Principles of Naval Architecture Second Revision

Volume I • Stability and Strength

Edward V. Lewis, Editor

1988

Published by
The Society of Naval Architects and Marine Engineers
601 Pavonia Avenue
Jersey City, NJ

Copyright © 1988 by The Society of Naval Architects and Marine Engineers.

It is understood and agreed that nothing expressed herein is intended or shall be construed to give any person, firm, or corporation any right, remedy, or claim against SNAME or any of its officers or members.

Library of Congress Catalog Card No. 88-60829
ISBN No. 0-939773-00-7
Printed in the United States of America
First Printing, April, 1988

Introduction

A Word From the President

The original version of this book, *Principles of Naval Architecture*, was first published by the Society in 1939. Editors H. E. Rossell and L. B. Chapman stated that the purpose of the work was to "adequately cover the field of naval architecture in one text." This they did, in two volumes, serving the Society's students and members more than adequately for nearly 30 years.

The First Revision was published in 1967, with John P. Comstock serving both as Chairman of the Control Committee and Editor. It consisted of one volume containing 11 chapters and an Appendix. Continuing changes in naval architecture, such as technical practices, new criteria and regulations regarding damage stability and ship strength, new knowledge about ocean waves and seakeeping, and the use of computers, prompted the Society to undertake the Second Revision in 1978. President Robert T. Young appointed John J. Nachtsheim as Chairman of the Control Committee, and Professor Edward V. Lewis was named Editor. Serving on the Control Committee, charged with the important tasks of choosing the authors and review of the chapters, were Thomas M. Buermann, William A. Cleary, Richard B. Couch, Jerome L. Goldman, Jacques B. Hadler, Ronald K. Kiss, Donald P. Roseman, Stanley G. Stiansen and Charles Zeien. This Second Revision of *Principles of Naval Architecture* (PNA) is the result of this Committee's work.

Even though the First Revision chapters on tonnage admeasurement, load line assignment and launching were removed from PNA to the 1980 edition of *Ship Design and Construction*, the remaining PNA chapters were so enlarged by new material that the decision was made to expand the Second Revision into three volumes.

Only the authors and the editors can appreciate the time and difficulties involved in writing, reviewing and editing this mass of knowledge into suitable form for publication. The work of these people, who are esteemed in their respective fields, has been as selfless as it is priceless to our Society and its membership. The Society is deeply indebted to the authors and to the tireless reviewers of the Control Committee.

To quote the late Matthew G. Forrest, a Past President of the Society, "The Society hopes that this First Revision of *Principles of Naval Architecture* will prove to be as useful, both to students and to those engaged in the practice of the profession, as the original edition proved to be."

I could not say it any better regarding the Second Revision.

EDWARD J. CAMPBELL
President, SNAME

Foreword

This revision of *Principles of Naval Architecture* began in 1978. It had been only eleven years since the prior revision but that time span was an explosive one in maritime technology. It was during this period that containerships became a commercial reality as did barge-carrying ships. Tankers of unprecedented size, some exceeding a half-million deadweight tons, became the norm. Roll-on, roll-off ships were built, as were liquefied natural gas carriers. Heavy lift ships appeared. Tug-barges, the size of ships, joined with ingenious mechanical linkages went into ocean service. The *Manhattan* was ice-strengthened and traveled north of Canada to Alaska. Off-shore drilling rigs of unique shapes and forms went to work in the most severe sea conditions imaginable. Concern for the sea as an important element of the environment became real after the *Torrey Canyon* broke up, spilling 100,000 tons of oil.

In the United States, the passage of the 1970 Merchant Marine Act provided a tremendous stimulus to merchant shipbuilding, ship operations, and maritime research.

A worldwide upsurge in these same activities resulted from a very healthy global maritime economy. The sum total of these stimulating activities provided the impetus for expanded design technology, enhanced shipbuilding productivity measures, and extremely creative maritime research activity.

The capture of these technological advances in this revision of *Principles of Naval Architecture* was the goal of the Control Committee.

It is our hope that we have done that. Our authors and Control Committee members were chosen for their extensive backgrounds as well as their involvement in these rapidly growing fields. In fact, one of the greatest continuing difficulties as the book progressed was in deciding where to divide "research" and "principles" in determining which material would be included in many of the chapters. I hope we have done that well. I do most sincerely thank the members of the Control Committee, the authors, the headquarters staff, and particularly our editor, Ned Lewis for their efforts. I hope you, the reader, will benefit from their most commendable, professional contributions.

The surge in maritime economic well-being ended later in the time period between revisions with the Arab oil embargo, the resulting crash in the world economy, and the precipitous drop in trade between nations . . . too many ships chasing too little cargo. With that decline came a corresponding decrease in technological growth. Survival, not growth, became the watchword. The most telling example of the depth of the decline since then is the fact that, as this Foreword is being written, not one merchant ship is on order or under construction in the United States. Maritime research funds, throughout the world, have become an endangered specie.

When maritime activity will again emerge, when technological growth again becomes a competitive necessity, no one can say. Until then we can at least take heart in knowing that this revision of *Principles of Naval Architecture* is a reflection of the latest technology, having taken advantage of probably the single most productive brief period of growth, from a maritime technology viewpoint, in the history of our profession.

Let's hope the current valley of worldwide maritime inactivity won't last for too long. Let's hope for better times, further technological growth, and the need once more, not too far away, for the next revision of *Principles of Naval Architecture*.

JOHN J. NACHTSHEIM
Chairman, Control Committee

Preface

The aim of this second revision (third edition) of the Society's successful *Principles of Naval Architecture* was to bring the subject matter up-to-date through revising or rewriting areas of greatest recent technical advances, which meant that some chapters would require many more changes than others. The basic objective of the book, however, remained unchanged: to provide a timely survey of the basic principles in the field of Naval Architecture for the use of both students and active professionals, making clear that research and engineering are continuing in almost all branches of the subject. References are to be included to available sources of additional details and to ongoing work to be followed in the future.

The preparation of this third edition was simplified by an earlier decision to incorporate a number of sections into the companion SNAME publication, *Ship Design and Construction*, which was revised in 1980. The topics of Load Lines, Tonnage Admeasurement and Launching seemed to be more appropriate for the latter book, and so Chapters V, VI, and XI became IV, V and XVII respectively, in *Ship Design and Construction*. This left eight chapters, instead of 11, for the revised *Principles of Naval Architecture*.

At the outset of work on the revision, the Control Committee decided that the increasing importance of high-speed computers demanded that their use be discussed in the individual chapters instead of in a separate Appendix as before. It was also decided that throughout the book more attention should be given to the rapidly developing advanced marine vehicles.

In regard to units of measure, it was decided that the basic policy would be to use the International System of Units (S.I.). Since this is a transition period, conventional U.S. (or "English") units would be given in parentheses throughout the book. This follows the practice adopted for the Society's companion volume, *Ship Design and Construction*. The U.S. Metric Conversion Act of 1975 (P.L. 94-168) declared a national policy of increasing the use of metric systems of measurement and established the U.S. Metric Board to coordinate voluntary conversion to S I. The Maritime Administration, assisted by a SNAME Ad Hoc Task Group, developed a *Metric Practice Guide* to "help obtain uniform metric practice in the marine industry," and this guide was used here as a basic reference. Following this guide, ship displacement in metric tons (1000 kg) represents mass rather than weight. (In this book the familiar symbol, Δ, is reserved for the displacement mass). When forces are considered, the corresponding unit is the kilo-Newton (kN), which applies, for example, to resistance and to displacement weight (symbol W, where $W = \Delta g$) or to buoyancy forces. (See Chapter I.) When conventional or English units are used, displacement weight is in the familiar long ton unit (2240 lb), which numerically is $1.015 \times$ metric ton. A conversion table also is included with the symbols and abbreviations or Nomenclature at the end of this volume.

This first volume of the third edition of *Principles of Naval Architecture*, comprising Chapters I through IV, covers almost the same subject matter as the first four chapters of the preceding edition. Thus, it deals with the essentially static principles of naval architecture, leaving most dynamic aspects to the remaining volumes. Chapter I deals with the graphical and numerical description of hull forms and the calculations needed to deal with problems of flotation and stability that follow. Chapter II considers stability in normal intact conditions, while Chapter III discusses flotation and stability in damaged conditions. Finally, Chapter IV deals with principles of hull structural design, first under static calm water conditions, and then introducing the effect of waves which also is covered more fully in Volume III, Chapter VII on Seakeeping.

These first four chapters were found to require less revision than those dealing, for example, with maneuverability and motions in waves. The latter required more time than anticipated. Some of the principal changes may be noted:

In Chapter I there is some rearrangement and change of emphasis. A few additions were made, such as developable lines and a containership, as well as a conventional cargo ship, as examples.

(Continued)

PREFACE

In Chapter II more attention is given to stability curves and to criteria for acceptable stability based on them.

In Chapter III more space is allotted to standards of flooding and damage stability, with emphasis on new probability-based international regulations.

Finally, Chapter IV has been extensively rewritten to cover new probabilistic techniques for dealing with loads and structural analysis methods concerned with ultimate strength. Several sections, including 3.3, Calculation of section modulus, and 3.14, Stress concentrations, were reproduced without change from the earlier edition.

February 1988

EDWARD V. LEWIS
Editor

Table of Contents
Volume I

	Page		Page
Introduction	iii	Editor's Preface	v
Foreword	iv	Acknowledgments	viii

Chapter 1 — SHIP GEOMETRY

NORMAN A. HAMLIN, Professor, Webb Institute of Naval Architecture

1. Ships' Lines ... 1
2. Displacement and Weight Relationships ... 16
3. Coefficients of Form ... 18
4. Integrating Rules and Methods ... 22
5. Hydrostatic Curves and Calculations ... 31
6. Bonjean Curves ... 44
7. Wetted Surface ... 47
8. Capacity ... 51

Chapter 2 — INTACT STABILITY

LAWRENCE L. GOLDBERG, University of Maryland

1. Elementary Principles ... 63
2. The Weight Estimate ... 69
3. Metacentric Height ... 71
4. Curves of Stability ... 78
5. Effect of Free Liquids and Special Cargoes ... 93
6. Effect of Changes in Weight on Stability ... 102
7. Evaluation of Stability ... 106
8. Drafts Trim and Displacement ... 115
9. The Inclining Experiment ... 122
10. Submerged Equilibrium ... 128
11. The Trim Dive ... 134
12. Methods of Improving Stability, Drafts and List ... 135
13. Stability when Grounded ... 136
14. Intact Stability of Unusual Ship Forms ... 138

Chapter 3 — SUBDIVISION AND DAMAGE STABILITY

GEORGE C. NICKUM, President, Nickum & Spaulding Associates

1. Introduction ... 111
2. Fundamental Effects of Damage ... 146
3. Subdivision and Damage Damage Stability Calculations ... 149
4. Manual Subdivision and Damage Stability Calculations ... 152
5. Subdivision and Damage Stability Calculation by Computer ... 176
6. Definitions for Regulations ... 178
7. Subdivision and Damage Stability Criteria ... 180
8. Alternate Equivalent Passenger Vessel Regulations ... 194

Chapter 4 — STRENGTH OF SHIPS

J. RANDOLPH PAULLING, Professor, University of California, Berkeley

1. Introduction ... 205
2. Ship Structural Loads ... 208
3. Analysis of Hull Girder Girder Stress and Deflection ... 233
4. Load Carrying Capability and Structural Performance Criteria ... 275
5. Reliability of Structures ... 290

Nomenclature ... 301
Index ... 305

Acknowledgments

All of the authors and the Editor first wish to acknowledge their indebtedness to the authors of the corresponding chapters of the preceding edition. The former have made extensive use of the original text and figures. The preceding authors were the late W. Selkirk Owen and the late John C. Niedermair (Chapter I), Charles S. Moore (Chapter II), James B. Robertson, Jr. (Chapter III) and Donald F. MacNaught (Chapter IV). The Control Committee, under the chairmanship of John J. Nachtsheim provided essential guidance, as well as valuable assistance in reviewing early drafts of the manuscript. Many members of the Committee provided extra help in areas of their particular expertise. Individual authors' acknowledgments follow.

Norman A. Hamlin—as well as the Editor—wishes to thank Webb Institute of Naval Architecture for allowing him to devote some of his time to work on Chapter I and for furnishing needed secretarial assistance. Prof. Hamlin also appreciates the help of a number of individuals and former Webb Institute students at MARAD, Coast Guard, NAVSEA, American Bureau of Shipping and shipyards and design agencies—in particular, Kevin H. Calhoun, George H. Levine, Ronald K. Kiss (member of Control Committee), James L. Mills, Jr., and Francis J. Slyker.

Lawrence L. Goldberg (Chapter II) acknowledges helpful information received from William A. Cleary, U.S. Coast Guard (member of Control Committee) and text material for Section 14, Intact Stability of Unusual Ship Forms, from George Wachnik, DTNSRDC, Daniel Savitsky, Director of the Davidson Laboratory, E.G.U. Band and David Lavis (of Band, Lavis and Associates) and Robert G. Tucker, current head of the Stability Branch, NAVSEA.

George Nickum (Chapter III) acknowledges helpful information and comments received from William A. Cleary, U.S. Coast Guard (member of Control Committee).

J. Randolph Paulling (Chapter IV) acknowledges the assistance of John F. Dalzell, DTNSRDC, in providing text material and helpful comments, particularly on Sections 2.7–2.10 dealing with long-term probabilities. He also wishes to acknowledge the general guidance and appreciation for ship structural analysis received through many years of close association with H. A. Schade of the University of California. A number of other individuals provided invaluable assistance through personal discussion and commentary on all or part of the chapter. In this regard, the assistance of Alaa Mansour of the University of California, Douglas Faulkner of the University of Glasgow, C. S. Smith of AMTE, Dunfermline, Scotland, Stanley Stiansen, Donald Liu and H. Y. Jan of the American Bureau of Shipping are gratefully acknowledged. Especial thanks are in order for the assistance of his student, Jan Otto DeKat, who performed the computations and prepared the plots of structural loading contained in Fig. 8.

Finally, the Editor wishes to thank the authors for their fine work and for their full cooperation in making suggested revisions. He acknowledges the indispensable efforts of Trevor Lewis-Jones in doing detailed editing and preparing text and figures in proper format for publication.

CHAPTER I

Norman A. Hamlin | **Ship Geometry**

Section 1
Ships' Lines

1.1 Delineation and Arrangement of Lines Drawing. The exterior form of a ship's hull is a curved surface defined by the lines drawing, or simply "the lines." Precise and unambiguous means are needed to describe this surface, inasmuch as the ship's form must be configured to accommodate all internals, must meet constraints of buoyancy, stability, speed and power, and seakeeping, and must be "buildable." Hence, the lines consist of orthographic projections of the intersections of the hull form with three mutually perpendicular sets of planes, drawn to a suitable scale.

Fig. 1 shows a lines drawing for a single-screw cargo-passenger ship.

The profile or *sheer plan* shows the hull form intersected by the centerplane—a vertical plane on the ship's centerline—and by buttock planes which are parallel to it, spaced for convenient definition of the vessel's shape and identified by their distance off the centerplane. The centerplane intersection shows the profile of the bow and stern. Below the profile is the *half-breadth* or *waterlines plan*, which shows the intersection of the hull form with planes parallel to the horizontal baseplane, which is called the base line. All such parallel planes are called waterline planes, or waterplanes. It is convenient to space most waterplanes equally by an integral number of meters (or feet and inches), but a closer spacing is often used near the baseline where the shape of hull form changes rapidly. DWL represents the design waterline, near which the fully loaded ship is intended to float. All waterlines are identified by their height above the baseline.

The *body plan* shows the shapes of sections determined by the intersection of the hull form with planes perpendicular to the buttock and waterline planes. In Fig. 1 this is shown above the profile, but it might otherwise be drawn to the right or left of the profile, using a single extended molded baseline, depending upon the width and length of paper being used. Alternatively, the body plan is sometimes superimposed on the profile, with the body plan's centerplane midway between the ends of the ship in profile view. Planes defining the body plan are known as body plan stations. They are usually spaced equally apart, such that there are 10 spaces—or multiples thereof—in the length of the ship, but with a few extra stations at the ends of the ship at one half or one quarter this spacing.

Most ships are symmetrical about the centerplane, and the lines drawing shows waterlines in the half-breadth plan on only one side of the centerline. Asymmetrical features on some ships, such as overhanging flight decks on aircraft carriers, must be depicted separately. Correspondingly, the body plan shows sections on one side of the centerline only—those in the forebody on the right hand side and those in the afterbody on the left. By convention in the U.S., the bow of the ship is shown to the right. With the arrangement of the lines as shown in Fig. 1, the drawing represents a case of first angle projection in descriptive geometry.

The lines in Fig. 1 represent the *molded surface* of the ship, a surface formed by the outer edges of the frames, or inside of the "skin," in the case of steel, aluminum and wooden vessels. In the case of glass reinforced plastic vessels, the molded surface is the outside of the hull. (The term molded surface undoubtedly arose from the use of wooden "molds" set up to establish a surface in space to which frames could be formed when wooden vessels were being built).

The shell plating of a steel or aluminum ship constitutes the outer covering of the molded surface. The shell plating is relatively thin and is formed of plates that are usually of varying thickness, causing some unevenness, although the molded surface is generally smooth and continuous.

The thickness of planking of a wooden boat is relatively larger than the shell thickness of a steel vessel, and it is the usual practice to draw the lines of a wooden boat to represent the surface formed by the outside of the planking, since this gives the true external form. However, for construction purposes it is necessary to deal with the molded form, and therefore it is not unusual to find the molded form of wooden vessels delineated on a separate lines drawing.

In the sheer plan of Fig. 1, the base line, representing the bottom of the vessel, is parallel to the DWL, showing that the vessel is designed for an "even-keel" condition. Some vessels—especially tugs and fishing vessels—are often designed with the molded keel line raked downward aft, giving more draft at the stern than the bow when floating at the DWL; such vessels are said to have a designed *drag to the keel.*

1.2 Perpendiculars; Length Between Perpendiculars. A vertical line in the sheer plan of Fig. 1 is drawn at the intersection of the DWL, which is often the estimated summer load line (defined subsequently), and the forward side of the stem. This is known as the *forward perpendicular,* abbreviated as FP. A slight inconsistency is introduced by this definition of FP in that the forward side of the stem is generally in a surface exterior to the molded form by the thickness of contiguous shell plating—or by the stem thickness itself if the stem is of rolled plate.

A corresponding vertical line is drawn at the stern, designated the *after perpendicular* or AP. When there is a rudder post the AP is located where the after side of the rudder post intersects the DWL. In Fig. 1 the AP is drawn at the centerline of the rudder stock, which is the customary location for merchant ships without a well defined sternpost or rudder post. In the case of naval ships, it is customary to define the AP at the after end of the vessel on the DWL. Such a location is also sometimes chosen for merchant vessels—especially vessels with a submerged stern profile extending well abaft the rudder. Fig. 2 shows the various locations of the AP here described.

An important characteristic of a ship is its length between perpendiculars, sometimes abbreviated LBP or L_{pp}. This represents the fore-and-aft distance between the FP and AP, and is generally the same as the length L defined in the American Bureau of Shipping *Rules for Building and Classing Steel Vessels* (Annual)[1]. However, in the *Rules* there is included the proviso that L, for use in the Rules, is not to be less than 96 percent and need not be greater than 97 percent of the length on the summer load line. The summer load line is the deepest waterline to which a merchant vessel may legally be loaded during the summer months in certain specified geographical zones. Methods for determining the summer load line are covered in the discussion of freeboard in *Ship Design and Construction* (Taggart, 1980).

When comparing different designs, a consistent method of measuring ship lengths should be used. Overall length is invariably available from the vessel's plans and LBP is usually also recorded. However, for hydrodynamic purposes, length on the prevailing waterline may be significant; alternatively, an "effective length" of the underwater body for resistance considerations is sometimes required.

One useful method of determining the after end of effective length is to make use of a sectional area curve, whose ordinates represent the underwater cross sectional area of the vessel up to the DWL at a series of stations along its length. (See Section 1.7.) The effective length is usually considered as the overall length of the sectional area curve. However, if the curve has a concave ending, a straight line from the midship-cross-sectional area can be drawn tangent to the curve, as shown in Fig. 3. The intersection of this straight line tangent with the baseline of the graph may then be considered to represent the after end of the effective length. On many single-screw designs it has been found that the point so determined is close to the location of the AP. Such an effective length ending might then be used in calculating hull form coefficients, as discussed in Section 3. A similar definition for the forward end of effective length might be adopted for ships with protruding bulbous bows extending forward of the FP.

It is important that in all calculations and measurements relating to length, the method of determining the length used, and the location of its extremities be clearly defined.

1.3 Midship Section; Parallel Middle Body. An important matter for any ship is the location and shape of the midship cross section, generally designated by the symbol ⊗, which was originally used to indicate the fullest cross section of the vessel. In some of the early sailing ships this fullest section was forward of the midlength, and in some high-speed ships and sailing yachts, the fullest section under water is somewhat abaft the midlength. In any case, the usual practice in modern commercial vessels of most types is to locate ⊗ halfway between the perpendiculars, while in naval ships it is usually midway between the ends of the DWL.

In many modern vessels, particularly cargo vessels, the form of cross section below the DWL amidships extends without change for some distance forward and aft, usually including the midship location. Such vessels are said to have parallel middle body. The ship in Fig. 1 has no parallel middle body, but the form of section under water changes but slightly for small distances forward or abaft the fullest section, which is located amidships.

1.4 Body Plan Stations; Frame Lines; Deck Lines. In order to simplify the calculation of underwater form characteristics, it is customary to divide the LBP into 10—or 20, or 40—intervals by the body plan planes. The locations of these planes are known as body plan stations, or simply stations, and are indicated by straight lines drawn in the profile and half-breadth plans at right angles to the vessel's baseline and centerline, respectively. The intersections of these planes

---
[1] Complete references are listed at end of chapter.

SHIP GEOMETRY

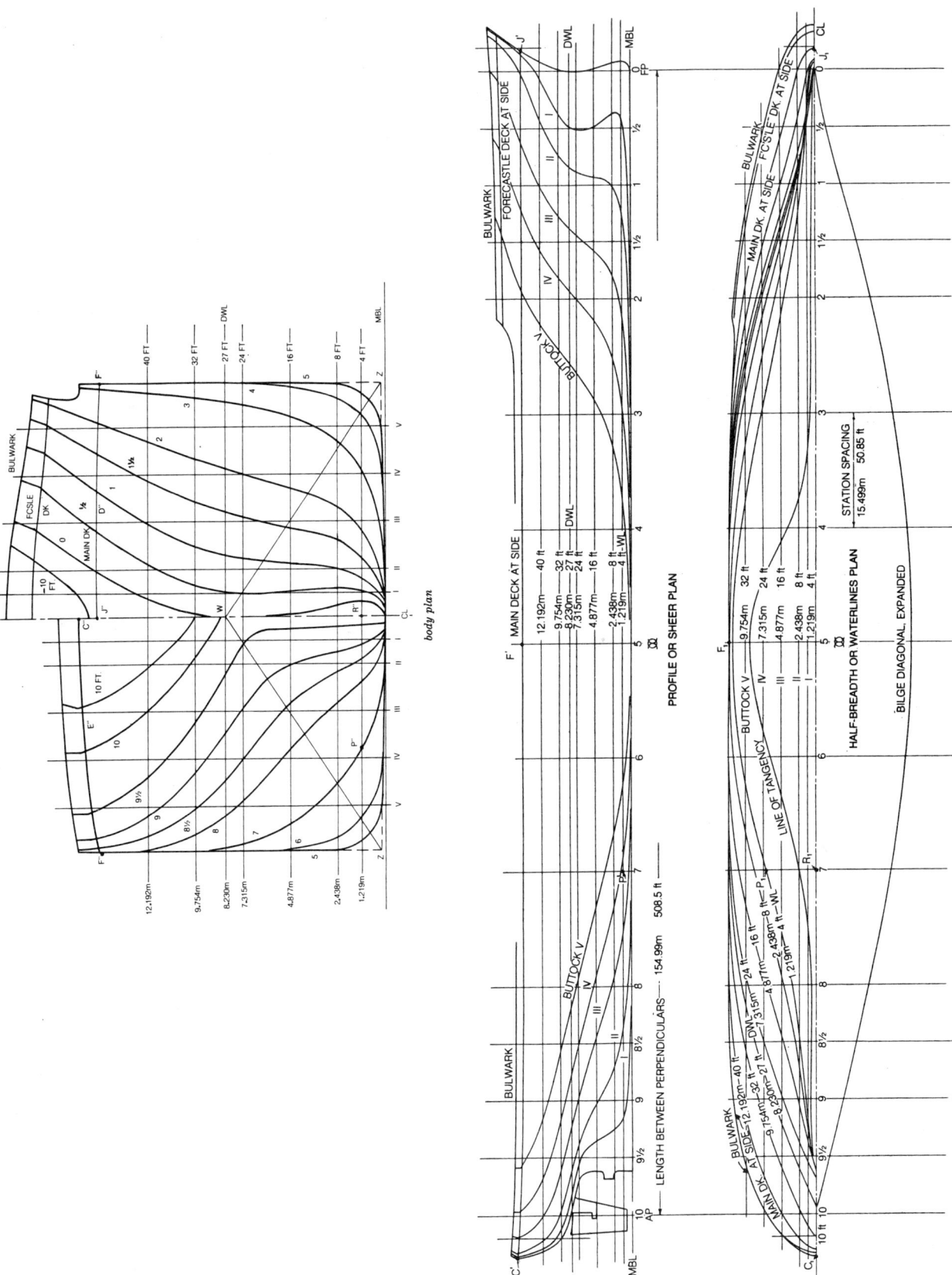

Fig. 1 Lines drawing

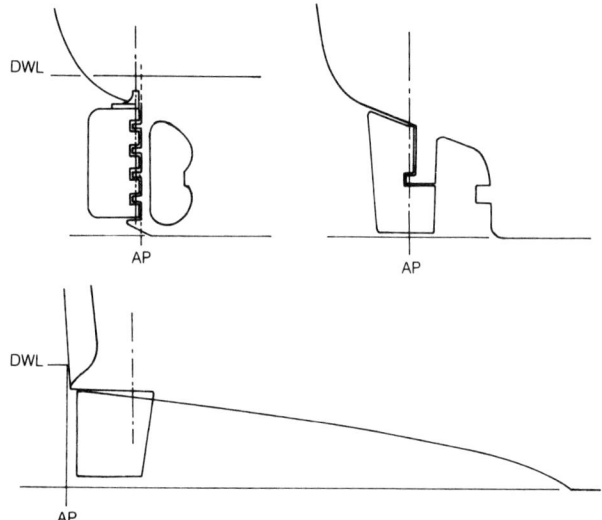

Fig. 2 Alternative locations of after perpendicular

with the molded form appear in their true shape in the body plan.

Body plan stations are customarily numbered from the bow, with the FP designated as station 0. In Europe and Japan, however, station 0 is often located at the AP, with station numbering from aft forward. For the ship shown in Fig. 1, station No. 10 represents the stern extremity of the vessel for calculations relating to the underwater body. It will be noted that additional stations are drawn midway between stations 0 and 1, and 9 and 10, and sometimes between 1 and 2, and 8 and 9, as well. This is done to better define the vessel's form near the ends where it may change rapidly for small longitudinal distances.

Additional stations are often also shown forward of the FP and abaft the AP. These may receive letter or distance designations from the perpendiculars, or a continuation of the numbering system equivalent to that used in the remainder of the ship, as negative numbers forward of the FP and numbers in excess of 10 (or 20, etc.) abaft the AP.

Body plan station planes are not to be confused with planes at which the vessel's frames are located, although frames are normally located in planes normal to the baseplane and longitudinal centerplane, which are therefore parallel to body plan station planes. Frames are normally spaced to suit the structure and arrangement of internals and their location is not dependent upon station plane locations. On some naval ships, frame spacing is an integral number of feet or one meter. Frame locations are usually chosen early in the design of a ship, and it is customary to show them on arrangement drawings and frequently also on final lines drawings. Therefore, frame locations, and their spacing, must be clearly stated. A body plan at frame locations is frequently drawn to assist the shipyard in fabricating the frames.

Frames are generally numbered with integer numbers, either starting at the FP and increasing aft, or at the AP and increasing forward. The latter practice is customary in tankers. In some instances, particularly naval ships, frames have been identified by the distance of the frame plane in meters from the FP.

A frame plane establishes a molded line, or surface, which will be coincident with the plane of either the forward or after edge of the frame. The location of frames, either forward of or abaft the frame line, should be clearly stated on relevant drawings.

The outline of the ship is completed in the sheer plan by showing the line of the main deck at the side of the ship, and also at the longitudinal centerline plane whenever, as is usual, the deck surface is crowned or *cambered*, i.e., curved in an athwartship direction with convex surface upwards, or sloped by straight lines to a low point at the deck edge. A ship's deck is also usually given longitudinal sheer; i.e., it is curved upwards towards the ends, usually more at the bow than at the stern. In case the sheer line of the deck at side curves downward at the ends, the ship is said to have reverse sheer.

Similarly, lines are shown for the forecastle, bridge, and poop decks when these are fitted; sometimes decks below the main deck are also shown. All such deck lines generally designate the molded surface of the

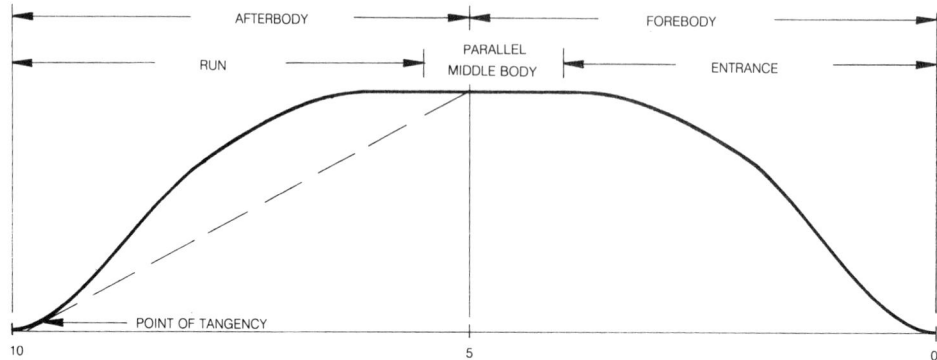

Fig. 3 Geometry of sectional area curve

SHIP GEOMETRY

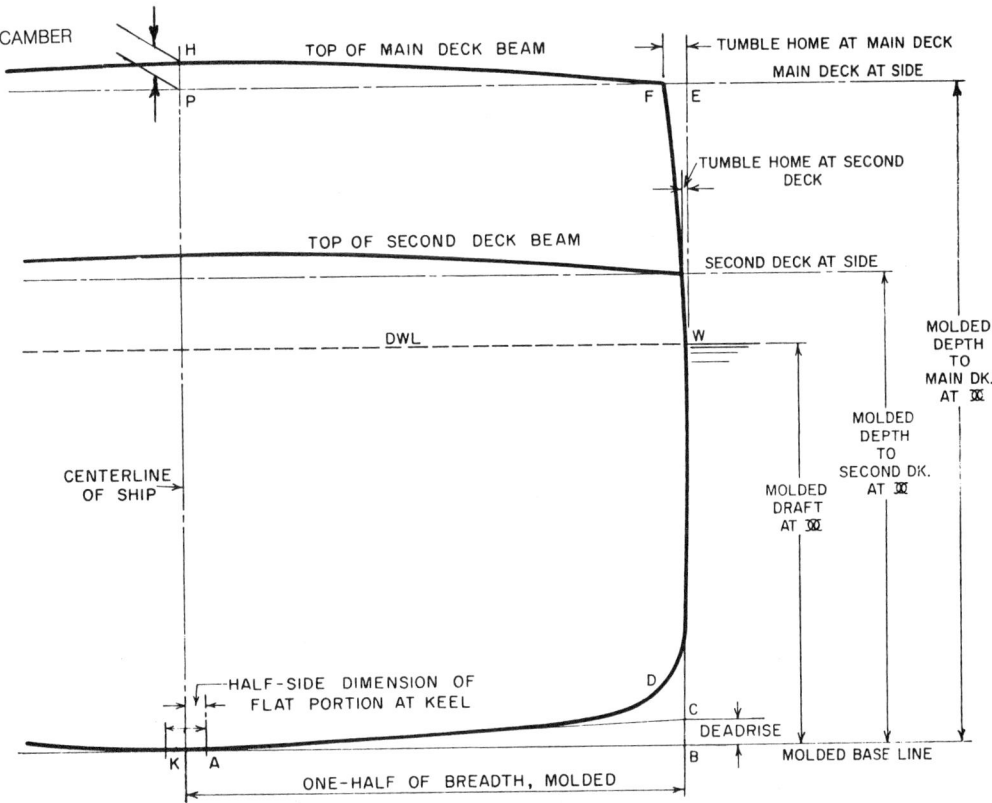

Fig. 4 Midship section, molded form

respective deck; i.e., the surface at the top of the deck beams, and are consequently referred to as molded deck lines at side or at center as the case may be.

In the lines drawing, Fig. 1, the curve of the main deck at the side is projected into the sheer plan as the curve $C'F'J'$ and into the body plan as $J''D''F''$ for the fore body and as $F'''E'''C'''$ for the after body, and also into the half-breadth plan as the curve $C_1F_1J_1$, which is known as the half-breadth line of the main deck.

Through the point where the molded sheer line of the main deck at side intersects the midship station in the sheer plan, there may be drawn a level line called the molded depth line of main deck at side. At any particular station, the vertical distance between this line and the sheer line of deck at side is known as the *sheer* of the deck at that station. The sheer of a deck would, therefore, be zero at the midship station, and it may be zero for an appreciable distance either forward of or abaft amidships. Of particular interest are the values of sheer at FP and AP.

The sheer line of deck at side in some vessels, particularly yachts, may dip below the level of the molded depth line at side. This usually occurs, if at all, in the region immediately abaft amidships, and the sheer of the deck in such a region is measured below the level of the molded depth line at side and is considered to have a negative value.

The molded lines of the principal transverse bulkheads are sometimes also shown on final drawings.

1.5 Molded Base Line; Molded Dimensions. The molded base line, drawn in the sheer plan and body plan as a straight horizontal line, represents an important reference datum, both for design and construction purposes. The line, in fact, represents a plane in space to which many vertical heights are referred. It also represents the bottom of the vessel's molded surface, and so is coincident with the top surface of the flat plate keel on most straight-keel ships with a single thickness of shell plating.

In the event the keel line of a ship is straight, but the vessel has a designed drag to the keel, it usually slopes downward aft. In this case the molded base line may mark the bottom of the molded surface amidships, or at the AP. When drawing the lines for such a vessel, the bottom of the molded surface is shown as a raked line.

In the event the vessel is designed with an external hanging bar keel, extending below the shell plating surface, the bottom of keel is drawn in the sheer plan to complete the lower contour of the vessel. However, on most other ships, only the bottom of the molded surface is drawn.

In the case of ships with "in and out" riveted plating, the keel plate is usually an "out" strake and the bottom of keel is then below the molded base line by not only its own thickness but that of the first outboard, or garboard strake, as well.

The molded depth of a vessel is the vertical distance

between the molded base line and the molded depth line of the uppermost deck at side as shown in Fig. 4.

The distance from K to B in Fig. 4 is one-half of the important dimension known as the molded beam or molded breadth of the vessel, which is normally a maximum at the midship station.

1.6 Characteristics of the Sections. In Fig. 4 from the point A the molded line of the bottom of the midship section extends towards the side in a straight line AC. This line often is inclined upwards slightly and intersects, at the point C, the vertical line EB drawn tangent to the widest part of the underwater body.

The line AC is known as the floor line, and the distance $\overline{BC}$ is referred to variously as the *deadrise*, rise of floor, or rise of bottom. For the ship shown in Fig. 1, the deadrise is 0.305 m.

The point K in Fig. 4 at the vessel's centerline is at the lowest part of the molded surface and the distance $\overline{KA}$ is the *half-side* dimension of the flat portion of the molded surface in the vicinity of the keel i.e., to the beginning of the deadrise. This half-side dimension is small in vessels having a hanging bar keel, being simply the half-thickness of the bar forming the keel, but in vessels having a dished-plate keel it will be considerably more, depending upon the size of the ship. It does not apply at all to ships with no deadrise.

The curved portion of the section, as at D, which joins the floor line with the side, is known as the *turn of bilge* and may be further described as a "hard" or as an "easy" turn of bilge, where hard refers to a small radius of curvature. The turn of bilge throughout the parallel middle body is usually, but not necessarily, a circular arc, and the radius of this curve is known as the *bilge radius*.

The molded line of the side above the waterline sometimes extends inboard somewhat to meet the line of the top of the main deck beam. In Fig. 4 this intersection is at the point F. The horizontal distance $\overline{EF}$ is known as *tumble home* at the deck. The opposite of tumble home is known as *flare*, and it is measured in a similar way.

A horizontal line through F in Fig. 4 meets the centerline of the section at P; the distance PH is called *camber* or *round of beam*. The camber curve may be an arc of a circle, a parabola, or several straight lines. Standard past practice has been to provide about 2 percent of the total breadth of the ship as camber amidships, and then to use the camber curve so determined as applicable to all other fore and aft locations. The use of camber accomplishes the important function of assuring that rain water and water shipped aboard will drain off readily.

1.7 Sectional Area Curve. A fundamental drawing in the design of a ship—particularly relative to resistance—is the sectional area curve, shown in Fig. 3 for a ship with some parallel middle body. The sectional area curve represents the longitudinal distribution of cross sectional area below the DWL. The ordinates of a sectional area curve are plotted in distance-squared units. Inasmuch as the horizontal scale, or abscissa, of Fig. 3 represents longitudinal distances along the ship, it is clear that the area under the curve represents the volume of water displaced by the vessel up to the DWL, or volume of displacement.

Alternatively, the ordinate and abscissa of the curve may be made non-dimensional by dividing by the midship area and length of ship, respectively. In either case, the shape of the sectional area curve determines the relative "fullness" of the ship (See Section 3).

The presence of parallel middle body is manifested by that portion of the sectional area curve parallel to the baseline of the curve. The shoulder is defined as the region of generally greater curvature (smaller radius of curvature) where the middle body portion of the curve joins the inward sloping portions at bow or stern.

The centroid of the vessel's sectional area curve is at the same longitudinal location as the center of buoyancy, LCB, and the ratio of the area under the sectional area curve to the area of a circumscribing rectangle is equal to the *prismatic coefficient*, C_p (See Section 3).

Fig. 3 also shows the customary division of the underwater body into *forebody* and *afterbody*, forward of and abaft amidships, respectively. *Entrance* and *run*, which represent the ends of the vessel forward of and abaft the parallel middle body, are also shown.

1.8 Molded Drafts; Keel Drafts; Navigational Drafts; Draft Marks. In general, the amount of water a vessel draws, or *draft*, is the distance measured vertically from the waterline at which the vessel is floating to its bottom. Drafts may be measured at different locations along the length. They are known as molded drafts if measured to the molded baseline; keel drafts if measured to the bottom of the keel. Mean draft is defined as the average of drafts forward and aft.

Ships are customarily provided with draft marks at the ends and amidships, arranged in a plane parallel to station planes and placed as close to the perpendiculars as practical. These draft marks are for the guidance of operating personnel, and therefore the drafts indicated should be keel drafts. The marks are painted in a readily visible color to contrast with the color of the hull. Arabic numerals are usually used on merchant vessels, although Roman numerals also appear on some naval ships, particularly in way of appendages that extend below the baseline. The bottom of the numeral is located at the indicated waterline. For many years it has been the practice to use numerals 6 inches high and to mark the drafts in feet at every foot above the keel. Thus, if one were to see the numeral half immersed, the prevailing draft would be three inches deeper than the half-immersed number in ft.

With the ultimate conversion to the metric system in the United States a reasonable practice would seem to be that adopted by Australian maritime authorities (Australian Dept. of Transport, 1974). This provides that drafts be shown in meters at every meter in Arabic numerals, followed by M. Intermediate drafts are shown at every 0.2 m (2 decimeters), but only the numerals 2, 4, 6 and 8 are shown, with no decimeter designation. All numerals are to be one decimeter high. Thus, draft marks between 11 and 12 meters would show,

12M
8
6
4
2
11M

The difference between drafts forward and aft is called *trim*. If the draft aft exceeds that forward, the vessel is said to have trim by the stern. An excess of draft forward causes trim by the bow—or trim by the head. When trim is determined by reading the draft marks and the angle of inclination or the displacement of the vessel is to be determined, it is important to account for the specific fore and aft location of the marks.

Some vessels are designed with local projections below the keel of a permanent nature—for example, sonar transducer housings (domes), and the propeller blade tips of some naval vessels. It is important that operating personnel be well aware of the distance below the keel to which such projections extend. Navigational drafts—which represent the minimum depth of water in which the vessel can float without striking the bottom—would exceed keel drafts by this distance.

1.9 Diagonals; Types of Intersecting Planes. The shape of curves shown by the stations, buttock lines and waterlines do not necessarily convey the shape of hull form as one might wish to see it, and the designer need not be limited to use of these planes. Additional planes with which the hull form is sometimes intercepted are diagonal planes, which are planes normal to station planes, but inclined with respect to the baseplane and the longitudinal centerplane. Such a plane appears as a straight line in the body plan. The inclination of a diagonal plane is generally chosen so that it is approximately normal to the body plan sections.

It is customary to show the resulting intercept curve, called a *diagonal*, below the half-breadth view in the lines drawing. This practice has been followed in Fig. 1. Thus, the expansion of the diagonal is a plot of distance from the point W on the ship's C.L. in the body plan to the points where ZW crosses each station. The particular diagonal shown in Fig. 1 is called a *bilge diagonal*, inasmuch as it intersects the bilge. Point W is at the DWL on the vessel's centerline, and point Z marks the intersection of the vessel's half-beam line and deadrise line.

Projections showing the intersections of diagonal planes with the molded surface are generally omitted in the half-breadth plan and the profile.

1.10 Cant Frame Lines. On some types of vessels, it is found that near the ends of the vessel, the inclination of the ship's surface to the planes of transverse frames becomes so great as to require these planes to be moved to a position more nearly normal to the surface, so that the frame when so constructed may give a better support to the surface in its vicinity. In the event the plane of the frame remains normal to the baseplane, the trace of the plane in the centerplane appears as a line perpendicular to the molded baseline in the sheer plan, and the frame is called a *single cant frame*, or simply *cant frame*.

Frames are also occasionally placed in planes normal to the longitudinal centerplane, but inclined to the baseplane, whereby the trace of the plane in the baseplane, as seen in the half-breadth plan, is perpendicular to the vessels' centerline. The term *inclined frame* has been applied to this case.

Double cant frames lie in planes which are neither normal to the longitudinal centerplane nor the baseplane. Determining the trace of such a plane in the molded surface is an exercise in descriptive geometry, and such frames are rarely used.

1.11 Fairness: Fairing of Lines. It is of interest to note certain features of hull form shown by the lines in Fig. 1. The lower waterline shapes near the bow and stern are drawn with some *hollow*—that is, they are concave. Similarly, the body plan sections and buttocks are hollow, generally, in the vicinity of the DWL and particularly in the afterbody. Thus the shape of the lines in one view is reflected in the other views, and vice versa.

With the exception of deliberate discontinuities at the stem, knuckles, chines, transom corners, etc., the shape of a vessel's exterior form below the deck is virtually always designed as a fair surface. A fair surface is defined as one that is smooth and continuous, and which has no local bumps or hollows, no hard spots and a minimum of points of inflection. Localized flat spots between areas of the surface with curvatures of equal sign are generally considered unfair, unless they occur as part of the bottom or sides, especially with parallel middle body. Mathematically, the property of fairness of surface might be thought of as that of continuity in a plot of curvature, or radius of curvature, of the intersection of any plane with the surface. Inasmuch as waterlines, buttocks, station lines and diagonals all represent the intersection of planes with the molded surface, it may be seen that a fair hull form will be characterized by fairness in these curves; correspondingly, it is usually assumed that if these curves are fair, then so will the hull form. In general, discontinuities in the first derivative, indicating abrupt changes in slope, occur at knuckle lines. Other sudden changes in curvature, indicated by discontinuities in the second derivative, are considered to show unfairness. A common situation on ships with parallel middle body is a bilge of constant radius, r, connecting to flat bottom and/or side, with a change in curvature of the transverse section from $1/r$ to 0 at the point of tangency. Although such a section is not fair, its shape is not necessarily disadvantageous. It can be made fair if desired by easing the transition in curvature. On the other hand, continuity in both first and second derivatives does not guarantee fairness, inasmuch as the achievement of fairness has always been and probably will continue to be a matter of opinion or judgment.

An additional condition implied by the term fairness is that of consistency, that is, each projection of any point on the surface onto the corresponding reference plane must agree with the locations of its other projections. For example, consider a point P to be on the surface of the ship in Fig. 1 at station 7 and 4 ft (1.22 m) above the molded base line. This point would be shown in the sheer plan at P'. Its location in the body plan would be on transverse section 7, and on the 4-ft WL. The horizontal distance of the point P from the ship's centerplane would be determined by the distance in the body plan of the point P'' from the ship's centerline, as $\overline{P''R''}$. The point P_1 in the half-breadth plan would be at the ordinate for station 7 and on the 4-ft WL and its distance from the ship's centerline would be $\overline{P_1R_1}$ as shown in that plan. A test of consistency of the point P would be that the distance $\overline{P_1R_1}$ in the half-breadth plan must equal $\overline{P''R''}$ in the body plan.

In case the point P had been originally selected on the surface at a location where no transverse section, waterline, or buttock already existed, a check of fairness would require one to introduce any two of these three types of intersecting planes through the point, find the corresponding projections of the lines of intersection and proceed as before.

The process of fairing a set of lines is invariably an iterative, or cut and try one, requiring patience and perseverance. It consists essentially of investigating the fairness or suitability of each line of the vessel in succession. It often happens that, after testing and accepting a number of lines, the next line to be considered will require changes to be made to it that will be so far-reaching as to affect some of the lines previously accepted. It then becomes necessary to make whatever changes seems best, all things considered, and to proceed anew through the same fairing steps as before. Usually several such difficulties have to be overcome successively before the whole fairing process is completed. Thus, the process may be laborious.

Fairing lines for a new ship design is normally accomplished at least twice—first in the design phase, and second in the construction phase, at which time the lines are faired either full-scale, on the mold loft floor, or in the optical detailing room to a scale of 1/10 or 1/20 of full size, or by computer as discussed in Section 1.16. In the design phase, there is greater freedom to make changes and to achieve hull form features which the designer favors. Curves are usually drawn using a combination of free hand sketching, ship curves and flexible battens (or splines) held by batten weights ("ducks").

Waterlines are usually drawn by the last of these methods. A uniform batten will be fair between a pair of ducks, but it can be forced into an unfair overall curve by the ducks. Hence, a customary method of fairing or smoothing is to adjust the ducks—and hence the batten defining the waterline—until any one of the ducks can be removed without the batten moving. This is intended to assure that changes in curvature are made gradually.

In the final design or construction phase, the lines are reasonably well defined at the start. The process of fairing is more localized and directed at achieving consistency among the various views. However, the larger scale used in this case is intended to assure that local deviations, which may not have been evident in the earlier small-scale design phase, will be eliminated.

1.12 Developing a Set of Lines. The development of a set of lines presupposes a tentative (or final) selection of suitable hull dimensions, coefficients (section 3), LCB, sectional area curve (Fig. 3) and design waterline. This selection is based on considerations of displacement, capacity, trim, stability, resistance and propulsion, all of which are discussed in other chapters, as well as in the chapter on Mission Analysis and Basic Design, *Ship Design and Construction* (Taggart, 1980). Fig. 5 is a generalized plot whereby the offsets of a sectional area curve may be drawn to fit prescribed hull features (prismatic coefficients and LCB.) In order to use Fig. 5, one enters Fig. 5a with LCB and total C_P to get C_{PA} and C_{PF}; these are then used in Fig. 5b to find the sectional area curve offsets.

Given the desired hull characteristics, the process of drawing and fairing a preliminary small-scale set of lines generally begins with fixing the profile of the vessel in the centerplane, the design waterline and deck line in the half-breadth plan, and the midship body plan section. Intermediate sections may next be sketched in to satisfy the pre-determined sectional area curve, often by reference to previous designs and typical hull forms (SNAME Hydrodynamics Committee, 1966). A few additional waterlines, between the deck and the DWL, and between the DWL and the baseline, are then drawn in the half-breadth view using half-breadths at the stations and making as small and as few changes as possible in these. The sections in the body plan are then changed to achieve consistency with the waterline half-breadths, and section areas checked. A few buttocks are then drawn in and checked and the process repeated. Alternatively, diagonals, rather than waterlines, are preferred by some designers as a fairing medium, and are used to check the consistency of section shape variation from station to station before buttocks and intermediate waterlines are drawn. Liberal use of the eraser is required, the drawing frequently being made on the back of transparent cross-section paper, chosen so that the grid of the paper matches the grid of waterlines, buttocks and

SHIP GEOMETRY

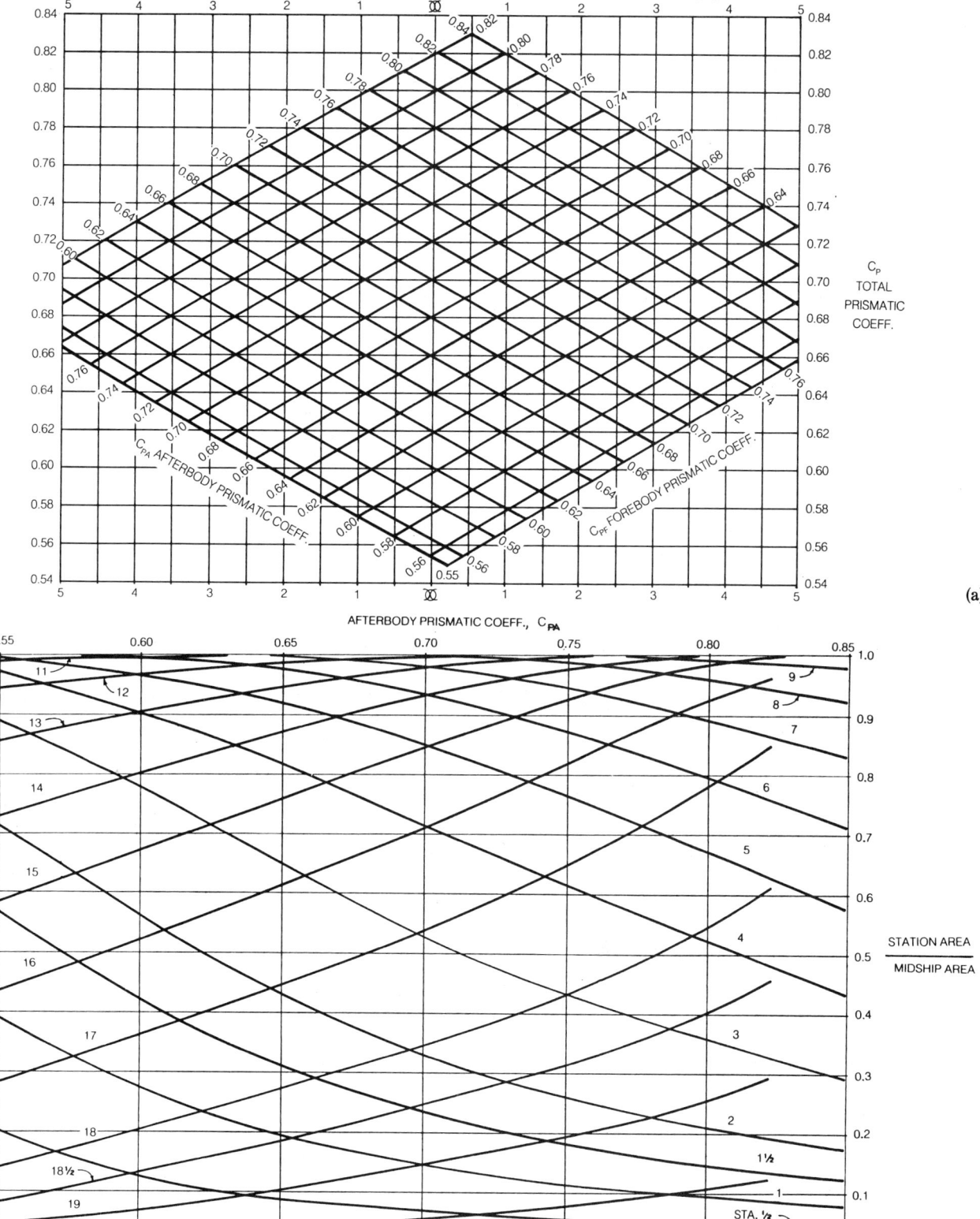

Fig. 5 Generalized plot of sectional areas, including forebody and afterbody prismatic coefficient as functions of longitudinal center of buoyancy (a & b)

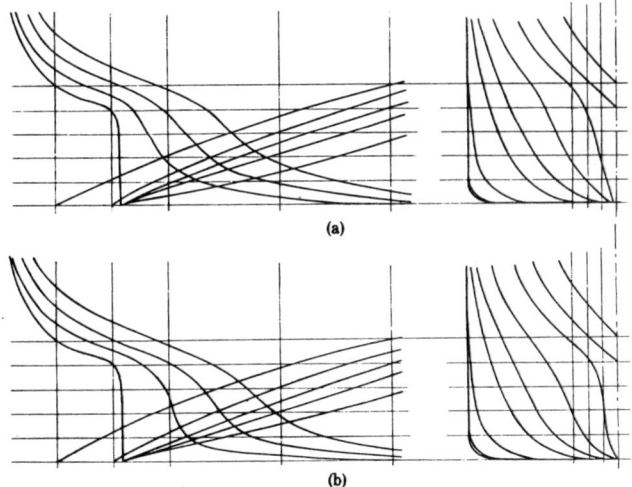

Fig. 6 Typical original and modified stern lines

stations desired. Because of the flatness of angle of intersection of buttocks and waterlines on narrow, fine-lined ships at the quarter length, it is sometimes the practice to foreshorten station spacing in the profile and half breadth plan to assist in fairing.

As the ship design progresses, one or more larger scale lines drawings must be prepared and faired with increasing precision. In the fairing process, some general guides should be remembered. For example, the general shape of buttock endings, particularly for buttocks near the ship's centerline, must reflect the shape of body plan stations, if a gradual and progressive hange of waterline slope is to be achieved. Fig. 6A illustrates a case in which (a) this guide was not followed, and (b) the lines might be modified to suit the guide.

Nine of the often conflicting considerations involved in developing a set of lines, other than those of resistance and propulsion discussed in detail in Chapter V, may be noted here:

(a) Generous clearances around the propeller tend to reduce vibration excitation forces, but a large diameter propeller tends to improve populsive efficiency and hence to reduce required shaft horsepower, assuming the propeller design is not restricted in RPM.

(b) A large amount of "fin" area aft, both fixed and movable, tends to promote directional stability. Generous movable area (rudder area) tends to improve the ability to initiate and recover from turns.

(c) A small bilge radius, together with a bilge keel right at the turn of the bilge, tends to increase roll damping. However, wetted surface, and hence frictional resistance, tend to be increased by a small bilge radius.

(d) V-sections are generally favorable to stability and seakeeping performance, but are often objectionable from the viewpoint of resistance and/or propulsion.

(e) Ships which must operate in heavy weather may experience slamming on the flat of bottom forward unless large deadrise angles are used and the extent of flatness is minimized. However, a long straight flat keel is desirable from drydocking considerations.

(f) Generous flare forward, sometimes with a gently sloped longitudinal knuckle well above the waterline, may be used instead of an increase in freeboard forward in achieving dry decks when in a seaway.

(g) Ships with bulbous bows may experience damage to the bulb from anchor handling unless the bow in way of the hawsepipe is flared out sufficiently to allow an unobstructed drop from the pipe extremity, taking into account the possibility of the ship's rolling to the opposite side.

(h) Hull surfaces composed of portions of cylinders and cones—i.e., developable surfaces—are more easily fabricated than surfaces of compound curvature, but may incur added resistance. See section 1.14.

(i) Excessive waterline angles forward of the propeller should be avoided, as well as blunt waterline endings, since they may promote separation in the flow, especially in the case of very full, slow-speed vessels. Such separation tends to cause propeller-excited vibration, as well as greater resistance and less efficient propulsion.

1.13 Offsets. In the process of building a ship, some means must be devised for determining the shapes of the frames with greater precision than can be obtained directly from the usual lines drawings. It has been the practice in most shipyards in the past, to attain the necessary accuracy, to redraw and refair the lines to full size on a large wooden floor located in a space known as the mold loft. The mold loftsmen were supplied sufficient information to enable certain portions or the whole of the vessel's lines to be drawn full size, often in contracted form, i.e., with all breadths and heights full size but with lengths reduced. These operations, complete with refairing as necessary, are known as laying off or laying down the lines.

For laying off a mold loftsman needs not only the lines drawing but also a list of the measurements he must use in locating points through which the various curves are to be drawn. Consider a waterline in the half-breadth plan and suppose that the distance on each station from the vessel's centerline to the waterline were measured. Such measurements are known as *offsets*, and by their use the loftsman can lay off the necessary points on the floor through which the required curve can be drawn in a fair line by using long flexible wooden battens. For a buttock line in the sheer plan, the offsets would be given as heights above the molded base line at each station. If in the English system, these would be in feet, inches, and eighths (or sixteenths) of an inch. It is expected that as shipyards in the U. S. convert to the metric system, full-scale offsets will be recorded to the nearest millimeter—that is, three decimal places after the meter—inas-

Table 1—Typical Table of Offsets
Halfbreadths, m

Station	Half Siding	Bottom tangent	4-ft WL 1.219 m	8-ft WL 2.438 m	16-ft WL 4.877 m	24-ft WL 7.315 m	27-ft WL 8.230 m	32-ft WL 9.754 m	Station
0, FP	0	—	0.759	0.581	0.108	—	—	0.133	0, FP
½	0.394	—	1.308	1.432	1.270	1.172	1.245	1.613	½
1	0.483	—	1.968	2.438	2.730	2.962	3.140	3.610	1
1½	0.571	—	2.978	3.848	4.626	5.102	5.359	5.886	1½
2	0.660	—	4.324	5.534	6.575	7.315	7.597	8.093	2
3	"	0.860	7.509	8.909	10.173	10.792	10.956	11.195	3
4	"	3.832	10.293	11.208	11.830	11.986	12.007	12.033	4
5	"	9.144	11.417	11.916	12.039	12.039	12.039	12.039	5
6	"	6.268	10.344	11.338	11.983	12.039	12.039	12.039	6
7	"	2.324	6.833	8.490	10.627	11.703	11.899	12.033	7
8	"	0.679	3.314	4.423	6.788	9.458	10.271	11.246	8
8½	"	0.660	2.207	2.896	4.518	7.306	8.417	9.976	8½
9	0.660	—	1.445	1.778	2.508	4.677	5.962	7.973	9
9½	0.432	—	0.549	0.568	0.600	1.553	3.057	5.410	9½
10, AP	—	—	—	—	—	—	—	2.130	10, AP
10-ft aft (3.048m)	—	—	—	—	—	—	—	—	10-ft aft (3.048m)

Table 1 (continued)

	Halfbreadths, m			Buttock Heights, m					
Station	40-ft WL 12.192 m	Main Deck	Foc'sle Deck	I 4-ft 1.219 m	II 8-ft 2.438 m	III 16-ft 4.877 m	IV 24-ft 7.315 m	V 32-ft 9.754 m	Station
0, FP	0.879	2.337	4.477	12.872	14.967	—	—	—	0, FP
½	2.775	4.483	6.674	0.911	11.586	15.335	—	—	½
1	4.823	6.518	8.477	0.378	2.438	12.284	15.888	—	1
1½	6.988	8.404	9.934	0.178	0.787	6.401	12.808	17.066	1½
2	8.979	9.966	11.011	0.083	0.368	1.654	7.315	14.167	2
3	11.484	11.716	11.944	0.019	0.089	0.359	1.111	3.810	3
4	12.039	12.039	—	0.016	0.048	0.117	0.251	0.851	4
5	"	"	—	"	"	0.111	0.178	0.279	5
6	"	"	—	"	"	0.111	0.213	0.835	6
7	12.039	"	—	0.016	0.048	0.394	1.524	3.708	7
8	11.932	12.039	—	0.044	0.517	2.953	5.347	7.630	8
8½	11.389	11.890	—	0.175	1.600	5.264	7.325	9.503	8½
9	10.252	11.370	—	0.676	4.712	7.461	9.223	11.510	9
9½	8.236	10.001	—	7.074	7.852	9.389	11.271	14.478	9½
10, AP	4.861	6.826	—	9.093	9.989	12.211	—	—	10, AP
10-ft aft (3.048m)	2.658	4.553	—	10.598	11.919	—	—	—	10-ft aft (3.048m)

much as one millimeter = 0.0397 in. ≈ 1/25 in., very nearly.

A complete set of offsets for the various lines of the vessel, arranged in tabular form, is known as a table of offsets. The typical example given in Table 1 applies to the ship shown in Fig. 1. Sometimes such tables are included on the lines plan. The offsets originally supplied to the loftsmen are usually marked "preliminary." After the lines have been faired on the loft floor, another set of offsets, known as the "returned" or "finished" table of offsets is usually lifted from the floor and returned to the drafting office. This finished set should include offsets lifted for every frame station throughout the ship, in addition to those for the lines stations.

Fairing ship lines on a mold loft floor is time consuming and requires substantial amounts of floor space. To overcome these disadvantages, some shipyards in the 1950s began to have the preliminary lines redrawn and refaired to a scale of one-tenth of full size, the work being done on large drawing tables with precise drafting instruments. Originally 1/10-scale drawings of structural parts were made and photographically reduced to 1/100 or less of full scale. The photographic negatives were then projected optically full-size onto the plates for marking and cutting. Later

12 PRINCIPLES OF NAVAL ARCHITECTURE

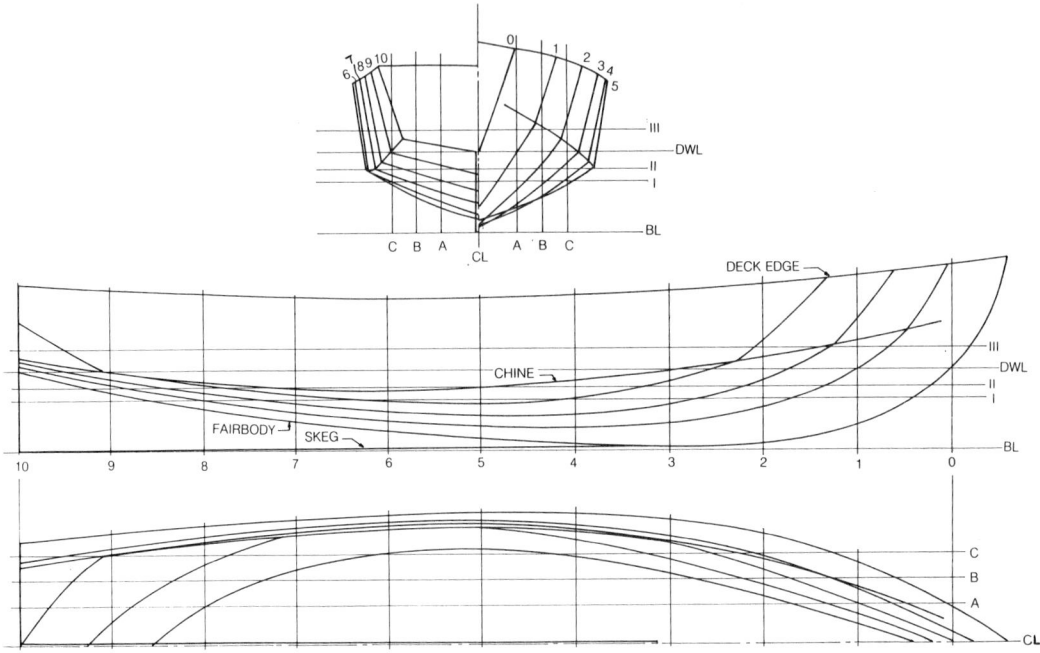

Fig. 7 Lines of small developable surface vessel

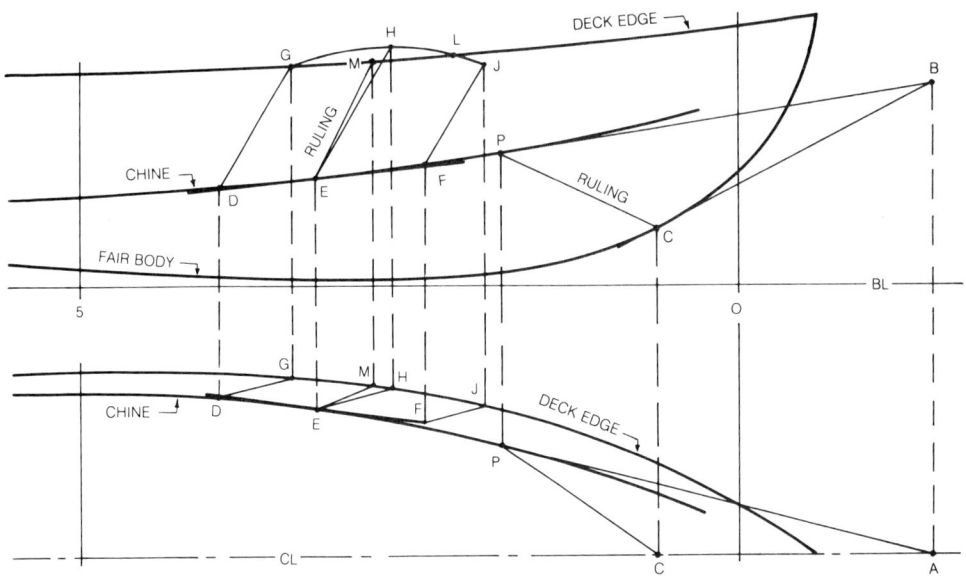

Fig. 8 Construction drawing for developable surface lines

To find ruling between deck edge and chine:
1. Draw $\overline{DF}$, tangent to chine at E.
2. Draw $\overline{DG}$, $\overline{EH}$ and $\overline{FJ}$ in half-breadth plan parallel to each other at arbitrary angle.
3. Project G to deck edge in profile.
4. Find points M and J in profile as projections from half-breadth where $\overline{DG}$, $\overline{EH}$ and $\overline{FJ}$ are parallel.
5. Draw curve GHJ in profile cutting deck edge at L.
6. M is midpoint of GL and is end of ruling $\overline{EM}$
Plane DGJF is tangent to chine at E.

Curve GHLJ is the intersection of plane DGJF with a cylindrical surface with vertical elements through deck edge.

To find ruling between stem profile and chine:
1. Draw $\overline{PA}$ and $\overline{PB}$ tangent to chine at P.
2. Project point A in half breadth at CL, to point B in profile.
3. Draw $\overline{BC}$ tangent to stem profile at C, giving end C of ruling $\overline{PC}$.

the 1/10-scale drawings were used directly as templates for optically controlled burning machines. Offsets from the 1/10-scale lines were considered to be the finished offsets.

During the 1960s and 1970s there was rapid development of the use of computers, and most large shipyards now use computers as an aid to the entire process of fairing the lines (section 1.16), computing the offsets, and preparing numerical control for automatic flame cutting. See Chapter XVI, *Ship Design and Construction* (Taggart, 1980) for further details.

1.14 Developable and Straight-Frame Lines. Ship hull forms as traditionally designed are composed of surfaces of compound curvature, such that the intersection of any plane with the surface will form a curved line. A simpler type of curvilinear surface is one on which the intersection of certain planes with the surface form straight lines—called *rulings* or *elements*—which never cross each other. Such a surface is known as a developable surface because it is possible to unbend or unroll the surface and flatten it into a plane. Hence, it will usually be a portion of cylinders or of cones. Correspondingly, it is possible to form a developable surface from a plane surface, such as a sheet of paper or a sheet of steel, by bending it in only one direction along successive rulings.

Hull forms composed entirely of developable surfaces have been successfully designed, particularly for smaller vessels, but they have potential value for larger vessels as well. The different surfaces usually connect at *chines*, or curved knuckle lines, which should be oriented to follow lines of flow as much as possible.

According to the method described by Kilgore (1967), a developable surface can be formed to include two arbitrary curves in space, provided the curvature of the projections of the two curves on the planes of a Cartesian system always have the same sign. However, this is not a necessary condition and the drafter must rely on experience and sometimes on trial. Michelsen, in discussing Kilgore (1967), gives methods by which the existence of a developable surface between two space curves may be checked.

Chine and deck edge lines usually meet the constant curvature sign condition. When they do not, test construction lines may be drawn according to the method outlined to see if rulings can be determined. If none are found, one or both of the curves may be modified and rechecked. In general, it may be said that straight line stems and points of inflection in the chine and deck edge should be avoided. Points of inflection should also be avoided in the intersection of the bottom surface of the vessel with the longitudinal centerplane, defined as *fairbody* line (Kilgore, 1967).

Fig. 7 shows the lines for a small single-chine developable surface vessel. It will be seen that body plan stations, especially forward between the fairbody line and the chine, are slightly convex when seen from the exterior, which is characteristic of developable surface lines in general.

The construction drawing, Fig. 8, shows how rulings on the side between deck edge and chine may be drawn. A ruling is not only a straight line element of the developable surface, but also lies in a plane tangent to it. The construction at the bow shows how rulings are found between the stem profile and the chine. Rulings in the bottom surface may be found in a manner analagous to that used between the chine and deck edge. Kilgore (1967) provides the basic theorems which govern the determination of rulings and the uniqueness of the resulting developable surface. A useful feature of rulings is that tangents to buttocks in the profile view at a single ruling are parallel to each other, and tangents to waterlines in the halfbreadth plan at a single ruling are parallel to each other.

The process of drawing lines for a developable hull form is, therefore, one of finding rulings between chine and deck edge, between pairs of chines, and between chine and fairbody line. Once the rulings are found, it is a relatively simple task to find stations, waterlines and buttocks using normal projection techniques, inasmuch as the rulings are defined in two views.

Fig. 9 shows the body plan of a comparable straight-frame vessel and it may be seen that differences, compared with the developable surface vessel, are relatively small. Thus, the choice of designing a vessel with one system or the other may well depend upon comparing the difficulty of forming curved frames—together with the ease of plating the developable shell—with the ease of forming straight frames—together with the difficulty of plating a warped shell.

1.15 Methodical Lines. If the naval architect wishes to draw lines representing a particular type of ship, there are in the open literature several *methodical series* of hull forms which permit offsets to be developed directly, without the necessity of going through the fairing process. By a methodical series is

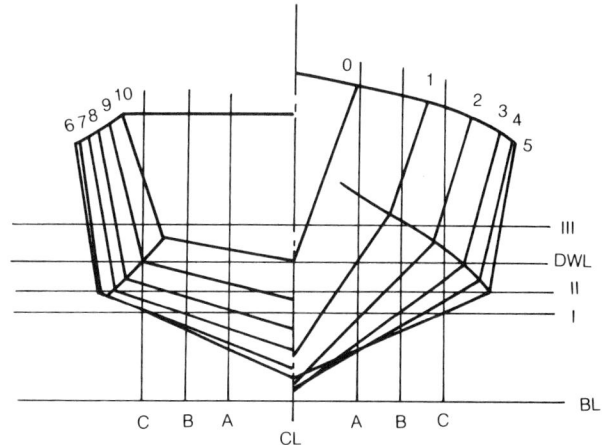

Fig. 9 Straight frame body plan of vessel

meant a group of uniquely related forms whereby specific offsets may be obtained from charts or tables for given arbitrary input characteristics, especially prismatic coefficient, length-to-breadth ratio, volumetric coefficient, and position of the longitudinal center of buoyancy. Particular examples are the Series 60 single screw merchant ship forms (Todd, et al, 1957), Taylor's Standard Series (Taylor, 1943), Townsend's seakeeping series of single screw forms (Townsend, 1967), the MARAD low L/B series (1987), and the Webb trawler series (Ridgely-Nevitt, 1963). Furthermore, by following methods outlined in Section 3.3, the lines of any specific hull form may be transformed in a methodical way to suit arbitrary hull form characteristics.

The constraints of end details—stern frame ending, size and location of rudder, stem profile, etc.—with which such methodically chosen lines are endowed often leads to "tailoring" of the resulting lines at the ends, which may require refairing substantial portions of the hull surface.

1.16 Use of Computer in Lines Definition; Mathematical Lines. Among the most useful applications of digital computers in naval architecture, giving direct geometrical answers, are:

(a) Determining lines and offsets to suit arbitrary hull form characteristics derived from a prescribed parent form.

(b) Final fairing and determination of closely spaced frame offsets for shipyard use based upon widely spaced preliminary design offsets.

The first of these applications provides the capability to carry out by computer and in a more general fashion what Adm. D. W. Taylor began prior to World War I, that is, to design ship lines mathematically. By the method of Taylor (1915) waterlines and sectional area curves took the form of a 5th order curve, separately for forebody and afterbody,

$$y = tx + ax^2 + bx^3 + cx^4 + dx^5,$$

where t, a, b, c and d are constants.

By suitable transformation, this equation was rewritten as,

$$y = C_y + PC_p + tC_t + \alpha C_\alpha,$$

where P is the waterplane area coefficient (for a water line), or prismatic coefficient (for a sectional area curve) of forebody or afterbody, t is the tangent of the curve at bow or stern, and α is a function of the second derivative of the curve at amidships ($x = 1.0$). A simple table was provided giving values of the coefficients, C, which were fixed for each body plan station. The resulting curves had at most one point of inflection. This method was used to draw the lines for Taylor's Standard Series (of ship models used in resistance tests—see Chapter V). Taylor (1915) noted that, "practically all U.S. naval vessels designed during the last ten years have had mathematical lines."

During the intervening 65 years, the use of mathematical ship lines appears to have declined until the advent of computers. A number of successful attempts have now been reported (Fuller, et al, 1977 and Söding, et al, 1977, for example) where ship lines in keeping with hull forms favored today have been produced and plotted with the aid of the computer. Polynomials of higher order than used by Taylor have been used for waterlines and sectional area curves, with particular attention taken to avoid unwanted points of inflection. However, unless some adjustment is done to the end profiles, the resulting hull forms are endowed with waterline endings or stern profiles that may not satisfy the user. Kuiper (1970) presented a method whereby the design waterline is expressed as two eight-term polynomials, one for forebody and one for afterbody, which are easily determined using the basic hull form characteristics. However, to define the hull form above and below the design waterline requires the use of seventeen form parameters which must be defined at all drafts, for forebody and afterbody.

Recently somewhat different computer techniques have been developed to assist in the early stage of lines development. For example, a Ship Hull Form Generator Program (HULGEN) was developed in the Ship Design Division of NAVSEC. (Fuller, et al, 1977). The key to this program is the use of polynomials in various combinations to build up a line-for-line definition of the hull form that is remarkably fair. The strength of the program is the user-oriented interactive-graphics method of data input, display and modification. Results of variations of parameters can be viewed instantly, or the hull form can be stretched and distorted into shapes to maintain those parameters.

The second application is that of final fairing of preliminary lines, which necessarily embodies judgment, in that the drafter's eye and opinion ultimately determine fairness. In this process, the drafter, or the mold loftsman, is faced with the problem of passing a curve through a set of points, usually equally spaced along a reference axis, and satisfying himself that the curve is smooth, with a minimum number of points of inflection and with curvature varying in a gradual way. In order to achieve fairness, the curve may have to miss some of the points by small amounts. Also, for consistency, interesecting curves in other views which contain these points must be checked and adjusted.

As previously noted (1.11), battens or splines are commonly used in drawing such curves, with batten weights (ducks) positioned to hold the batten at or near the given points. Therefore, computerized representations of ship lines often make use of the equations for spline curves. The bending induced in the batten by the ducks is describable by the theory of bending of a simple weightless beam with concentrated loads or supports at a series of discrete points, corresponding to the points of duck restraint. It is shown in Strength of Materials texts that the deflection of such a beam is given by polynomials no higher than the

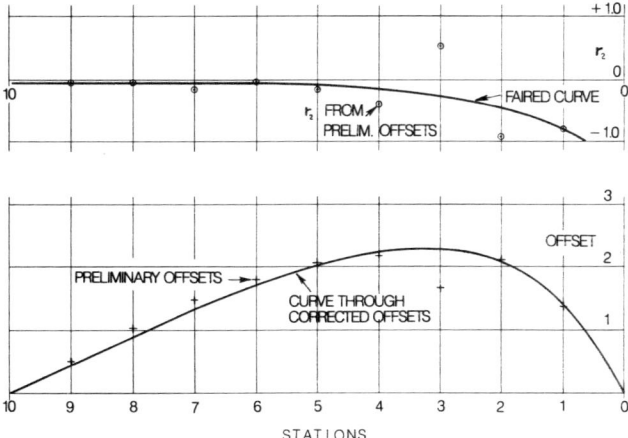

Fig. 10 Unfair and faired rudder section

third order, that is, by cubic functions of the x-dimension parallel with the beam. Such cubics have continuity in first and second derivatives ($y' = dy/dx$, $y'' = d^2y/dx^2$) at the points of application of the concentrated loads.

An assumption made in developing the deflection equation for a simple beam is that the deflections are small, and the beam assumes only small angles to the x-axis. Under these conditions the differential equation of bending is

$$M(x)/EI = y''/[1+(y')^2]^{3/2}.$$

This can be linearized by assuming that $(y')^2 = 0$, inasmuch as y' is small; thus,

$$M(x)/EI \approx y''$$

and y'' becomes a linear function of x. A close approximation to y'', at point n for example, is given by the second difference r_2, where

$$r_2 = (y_{n+1} - 2y_n + y_{n-1})/h^2,$$

and h is the spacing between any pair of equally spaced points. Inasmuch as r_2 is quite sensitive to changes in curvature, it is apparent that by adopting r_2 values from a smooth curve, and by adjusting offsets to match the faired r_2 values, a curve through the adjusted offsets will usually be quite fair.

For illustrative purposes, Fig. 10 plots rough—and obviously unfair—points representing preliminary offsets of a rudder section. Also shown is a plot of r_2 from the given offsets, and a smooth curve interpreting the plot but missing some of the points.

The final faired rudder profile curve has been obtained from the smooth curve of r_2, beginning at the nose of the section and working aft, but with the addition of two additional linear corrections, first to make the tail of the section sharp, and second, to make the average value of the faired offsets equal to the mean value of the given offsets.

The spline curve representation of ship lines by the differential equation of the deflection of a simple beam may become unrealistic when the slope y' of the line being represented becomes so large that it cannot be assumed equal to zero. Thus, most ships' waterlines can readily be represented by spline curve equations over most of the length of the ship. However, such is not the case for body plan stations, nor for many buttocks, especially near the ends of the ship, where steep slopes are often met. In order to overcome this problem, some early attempts to define ship lines with the aid of a computer required that the coordinate reference axes be rotated. More recently, lines have been expressed as parametric spline curves, by which the curves are defined by a parameter s, rather than directly by x,y coordinates. The parameter s is defined as the cumulative length of segments of the line from the start point up to the point in question (IIT Research Institute, 1980).

A computer program in which this representation is used is HULDEF, developed by the U.S. Navy for design use but now extended and made available to a number of shipyards in the U.S. for final hull form definition in the construction phase. HULDEF is said to be economical of computer time, and has been made compatible with other computer-based hull production programs. By HULDEF lines along the hull are developed from the given input waterlines and buttocks into *iso-girth* lines, formed by taking fixed percentages of the girthed length around each body plan station from centerline to deck edge (or chine) all along the hull from the tip of the bow to the stern. The lines are mathematized as parametric spline curves. The HULDEF system has been provided with interactive graphics capability so that the operator can readily display curves, first differences, and second differences, and can fair these on the scope to suit his own idea of fairness. This puts the fairing capability under the control and judgment of the operator just as it has been in the past under the control of the traditional drafter or mold loftsman. However, the previous time consuming operation of drawing the line—on a drafting table or on the mold loft floor—is no longer needed (Fuller, et al, 1977). Other similar systems are in use in some U.S. shipyards and abroad.

Computer applications in hydrostatic calculations are discussed in Section 5.16.

Section 2
Displacement and Weight Relationships

2.1 Archimedes' Principle. The fundamental physical law controlling the static behavior of a body wholly or partially immersed in a fluid is known as Archimedes' Principle which, as normally expressed, states that a body immersed in a fluid is buoyed up by a force that equals the weight of the displaced fluid.[2] Thus, the weight is considered to be a downward force that is proportional to the body's mass; the equal buoyant force is proportional to the mass of the displaced fluid.

Consider a body of fluid such as water, with a free surface, at rest. The fluid is of constant mass density, ρ (i.e., mass per unit volume). At any point P, a distance t below the free surface, the mass of fluid above the point is $\rho A t$, where A is the cross sectional area parallel to the free surface of the column of fluid. In general, a fluid cannot support shear forces. Therefore, if the fluid be in a state of static equilibrium, it is necessary that equal forces be experienced in all directions at any such point. Since the gravitational force resulting from the mass of the fluid above is equal to its mass $\times g$, the pressure force experienced by the fluid at that point is $\rho g A t$—or the weight of the column of fluid above P.

If a rigid body is afloat in the water in static equilibrium, Fig. 11, a consequence of the above reasoning is that the same pressure forces are directed normal to the surface of the body. The integration of the vertical component of all such pressures experienced by the surface S of the body is the buoyant force,

$$\sum_{o}^{S} \rho g t \cos \alpha \, \delta s,$$

where α is the inclination of any part of S from the horizontal. But $\sum_{o}^{S} t \cos \alpha \, \delta s$ represents the volume of the body beneath a plane coincident with the free surface. The weight of fluid would occupy this volume in the absence of the body is identically equal to ρg multiplied by the volume.

For the body to be in equilibrium, the integration of upward components of hydrostatic pressures over the surface of the body, or buoyancy, must be exactly balanced by the gravitational force of the body's mass, directed downward, i.e., its weight. Therefore, the weight of a ship and its contents is equal to the weight of displaced water, or displacement. Likewise, the mass of a ship and its contents is equal to the mass

---
[2] First stated about 250 BC by Archimedes, Greek mathematician and inventor (C.287-212BC).

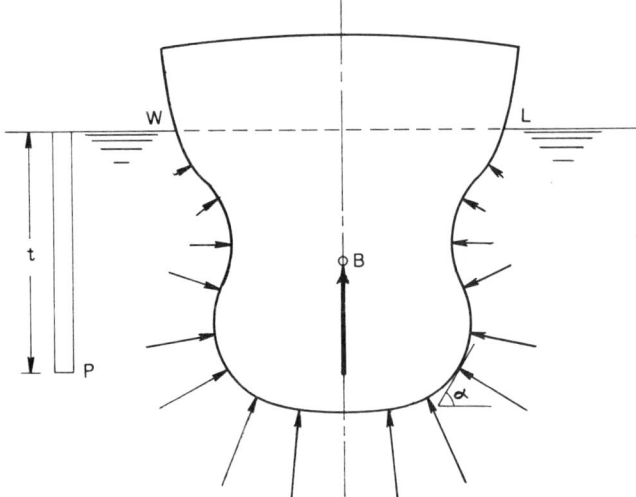

Fig. 11 Buoyant forces on a floating body

of displaced water. Hence, displacement can be expressed in either weight or mass units.

It is evident that a fully immersed rigid body, such as a submarine, also experiences an upward buoyant force equal and opposite to the weight of the water it displaces. A totally submerged body may weigh either more or less than the displaced water. For the body to be in equilibrium in its submerged position, it would have to receive, in the first case, an additional upward force, and in the second case, an additional downward force. When submerged and not resting on the bottom, a body may remain stationary, without rising or falling, only in the unusual case when its mass exactly equals the mass of the water it displaces.

2.2 Displacement and Center of Buoyancy. The volume of the underwater portion of a vessel may be calculated by methods outlined in Sections 4 and 5. The result is known as the volume of displacement, ∇, up to the waterline at which the vessel is floating.

If we know the mass density of the water, ρ, in which a ship is floating, we can calculate the weight of the displaced fluid, or the displacement weight, W,

$$W = \rho g \nabla. \qquad (1)$$

By Archimedes' principle this weight is equal to the weight of the ship and its contents. In inch-pound (or "English") units, W is in long tons if ∇ is in ft^3 and $\rho g = 1/35.9$ long tons/ft^3 (62.4 lb/ft^3) in fresh water (FW) or $\rho g = 1/35.0$ long tons per ft^3 (64.0 lb/ft^3) in salt water (SW); i.e.,

$W = \nabla/35.9$ or $\nabla/35.0$ long tons (2240 lb per ton) in FW or SW, respectively.

In SI (*Système International*), the above expression for displacement weight (Eq. 1) applies if units of force are newtons (with ρ in kg/m³) or kilonewtons (with ρ in t/m³). In FW the value of ρg is approximately 9.81 kN/m³ ($\rho = 1.0$ t/m³) and in SW ρg is 10.06 kN/m³ ($\rho = 1.026$ t/m³). Such units are common in resistance and propulsion calculations (Chapter V).

However, adherance to the SI system obliges one to think of ship displacement, Δ, in mass units, rather than weight (force) units, with the unit of mass being a multiple of grams, such as a kilogram (1000 grams), or a metric ton (1000 kilograms) t, sometimes written as "tonne."[3] Hence, in the SI system, mass displacement,

$$\Delta = \rho \nabla, \qquad (2)$$

where Δ is in metric tons, ∇ is in m³, $\rho = 1.00$ t/m³ (equal to kg/L) in FW and $\rho = 1.026$ t/m³ in SW. For the above relationship to be true in inch-pound units, Δ would have to be expressed in lb-sec²/ft or in the seldom-used *slugs* and ρ in lb-sec²/ft⁴ or in slugs/ft³.

Since the mass density of fresh water is 1.0 kg/L or 1.0 t/m³, density is numerically the same in SI units as specific gravity, γ (at standard temperature). Hence, it may be more convenient when using SI units to use,

$$\Delta = \gamma \nabla. \qquad (3)$$

Again this is true in inch-pound units only if Δ is in lb-sec²/ft or in slugs.

Sometimes naval architects prefer (as in Chapter II) to make use of the reciprocal of density, or specific volume, δ (volume per unit mass), in their calculations. For fresh water, of course, $\delta = 1.00$ m³/t; for salt water $\delta = 1/1.026 = 0.975$ m³/t.

We shall in general consider ship displacement in units of metric tons of mass, where one metric ton is equal to the mass of one cubic meter of fresh water (at standard temperature), i.e., $\rho = 1.0$ m³/t. It should be noted that one m³ of fresh water in inch-pound units is 2204 lb or 0.9839 long ton (2240 lb/ton). Hence, it can be seen that one SI ton is roughly equivalent to a long ton of weight (1.6 percent error) in the inch-pound system. The term weight will often be used loosely to mean either weight in tons (lb) or mass in metric tons (kg.).

The centroid of the underwater portion of a vessel may be calculated by the principle of moments, using methods also outlined in Sections 4 and 5. The centroid is called the center of buoyancy. It represents a point through which the vertical buoyant vector is considered to pass, i.e., point B in Fig. 11.

2.3 Effect of Density of Medium. A decrease in the density of the fluid in which a vessel floats requires an increase in the volume of displacement ∇ in order to satisfy static equilibrium requirements. Therefore, a ship moving from salt water to fresh water, for example, experiences an increase in draft, δT. This increase can be calculated by equating the increase in displacement volume to the volume of a layer of buoyancy of uniform thickness, δT, distributed over the original load waterplane. The increase in displacement volume,

$$\nabla_F - \nabla_S = \nabla_S \frac{\rho_S}{\rho_F} - \nabla_S = \nabla_S \left(\frac{\rho_S}{\rho_F} - 1\right)$$

where subscript S refers to salt water, subscript F to fresh water. But, on the assumption that the ship is "wall-sided," the equal layer of buoyancy is,

$$\nabla_F - \nabla_S = A_{WP} \cdot \delta T$$

Hence,

$$A_{WP} \cdot \delta T = \nabla_S \left(\frac{\rho_S}{\rho_F} - 1\right)$$

and the increase in draft is,

$$\delta T = \frac{\nabla_S}{A_{WP}} \left(\frac{\rho_S}{\rho_F} - 1\right) = \frac{\nabla_S (\gamma_S - 1)}{A_{WP}} \qquad (4)$$

where ∇ is displacement volume, ρ is mass density, A_{WP} is waterplane area, and $\frac{\rho_S}{\rho_F} = \gamma_S$ is specific gravity.

The centroid of the underwater body may shift, both vertically and longitudinally, with such a change in medium. In particular, an increase in draft as a result of a decrease in fluid density causes the vertical location of the center of buoyancy to rise with respect to the keel as a result of the increase in displacement volume, ∇.

When a ship becomes partially supported by mud of mass density ρ_M, the volume of displacement must decrease to the point where the sum of products of volume of displacement in the medium multiplied by the density of the medium equals the weight of the vessel. Correspondingly, the center of buoyancy may be found, using methods in Sects. 4 and 5 by calcu-

---

[3] Since this edition of *Principles of Naval Architecture* incorporates the transition from inch-pounds to SI units, reference will in general be made to both systems. The distinction, however, should always be borne in mind between inch-pound weight units and SI mass units.

lating the buoyant moment as the sum of products of buoyancy from each medium, multiplied by the distance to the centroid of each volume.

It is important to use the correct density of the water in making displacement calculations. There is about a 2½ percent difference between the density of fresh water, as in the Great Lakes, and the salt water of the oceans. The water in some rivers and harbors and off the mouth of estuaries is usually brackish, and its density may vary considerably with the tides. When draft readings are taken to determine displacement, samples of the water should be taken at the same time in order to determine its density.

In principle, since the density of water changes slightly with temperature, a correction should be made to account for any differences from an agreed upon temperature standard. Furthermore, the temperature coefficient of expansion of steel may influence the volume of displacement of a steel vessel slightly up to any waterline if the temperature of the steel differs significantly from the standard.

2.4 Displacement vs. Weight Estimate. When preparing the design for a proposed ship, a careful estimate of its total weight and position of its center of gravity should be made, as discussed in Chapter II, Section 2. The total weight thus estimated may be compared later with the total displacement obtained from draft readings after the ship is afloat. If differences occur, as is usually the case, the error is assumed to be in the weight estimate. A check of the accuracy of the weight estimate for the vessel in its partially completed condition is usually first obtainable immediately after launching. As previously noted, these "weights" may be given in mass units.

Section 3
Coefficients of Form

3.1 General. In comparing ships' hull forms, displacements and dimensions, a number of coefficients are used in naval architecture. These coefficients are useful in power estimates and in expressing the fullness of a ship's overall form and those of the body plan sections and waterlines. Table 2 lists coefficients and particulars for a number of typical vessels, which will be found helpful in understanding the significance of the coefficients defined below.

Section 3 and Table 2 define and discuss the Block Coefficient, Midship Coefficient, Waterplane Coefficient Vertical Prismatic Coefficient, and Volumetric Coefficient. Table 2 also gives the general geometrical characteristics of 19 types of ships, ranging from a large, high-speed passenger liner capable of 33 knots sustained sea speed to a naval dock ship 171 m (555 ft) in length.

3.2 Definitions and Uses of Coefficients. (a) *Block Coefficient, C_B.* This is defined as the ratio of the volume of displacement ∇ of the molded form up to any waterline to the volume of a rectangular prism with length, breadth and depth equal to the length, breadth and mean draft of the ship, at that waterline.

Thus,

$$C_B = \frac{\nabla}{L \cdot B \cdot T}$$

where L is length, B is breadth and T is mean molded draft to the prevailing waterline. Practice varies regarding L and B. Some authorities take L as LBP, some as LWL, and some as an effective length, as discussed in Section 1.2. B may be taken as the molded breadth at the design waterline and at amidships, the maximum molded breadth at a selected waterline (not necessarily at amidships), or according to another standard. Most merchant ships have vertical sides amidships, with upper waterlines parallel to the centerline, thereby removing possible ambiguity in B.

Values of C_B at design displacement may vary from about 0.36 for a fine high-speed vessel to about 0.92 for a slow and full Great Lakes bulk carrier.

(b) *Midship Coefficient, C_M.* The midship section coefficient, C_M, sometimes called simply midship coefficient, at any draft is the ratio of the immersed area of the midship station to that of a rectangle of breadth equal to molded breadth and depth equal to the molded draft amidships.

Thus,

$$C_M = \frac{\text{Immersed area of midship section}}{B \cdot T}$$

Values of C_M may range from about 0.75 to 0.995 for normal ships, while for vessels of extreme form with a slack bilge and a hollow *garboard* area (immediately outboard of the keel) amidships, C_M might be as low as 0.62. In some cases vessels have been

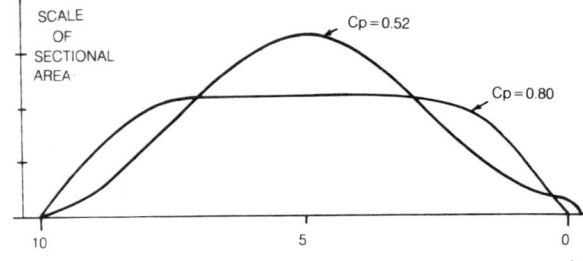

Fig. 12 Sectional area curves for different prismatic coefficients

built with bulges or *blisters* below the design waterline. Assuming B is taken at the prevailing waterline, then C_M may be greater than unity on such vessels.

(c) Prismatic Coefficient, C_P. The prismatic coefficient, sometimes called longitudinal prismatic coefficient, or simply longitudinal coefficient, gives the ratio between the volume of displacement ∇ and a prism whose length equals the length of the ship and whose cross section equals the midship section area.

Thus

$$C_P = \frac{\nabla}{L \times \text{immersed area of midship section}}$$
$$= \frac{\nabla}{L \cdot B \cdot T \cdot C_M} = \frac{C_B}{C_M}.$$

The term longitudinal coefficient was originated and used by Adm. D. W. Taylor (1943) for the reason that this coefficient is a measure of the longitudinal distribution of a ship's buoyancy. If two ships with equal length and displacement have different prismatic coefficients, the one with the smaller value of C_P will have the larger midship sectional area ($B \cdot T \cdot C_M$) and hence a larger concentration of the volume of displacement amidships. This is clearly shown by Fig. 12, which compares the sectional area curves for two different vessels. The ship with the smaller C_P is also characterized by a protruding bulbous bow, which causes the swelling in the sectional area curve right at the bow, and its extension forward of Station O.

Prismatic coefficient is a frequently used parameter in studies of speed and power (Chapter V). Usual range of values is from about 0.50 to about 0.90. A vessel with a low value of C_P (or C_B) is said to have a fine hull form, while one with a high value of C_P has a full hull form.

(d) Waterplane Coefficient, C_{WP}. The waterplane coefficient is defined as the ratio between the area of the waterplane A_{WP} and the area of a circumscribing rectangle. Thus,

$$C_{WP} = \frac{A_{WP}}{L \cdot B}$$

As with the other coefficients, the length and breadth are not always taken in a standard way. The coefficient may be evaluated at any draft. The values of C_{WP} at the DWL range from about 0.65 to 0.95, depending upon type of ship, speed, and other factors.

(e) Vertical Prismatic Coefficient, C_{VP}. This coefficient is the ratio of the volume of a vessel's displacement to the volume of a cylindrical solid with a depth equal to the vessel's molded mean draft and with a uniform horizontal cross section equal to the area of the vessel's waterplane at that draft. This ratio is analogous to the prismatic or longitudinal coefficient, except that the draft and area of waterplane have been substituted for the vessel's length and area of midship section. The vertical prismatic coefficient of fineness is designated as C_{VP} and written as follows:

$$C_{VP} = \frac{\nabla}{C_{WP} \times L \times B \times T} = \frac{C_B}{C_{WP}}$$

(f) Volumetric Coefficient, C_V. This coefficient (or fatness ratio) is defined as the volume of displacement divided by the cube of one tenth of the vessel's length, or

$$C_V = \nabla/(L/10)^3$$

In essence, it is the dimensionless equivalent of displacement-length ratio, $\Delta / \left(\frac{L}{100}\right)^3$ frequently used in the past, where Δ is ship displacement in long tons in salt water, and L is ship length in feet. These coefficients express the displacement of a vessel in terms of its length. Ships with low volumetric coefficients might be said to be "thin", while those with a high coefficient are "fat." Values of the volumetric coefficient range from about 1.0 for light, long ships like destroyers, to 15 for short heavy ships like trawlers.

(g) Ratios of Dimensions. The three principal dimensions of the underwater body are sometimes referred to in ratio form. These are noted below, with approximate ranges for each:

Ratio of length to breadth = L/B Approx. range 3.5 to 10.
Ratio of length to draft = L/T Approx. range 10 to 30.
Ratio of breadth to draft = B/T Approx. range 1.8 to 5.

In view of the confusion which can arise when different definitions of dimensions—especially length—are used by different designers in forming the above coefficients and ratios, it is suggested that length between perpendiculars—on single-screw ships—and molded breadth at the design waterline and at amidships be used in forming these ratios. The length on the DWL is preferred for twin-screw ships (see Section 1.2). The definitions adopted should always be specified.

3.3 Geometrical Modification to Lines. It frequently happens during the design of a ship that unexpected requirements which were not foreseen necessitate a change in dimensions without changing the coefficients of form. Examples are an increase in breadth to provide greater stability, a decrease in design draft to allow entering a port with restricted water depth, or an increase in length to reduce wavemaking resistance. If this should happen after preliminary lines are faired, it seemingly requires that a

Table 2—Geometrical Characteristics of Typical Ships

	1 Pass. Liner	2 Cargo-Pass. Ship	3 Container Ship	4 Container Ship	5 Gen. Cargo Ship	6 Barge Carrier	7 Roll on/Roll off Ship	8 Bulk Carrier	9 Gt. Lakes Ore Carrier
Length overall, m	301.75	166.60	262.13	185.93	171.80	272.29	208.48	272.03	304.80
Length between perpendiculars, L_{pp}, m	275.92	154.99	246.89	177.09	171.80	243.03	195.07	260.60	301.30
Length for coefficients, L, m	286.99	154.05	246.89	176.78	158.50	247.90	195.07	260.60	301.30
Molded depth to strength dk., m	22.63	14.66	20.12	16.61	13.56	18.29	21.18	19.05	14.94
Molded breadth, B, m	30.94	24.08	32.23	23.77	23.16	30.48	31.09	32.23	31.88
Molded draft for coeffs., T, m	9.65	8.23	10.67	8.23	8.23	8.53	9.75	13.96	7.85
Molded displacement, Δ, S.W., t	46,720	18,250	50,370	22,380	18,970	38,400	34,430	100,500	71,440
Block coefficient, C_B	0.532	0.583	0.579	0.630	0.612	0.582	0.568	0.836	0.924
Midship coefficient, C_M	0.953	0.967	0.965	0.975	0.981	0.922	0.972	0.996	0.999
Prismatic coefficient, C_P	0.558	0.603	0.600	0.646	0.624	0.631	0.584	0.839	0.924
Waterplane coefficient, C_W	0.687	0.725	0.748	0.675	0.724	0.765	0.671	0.898	0.975
Vertical prismatic coeff., C_{PV}	0.774	0.807	0.774	0.851	0.845	0.762	0.846	0.931	0.948
Longitudinal center of buoyancy from midship, %L	Amids.	Amids.	−1.1	−1.2	−1.5	−1.6	−2.4	+2.5	+0.5
Bulb area, % midship area	2.0	2.5	8.3	4.0	4.0	5.6	9.7	10.7	0
Volumetric coefficient, $(\nabla/L^3) \times 10^3$	1.93	4.87	3.26	3.95	4.65	2.46	5.18	5.54	2.55
L/B	9.28	6.40	7.94	7.44	6.84	8.13	6.27	8.09	9.45
B/T	3.21	2.93	2.91	2.89	2.81	3.57	3.19	2.31	4.06
Shaft horsepower, normal	158,000	18,000	43,200	19,250	17,500	32,060	37,000	24,000	14,000
Sea speed, knots	33	20	25	20	20	22	23	16.5	13.9
Froude number	0.320	0.265	0.261	0.427	0.261	0.229	0.270	0.168	0.132
Number of propellers, rudders	4,1	1,1	1,1	1,1	1,1	1,1	1,1	1,1	2,2

	10 Crude Oil Carrier	11 Petroleum Prods. Tanker	12 LNG Tanker	13 Off-Shore Supply Vessel	14 Double-ended Ferry	15 Fishing Trawler	16 Arctic Ice-breaker	17 Naval Replen. Ship	18 Naval Frigate	19 Naval Dock Ship
Length overall, m	335.28	201.47	285.29	56.46	94.49	25.65	121.62	236.37	135.64	170.99
Length between perpendiculars, L_{pp}, m	323.09	192.02	273.41	53.19	91.59	23.04	106.98	234.70	124.36	164.59
Length for coefficients, L, m	323.09	192.02	273.41	53.19	91.59	23.75	107.29	234.70	124.36	164.59
Molded depth to strength dk., m	26.21	13.79	24.99	4.27	6.30	3.33	13.18	17.07	9.14	13.41
Molded breadth, B, m	54.25	27.43	43.74	12.19	19.81	6.71	23.77	32.72	13.74	24.99
Molded draft for coeffs., T, m	20.39	10.40	10.97	3.35	3.81	2.53	8.53	11.58	4.37	5.41
Molded displacement, Δ, S.W., t	308,700	43,400	97,200	1472.	2760.	222	10,900	52,140	3390	12,850
Block coefficient, C_B	0.842	0.772	0.722	0.660	0.392	0.538	0.488	0.569	0.449	0.563
Midship coefficient, C_M	0.996	0.986	0.995	0.906	0.732	0.833	0.853	0.987	0.741	0.933
Prismatic coefficient, C_P	0.845	0.784	0.726	0.729	0.534	0.646	0.572	0.577	0.605	0.603
Waterplane coefficient, C_W	0.916	0.854	0.797	0.892	0.702	0.872	0.740	0.734	0.727	0.720
Vertical prismatic coeff., C_{PV}	0.919	0.904	0.906	0.740	0.558	0.617	0.660	0.779	0.618	0.782
Longitudinal center of buoyancy from midship, %L	+2.7	+1.9	Amids.	−0.3	Amids.	−1.7	+1.3	−0.9	−1.4	−1.4
Bulb area, % midship area	0	0	9.7	0	0	16.2	0	10.0	0	2.0
Volumetric coefficient, $(\nabla/L^3) \times 10^3$	5.96	5.98	4.64	9.53	3.51	3.54	8.97	3.9	1.7	2.8
L/B	5.96	7.00	6.25	4.35	4.62	2.65	4.51	7.17	9.05	6.59
B/T	2.66	2.64	3.99	3.33	5.20	2.65	2.79	2.82	3.14	4.62
Shaft horsepower, normal	35,000	15,000	34,400	3,740	7,000	500	18,000	100,000	40,000	22,900
Sea speed, knots	15.2	16.5	20.4	12	16.1	10.7	18	26	30	21.5
Froude number	0.139	0.196	0.203	0.270	0.276	0.361	0.285	0.279	0.442	0.275
Number of propellers, rudders	1,1	1,1	1,1	2,2	2,0	1,1	3,1	2,2	1,1	2,2

1) Vessel 10 has a cylindrical bow. 2) Vessel 14 has vertical axis propellers and a fixed skeg at each end.

completely new set of lines be drawn and faired. However, by making systematic changes in the offsets, it may be possible to accomplish the desired transformation without disturbing the fairness of the lines, and without necessitating complete recalculation of the curves of form.

For example, a simple respacing of body plan stations leads to an elongation or shortening of the lines at constant breadth and draft, with displacement changing in direct proportion to the station spacing; the form coefficients C_B, C_P, C_M, C_{WP} and C_{VP} will not change, and the fairness of lines will be preserved. Of the curves of form, changes will be experienced only in those quantities which depend upon length and displacement, including C_V. Correspondingly, an increase in waterline spacing leads to a proportionate change in displacement with no change in C_B, C_P, C_M, C_{WP} and C_{VP}. Those curves of form which are dependent upon displacement and draft are the only ones which will change. Similar conclusions are reached insofar as changes in buttock spacing—that is, changes in halfbreadth—are concerned.

The combined effect of two or more of such changes is multiplicative. For example, if the length of the vessel were to increase 10 percent by an increase in station spacing, the breadth were to increase 5 percent by an increase in halfbreadths, and the draft were to decrease 8 percent by a reduction in waterline spacing, the resulting volume of displacement ∇_2 would be obtained from ∇_1, the initial volume of displacement, by,

$$\nabla_2 = 1.1 \cdot 1.05 \cdot 0.92 \cdot \nabla_1 = 1.0626\, \nabla_1$$

A new body plan, waterlines plan and profile could be drawn directly, in which new longitudinal distances x_2 are obtained from old longitudinal distances x_1 by $x_2 = 1.1\, x_1$; new halfbreadths y_2 are obtained from old halfbreadths y_1 by $y_2 = 1.05\, y_1$; etc.

Changes in the more important curves of form, defined in Section 5, would give,

$$\overline{KB}_2 = 0.92\, \overline{KB}_1$$
$$TP\, cm_2 = 1.1 \cdot 1.05 \cdot TP\, cm_1$$
$$LCF_2 = 1.1\, LCF_1$$
$$LCB_2 = 1.1\, LCB_1$$
$$\overline{KM}_2 = \overline{KB}_2 + \overline{BM}_2$$
$$= 0.92\, \overline{KB}_1 + \left(\frac{1.1 \cdot (1.05)^3}{1.1 \cdot 1.05 \cdot 0.92}\right)\overline{BM}_1$$
$$\overline{KM}_{L_2} = 0.92\, \overline{KB}_1 + \left(\frac{(1.1)^3 \cdot 1.05}{1.1 \cdot 1.05 \cdot 0.92}\right)\overline{BM}_{L_1}$$

Wetted surface, which depends upon girthed distances, does not vary in a simple manner and would have to be recomputed for the transformed design.

Methods have been developed (Rawson & Tupper, 1983) to estimate modifications to the geometrical quantities on the basis of partial derivatives. Inasmuch as these methods assume infinitesimal changes in the independent variables, L, B, etc., they may lead to inaccuracies in practical use. On the other hand, direct calculations to find the transformed quantities are by their nature both exact and correct, and therefore they are recommended.

A traditional and practical way of shifting the LCB of a new design without changing displacement is known as the method of *swinging stations*. Fig. 13 shows the sectional area curve of a ship and the centroid of the area under the curve, the latter having been found from both axes ($\bar{x}$ and $\bar{y}$). If the centroid now be moved forward (or aft) a distance $\delta\bar{x}$, and a straight line be drawn through the shifted position and original base, it will establish an angle γ by which all points on the curve may be similarly shifted so that the desired shift of LCB occurs. Any original body plan station such as station 3 must then be shifted by distance δx. This allows one to find the shift of any offset (height or halfbreadth) forward or aft directly from the transformed sectional area curve. Hence, the waterlines and profile views on the lines plan may be redrawn without refairing being required. From the redrawn waterlines and profile a new body plan, with equally spaced stations, may then be constructed.

A somewhat similar transformation can be done to the separate ends of a sectional area curve with some parallel middle body if one wishes to change the fullness of the design. Let us suppose the forebody of a given sectional area curve has a prismatic coefficient of C_{PF1}, but it is desired to increase this by respacing stations to gain more displacement. The new forebody prismatic is to be C_{PF2}. Thus $\delta C_{PF} = C_{PF2} - C_{PF1}$. Then it can be shown (Lackenby, 1950) that if x_1 is the dimensionless distance from the left-hand axis of the curve, where x_1 lies between 0 and 1.0, the shift forward δx to give the required new prismatic coefficient of the forebody is obtained from,

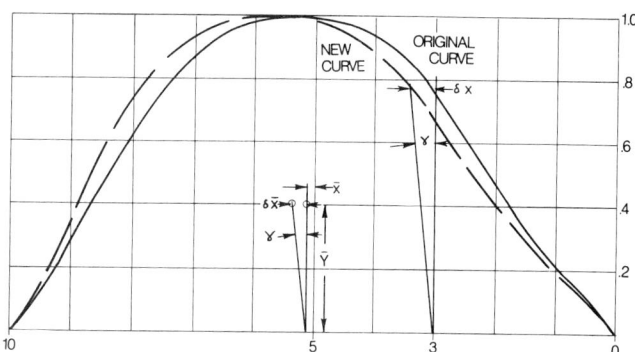

Fig. 13 Swinging stations method of modifying sectional area curve

$$\delta x/(1-x_1) = \delta C_{PF}/(1-C_{PF1}),$$

or

$$\delta x = \delta C_{PF}[(1-x_1)/(1-C_{PF1})]. \quad (5)$$

This procedure, which is known as the *one-minus-prismatic* rule, is illustrated in Fig. 14 (Lackenby, 1950). Having modified the sectional area curve in the indicated way, body plan stations must now be shifted the indicated amount. Thus, the waterlines and profile views in the entrance may be redrawn, with a new body plan for the forebody to suit equally spaced stations. It should be noted, however, that having first transformed the forebody a similar transformation of the afterbody in general leads to a combined longitudinal center of buoyancy of the entire ship which will differ from that of the basic ship before the transformation.

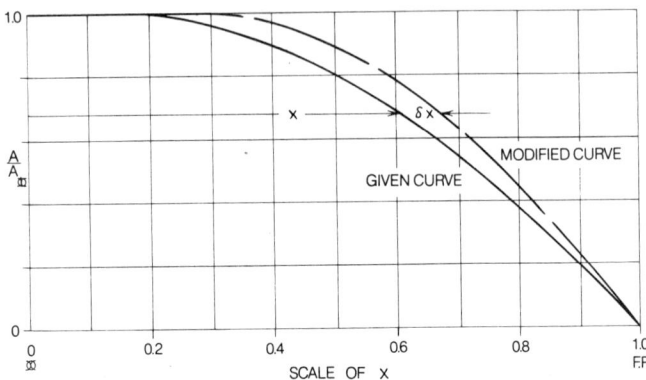

Fig. 14 One-Minus-Prismatic method of modifying sectional area curve

Söding, et al (1977) show an extensive transformation of an existing containership design to a design of widely different particulars following generally the methods of Lackenby (1950).

Section 4
Integrating Rules and Methods

4.1 General. For a variety of reasons, it is necessary to be able to calculate areas, centroids, volumes—and other geometrical characteristics—of a ship's form when floating at any prescribed waterline. Areas of the immersed cross sectional area at each body plan station and of each waterplane are of particular interest, not only for their own sake but because—as will be shown later—volumes can be calculated from areas. Because of the symmetry of the two sides of most vessels, most of these calculations need be performed for only one side of the ship and then multiplied by 2.

Each of the half transverse sections, or half waterplanes, form a closed curve, such as $OABD-GH$ in Fig. 15. The area enclosed may be found by integral calculus, provided $AB-G$ is a curve whose mathematical equation is known. Inasmuch as most ship curves are not mathematical curves, it is customary to approximate the area by numerical integration.

An important property of such a closed curve is its centroid, which is located at a distance from the axis OY equal to $\bar{x}$, where $\bar{x}$ is the quotient of the first moment of the area about axis OY divided by the area itself. If the curve $OABD-GH$ were to represent a thin lamina of uniform density and of constant thickness, then the centroid would represent the location of its center of mass (generally known as center of gravity).

4.2 Formulas For Area, Moment, Centroid, Moment of Intertia, and Gyradius. In Fig. 15 the area enclosed by the x and y-axes and the curve $ABDG$ may be considered as comprised of many small rectangles such as $NBPQ$, of dimensions y and δx, where δx is very small. Using methods of the calculus, we may derive expressions for the area of the curvilinear figure and for various properties of the area.

(a) Areas. Let δA be the area of the elementary rectangle $NBPQ$. Then $\delta A = y\delta x$, and the entire area under the curve, A, is given by the summation of all such elementary areas, or,

$$A = \Sigma \delta A = \Sigma y \delta x.$$

Putting this in the form of a definite integral between the limits 0 and H,

$$A = \int_0^H y\,dx. \quad (6)$$

(b) Moments and Centroids. Let δM_ℓ be the first moment of the area of the elementary rectangle $NBPQ$ about axis OY. Then $\delta M_\ell = (\delta A)x = xy\delta x$. Hence, the moment of the entire area under the curve about axis OY may be written as $M_\ell = \Sigma xy\delta x$, which may be expressed as the definite integral,

$$M_\ell = \int_0^H xy\,dx. \quad (7)$$

The distance $\bar{x}$ of the centroid of the area from axis OY is given by the quotient of moment about OY divided by area or,

$$\bar{x} = \frac{\int_0^H xy\,dx}{\int_0^H y\,dx}. \quad (8)$$

Let δM_t be the first moment of the elementary area $NBPQ$ about the baseline OX. Then

$$\delta M_t = (\delta A)\frac{y}{2} = \frac{y^2}{2}\delta x.$$

The moment of the entire area about the baseline becomes,

$M_t = \frac{1}{2}\Sigma y^2 \delta x$, or in the form of an integral,

$$M_t = \frac{1}{2}\int_0^H y^2 dx. \quad (9)$$

The distance $\bar{y}$ of the centroid of the area from the baseline OX is the quotient of moment about the baseline divided by area, or,

$$\bar{y} = \frac{\frac{1}{2}\int_0^H y^2 dx}{\int_0^H y\,dx}. \quad (10)$$

c Moments of Inertia and Gyradii. Let δI_ℓ be the second moment, or moment of inertia, of the area of the elementary rectangle $NBPQ$ about axis OY. Then $\delta I_\ell = (\delta A)x^2 = x^2 y \delta x$. Hence the moment of inertia of the entire area under the curve about OY, I_ℓ, is,

$$I_\ell = \Sigma x^2 y \delta x \text{ or } I_\ell = \int_0^H x^2 y\,dx. \quad (11)$$

The gyradius r_ℓ of the area about axis OY is given by the square root of the quotient of moment of inertia divided by area, or,

$$r_\ell = \sqrt{\frac{\int_0^H x^2 y\,dx}{\int_0^H y\,dx}}. \quad (12)$$

If $I_{g\ell}$ be the longitudinal moment of inertia of the area under the curve about a transverse axis through the centroid (axis parallel to the Y-axis), we have by the parallel axis principle of mechanics, $I_{g\ell} = I_\ell - A\bar{x}^2$.

The area under the curve AG may also be considered as comprised of many small squares such as $\delta x \delta y$, Fig. 15. Then let δI_t be the second moment, or moment of inertia, of the area of the elementary square about the baseline OX. But $\delta I_t = \delta x \delta y \cdot y^2$. Thus the moment of inertia of the entire area under the curve about the baseline I_t may be written as $I_t = \Sigma\Sigma \delta x \delta y \cdot y^2$, or,

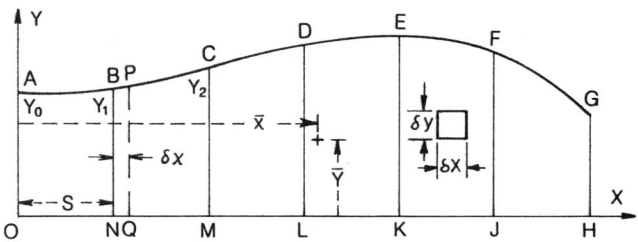

Fig. 15 Curve to be integrated

$$I_t = \int_0^H \int_0^y y^2\,dy\,dx.$$

Since

$\int_0^y y^2 dy = \frac{1}{3}y^3$, then $I_t = \frac{1}{3}\int_0^H y^3 dx. \quad (13)$

The gyradius r_t of the area about the baseline OX is given by,

$$r_t = \sqrt{\frac{\frac{1}{3}\int_0^H y^3 dx}{\int_0^H y\,dx}}. \quad (14)$$

In order to evaluate these integrals, naval architects again overcome the limitation that most ship lines are not represented by mathematical formulas by utilizing approximate rules of integration. A rule of integration assumes that the curve to be integrated is closely approximated by a mathematical curve that has the same offsets (or ordinates) as the actual ship curve at a series of stations. The desired integrals are then approximated by taking the sum of products of offsets and particular multipliers developed for each rule and multiplying the sum by an integrating factor, as described in the following subsections.

4.3 Trapezoidal Rule. In Fig. 15 each portion of the curve AB—G between pairs of ordinates as AB, BC, etc. is considered to be approximated by a straight line through each pair of points. If the spacing between each pair of ordinates is s, then the area of trapezoid $OABN = s\cdot\frac{1}{2}(y_0 + y_1)$, the area of trapezoid $NBCM = s\cdot\frac{1}{2}(y_1 + y_2)$ and the area of trapezoid $JFGH = s\cdot\frac{1}{2}(y_{n-1} + y_n)$. If the areas of all n trapezoids are added, their combined area, and the approximate area A under the curve is,

$$A = s\left(\tfrac{1}{2}y_0 + y_1 + y_2 + \ldots + y_{(n-1)} + \tfrac{1}{2}y_n\right) \quad (15)$$

This is known as the *trapezoidal rule* for area.

Rules for moments of the area based upon the trapezoidal rule may be derived. Thus, the moment M_ℓ of the combined area of all the trapezoids about axis OY is,

$$M_\ell = s^2 \left[\frac{1}{6} y_0 + y_1 + 2y_2 + \ldots \right.$$
$$\left. + (n-1)y_{(n-1)} + \left(\frac{3n-1}{6}\right) y_n \right]. \quad (16)$$

Correspondingly, the moment of inertia I_ℓ of the combined area of all the trapezoids about the OY axis is,

$$I_\ell = s^3 \left[\frac{1}{12} y_0 + \frac{7}{6} y_1 + \frac{25}{6} y_2 + \ldots \right.$$
$$\left. + \left(\frac{6n^2 - 12n + 7}{6}\right) y_{(n-1)} + \left(\frac{6n^2 - 4n + 1}{12}\right) y_n \right]. \quad (17)$$

The trapezoidal rule may be adapted to give transverse moment M_t and transverse moment of inertia I_t, but the expressions are complicated by the presence of products of the ordinates $y_0 y_1$, $y_1 y_2$, etc. for M_t and $y_0^2 y_1$, $y_0 y_1^2$, $y_1^2 y_2$, $y_1 y_2^2$, etc. for I_t. To overcome this complexity the squares and cubes of the ordinates as given by the integrals in Sect. 4.2 are sometimes weighted by the trapezoidal area rule multipliers to give rough approximations of moment and moment of inertia about the x-axis.

Owing to the straight line approximation inherent in the trapezoidal rule, a closer spacing of ordinates is needed to approach the same level of accuracy for area obtainable with other rules described later, and its application is limited in naval architectural calculations to finding areas. In the case of a convex curve with no point of inflection, the area found by the trapezoidal rule is always less than the true area.

4.4 Simpson's First Rule. This rule, and that to follow in 4.5, are part of a group of rules known as Newton-Cotes Rules. *Simpson's First Rule* rigorously integrates the area under a curve of the type $y = a + bx + cx^2$, which is a second order parabola, or polynomial of degree 2, by applying multipliers to groups of three equally spaced ordinates. That is, if the portion of the curve in Fig. 15 extending from A to C is parabolic, and the ordinates y_0, y_1, and y_2 are equally spaced, then the area found by Simpson's First Rule is precisely correct. Inasmuch as many ship curves are not dissimilar to the parabola, the area so found is a close approximation to that of the ship, and the rule is widely used in naval architecture.

The rule may be derived by assuming the area is given by the expression, $A = k_0 y_0 + k_1 y_1 + k_2 y_2$. Given the mathematical form of the curve, $(y = a + bx + cx^2)$ and ordinates at spacing s, then $y_0 = a$, $y_1 = a + bs + cs^2$ and $y_2 = a + 2bs + 4cs^2$. Putting the three y values into the expression for A, an equation for the coefficients a, b and c results. But A is also equal to the definite integral,

$$A = \int_0^{2s} y\,dx = \int_0^{2s} (a + bx + cx^2)\,dx$$
$$= 2as + 2bs^2 + \frac{8}{3} cs^3.$$

Equating the two expressions for A, we may set the coefficients of a, b, and c equal to each other. There are three resulting equations in the three unknowns, k_0, k_1 and k_2, which may be solved simultaneously. This gives,

$$k_0 = \frac{s}{3}, \; k_1 = \frac{4s}{3}, \; k_2 = \frac{s}{3}.$$

The curve to be integrated must be divided into an even number of spaces by equally spaced ordinates. The multipliers for even numbered ordinates are then found on the assumption that each such ordinate represents the termination of one parabolic curve and the initiation of another. Knuckles in the curve are allowed at these ordinates. Hence the multiplier for such even numbered ordinates (except for the first and last) is 2, giving the following form of the rule,

$$A = \frac{s}{3}[y_0 + 4y_1 + 2y_2 + 4y_3 \ldots$$
$$+ 4y_{(n-1)} + y_n] \quad (18)$$

where n is even. In order to simplify the multipliers when using the First Rule, it is not uncommon to divide them by 2, in which case they are known as *half multipliers*. The final integration is then found by multiplying the summed products of ordinates and multipliers by an additional factor of 2, so that the integrating factor becomes $2s/3$.

Simpson's First Rule may be adapted to the calculation of longitudinal moment M_ℓ and longitudinal moment of inertia I_ℓ in a similar way to that used for finding a formula for area, with the assumption that the ordinates of a 2nd order parabolic curve are xy, and x^2y, respectively.

In practice, it is customary to perform calculations for area, longitudinal moment, and longitudinal moment of inertia using Simpson's First Rule by means of tables such as Table 6, described in sections 5.3, 5.4, and 5.5. Separate columns are provided in the table for the ordinates, for Simpson's Multipliers, for levers (for longitudinal moment), for the squares of levers (for longitudinal moment of inertia), and for the products of the ordinates times the levers times Simpson's Multipliers, etc. For simplicity, the levers are usually non-dimensionalized by dividing by the station spacing, s. When this is done, the tabular calculations for M_ℓ and I_ℓ (axis of moments at origin, often located amidships) may also be expressed by the following formulas, which may be found more appropriate for computer programming, taking the axis at $x = 0$,

$$M_\ell = \frac{s^2}{3}[4y_1 + 4y_2 + 12y_3 + \ldots +$$
$$2(n-2)y_{(n-2)} + 4(n-1)y_{(n-1)} + ny_n]$$
where n is even. (19)

$$I_\ell = \frac{s^3}{3}[4y_1 + 8y_2 + 36y_3 + \ldots +$$
$$2(n-2)^2 y_{(n-2)} + 4(n-1)^2 y_{(n-1)} + n^2 y_n]$$
where n is even. (20)

It will be noted that there are no y_0 terms above because the axis for moments is at $x = 0$ where $y = y_0$ and the lever arm is zero.

If similar derivations are applied to the determination of formulas for transverse moment of area M_t and transverse moment of inertia of area I_t it will be found that expressions of the form,

$$M_t = k_0 y_0^2 + k_1 y_1^2 + k_2 y_2^2 \text{ and}$$
$$I_t = k_0 y_0^3 + k_1 y_1^3 + k_2 y_2^3,$$

cannot be solved, owing to an excess of equations. This results from the presence of cross products of the ordinates, as noted in Sect. 4.3. Nevertheless, Simpson's First Rule is routinely applied to the calculation of transverse moment of area, and transverse moment of inertia of area, by weighting the squares and cubes of ordinates by Simpson's area multipliers, and in accordance with the integrals in Section 4.2. This is equivalent to assuming that the ordinates of the 2nd order parabola are y^2, and y^3, respectively.

Therefore, in the event the squares of the ordinates of the curve to be integrated, or the cubes of the ordinates, respectively, followed a 2nd order parabolic curve, the integration for transverse moment, and for transverse moment of inertia, by Simpson's First Rule, would be precisely correct.

Table 6 includes columns for the cubes of ordinates and for the products of these times Simpson's Multipliers, in order to calculate transverse moment of inertia (about the ship's centerline) by Simpson's First Rule.

It may be shown that Simpson's First Rule also precisely integrates the area under a third order parabolic curve of the form,

$$y = a + bx + cx^2 + dx^3,$$

which passes through the three given ordinates. Hence, Simpson's First Rule is accurate enough for most ship problems.

4.5 Simpson's Second Rule. This rule correctly integrates the area under a third order parabolic curve, or polynomial of degree 3, when four equally spaced ordinates are provided. The derivation of appropriate Simpson's multipliers is achieved using similar steps to those outlined in Sect. 4.4. It may be shown that if we assume $A = k_0 y_0 + k_1 y_1 + k_2 y_2 + k_3 y_3$, then, $k_0 = k_3 = 3s/8$, and $k_1 = k_2 = 9s/8$ where s is the station spacing. Thus, in general, the area A under an arbitrary curve by *Simpson's Second Rule* is

$$A = \frac{3s}{8}[y_0 + 3y_1 + 3y_2 + 2y_3 + 3y_4 + 3y_5$$
$$+ \ldots + \ldots + 2y_{(n-3)} + 3y_{(n-2)}$$
$$+ 3y_{(n-1)} + y_n], (n = 3, 6, 9 \ldots \text{etc.}) \quad (21)$$

As with Simpson's First Rule, a separate parabolic curve is assumed between the extremities of each group of intervals—three intervals in the case of the Second Rule—and knuckles in the curve to be integrated are permitted at these points.

Simpson's Second Rule may be applied to the calculation of longitudinal moment of area M_ℓ and longitudinal moment of inertia of area I_ℓ by combining $3s^2/8$ and $3s^3/8$, respectively, with the Simpson's multipliers for area, together with non-dimensionalized levers and levers squared, respectively, as done when using the First Rule. The resulting M_ℓ and I_ℓ are not rigorously correct for a parabolic curve of the third order, but are routinely used. The resulting errors are quite small, in general.

The accuracy of transverse moment of area M_t and transverse moment of inertia of area I_t when calculated by Simpson's Second Rule is subject to the same limitations as apply to the First Rule. However, the rule is routinely used for these purposes.

4.6 Single Interval Rules. These rules allow one to find area under the curve, A, longitudinal moment of area M_ℓ, and longitudinal moment of inertia of area I_ℓ about axis OY for a single interval between the first two ordinates of a second order parabola of the form $y = a + bx + cx^2$ when the curve is defined by three equally-spaced ordinates with spacing s.

Consider Fig. 15. The *five, eight, minus one* rule states that the area A between ordinates y_0 and y_1 is,

$$A = \frac{s}{12}(5y_0 + 8y_1 - y_2). \quad (22)$$

The *three, ten, minus one* rule states that the longitudinal moment M_ℓ of the area between y_0 and y_1 about axis OY is,

Table 3—Newton-Cotes Rules
Ordinates equally spaced, with end ordinates at ends of curve

Number of Ordinates	Multipliers for ordinate numbers								
	1	2	3	4	5	6	7	8	9
2	1/2	1/2							
3	1/6	4/6	1/6						
4	1/8	3/8	3/8	1/8					
5	7/90	32/90	12/90	32/90	7/90				
6	19/288 = 0.0660	75/288 = 0.2604	50/288 = 0.1736	50/288 = 0.1736	75/288 = 0.2604	19/288 = 0.0660			
7	0.0488	0.2571	0.0321	0.3238	0.0321	0.2571	0.0488		
8	0.0435	0.2070	0.0766	0.1730	0.1730	0.0766	0.2070	0.0435	
9	0.0349	0.2077	−0.0327	0.3702	−0.1601	0.3702	−0.0327	0.2077	0.0349

Area = Σ (Multipliers × Ordinates) × distance between end ordinates, R

$$M_\ell = \frac{s^2}{24}(3y_0 + 10y_1 - y_2). \qquad (23)$$

A similar rule may be derived for longitudinal moment of inertia I_ℓ of the area between y_0 and y_1 about axis OY. It might be called the *seven, thirty six, minus three* rule and is,

$$I_\ell = \frac{s^3}{120}(7y_0 + 36y_1 - 3y_2). \qquad (24)$$

These rules are exact for the 2nd order parabolic curve assumed.

4.7 Higher Order Curves. In the event a curve is believed to be more closely approximated by a higher order parabola, or polynomial of higher degree, the Newton-Cotes multipliers may be used, but a greater number of equally spaced ordinates is needed in way of that portion of the curve over which the defining parabola is assumed to hold.

Thus, five equally spaced ordinates are needed to define a curve in the form $y = a + bx + cx^2 + dx^3 + ex^4$. It may be shown that the area A under such a curve is given by,

$$A = \frac{s}{45}[14y_0 + 64y_1 + 24y_2 + 64y_3 + 14y_4]. \qquad (25)$$

By combining end ordinates for two or more groups of four equal intervals, a rule analagous to Simpson's First or Second Rule may be devised.

Based upon Miller (1963), multipliers for higher order curves would be as shown in Table 3. In each case, the area under the curve would be the product of the distance between end ordinates R and the sum of products of multipliers and ordinates. It may be noted that the sum of the ordinates in Table 3 equals 1.0 for each polynomial.

4.8 Half-Spaced Ordinates. Near the ends of a ship it is customary to introduce additional body plan stations midway between pairs of the normal 10 or 20 stations in the length between perpendiculars. This is done to better define the hull form in these regions, as it is usually changing more rapidly with longitudinal distance than near midship. In order to improve the accuracy of integration, one may take advantage of offsets at such half-spaced stations—i.e., stations ½, 1½, 18½ and 19½ in a 20-station length. The foregoing rules of integration may be easily modified to this end. The modification implies that the distance over which the curve to be integrated may be assumed to match the hypothetical curve is cut in half.

In the above case, assuming Simpson's First Rule, each separate parabolic curve would be assumed to extend from station 0 to 1, 1 to 2, 18 to 19 and 19 to 20, while in the middle portion of the ship separate parabolas would be considered to extend from station 2 to 4, 4 to 6 . . . 16 to 18. In order to accommodate this combination of spacings, the Simpson's multipliers are reduced to one half their normal values in way of the half stations.

Fig. 16 shows the arrangement of half stations at the end of a 10-station ship, along with the Simpson's half multipliers appropriate to the First Rule.

An important consideration in any ship calculation is whether the area (or quantity being integrated) is complete, or applies to one side of the ship only. If the latter, a factor of 2 must be introduced into the calculation to obtain the total for the ship.

4.9 Tchebycheff's Rules. The Tchebycheff Rules use varying numbers of ordinates located at irregular intervals along the base line, spaced in such a way that the sum of the ordinates is directly proportional to the area under the curve. The curve to be integrated is assumed parabolic, i.e., $y = a + bx + cx^2 \ldots + kx^n$. The number of ordinates needed is the same as the order of parabola assumed. The length of curve is taken as $2s$, as in Fig. 17. The validity of the rule is based upon the location of the ordinates, which are symmetrically disposed about the middle, such as y_1 and y_2 at locations x and $-x$.

Consider a second order parabola, with origin at 0,

as shown in Fig. 17. Assume the area under the curve from D to E is given by,

$$A = p(y_1 + y_2),$$
$$y_1 = a - bx + cx^2, \quad y_2 = a + bx + cx^2$$
$$\therefore y_1 + y_2 = 2a + 2cx^2$$

Hence, $A = 2pa + 2pcx^2$.

$$\text{But } A = \int_{-s}^{s} y\,dx = \left[ax + b\frac{x^2}{2} + c\frac{x^3}{3} \right]_{-s}^{s}$$
$$= 2as + 2c\left(\frac{s^3}{3}\right).$$

Equating coefficients of a and c,

$$2p = 2s \therefore p = s;$$
$$2px^2 = 2\left(\frac{s^3}{3}\right) \therefore x^2 = \frac{s^3}{3p} = \frac{s^2}{3} \text{ and}$$
$$x = \frac{s}{\sqrt{3}} = 0.57735s.$$

Then the area,

$$A = s(y_1 + y_2), \tag{26}$$

where y_1 and y_2 are at locations $\pm 0.57735s$ from the origin.

Table 4 shows the locations of ordinates for numbers of ordinates up to 10. It will be noted that in the case of an odd number the middle ordinate is at the middle of the curve, or origin of Fig. 17. In each case, the area under the curve is found as the average length of ordinate $\frac{1}{n}(y_1 + y_2 \ldots + y_n)$ multiplied by the length of the base line $2s$.

4.10 Integration For Arbitrarily Spaced Ordinates. It frequently happens that when integrating a ship's waterplane, the extremities of the curve, either at bow or stern, do not fall at integral stations—in particular the FP or AP—but instead the curve originates either forward of or abaft the perpendicular.

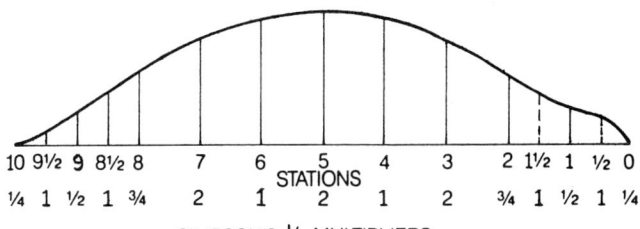

Fig. 16 Half-spaced ordinates

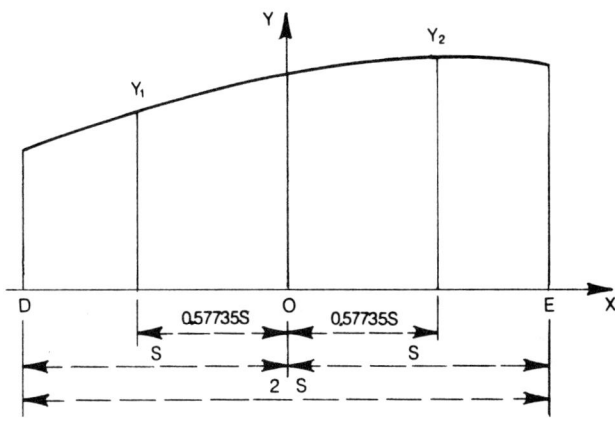

Fig. 17 Integration by Tchebycheff Rule

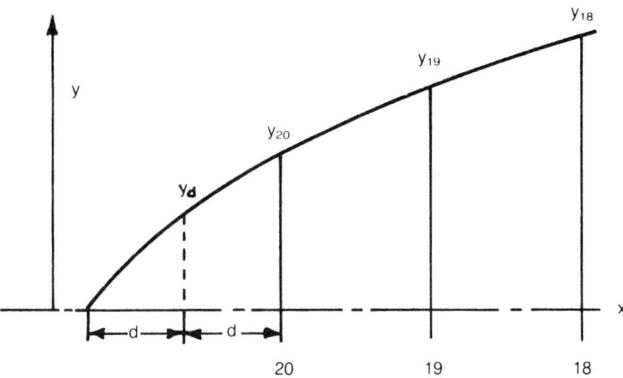

Fig. 18 Curve ending beyond perpendicular

Table 4—Spacing of Tchebycheff's Ordinates

Number of ordinates used	Positions of ordinates from middle of base, in fractions of half-length (s) of base
2	0.5773
3	0 0.7071
4	0.1876 0.7947
5	0 0.3745 0.8325
6	0.2666 0.4225 0.8662
7	0 0.3239 0.5297 0.8839
8	0.1026 0.4062 0.5938 0.8974
9	0 0.1679 0.5288 0.6010 0.9116
10	0.0838 0.3127 0.5000 0.6873 0.9162

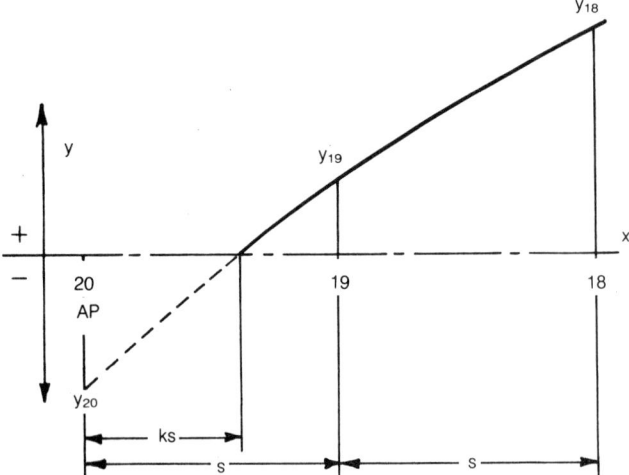

Fig. 19 Curve ending before perpendicular

Fig. 18 shows such a curve extending abaft the AP, or station 20. To handle this situation, one may add and measure an additional ordinate midway between AP and 0-ordinate and add a Simpson's First Rule integration to that for the waterline as a whole, using methods in Sect. 4.11.

In case the curve terminates forward of AP, as in Fig. 19, the curve may be extended by the dotted line and a fictitious negative ordinate, or ordinates, such as y_{20}, may be read. Assuming the extremity of the curve is at ks, where s is the station interval, and the curve is a second order parabola, it may be shown that the area A under the real part of the curve abaft station 18 is given by,

$$A = \frac{s}{3}\left[\left(1 - 3k + \frac{9}{4}k^2 - \frac{1}{2}k^3\right)y_{20} + \left(4 - 3k^2 + k^3\right)y_{19} + \left(1 + \frac{3}{4}k^2 - \frac{1}{2}k^3\right)y_{18}\right] \quad (27)$$

This form might be known as the *partial area rule*. Depending upon whether k is less or more than unity, either y_{20} or y_{20} and y_{19} would be negative.

As an example, in Fig. 19, let $k = 0.3$. Then the calculation for area under the curve A from its extremity to station 18 would be,

$$A = \frac{s}{3}[0.289y_{20} + 3.757y_{19} + 1.054y_{18}]. \quad (28)$$

It may be seen that this formula reverts to Simpson's First Rule when $k = 0$. The *five, eight, minus one rule* multipliers result when $k = 1.0$.

Additional integrating rules could be derived using non-equally-spaced stations, with multipliers dependent upon the specific station locations chosen and predicated on the use of a parabolic curve, but there has in the past been little need for these.

4.11 Combining Rules for Any Number of Ordinates. Ship curves are sometimes divided by a number of equally spaced ordinates which are incompatible with the number needed for integration by either Simpson's First or Second Rules—for example, in the case of 5, 7, 9 or 11 intervals, defined by 6, 8, 10 and 12 ordinates. However, these cases are readily handled by a combination of the two rules. In case both rules are used to integrate such a curve, one may perform a separate integration for each portion of the curve, or a single integration for the entire curve, in which case Simpson's multipliers for the "secondary" portion are modified to suit the integrating factor appropriate to the "primary" portion.

For example, assume there are eight ordinates and the area is to be found using the First Rule over the primary portion. Second Rule multipliers must then be multiplied by the factor $\frac{3s}{8} \div \frac{s}{3} = \frac{9}{8}$. The calculation takes the form shown by Table 5. The integrating factor is $\frac{s}{3}$, appropriate to the First Rule. Also shown in Table 5 are multipliers when the primary portion is integrated using the Second Rule.

4.12 Polar Integration. Whereas most ship curves are defined in rectangular coordinates, there are cases where polar coordinates are more convenient. For example, in calculations relating to static stability, portions of transverse sections of a ship might be wedge shaped such as $OACO$ in Fig. 20.

The elementary cross-hatched, four-sided figure in Fig. 20 has sides of length $r\delta\theta$ and δr. Thus, the area of the sector $OACO$ is,

$$A = \sum\sum(r\delta\theta)\delta r = \int_0^\rho \int_{\theta_1}^{\theta_2} (rd\theta)dr$$

$$= \frac{1}{2}\int_{\theta_1}^{\theta_2} \rho^2 d\theta. \quad (29)$$

Here ρ is the distance from the origin 0, i.e. the value of r, to any point P in the curved side of the figure, and the angle θ is in radians.

For a given wedge-shaped figure, the foregoing integration may be performed by any of the practical rules for integration previously described. The quantities ρ^2 and $d\theta$ above are analogous to the quantities y and dx, respectively, in the equation for the area of a typical curvilinear figure in rectangular coordinates; $\overline{OA}$ and $\overline{OC}$ are the end radial distances corresponding to the end ordinates $\overline{OA}$ and $\overline{HG}$ of Fig. 15. The angle AOC is analogous to the length of base $\overline{OH}$ in Fig. 15. The angle AOC is divided by radial lines through O into a suitable number of equal parts. The length of each radial line is squared, and thereafter is treated in the same manner as an ordinate; that is, by applying the proper ordinate multiplier as required by the particular integrating rule that is being used. In polar integration for area, the factor 1/2 that appears before the sign of integration must be used, whereas no such fractional factor exists in front of the integral $\int y\,dx$ for the area of a figure determined by rectangular coordinates. Also, in polar integration, the common interval is the angular distance in radians between the adjacent radial lines. It is analogous to the linear common interval s of rectangular integration.

In Fig. 20, the centroid of the small elementary triangle POQ is at a distance from O of $(2/3)\rho$. The moment of this elementary triangle about any axis in the plane may be obtained by multiplying its area by the distance of the centroid from that axis. Thus, the distance of the centroid from OY is $(2/3)\rho\cos\theta$ and the moment of the elementary triangle POQ about OY is,

$$\frac{1}{2}\rho^2 d\theta \cdot \frac{2}{3}\rho \cos\theta = \frac{1}{3}\rho^3 \cos\theta\, d\theta.$$

The moment M_{oy} of the figure $OACO$ about OY is

$$M_{oy} = \frac{1}{3}\int_{\theta_1}^{\theta_2} \rho^3 \cos\theta\, d\theta. \tag{30}$$

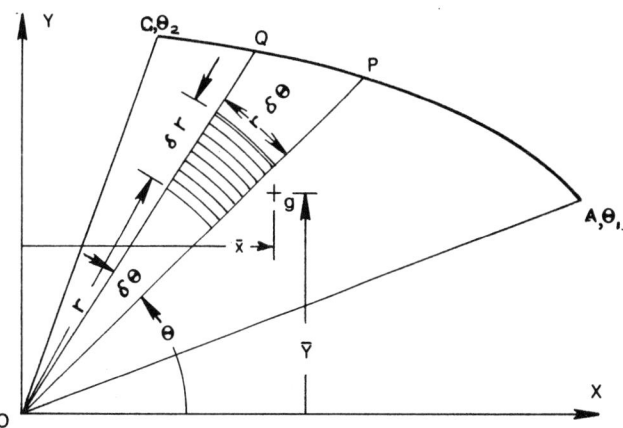

Fig. 20 Curve for integration by polar coordinates

The distance $\bar{x}$ from OY of the centroid, g, of the figure $OACO$ may be obtained by dividing the moment of $OACO$ about OY by the area. Thus distance $\bar{x}$ of g from OY is,

$$\bar{x} = \frac{\dfrac{1}{3}\int_{\theta_1}^{\theta_2} \rho^3 \cos\theta\, d\theta}{\dfrac{1}{2}\int_{\theta_1}^{\theta_2} \rho^2\, d\theta}. \tag{31}$$

Table 5—Typical Integrations Using Combined Simpson's First and Second Rules
First Rule as Primary Rule

Sta.	Ordinate	SM	Prod.
0	y_0	1	y_0
1	y_1	4	$4y_1$
2	y_2	2	$2y_2$
3	y_3	4	$4y_3$
4	y_4	17/8	$(17/8)y_4$
5	y_5	27/8	$(27/8)y_5$
6	y_6	27/8	$(27/8)y_6$
7	y_7	9/8	$(9/8)y_7$
			Σ Prod.

Area under curve $= \dfrac{s}{3} \times \Sigma$ Prod. where s is station spacing.

Second Rule as Primary Rule

Sta.	Ordinate	SM	Prod.
0	y_0	1	y_0
1	y_1	3	$3y_1$
2	y_2	3	$3y_2$
3	y_3	17/9	$(17/9)y_3$
4	y_4	32/9	$(32/9)y_4$
5	y_5	16/9	$(16/9)y_5$
6	y_6	32/9	$(32/9)y_6$
7	y_7	8/9	$(8/9)y_7$
			Σ Prod.

Area under curve $= \dfrac{3s}{8} \times \Sigma$ Prod., where s is station spacing.

Similarly, the moment of the figure $OACO$ about OX, represented by the symbol M_{ox}, may be written,

$$M_{ox} = \frac{1}{3}\int_{\theta_1}^{\theta_2} \rho^3 \sin\theta \, d\theta. \tag{32}$$

Also, the distance from OX of the centroid of the entire figure, $\bar{y}$ may be obtained by dividing M_{ox} by the area. Thus distance of g from OX is,

$$\bar{y} = \frac{\frac{1}{3}\int_{\theta_1}^{\theta_2} \rho^3 \sin\theta \, d\theta}{\frac{1}{2}\int_{\theta_1}^{\theta_2} \rho^2 \, d\theta}. \tag{33}$$

The integration indicated by equations for M_{oy} and M_{ox} may be done in the same way as that previously described for integrating the area equation.

In order to find moment of inertia, about the x-axis for example, it is convenient again to think of the elementary cross hatched portion shown in Fig. 20. Thus, the moment of inertia of the sector $OACO$ about the OX axis is,

$$I_x = \sum\sum (r\delta\theta)\delta r(r\sin\theta)^2$$
$$= \int_O^\rho \int_{\theta_1}^{\theta_2} \sin^2\theta \, d\theta r^3 dr = \frac{1}{4}\int_{\theta_1}^{\theta_2} \rho^4 \sin^2\theta \, d\theta \tag{34}$$

In performing a Simpson's Rule integration for moment, or for moment of inertia, based upon polar coordinates, $(1/3)\rho^3 \sin\theta$, and $(1/4)\rho^4 \sin^2\theta$ respectively would replace the $(1/2)y^2$ and $(1/3)y^3$ terms in similar calculations based upon rectangular coordinates, described in Section 4.2.

4.13 Mechanical Integration. For many years there have been available mechanical instruments allowing the important geometrical properties of any plane curve to be determined without the necessity of reading ordinates and performing a calculation. That is, the final results are obtained directly from dials on the instrument. Unfortunately, the only way to verify the results of such a determination is to repeat the operation. Mechanical integrators are, in effect, a form of analog computer.

The planimeter is used to find the area of any closed curve; the integrator to find the moment of an area about a chosen axis, and sometimes also the moment of inertia. With the integraph the area of any figure may be obtained from its integral curve, which is drawn by the instrument. With a map measurer, the perimeter of any figure, or any part of it may be determined. Of these, the planimeter and the integrator are those most commonly used. Mechanical integration is particularly useful in checking the results obtained by calculations, and also in obtaining quickly, if only approximately, many of the quantities that are needed in the early stages of the design of a vessel.

Today the most common method of calculation is the use of electronic digital computers that employ numerical methods. See section 5.16. Many of the calculations can be performed on a hand-held programmable calculator.

4.14 The Planimeter. The planimeter is an instrument for finding the area of any plane figure. A perspective view of a usual and typical form, known as a polar planimeter[4], is shown in Fig. 21.

The area to be found is bounded by the closed curve $PBEFP$. Any portion of the enclosing line may be straight, curved, or irregular. The planimeter has a tracing point P at one end of a moving bar PA; in operation this tracing point is moved by hand so as to trace entirely around the closed curve. Any point on the curve may be selected from which to start, and the motion is usually in a clockwise direction and continues until the tracing point arrives back at the starting point. The other end of the moving bar PA is jointed at A to a weighted link AO, and this link is free to rotate about the point O. During the operation of the planimeter, the point O, which is located at a needle point on the instrument, is fixed in position on the plane of the table or paper on which the given area is drawn, but O should be outside of the given area.

Attached to the bar PA, and parallel to it, is a shaft on which is mounted the measuring wheel R, which rests on the table or plane of the given figure. The circumferential edge of R is thin so as to insure almost a single-point contact with the horizontal plane. This point of contact, the tracing point P, and the support wheel W constitute the three points of support of the bar BA assembly upon the table. In operation, the

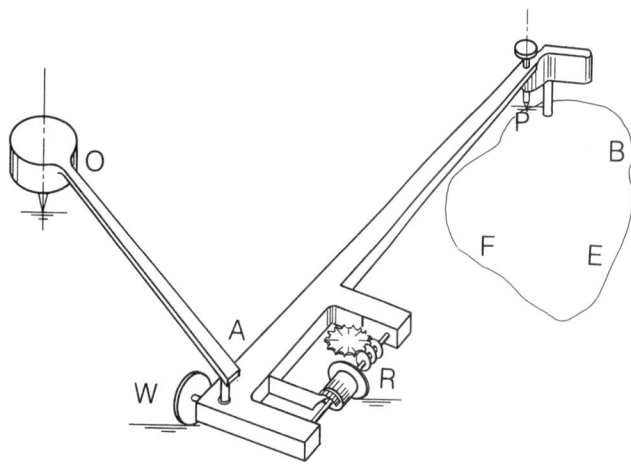

Fig. 21 The planimeter

---

[4] The planimeter, as well as the integrator (4.15), were invented by Professor Jacob Amsler in Switzerland about 1856.

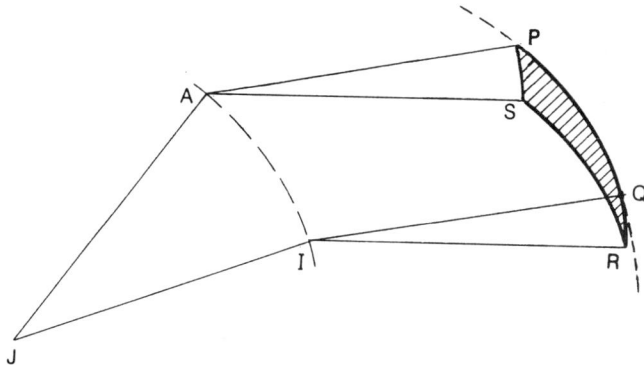

Fig. 22 Integration by planimeter

movement of the assembly can rotate about axis A. The wheel R carries a scale which gives the area reading for the figure $PBEFP$ after it has been traced.

Consider the shaded area $PQRS$ in Fig. 22. Arc PQ is circular, obtained by moving the tracing point at P to Q so that the moving bar IQ remains parallel to AP. Arc QR is circular about I as center and arc SP is circular about A as center. Arc RS is circular, obtained by moving the tracing point from R to S so that the moving bar remains parallel to IR. Since rotation of the wheel results only from motion normal to the moving bar, the increase in revolutions from movement of the tracing point from P to Q is proportional to the area of $APQI$, and the decrease in revolutions in moving from R to S is proportional to the area of $ASRI$. Since sectors APS and IQR are of equal area, tracing arcs QR and SP cancel out, and the cross hatched area $PQRS$ equals the area of $APQI$ minus the area of $ASRI$. Hence, the difference in wheel revolutions before and after tracing right around $PQRS$ is proportional to its area. By approximating any closed curve by an increasingly large number of smaller closed curves generated in the same manner as $PQRS$, we may approach the arbitrary curve as closely as we please, which shows that the area within any closed curve is proportional to the difference in wheel revolutions as a result of tracing right around the curve.

In each actual use it is wise to calibrate the instrument by tracing a rectangle of known area and obtaining a calibration factor.

4.15 The Integrator. The integrator is an instrument for obtaining the area of any plane closed figure and the moment of that area about a chosen axis. In most types of integrators means are also provided for obtaining the moment of inertia of the given area about the same axis. Since it is used primarily for obtaining cross curves of stability, it will be described and discussed in Chapter II.

Section 5
Hydrostatic Curves and Calculations

5.1 Curves of Form. It is customary in the design of a ship to calculate and plot as curves a number of hydrostatic properties of the vessel's form at a series of drafts. Such curves are useful in loading and stability studies during the design phase. Large scale plots of these curves for a newly built ship are then made for the assistance of the vessel's operating personnel. Such curves are known as the vessel's *curves of form*, or synonymously, *hydrostatic curves*. Fig. 23 shows the curves of form for the vessel shown in Fig. 1.

Curves of form are generally drawn on a large sheet of graph paper with all curves plotted against a vertical scale of draft, and with the bottom of the vessel (zero draft) at the foot of the sheet. In order to avoid showing a separate scale for each curve, one horizontal scale of units may be provided, together with separate conversion factors for most of the curves. The practice of providing a horizontal scale of inches, instead of units, often followed in the past, is not recommended in view of the possibility of reproducing the curve sheet at a scale different from that of the original.

Final curves of form as furnished for use by ship's personnel are usually plotted against drafts measured to the bottom of the keel. However, it is not uncommon in the design stage to plot the curves against molded drafts.

The curves of form are customarily calculated with the ship in an even keel condition (no trim). The draft scale is identified as mean draft, and it is assumed that the effect of trim at constant mean draft on most of the plotted quantities is small. This is equivalent to assuming that the vessel is wall-sided—that is, section shapes in way of the prevailing waterline are vertical. The effect of trim is often shown, however, by auxiliary curves.

The range of drafts to which the curves are plotted should extend from below the lightest possible operational draft to the deepest possible draft. The displacement curve should extend down to the origin, in order to provide information for calculating the height of the center of buoyancy, as described in Section 5.10.

5.2 Calculations Required. Calculations of hydrostatic properties of the ship's hull require application of the methods of integration described in Section 4. The calculations take three forms: integrations of

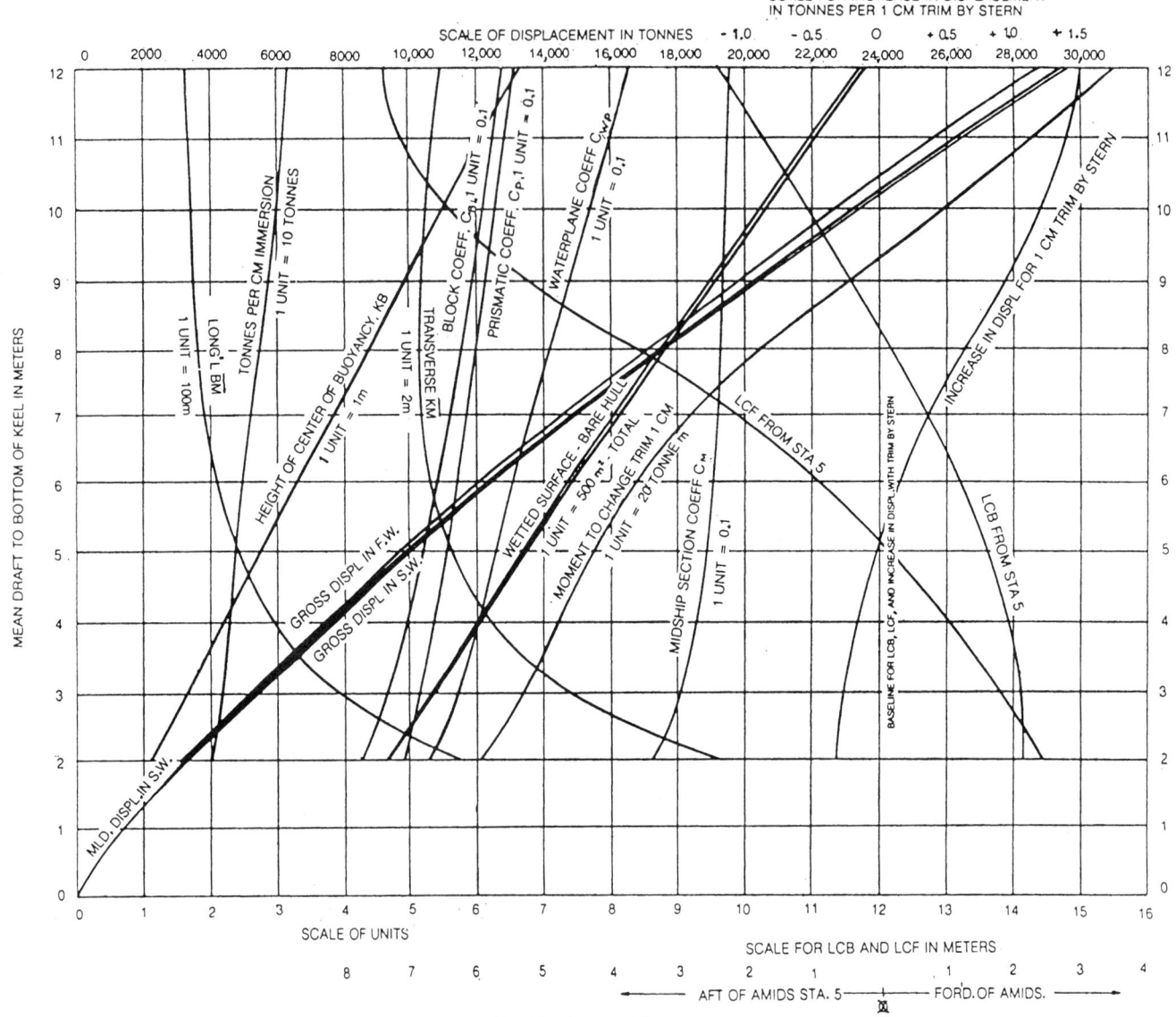

Fig. 23 Curves of form

plane areas to find quantities which are only area related, integrations to find volumes and related quantities, integrations which are accomplished for an area, but for which the answer required also makes use of a volume.

Where calculations for volume of displacement are needed, the integration may be done with either longitudinal distance or vertical distance as the independent variable. Both such integrations are sometimes done for the same volume, to serve as a check.

In all such integrations, the ship's offsets are used. These should always be recorded, and printed out when computer calculations are undertaken.

A formalized tabular method of performing and recording the calculations makes use of a *displacement sheet*, which is a large printed form with spaces for entering all offsets used, multipliers for the rule of integration adopted, and products of these. The displacement sheet forms a permanent record of the calculations, but has the disadvantage of lack of calculational flexibility, and the large sheet may be awkward to file. The majority of such calculations are now accomplished by programmed digital computers. The procedure in this case should include a careful check of the input data, and an assurance that the calculation routines inherent in the computer program are understandable to the user, have adequate precision, and are compatible with the details of the particular form to be integrated.

5.3 Area of Waterplane; Tons per Unit Immersion, Longitudinal Center of Flotation. The ship's waterplane area must be calculated at a sufficiently large number of waterlines to allow a well-defined curve to be drawn over the range of drafts needed. If at any such waterline the draft were to increase by a small amount with no change of trim, the volume of displacement would increase very nearly by an amount equal to the product of waterplane area and the increase in draft, or increase in immersion. The corresponding increase in displacement would be found by multiplying by the density of the water. In the past, an increase in draft of one inch was assumed, giving the *Tons per Inch Immersion* in salt water,

$$TPI = A_{WP}/12 \times 35 = A_{WP}/420 \quad (35)$$

A preferred quantity in the metric system of measurements is metric *Tons per cm Immersion (TPcm)*. Since the density of fresh water is 1 metric ton per m³, TP cm for a ship in fresh water would be,

$$TP\ cm = A_{WP}/100$$

where A_{WP} is waterplane area in m². Assuming the ship in salt water of density, $\rho = 1.025$ t/m³,

$$TP\ cm = \rho\ A_{WP}/100 = 1.025\ A_{WP}/100. \quad (36)$$

With a ship in brackish water, intermediate values of density should be used.

The *Center of Flotation*, which is the point in the waterplane at which a weight added to a vessel would produce parallel sinkage, with no change of trim or heel, is at the centroid of waterplane area. The longitudinal location, LCF, is found by calculating the longitudinal moment of waterplane area in conjunction with the calculation of area. Any axis of reference may be used to find the moment, such as the FP or AP. A midship axis is often preferred, in order to reduce the magnitude of numbers which result. Positive distance is customarily taken forward of amidships; negative

Table 6—Calculation of Waterplane Characteristics at 8.23m (27-ft) Waterline

Station	Half-breadth (m)	½SM	Prod.	Lever	Prod.	Lever²	Prod.	(Half-breadth)³	Prod.
0	0	0.25	0	5.0	0	25.0	0	0	0
½	1.245	1.0	1.245	4.5	5.603	20.25	25.211	1.93	1.93
1	3.140	0.50	1.570	4.0	6.280	16.0	25.120	30.96	15.48
1½	5.359	1.0	5.359	3.5	18.757	12.25	65.648	153.90	153.90
2	7.597	0.75	5.698	3.0	17.094	9.0	51.282	438.46	328.84
3	10.956	2.0	21.912	2.0	43.824	4.0	87.648	1315.09	2630.18
4	12.007	1.0	12.007	1.0	12.007	1.0	12.007	1731.03	1731.03
5	12.039	2.0	24.078	0	0	0	0	1744.90	3489.80
6	12.039	1.0	12.039	−1.0	−12.039	1.0	12.039	1744.90	1744.90
7	11.899	2.0	23.798	−2.0	−47.596	4.0	95.192	1684.73	3369.46
8	10.271	0.75	7.703	−3.0	−23.109	9.0	69.327	1083.52	812.64
8½	8.417	1.0	8.417	−3.5	−29.460	12.25	103.108	596.31	596.31
9	5.962	0.5	2.981	−4.0	−11.924	16.0	47.696	211.92	105.96
9½	3.057	1.0	3.057	−4.5	−13.756	20.25	61.904	28.57	28.57
10	0	0.25	0	−5.0	0	0	0	0	0
			$\Sigma_1 = 129.864$		$\Sigma_2 = -34.319$		$\Sigma_3 = 656.182$		$\Sigma_4 = 15009.00$

Station sp., $s = \dfrac{L}{10} = \dfrac{154.99}{10} = 15.499$ m

Waterplane area, $A_{WP} = \Sigma_1 \times \dfrac{4}{3} \times s = (129.864 \times 20.666) = 2{,}683.77$ m²

Waterplane coeff., $C_{WP} = A_{WP}/(L \times B) = 2683.77/(154.99 \times 24.078) = 0.719$

Tonnes per cm immersion $= 2{,}683.77 \times 1.025/100 = 27.51$ t (S.W.)

Long'l Center of Flotation $LCF = (\Sigma_2/\Sigma_1) \times s = (-34.319/129.864) \times 15.499 = 4.10$ m abaft Sta. 5

Long'l moment of inertia about Sta. 5 $= \Sigma_3 \times \dfrac{4}{3} \times s^3 = 656.182 \times \dfrac{4}{3} \times (15.499)^3 = 3{,}257{,}400$ m⁴

Long'l moment of inertia about LCF, $I_L = 3{,}257{,}400 - 2{,}683.77 \times (4.10)^2 = 3{,}212{,}300$ m⁴

Trans. moment of inertia, $I_T = \Sigma_4 \times \dfrac{4}{9} s = 15{,}009 \times 6.8884 = 103{,}390$ m⁴

Vol. of displacement, ∇ (from displacement curve) $= 17{,}845$ m³

Long'l $\overline{BM} = I_L/\nabla = 3{,}212{,}300/17{,}845 = 180.0$ m

Transverse $\overline{BM} = I_T/\nabla = 103{,}390/17{,}845 = 5.79$ m.

is abaft amidships. LCF is then moment divided by area.

Table 6 shows the tabular calculation for these quantities. The rule of integration used is Simpson's First Rule with half multipliers; the integrating factor for area, in accordance with section 4.2 (b) is $2 \cdot 2 \cdot \frac{s}{3}$. The results apply to the full waterplane (both sides of ship). The integrating factor for longitudinal moment is $2 \cdot 2 \cdot \frac{s^2}{3}$, since each ordinate is weighted by its dimensionless distance from amidships, in units of station spacing, s. Also included with Table 6 is the calculation for waterplane coefficient C_{WP}, discussed in Section 3.

$$C_{WP} = \frac{Waterplane\ area}{L \cdot B}.$$

Here L is the length between perpendiculars, although the actual length of the example waterplane exceeds LBP. Both the TPcm and LCF curves are useful for checking the correctness of input data, in that errors in offsets used in calculating these two curves are usually detectable in the uncharacteristic appearance of the curves.

5.4 Transverse Metacentric Radius; Height of Transverse Metacenter. The terms *transverse metacenter* and *transverse metacentric height* are defined and discussed in Chapter II. There it is shown that the vertical distance from the center of buoyancy to the transverse metacenter is called transverse metacentric radius $\overline{BM}$, where

$$\overline{BM} = I_T/\nabla, \qquad (37)$$

and I_T is transverse moment of inertia of entire waterplane area about the longitudinal centerline and ∇ is volume of displacement. Molded dimensions and volume are customarily used in this calculation (Section 5.7).

Table 6 includes the calculation for I_T, inasmuch as the waterplane halfbreadths needed are available from the waterplane area calculation. The integrating factor is the same as for waterplane area, but multiplied by 1/3, and halfbreadths must be cubed in accordance with Section 4.2.

Also shown with Table 6 is the calculation of $\overline{BM}$. Here the volume of displacement ∇ may be obtained by displacement calculations up to the same waterline, or may be read from the molded displacement curve.

The height of the transverse metacenter above the molded baseline is called $\overline{KM}_T$, or simply $\overline{KM}$. This is found by adding the height of the center of buoyancy $\overline{KB}$ to the metacentric radius $\overline{BM}$. That is,

$$\overline{KM} = \overline{KB} + \overline{BM}. \qquad (38)$$

The curves of form include transverse $\overline{KM}$, which is an important quantity for a ship from considerations of stability. It is important to distinguish between the height of metacenter $\overline{KM}$, which is a purely geometrical quantity, and the metacentric height, $\overline{GM}$, which involves the location of the ship's center of gravity, as discussed in Chapter II.

5.5 Longitudinal Metacentric Radius; Height of Longitudinal Metacenter. The terms *longitudinal metacenter* and *longitudinal metacentric height* are also discussed and defined in Chapter II. Longitudinal metacentric radius is there defined as $\overline{BM}_L$, where

$$\overline{BM}_L = I_L/\nabla, \qquad (39)$$

and I_L is longitudinal moment of inertia of entire waterplane area about a transverse axis through the longitudinal center of flotation LCF.

Table 6 also includes the calculation for longitudinal moment of inertia of the waterplane about amidships, which requires the waterplane halfbreadths. In performing the calculation, each ordinate must be weighted by the square of its distance from the reference axis. This is done nondimensionally in units of the common interval, or station spacing s. The integrating factor is,

$$2 \cdot 2 \cdot s^3/3.$$

In order to correct the longitudinal moment of inertia to a transverse axis through the LCF, the product of $A_{WP} \cdot (LCF)^2$ is deducted from the longitudinal moment of inertia about amidships, in accordance with section 4.2. This calculation is shown at the foot of Table 6.

The longitudinal metacentric radius $\overline{BM}_L$ is then calculated and the height of the longitudinal metacenter above the baseline $\overline{KM}_L$ is found, where

$$\overline{KM}_L = \overline{KB} + \overline{BM}_L. \qquad (40)$$

As in the case of the transverse metacenter, the height of the longitudinal metacenter $\overline{KM}_L$ should not be confused with *longitudinal metacentric height*, which is discussed in Chapter II.

5.6 Molded Displacement and Total Displacement. The displacement of a vessel is the product of underwater volume—or volume of displacement—and the density of the medium in which the vessel floats. Curves of form for an oceangoing vessel usually include three displacement curves; molded displacement in salt water, total or gross displacement in salt water, and total displacement in fresh water, Fig. 23. Of these, total displacement in salt water is probably the most useful to operating personnel of oceangoing ships. A scale of displacement in metric tons is usually provided at the top of the curve sheet.

The volume of the underwater portion of a steel

SHIP GEOMETRY

Table 7—Calculation of Displacement and Longitudinal Center of Buoyancy at 5m (16.41-ft) Waterline

Station	Area (m)²	½ SM	Prod	Lever (nondimens.)	Prod	
−0.07	0	0.0175	0	5.07	0	⎤ First
−0.035	3.0	0.07	0.21	5.035	1.1	⎦ Rule
0	4.2	0.2675	1.12	5.0	5.6	⎤
½	12.7	1.00	12.70	4.5	57.2	
1	22.6	0.50	11.30	4.0	45.2	
1½	35.1	1.00	35.10	3.5	122.9	
2	50.6	0.75	37.95	3.0	113.9	
3	83.3	2.0	166.60	2.0	333.2	First
4	106.1	1.0	106.10	1.0	106.1	Rule
5	113.7	2.0	227.40	0	0	
6	107.6	1.0	107.60	−1.0	−107.6	
7	81.4	2.0	162.80	−2.0	−325.6	
8	44.0	0.7813	34.38	−3.0	−103.1	⎦
8½	29.1	0.8438	24.30	−3.5	−85.1	⎤ Second
9	17.4	0.8438	14.68	−4.0	−58.7	⎦ Rule
9½	5.3	0.3138	1.66	−4.5	−7.5	⎤ First
9.565	3.3	0.13	0.43	−4.57	−2.0	⎦ Rule
9.63	0	0.0325	0	−4.63	0	
			$\Sigma_1 = 944.33$		$\Sigma_2 = 95.6$	

Sectional area curve extended beyond Stations 0 and 9½ to extremities, as shown by Fig. 24 and read at midpoint between last station and extremity. Simpson's Multipliers proportioned accordingly. Thus, at Station −0.035, ½ SM = ½ × 4 × 0.035 = 0.07; at Station 0, ½ SM = 0.25 + 0.0175 = 0.2675; at Station 8, ½ SM = $\frac{1}{2}\left(1.0 + \frac{9}{8} \times \frac{1}{2}\right)$ = 0.7813 (First and Second Rules); at Station 8½ and 9, ½ SM = $\frac{1}{2} \times \frac{9}{8} \times \frac{3}{2}$ = 0.8438; at Station 9½, ½ SM = $\frac{1}{2}\left(\frac{9}{8} \times \frac{1}{2} + \frac{0.065}{1.0}\right)$ = 0.3138; at Station 9.565, ½ SM = ½ × 4 × $\frac{0.065}{1.0}$ = 0.13.

Then ∇, volume of displacement = $\Sigma_1 \times \frac{2}{3} \times s$ = 944.33 × $\frac{2}{3}$ × 15.499 = 9,757 m³.

Displacement, Δ = 1.025 × 9757 = 10,000 t (SW.)

LCB = $\frac{\Sigma_2}{\Sigma_1} \times s = \frac{95.6}{944.33} \times 15.499$ = 1.57 m for'd of Station 5.

vessel is made up of the volume of the molded form (Section 5.7) plus the volume of the steel shell plating and other appendages, such as rudder, propeller, shaft bossings, sonar domes, bilge keels, etc. (Section 5.13). In a wooden vessel it is the volume to the outside of planking plus the volume of other appendages.

In a steel single-screw cargo vessel the volume of all appendages is usually slightly less than 1 percent of the molded volume, and the shell plating is by far the largest contributor, perhaps 0.75 percent. For very large ships, this percent tends to be lower—less than 0.5 percent for a typical large tanker. In multiple-screw vessels, the appendages constitute a greater percentage of the molded volume than in the case of single-screw vessels.

Of the various curves of form, the displacement curves are of particular importance. They are expected to be accurate, and are frequently utilized for the precise determination of displacement, as for example at the inclining experiment and deadweight check before delivery of a vessel.

Having calculated the displacement at different drafts, the coefficients of form discussed in Section 3 can be readily calculated. Curves of c_B, c_P, c_M, and c_{WP} are usually included among the curves of form.

5.7 Displacement and LCB. The calculation of a molded displacement curve requires that all portions of the vessel below the waterline of interest be included. This requires integration of volumes upward from the baseline. Should the vessel extend below the baseline, as from drag to the keel, a finite volume of displacement would exist at zero mean molded draft. The usual method is to calculate sectional areas directly and then to integrate them longitudinally. The longitudinal center of buoyancy for the waterline of interest is also conveniently found in this calculation.

Table 7 shows such a calculation up to one waterline for the example ship of Fig. 1. Simpson's First Rule

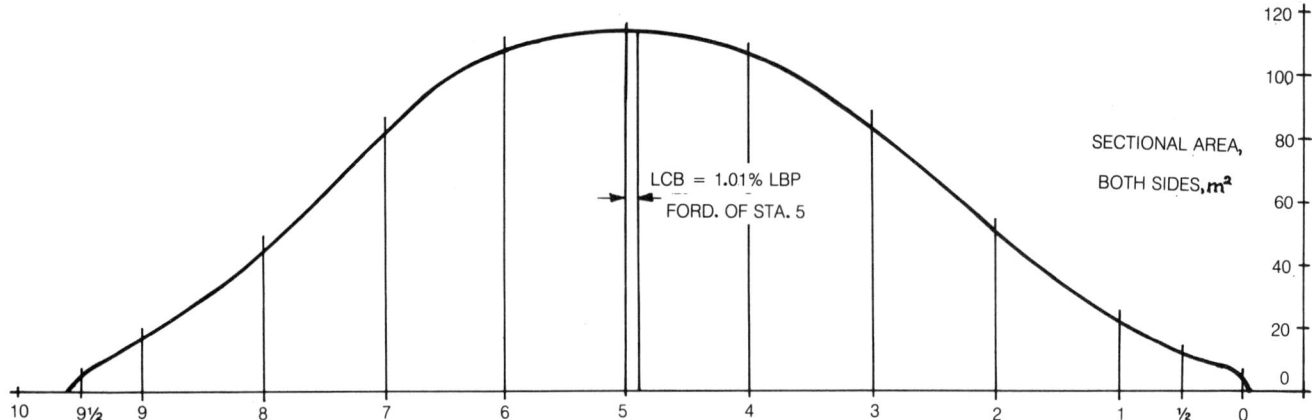

Fig. 24 Sectional area curve for 5 m waterline

is used as the primary rule, with half multipliers. In order to find the longitudinal moment of volume, the sectional areas are multiplied by their non-dimensional distances from amidships. The longitudinal center of buoyancy is then found as the quotient of moment divided by volume. In some cases the inclusion of appendages may have a significant effect.

The integrating factor for volume of displacement is simply $2 \cdot s/3$, inasmuch as the sectional areas used represent the total sectional area for both sides of the ship, as plotted for *Bonjean Curves*. In the case of longitudinal moment, the integrating factor is $2 \cdot s^2/3$. Here s is station spacing.

A substantial number of points are desired for the displacement curve. Most vessels change rapidly with draft at the lower waterlines. For example, a ship with small deadrise and no drag would have a large change in TP cm between zero mean draft and say the 0.5 m waterline.

In order to find the area to the lowest waterline one may use a planimeter, or else a number of closely-spaced waterlines together with numerical integration. The latter method is often used with digital computer calculations.

Calculations of volume and moment of volume for upper waterlines may be simplified somewhat if they are confined to successive "layers" of the underwater body above the uppermost waterline for which volume and moment calculations have already been completed. Thus, to calculate values up to the 5 m waterline, the "ordinate" to the curve to be integrated may be taken as the difference between areas for 5 m and 4 m waterlines. The volume of displacement increase so found would be added to that for the 4 m waterline.

As shown in Section 2.2, displacement is obtained from displacement volume by multiplying by the mass density of the liquid in which the ship is assumed to float. In SI units,

$$\Delta = \rho \nabla. \qquad (2)$$

In English units,

$$W = \rho g \nabla \text{ tons.} \qquad (1)$$

Correspondingly, to find LCB for the 5-m waterline, the longitudinal moment of volume between the 5 m and 4 m waterlines may be added to that for the 4 m waterline. LCB for the 5 m waterline is obtained by dividing total moment by total volume.

Fig. 24 shows the sectional area curve for the 5 m waterline, as well as its longitudinal centroid, which represents LCB at a molded draft of 5 m without trim.

In principle, waterplane areas may be found at several closely spaced, but low waterlines, and these areas integrated vertically to find displacement. However, greater accuracy is generally attainable by the longitudinal integration method.

Having calculated the displacement at different drafts, the coefficients of form discussed in Section 3 can be readily calculated.

5.8 Vertical Center of Buoyancy by Vertical Integration of Waterplanes. A basic feature of any vessel from the point of view of stability is the height of the center of buoyancy above the baseline, called $\overline{KB}$. It may be calculated by first finding the vertical moment of the volume of displacement above the baseline at any waterline.

In integral form, the moment up to draft T_P is,

$$\int_0^{T_p} T A_{WP} dT.$$

Then $$\overline{KB} = \frac{\int_0^{T_p} T A_{WP} dT}{\int_0^{T_p} A_{WP} dT} = \frac{\int_0^{T_p} T A_{WP} dT}{\nabla}. \qquad (41)$$

The calculation requires a curve of waterplane areas vs. draft, as shown in Fig. 25. The vertical location of the centroid of the area above this curve and below the waterline is identical with $\overline{KB}$ for any given draft.

For lower waterlines, a combination of the 5,8,-1 rule, giving volume of displacement, and the 3,10,-1 rule, giving moment of volume, may be utilized for integration, on the assumption that the plot of waterplane area against draft resembles a parabolic curve of the second order. Table 8 shows such a calculation for that portion of the vessel in Fig. 1 below the 1 m waterline. The integrating factor for volume of displacement is $s/12$ where s is waterline spacing. The integrating factor for moment of volume is $s^2/24$.

The same general procedure may be used for upper waterlines, but with the calculated volume and moment of volume, for the appropriate pairs of waterlines, added to corresponding values for the waterline below. Table 8 includes the calculations of volume of displacement and $\overline{KB}$ up to the 2 m waterline.

5.9 Vertical Center of Buoyancy by Integration of Displacement Curve. Another method of obtaining $\overline{KB}$ depends upon the fact that the curve of volume of displacement vs. draft is the integral of the curve of waterplane areas. Fig. 26 shows a curve of volume of displacement ∇ against draft T. The area A of the cross hatched section between the curve and the horizontal line at the prevailing draft T_p may be expressed as the sum of many small rectangles of area, $\delta \nabla (T_p - T)$. Thus,

Table 8—Calculation of Volume of Displacement and Height of Center of Buoyancy by Vertical Integration of Waterplane Areas

Height above baseline, m	Waterplane area, m²	Multiplier for volume	Product	Multiplier for moment	Product
0	194	5	970	3	582
1	1714	8	13712	10	17140
2	1976	−1	−1976	−1	−1976
			$\Sigma_1 = \overline{12706}$		$\Sigma_2 = \overline{15746}$
1	1714	5	8570	3	5142
2	1976	8	15808	10	19760
3	2137	−1	−2137	−1	−2137
			$\Sigma_3 = \overline{22241}$		$\Sigma_4 = \overline{22765}$

Values for 1 m draft

Volume of displ., $\nabla = \dfrac{S}{12} \times \Sigma_1 = \dfrac{1}{12} \times 12{,}706 = 1059 \text{ m}^3$

Moment of volume about baseline, $M_\nabla = \dfrac{S^2}{24} \times \Sigma_2$

$$= \dfrac{1^2}{24} \times \Sigma_2 = \dfrac{1^2}{24} \times 15{,}746 = 656.1 \text{ m}^4$$

Height of center of buoyancy, $\overline{KB} = \dfrac{M_\nabla}{\nabla} = \dfrac{656.1}{1059} = 0.62 \text{ m.}$

Values for 2 m draft

Added volume of displ., $\delta \nabla$ (1 m to 2 m) $= \dfrac{s}{12} \times \Sigma_3 = \dfrac{1}{12} \times 22{,}241 = 1853 \text{ m}^3$

Total volume, $\Sigma \nabla = \nabla + \delta \nabla = 1059 + 1853 = 2912 \text{ m}^3$

Moment of added volume, δM_∇ (1 to 2 m) about 1 m waterline

$$= \dfrac{s^2}{24} \times \Sigma_4 = \dfrac{1^2}{24} \times 22{,}765 = 948.5 \text{ m}^4$$

Moment of added volume about baseline $= 948.5 + 1853 (1 - 0) = 2801.5 \text{ m}^4$

Moment of total volume about baseline, $\Sigma M_\nabla = 656.1 + 2801.5 = 3457.6 \text{ m}^4$

Height of center of buoyancy, $\overline{KB} = \dfrac{\Sigma M \nabla}{\Sigma \nabla} = \dfrac{3457.6}{2912} = 1.19 \text{ m.}$

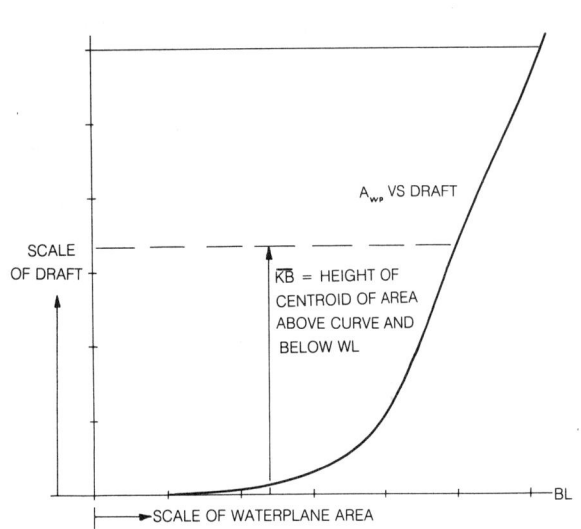

Fig. 25 Waterplane area vs. draft

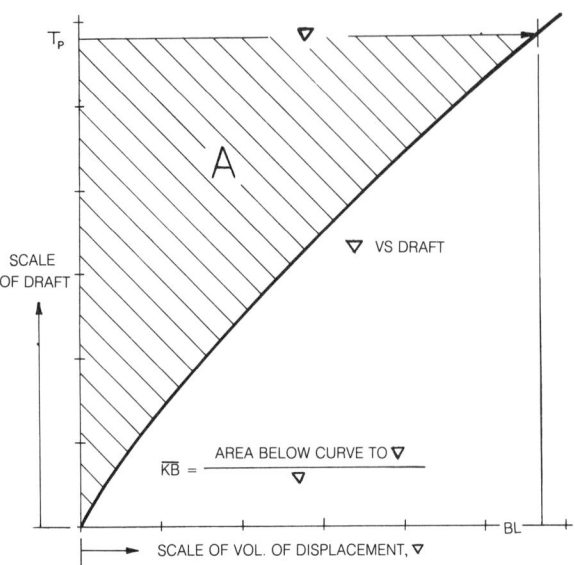

Fig. 26 Volume of displacement vs. draft

$$A = \sum(T_P - T)\delta\nabla = \int_o^{T_p}(T_P - T)d\nabla \text{ in the limit.}$$

But the volume of displacement $\nabla = \int_o^T A_{WP}dT$ and $d\nabla = A_{WP}dT$, where A_{WP} is waterplane area.

Hence, separating the integral into two parts and substituting for $d\nabla$,

$$A = \int_o^{T_p}(T_p - T)d\nabla = T_p\int_o^{T_p}d\nabla - \int_o^{T_p}Td\nabla$$
$$= T_p\int_o^{T_p}A_{WP}dT - \int_o^{T_p}A_{WP}TdT.$$

The first of these integrals is volume of displacement up to T_p. The second integral represents the moment of the volume of displacement about the baseline $M_{\nabla,o}$ or,

$$A = T_p\nabla - M_{\nabla,o}, \text{ and } M_{\nabla,o} = T_p\nabla - A.$$

Inasmuch as the area of the rectangle formed by T_p and ∇ at T_p is simply $T_p \cdot \nabla$, we see that to find the moment of the volume of displacement about the baseline, it is merely necessary to find the cross hatched area A and deduct it from the product of $T_p \cdot \nabla$. Alternatively, one can integrate the un-crosshatched area *under* the curve directly. The vertical center of buoyancy $\overline{KB} = \dfrac{M_{\nabla,o}}{\nabla}$ by definition, and therefore,

$$\overline{KB} = \frac{T_p\nabla - A}{\nabla} \qquad (42)$$

This provides a simple way of finding the vertical height of the center of buoyancy at any draft. In practice, a displacement curve must be available extending right to the baseline. The procedure to follow is to draw a vertical line at the desired displacement and carefully measure the area between the base line and the curve by a rule of integration or by planimeter. Then dividing the measured area by the displacement gives $\overline{KB}$.

It may be noted that the ton scale to which the displacement curve is usually plotted presents no problem here, inasmuch as the scale factor relating Δ to ∇ cancels when taking the quotient.

This method is considered the most accurate for finding $\overline{KB}$ at low drafts, but requires that the displacement curve at its lower end be carefully defined. Using longitudinal integration as described in Sec. 5.7, finding precise values of displacement at low waterlines should not pose a problem.

5.10 Approximate Formulas for Vertical Center of Buoyancy. In the initial stages of design, the height of the center of buoyancy may be required, yet a displacement curve is not available, precluding a calculation of $\overline{KB}$ by the method of Section 5.9. To this end, approximate formulas may be used as given below.

The Morrish formula (Morrish, 1892), also known as Normand's formula, gives the distance below the DWL of the center of buoyancy as,

$$\frac{1}{3}\left(\frac{T}{2} + \frac{\nabla}{A_{WP}}\right),$$

where T is molded mean draft, ∇ is corresponding volume of displacement and A_{WP} is corresponding waterplane area, all in a consistent system of units. The expression may be written more directly as,

$$\overline{KB} = \frac{1}{3}\left(\frac{5T}{2} - \frac{\nabla}{A_{WP}}\right). \qquad (43)$$

Experience has shown that for vessels of ordinary form the formula gives close approximation to the height of the center of buoyancy, not only for load draft but also for lighter drafts.

The formula may be derived from a diagram such as Fig. 27 in which KUP is a curve of waterplane areas plotted against molded drafts. LP equals A_{WP}, waterplane area at mean molded draft T, which equals KL. The area of the figure $LKUPL$ is equal to ∇, volume of displacement, and the centroid of this figure is at the same height above the baseline as the center of buoyancy. It is assumed that the centroid of the polygon $LKBPL$ is at the same height as the centroid of $LKUPL$.

Another approximation which gives values quite close to those of Fig. 23 is Posdunine's formula (Posdunine, 1925),

$$\overline{KB} = T\left(\frac{A_{WP}}{A_{WP} + \dfrac{\nabla}{T}}\right). \qquad (44)$$

The symbols have the same meaning as for the previous expression.

Experience with this formula indicates particularly good agreement with ships of high midship coefficient, C_M.

5.11 Change of Displacement with Trim. Curves of form are customarily calculated for an even keel (no trim) condition. It is shown in Chap II that when a vessel changes trim by a moderate amount because of the movement of a weight forward or aft, the draft remains constant at the center of flotation. Thus, if the center of flotation is abaft amidships, a weight shift causing trim by the stern will result in a reduction in mean draft. Therefore, if it be stipulated that mean draft remain constant, an increase in displacement results from trim by the stern in this case. The approximate increase in displacement in t per cm, (or tons per in.), of trim by the stern is (see Chapter II, 8.6),

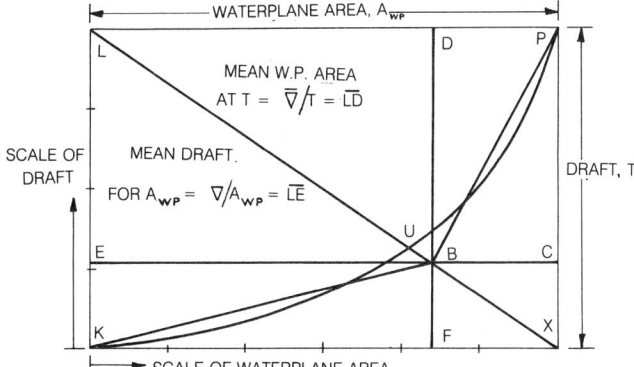

Fig. 27 Morrish Formula diagram

$$TPcm\ d/L \qquad (45)$$

or

$$TPI\ d/L \qquad (46)$$

where

d = distance LCF is abaft amidships,
L = length of ship between draft marks.

It is important that the direction of trim, by bow or stern, and the sign of the change in displacement, be clearly labeled on the curve.

As one of the curves of form, *change of displacement with trim* usually shares with longitudinal center of buoyancy, and longitudinal center of flotation, a separate reference axis from the displacement curve. If the LCF curve crosses amidships, the value of the change of displacement curve will be zero at the draft at which it crosses.

5.12 Moment to Change Trim. The moment necessary to change trim by a fixed quantity is an important characteristic of a vessel and one frequently used for loading studies. The *Moment to Change Trim 1 cm (MTcm)* may be found using principles outlined in Chap II. The expression is,

$$MTcm = \frac{\Delta \cdot \overline{GM_L}}{100\ L}, \qquad (47)$$

where Δ is ship displacement in metric tons, $\overline{GM_L}$ is longitudinal metacentric height = $\overline{KM_L} - \overline{KG}$ in m, L is length of ship between draft marks in m.

In English units, long tons and ft,

$$MT1'' = \frac{W\overline{GM_L}}{12L}. \qquad (48)$$

The value of $\overline{KM_L}$ is found as noted in Sect. 5.5 for any draft. The height of center of gravity $\overline{KG}$ will depend upon the loading condition of the ship. However, for most ships, the range of values of $\overline{KG}$ to be expected in service is a relatively small percentage of $\overline{GM_L}$ and it is sufficiently accurate to assume a standard and reasonable location. For many ships the height of the center of gravity is not far from the prevailing draft, and that location is sometimes chosen for the curves of form. However, for the example ship, the height of the center of gravity is assumed to be the same as the height of the center of buoyancy, so that $\overline{GM_L} = \overline{BM_L}$.

Inasmuch as it is directly proportional to displacement, it is evident that $MTcm$ will vary directly with water density. Thus, at any draft $MTcm$ should increase when in salt water, compared with its value in fresh water.

5.13 Displacement of Appendages. In order to find the total displacement of a ship, the displacement of the appendages—shell plating, rudder, propellers, bilge keels, bossings, etc.—up to any given waterline must be calculated and added to the displacement of

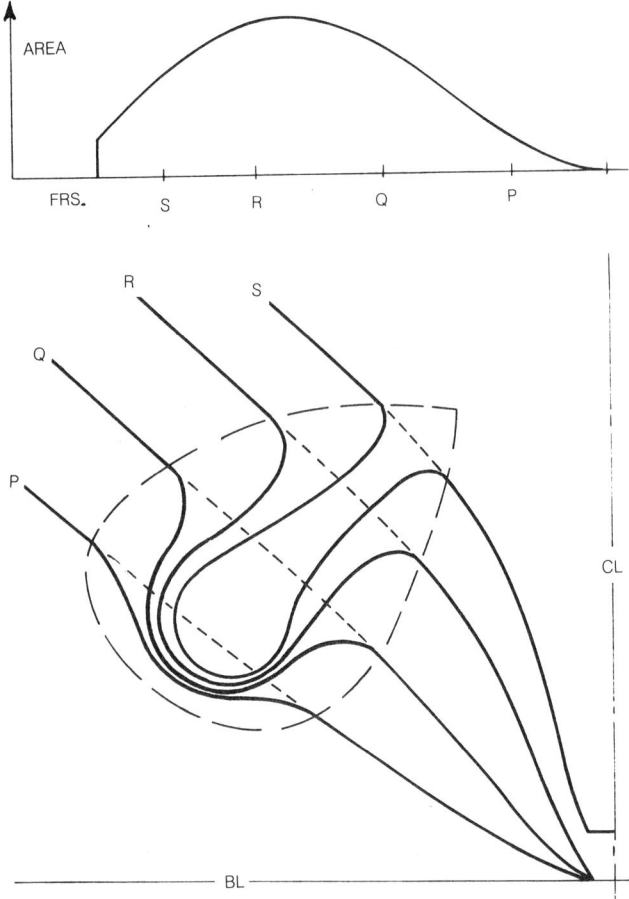

Fig. 28 Sections and sectional area curve of bossing

the molded form. The displacement of these appendages may be calculated from the shell expansion, midship section, and detail drawings of the various parts.

A typical example of such an appendage is the bossing on a twin screw ship. Fig. 28 shows transverse sections through such a bossing and the longitudinal distribution of areas associated with it, or sectional area curve. As in the case of the ship's molded form, the area of the sectional area curve represents the volume of displacement of the bossing.

Many ships built in recent years are fitted with transverse thrusters, which take the form of a cylindrical transverse free-flooding tunnel through the ship at a low waterline. These represent departures from the molded form of the ship, and may be accounted for as a "negative" appendage—that is, the net free-flooding volume of the tunnel should be deducted from the volume of other appendages. In case of naval ships with large sonar domes, the dome may be treated as part of the molded form, or alternatively as an appendage. The convention adopted should be clearly stated on the curve sheet.

In calculating the displacement of the shell plating, any strakes of "out" plating must be properly accounted for. For example, if the flat plate keel is an "out" strake, the volume between its inner surface and the molded form is also treated as appendage volume.

5.14 Curves of Form for Particular Types of Vessels. Certain specialized vessels may call for additional curves of form beyond those discussed in the foregoing. For example, bulk carrying vessels such as tankers and ore carriers can experience more than insignificant degrees of hull girder bending in still water as a result of concentrated weights. If the deflection of the hull girder can be predicted, this can be accounted for in hydrostatic characteristics, such as displacement for drafts read at bow and stern.

A simple but not unrealistic hull girder deflection curve is a second order parabola $y = ax^2$ where y is the deflection from the zero bending moment case, x is longitudinal distance from amidships, non-dimensionalized by the length of the ship, and a is vertical deflection at the ends of the ship, resulting in hog or sag, compared with amidships.

To find the increase in displacement with sag, waterplane halfbreadths must be weighted, first by the deflection, taken as $(a - y)$, and then by the rule of integration multipliers (Simpson's multipliers). Table 9 shows diagrammatically how the calculation may be performed, the increase in displacement with sag being that to be added to the displacement from drafts read at the ends of the vessel, assuming a straight keel.

In case the effect of trim on displacement is considered to be of unusual importance, trim correction curves may be calculated by assuming a series of trimmed waterlines, say for 1 m trim, 2 m trim, etc. by both bow and stern, and at a series of drafts. Such calculations are facilitated by the use of Bonjean curves, as discussed in Section 6. Resulting trim corrections in displacement are more accurate than can be obtained by the change in displacement with trim curve described in Section 5.11, inasmuch as they do not assume the vessel is wall-sided.

Unusual hydrostatic properties may be found for particular types of vessels, including floating drydocks, offshore mobile platforms, integrated tug-barges, and ships with large compartments which are occasionally free flooded, such as the dock area on float on-float off barge carriers. The curves of form in these cases may be characterized by knuckles at the draft at which large elements of buoyancy are immersed, or by several curves of the same quantity, depending upon the ship's condition. Careful thought is needed in analyzing such vessels. However, the calculation of displacement, and displacement related curves, may be more directly done when the underwater body is largely composed of simple geometrical bodies—cylinders, cones, prisms—rather than surfaces of compound curvature.

Table 9—Calculation for Finding Increased Displacement per Meter Sag

Sta.	Dimensionless distance from amidships (x/L)	Deflection, y m	$(y_{max}-y)$ m	HB m	SM	Product
0	0.5	1.00	0	y_0	½	0
½	0.45	0.81	0.19	$y_{½}$	2	0.38 $y_{½}$
1	0.4	0.64	0.36	y_1	1	0.36 y_1
1½	0.35	0.49	0.51	$y_{1½}$	2	1.02 $y_{1½}$
2	0.3	0.36	0.64	y_2	3/4	0.48 y_2
3	0.2	0.16	0.84	y_3	4	3.36 y_3
4	0.1	0.04	0.96	y_4	2	1.92 y_4
5	0	0	1.00	y_5	4	4.00 y_5
6	−0.1	0.04	0.96	y_6	2	1.92 y_6
7	−0.2	0.16	0.84	y_7	4	3.36 y_7
8	−0.3	0.36	0.64	y_8	3/4	0.48 y_8
8½	−0.35	0.49	0.51	$y_{8½}$	2	1.02 $y_{8½}$
9	−0.4	0.64	0.36	y_9	1	0.36 y_9
9½	−0.45	0.81	0.19	$y_{9½}$	2	0.38 $y_{9½}$
10	−0.50	1.00	0	y_{10}	½	0
						Σ

Assumed deflection, $y = 4 \times \left(\dfrac{x}{L}\right)^2$

Increased volume of displacement per meter sag $= \dfrac{2}{3} \times s \times \Sigma$ in m³ (both sides of ship), where s is station spacing in m.

Table 10—Condensed Summary of Curves of Form Values

Mean draft to bottom of keel, meters	Displacement in metric tons			LCB, long'l. center of buoyancy from ⊗, m	LCF, Long'l. center of flotation from ⊗, m	$\overline{KB}$, Center of buoyancy above baseline, m	Tons per cm immersion	Moment to change trim 1 cm, t-m
	Molded in salt water	Gross in salt water	Gross in fresh water					
2	3243	3273	3151	2.15 F	2.41 F	1.11	20.3	121
3	5347	5418	5245	2.11 F	1.83 F	1.65	21.9	134
5	9941	10033	9748	1.59 F	0.08 F	2.74	24.0	156
7	14932	15065	14658	0.73 F	2.11 A	3.82	26.0	184
9	20350	20462	19903	0.42 A	5.20 A	4.93	28.9	232
11	26317	26449	25707	1.73 A	7.08 A	6.08	30.8	286
12	29448	29611	28757	2.32 A	7.38 A	6.64	31.8	311

Mean draft to bottom of keel, meters	Increase in displacement for 1 cm trim by stern, t	$\overline{KM}$, Transv. metacenter above baseline m	$\overline{BM_L}$, Long'l. metacentric radius m	C_B Block coeff.	C_M Midship coeff.	C_P Prismatic coeff.	C_W Waterplane coeff.	Wetted surface in square meters	
								Bare Hull	Total
2	−0.306	19.4	579	0.427	0.864	0.494	0.529	2332	2339
3	−0.260	14.8	388	0.470	0.908	0.518	0.573	2669	2685
5	−0.023	11.3	243	0.522	0.944	0.553	0.629	3344	3377
7	+0.367	10.4	190	0.562	0.961	0.584	0.680	4013	4066
9	+0.951	10.4	177	0.593	0.969	0.612	0.743	4733	4794
11	+1.407	10.7	169	0.626	0.976	0.642	0.803	5478	5540
12	+1.507	11.0	164	0.641	0.978	0.658	0.830	5852	5920

Principal Ship Dimensions

Length overall, m	166.60	Design draft, molded, m	8.23
Length between perpendiculars, m	154.99	Displacement, molded, at design draft, s.w., tons	18,250
Length for coefficients, m	154.99	Bottom of keel below baseline, m	0.0254
Breadth, molded, m	24.08	Half siding at baseline, m	0.660
Depth, molded, to main deck amidships, m	14.66	Deadrise, m	0.305

1. Calculate, plot Bonjean curves (see Section 6).
2. Read Bonjean curves at desired drafts; integrate longitudinally to calculate, then plot at each draft: displacement, LCB; extend displacement curve to zero draft.
3. Read waterline half-breadths at each draft desired, calculate waterplane area, calculate, plot LCF, TPcm, C_{WP}; calculate I_T about ships centerline, calculate I_L about transverse axis through LCF at each draft.
4. Using displacement from (2) and I_T and I_L from (3), calculate, plot BM_L; calculate $\overline{BM}$, all at each draft.
5. Integrate displacement curve vertically to calculate $\overline{KB}$ at a series of drafts; plot $\overline{KB}$.
6. Calculate plot $\overline{KM}$ from $\overline{KB}$ and $\overline{BM}$, (4) and (5), at each draft.
7. Calculate $\overline{GM}_L$ using assumed $\overline{KG}$, $\overline{KB}$ from (5), $\overline{BM}_L$ from (4); using ship length between draft marks, calculate, plot MTcm at each draft.
8. Using TPcm, LCF (3), and ship length between draft marks, calculate coming plot increase in displ. for 1 cm trim by stern at each draft.
9. Read displacement from curve (2), calculate, plot C_B; calculate, plot C_M from midship Bonjean curve; calculate, plot C_P from C_M and C_B at each draft.
10. Read waterline half breadths at appropriate stations and drafts, calculate wetted surface over main body of ship at each draft using differential surface area method (see Section 7).
11. Girth stations at ends of ship, approximate wetted surface of ends; calculate, plot wetted surface of complete ship by adding wetted surface of main body from (10) at each draft.
12. Using waterplane areas from (3), integrate vertically to get displacement and $\overline{KB}$, check results obtained in (2) and (5).

Fig. 29 Sequence of Curves of Form calculations

5.15 Summary of Calculations. A concise summary of results should be furnished with the curves of form calculations to assist in plotting the curves and serve as a permanent record. Values should be listed at all drafts at which calculations are performed. Drafts should be spaced closely enough to allow drawing the curves without ambiguity. Table 10 is a condensed summarization of the plotted values in Fig. 23 for the example ship.

For most ship-shaped forms, the plotted curves will be fair, with few points of inflection. The presence of any unfairness usually indicates an error in the calculations—as from incorrect input data—or an error in plotting.

All of the data given in Table 10 apply to the molded form, with the exception of total displacement in fresh and salt water, and wetted surface, described in Section 7. Ordinarily, the other data, based on the molded form, are used without correcting for the effect of appendages, because such effects are so small as to be negligible as far as practical purposes are concerned—except in some cases for LCB. (Section 5.7).

Since final curves of form are for the benefit of operating personnel, the curve sheet furnished the ship owner should show drafts to the bottom of the keel, with the distance from the molded base line to the bottom of keel clearly noted.

5.16 Computer Applications. Calculations for curves of form require numerous repetitive calculations which were formerly performed on desk calculators. Consequently, one of the earliest applications of digital computers in naval architecture was in the calculation of hydrostatic properties. According to Lasky and Daidola (1977), in a survey of computer utilization by more than 200 ship design offices, the largest area of design activity to which computers are applied is hydrostatics, with about 80 percent of the responders so involved. There are known to be a large number of programs in existence for such hydrostatic calculations, and the already widespread application of such programs is expected to grow.

The basic input must be the overall dimensions of the ship and the offsets at a series of waterlines and stations together with a definition of the end profiles, stated or implied. It is a matter of judgment as to how many waterlines and body plan stations are needed to assure valid hydrostatic calculations. Clearly more closely spaced waterlines are needed when the vessel's form changes rapidly with draft, as normally applies at the lowest drafts, but the spacing should be appropriate to the rule of integration adopted. Figure 29 shows a possible sequential scheme for performing the various calculations. Fig. 30 shows a typical computer-generated body plan derived from stored offsets.

Some programs perform many of the hydrostatic calculations needed for intact and damage stability studies described in Chapters II and III, in addition to curves of form, inasmuch as the hull surface offsets, once defined by the computer, can serve as common input data. When the computer is linked to a cathode ray scope and/or a plotter, the results may be displayed for checking and/or hard copy curves can be

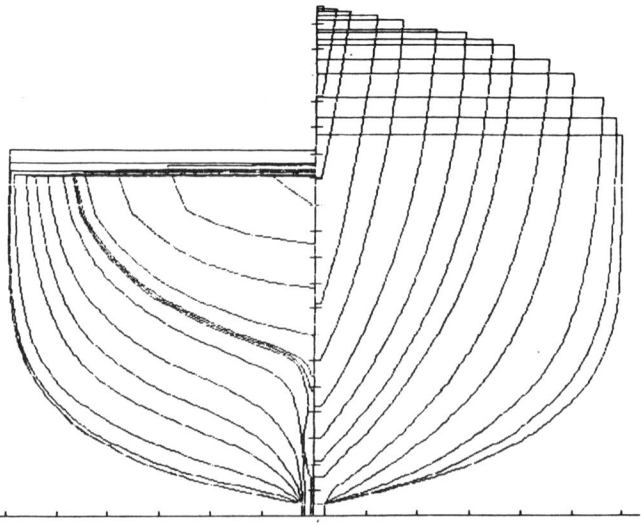

Fig. 30 Computer printout of body plan of small vessel in Fig. 31

plotted, in addition to appearing in the computer output in tabular form.

The U.S. Navy's Ship Hull Characteristics Program (SHCP) NAVSEA (1976) incorporates these features and is widely used in U.S. design offices and shipyards for both commercial and naval ships. It comprises a set of sub-programs which perform any or all of the following naval architectural calculations:

- Curves of form and Bonjean Curves.
- Longitudinal shear and bending moment (still water and in wave).
- Trim lines after flooding.
- Floodable length.
- Limiting drafts for survival after flooding.
- Curves of intact static stability and cross curves.
- Curves of static stability and cross curves, damaged.

Common input for the above are the hull offsets, which may be read from the lines drawing and entered by means of punched cards. Or a digitizer may be used to trace the body plan, read the points and enter them in the computer. A more economical procedure is to enter the offsets directly from the stored HULDEF data (Section 1.15).

When utilizing this program the user is not obliged to use the station or waterline location and spacing for which the lines have been drawn; rather, an odd number (minimum 3, maximum 41) of body plan stations are chosen. Each station must have a non-zero sectional area when fully immersed. Section offsets are specified at between 2 and 29 points for each body plan station which is assumed describable by a series of 2nd order curve segments. These segments are consistently taken between odd numbered waterlines, for integration by Simpson's First Rule.

From the offsets the Ship Data Table (SDT) is set up, which is the common data base for all of the sub-programs. It contains the following, calculated by Simpson's First Rule, for each point (waterline) on each section:

(a) Half-breadth and height above baseline; and cumulative properties (above baseline).
(b) Full section area (both sides).
(c) Transverse centroid of half section.
(d) Vertical centroid of section.
(e) Half-girth.

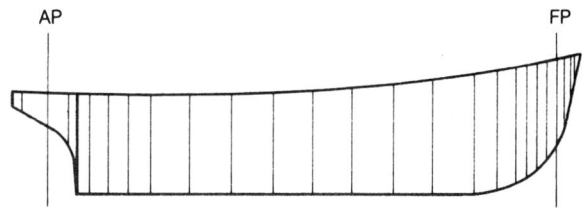

LBP OF SHIP = 12.5 STATIONS = 300 FT.
STATION SPACING = 24 FT.
STATION DISTANCE FROM F.P.
= STATION NO. x STATION SPACING.
STATIONS SHOWN:

−0.500	2.000	11.000
−0.417	3.000	11.500
−0.333	4.000	11.742
−0.167	5.000	11.771
0.000 (FP)	6.000	11.800
0.250	7.000	12.000
0.500	8.000	12.500 (AP)
0.750	9.000	13.000
1.000	10.000	13.417
1.500	10.500	

Fig. 31 Typical station spacing for integration by SHCP

The SDT can be printed out and/or retained in the computer memory.

In subsequent calculations Simpson's First Rule is also used between odd-numbered stations for longitudinal integration. In order to handle extremities of curves which do not terminate at selected stations, the program itself extrapolates the curve to the axis and determines one offset midway between the extremity so found and the first known offset.

The first subprogram produces the data for the usual curves of form for up to 21 waterlines and 7 trims. Calculated properties are presented in tabular form, and optional plots are provided of curves of form, waterlines and sectional areas (Bonjean curves).

Fig. 31 shows a typical spacing of stations for a small vessel integrated using the SHCP. Table 11 shows part of the output, in English units as customarily used by the U.S. Navy.

If computer output is to serve as the final record for use by ship's personnel, it is important that the data be presented in a completely clear way, bearing in mind the environment in which they will be used.

Table 11—Typical Output from Computer Calculation of Ship Hydrostatics using SHCP Program

```
SHIP-      SHCP SAMPLE SHIP S.S.SUSAN GAIL      SERIAL NUMBER-  717    DATE- 5/25/84
```

HYDROSTATICS - PART I   TRIM   0.000 FEET

	DRAFT	VOLUME	DISPLACEMENT	LCB	KB	WETTED SURFACE	PRISMATIC COEF	WPLANE COEF	WPLANE I COEF
	2.00	1722.	49.2	-9.56	1.03	2143.	0.759	0.590	0.372
	10.00	51593.	1474.1	8.37	6.81	12751.	0.547	0.580	0.410
	18.00	161991.	4628.3	6.41	11.90	20253.	0.599	0.666	0.502
	26.00	310907.	8883.1	4.30	16.81	26338.	0.644	0.738	0.578
	30.00	395435.	11298.2	3.22	19.20	29195.	0.665	0.770	0.607
DWL	36.00	531140.	15175.4	1.57	22.74	33410.	0.693	0.815	0.651
	44.00	725939.	20741.1	-0.80	27.39	39223.	0.729	0.875	0.707

HYDROSTATICS - PART II   TRIM   0.000 FEET

	DRAFT	WPLANE AREA	LCF	TPI	CIDDFTS	LONG. BM	TRNSV BM	LONG. KM	TRNSV KM	MT1
	2.00	1271.	0.12	3.03	-0.01	2965.5	2.00	2966.6	3.04	40.5
	10.00	10638.	7.38	25.33	-7.47	686.8	45.48	693.6	52.29	281.2
	18.00	16552.	3.75	39.41	-5.91	415.5	44.05	427.4	55.95	534.2
	26.00	20402.	0.31	48.58	-0.60	313.8	36.41	330.6	53.22	774.2
	30.00	21791.	-1.80	51.88	3.74	284.9	32.21	304.1	51.41	894.1
DWL	36.00	23401.	-4.77	55.72	10.64	251.9	26.84	274.6	49.58	1061.7
	44.00	25202.	-9.58	60.00	22.99	225.4	21.58	252.8	48.97	1298.9

SECTIONAL AREAS IN SQUARE FEET - PART 1   TRIM   0.000 FEET

	STATION DRAFT	-0.500	-0.417	-0.333	-0.167	0.000	0.250	0.500	0.750
	2.0	0.00	0.00	0.00	0.00	0.00	0.00	0.00	0.00
	10.0	0.00	0.00	0.00	0.00	0.00	0.00	0.00	5.28
	18.0	0.00	0.00	0.00	0.00	0.00	3.26	38.06	99.29
	26.0	0.00	0.00	0.00	0.00	5.25	69.01	164.73	289.98
	30.0	0.00	0.00	0.00	0.00	26.25	126.51	256.31	414.52
DWL	36.0	0.00	0.00	0.00	8.56	81.38	238.81	423.31	631.97
	44.0	0.00	0.00	0.00	55.55	190.27	431.89	691.69	967.61

SECTIONAL AREAS IN SQUARE FEET - PART 2   TRIM   0.000 FEET

	STATION DRAFT	1.000	1.500	2.000	3.000	4.000	5.000	6.000	7.000
	2.0	0.00	0.44	7.02	7.20	7.37	7.48	7.64	7.48
	10.0	26.06	90.64	144.83	217.50	266.63	293.07	309.75	308.99
	18.0	169.04	325.90	457.47	649.16	781.84	857.61	895.07	894.64
	26.0	411.40	670.16	868.00	1211.42	1426.43	1547.90	1598.99	1600.82
	30.0	563.29	873.90	1134.03	1522.66	1775.34	1916.85	1972.57	1973.82
DWL	36.0	822.79	1209.40	1551.72	2009.91	2314.97	2481.36	2542.69	2544.57
	44.0	1215.08	1699.35	2161.01	2686.58	3050.62	3242.45	3310.07	3311.95

Section 6
Bonjean Curves

6.1 Curves of Areas of Transverse Sections. Fig. 32(a) shows a typical transverse section through a ship on one side of the centerline, such as a body plan station. The area KCL_1W_1K from the baseline up to waterline W_1L_1 may be obtained by one of the rules of integration, or by planimeter or integrator. Twice the area, plotted to a convenient scale and at the same draft as W_1L_1, would appear as the point Q in Fig. 32 (b). Similarly, the area $KCLWK$ from the baseline up to WL could be obtained and would give the point P in Fig. 32(b). The curve $K'QPF'T$, Fig. 32 (b) thus represents the area of the full section on both sides of the centerline, from the baseline up to any waterline.

For wooden vessels, the half section should be taken to the outside of the planking, but for steel vessels, it is customary to draw the curve of areas for sections taken to the molded line.

In cases where vessels have unusually large appendages, it may be desirable to construct the curve of transverse section area with the inclusion of the shell thickness, corrected for the obliquity of the vessel's form, together with the cross sectional area of other appendages such as bilge keels. A longitudinal integration of such total cross section areas, together with the volume of appendages not intersected by the sections, would give the total displacement of the ship, but the calculation of the curves of cross sectional area would be too laborious for general use.

The curves of cross sectional area for all body plan stations are collectively called *Bonjean Curves*.[5] One of the principal uses of Bonjean Curves is determining volume of displacement of the ship at any level or trimmed waterline.

A convenient way to calculate Bonjean Curves is by the use of closely spaced waterlines at the lower levels and the use of the 5, 8, −1 rule described in Section 4.6. Table 12 shows the calculation for one point on a Bonjean Curve for the example ship. In the event the lowest contour of the station is considered to curve too sharply for a satisfactory parabolic approximation using available halfbreadths, a planimeter may be used. Further points on the Bonjean Curves may then be found, with the area between the next pair of waterlines added to that below. The moment of area of each section about the molded baseline may be found by an adaptation of the 3, 10, −1 rule. If these vertical moments be integrated longitudinally, one may find the moment of volume of displacement about the molded baseline, and hence $\overline{KB}$, the vertical height of center of buoyancy. The vertical moment of sections across the vessel up to any waterline may be useful in problems which arise in the case of a ship flooded throughout part of its length (Chapter III).

6.2 Construction of Bonjean Curves. Bonjean Curves may be plotted in either of two ways. Fig. 33 shows the curves for the ship shown in Fig. 1 plotted against a common scale of draft, with the cross sectional areas for stations in the forebody and amidships plotted to the right of the vertical axis and those for the afterbody plotted to the left. The draft scale may represent keel drafts, or molded drafts, but the distance from the molded baseline to the bottom of keel should be shown. Such a presentation has the advantage of compactness, and uses one scale of cross sectional area. It is convenient to show a contracted profile of the ship adjoining the curves.

An alternative plot is that shown by Fig. 34 in which a separate horizontal scale of cross sectional area is provided for each curve, and the curves are superposed on a contracted profile of the ship; in the latter case, the vertical axes coincide with the associated station lines in the profile. This arrangement is convenient for placing and locating trim lines on the profile, but has the disadvantage that the horizontal area scales for each station may be difficult to distinguish, one from the other, at areas of overlap. Draft scales corresponding to those on the ship should be shown at the appropriate locations on the profile.

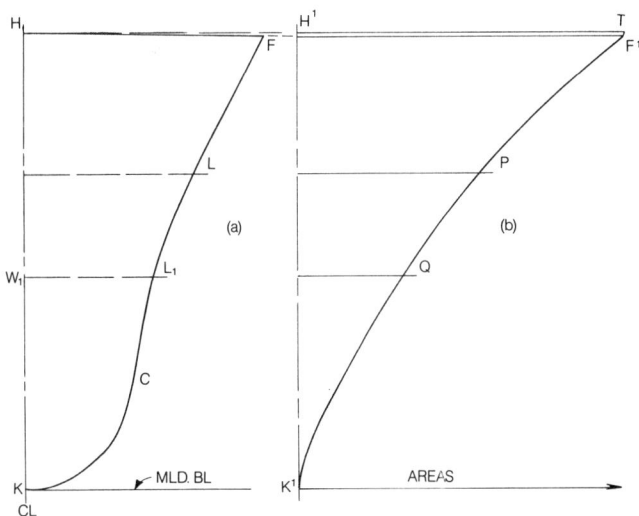

Fig. 32 Body plan section (a) and Bonjean Curve (b)

6.3 Uses of Bonjean Curves. As noted in Section 5.7, a standard method of calculating volume of displacement and LCB is by integrating transverse sectional areas. If the waterline at which the ship is floating is not for the even keel condition, Bonjean Curves are particularly useful. In the case of a trimmed waterline, the trim line may be drawn on the profile of the ship and drafts read at which the Bonjean Curves are to be entered. By drawing a straight line across the contracted profile of Figs. 33 and 34, the drafts at which the curves are to be read appear directly at each station.

Inasmuch as the curves of form are constructed for the ship in the even keel condition and most ships are not wall-sided, accurate hydrostatic characteristics for cases with a significant degree of trim are not in general obtainable from the curves of form and one must

Table 12—Calculation of Point on Bonjean Curve at Lower Level

(For Station 8 on Example Ship at 1m waterline)

Height above Baseline m	Halfbreadth m	Multiplier	Product
0	0.66	5	3.30
1	3.08	8	24.64
2	4.02	−1	−4.02
			$\Sigma = \overline{23.92}$

Transverse section area below 1m waterline

$$= \frac{s}{12} \times \Sigma = \frac{1}{12} \times 23.92$$

$$= 1.99 \text{ m}^2, \text{ one side of ship,}$$

$$= 3.99 \text{ m}^2, \text{ both sides of ship.}$$

---

[5] Named after a French naval engineer of the early nineteenth century.

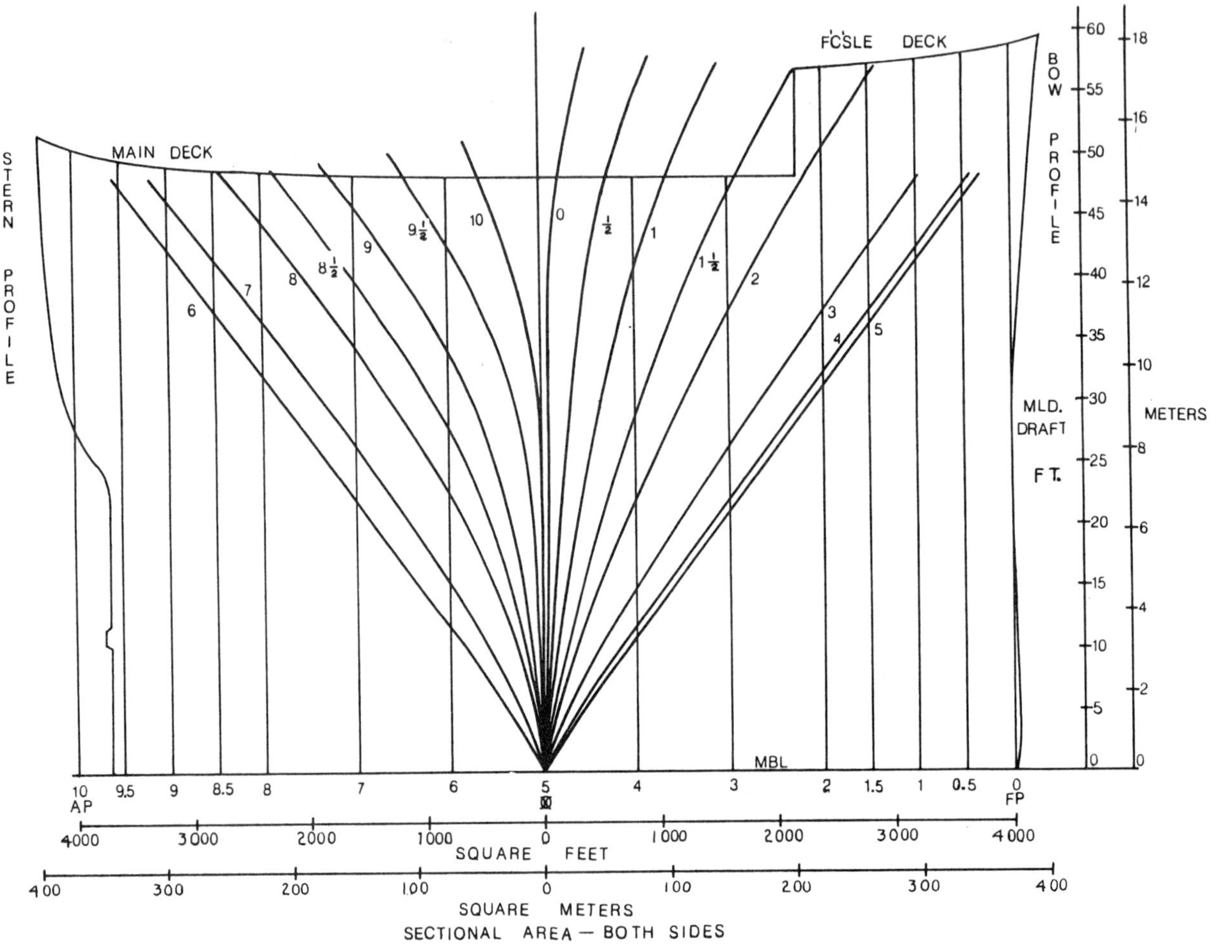

46 PRINCIPLES OF NAVAL ARCHITECTURE

Fig. 33 Bonjean Curves—common vertical axis

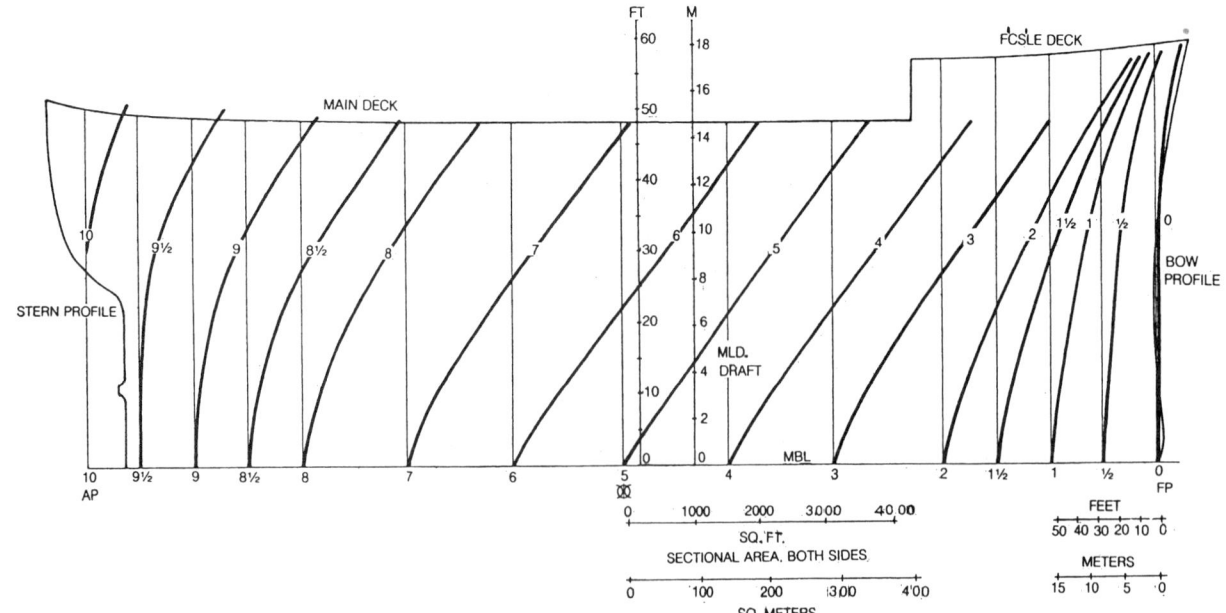

Fig. 34 Bonjean Curves—separate vertical axes

perform a complete longitudinal integration at the trimmed waterline (trim line) under consideration. The Bonjean Curves provide the basic input for such calculations. Cases where this is needed may occur in connection with launching, discussed in *Ship Design and Construction* (Taggart, 1980), and when the end compartments of a ship are flooded, as discussed in Chapter III. If the ship be considered in the crest or trough of a wave of known profile, as may be assumed for longitudinal strength calculations, again the displacement of the ship can be calculated, as discussed in Chapter IV.

Section 7
Wetted Surface

7.1 Definitions and Uses of Wetted Surface. For a vessel floating at a given waterline, the total area of its outer surface in contact with the surrounding water is known as its *wetted surface*. When estimating the frictional resistance to the motion of a vessel through the water, it is important to know the vessel's total wetted surface up to any waterline at which the vessel may operate. The subject of frictional resistance, and corrections to the wetted surface for the ship's wave profile, is treated in Chapter V.

The wetted surface may be used in estimating the amount of paint required to coat the vessel's bottom up to a given waterline. Also, the wetted surface below the waterline may be added to the area of the topsides above the waterline to obtain the total area of the shell plating. Thus, the approximate weight of the shell may be estimated as well as the paint required for it.

Wetted surface is customarily calculated at various waterlines for a new ship and appears as one of the curves of form. In some cases additions are made for appendages, such as stem, stern frame, rudder, propeller shaft bossing and bilge keels.

7.2 Calculation of Wetted Surface. The underwater surface of the molded form is the principal component of the total wetted surface of a ship. In a fine, high-speed, multiple-screw ship it may amount to 85 percent of the total wetted surface. In a full single-screw ship it may amount to 99 percent of the total. The wetted surface of the shell plating is virtually that of the molded form; therefore, calculations of the molded surface of the hull plus that of appendages extending beyond the shell may be considered the entire wetted surface.

The wetted surface of the molded form may be obtained by calculating various portions directly from the lines drawing, and estimating other portions closely. The calculation method has traditionally been that of drawing an expansion of the molded surface up to the desired waterline, and measuring the area enclosed by the expansion, it being assumed the area of the expansion is virtually that of the molded surface (See next section.).

At each transverse section of the body plan, the distance along the contour of the section from the centerline at the bottom up to any given waterline is known as the half-girth of the section up to that waterline. The half girth may be obtained by bending a thin flexible batten around the section, or a straight measuring scale or strip of paper may be placed in contact with the curve of the section at the starting point and thereafter kept in contact with and tangent to the curve at successive points, by rotating the strip of paper slightly with the paper held in place at the point of contact by the point of a pencil. The measuring scale or strip of paper should be rotated continually about its successive points of contact along the curve. Alternatively, a map measurer may be rolled along the section perimeter, following the curve carefully and noting the revolutions of the wheel; these may be interpreted, by reference to a calibrated scale, as girthed distance.

The half-girths of the various sections may be plotted as ordinates on their respective stations along a base line representing the length of the vessel. A fair curve passed through such points will enclose an area known as the transverse expansion of the molded surface of one side of the vessel up to the given waterline. Similarly, the transverse expansion of one side of the vessel's surface may be obtained between any two waterlines, or up to any deck line. Fig. 35 shows these transverse expansions up to several successive waterlines, constructed from the lines drawing, Fig. 1.

The area of the transverse expansion of a vessel's molded surface up to a given waterline may be calculated readily and considered as a first approximation to the true area of the vessel's molded wetted surface. This approximation is usually correct to within about 2 percent and for many purposes this degree of accuracy will suffice.

7.3 Graphical Corrections to Wetted Surface From Transverse Expansion. The molded surface is usually composed, in large part, of surfaces of compound curvature and so cannot be expanded into a plane. Hence, the area enclosed by the transverse expansion does not properly account for the obliquity of the vessel's lines. To overcome this inaccuracy, several alternative graphical corrections have been developed. Fig. 36 shows a portion of a half breadth view in the forebody

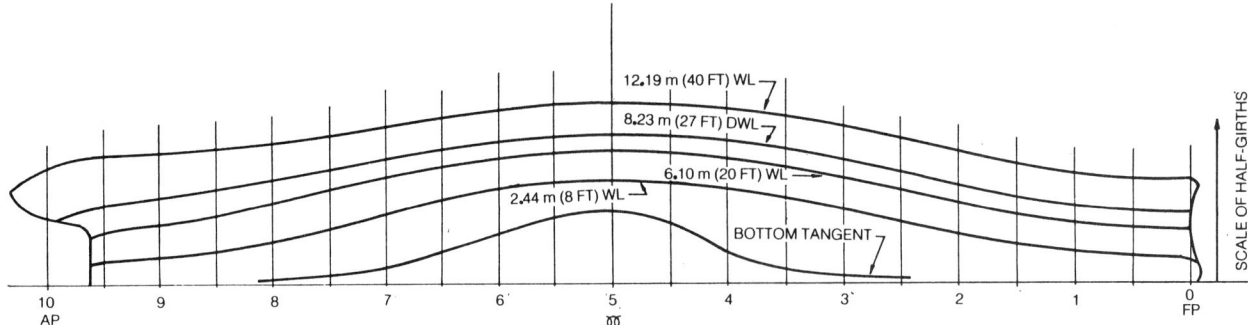

Fig. 35 Transverse expansion of half girths

of a ship, with the expansion of half girths shown above it. The 5 m waterline may be considered as giving the average slope of the 4 m and 6 m waterlines. By girthing the 5 m waterline, starting at amidships, station 5, and working forward, using either a thin, flexible batten or a strip of paper, the distances along the true shape of the waterline at which stations are crossed may be found. When these distances are laid out along the baseline of the expansion, it will be found that point 1 moves to point 1', point ½ moves to point ½', while the bow contour moves forward the maximum amount. The area under the dashed curve is then a closer approximation to the molded wetted surface of the vessel between the 4 m and 6 m waterlines than is the area under the uncorrected transverse expansion. This procedure is known as rectification of waterlines.

Another method, which is numerically equivalent to that of rectification of waterlines applies a secant correction to the half girths. Thus, in Fig. 36, the straight line CD, connecting the 5 m waterline half breadths at station 1½ and 2½, has a waterline angle ϕ which is practically the average angle of the average waterlines between the 4 m and 6 m waterlines, and in way of station 2. Thus, the length of the molded surface between these waterlines, and between stations 1½ and 2½ is quite close to CD. However, $CD = s \cdot \sec \phi$ where s is full station spacing. Thus, if the half girth at Station 2 be multiplied by $\sec \phi$, the area enclosed by the stations and line GHF in Fig. 36 will be a closer approximation to the true wetted surface than the area enclosed by the stations and line KTN. The secant method of modifying half-girths is probably more convenient for calculation purposes than is rectification of waterlines, inasmuch as station ordinate locations are not changed.

7.4 Integration of Differential Surface Areas. The foregoing methods are inherently approximate. The area of a transverse expansion is slightly less than the true surface area, while the rectification of waterline / secant half-girth correction method over compensates for the effect of obliquity on the transverse expansion, giving an area slightly too large. A more direct method of finding wetted surface is to integrate the differential surface area along the ship.

The elementary area of a curved three-dimensional surface, defined with respect to mutually perpendicular x, y, z axes, may be found as the product of elementary area in the x, z plane δx, δz and the secant of inclination of the surface from the x, z plane.

In vector analysis it is shown that for a surface $y = f(x,z)$, the angle β between a vector normal to the surface and the y axis is given by,

$$\cos \beta = \frac{1}{\sqrt{1 + \left(\frac{\partial f}{\partial x}\right)^2 + \left(\frac{\partial f}{\partial z}\right)^2}}.$$

Then

$$\sec \beta = \sqrt{1 + \left(\frac{\partial f}{\partial x}\right)^2 + \left(\frac{\partial f}{\partial z}\right)^2}.$$

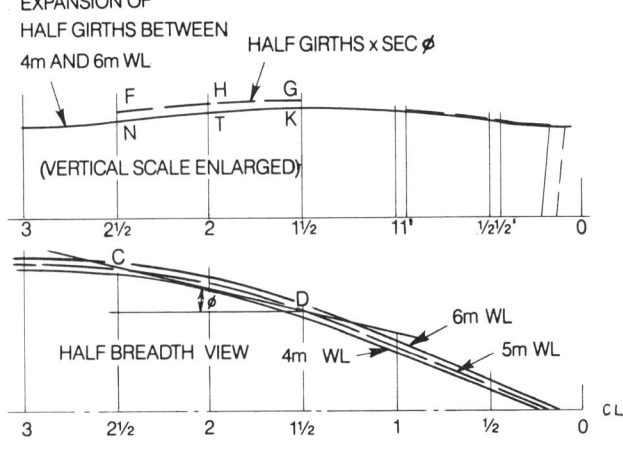

Fig. 36 Correction to transverse expansions for obliquity

Assume x distances are longitudinal distances along the ship, y distances are halfbreadths to the molded surface, and z distances are heights above the molded baseplane to the molded surface.

Consider the unit differential area δx, δz in Fig. 37 in the longitudinal centerplane of the vessel and project this to the molded surface, giving differential surface area δs.

Then

$$\delta s = \delta x \cdot \delta z \sqrt{1 + \left(\frac{\delta y}{\delta x}\right)^2 + \left(\frac{\delta y}{\delta z}\right)^2}, \text{ or area,}$$

$$S = \sum\sum \delta x\, \delta z \sqrt{1 + \left(\frac{\delta y}{\delta x}\right)^2 + \left(\frac{\delta y}{\delta z}\right)^2}, \text{ or} \quad (50)$$

$$S = \iint \sqrt{1 + \left(\frac{dy}{dx}\right)^2 + \left(\frac{dy}{dz}\right)^2}\, dx\, dz.$$

There seems to be no simple way of reducing this double integral to an integral in one variable, to permit calculating surface area by a single step rule of integration. We are obliged instead to find the area incrementally, the simplest and most useful way for a ship's molded surface being to calculate the surface between pairs of waterlines which are reasonably closely spaced. That is, let δz be a constant difference in draft and then integrate longitudinally using a rule of integration, such as Simpson's First or Second Rule to find the wetted surface between each pair of waterlines.

Table 13 illustrates a tabular calculation for finding wetted surface based upon differential surface area for the area between two waterlines of a tanker, and between stations 15 and 19 (20 stations LBP). The calculation makes use of Simpson's Second Rule as the primary rule of integration, together with Simpson's First Rule. It will be seen that the vertical station shape slope $\dfrac{\delta y}{\delta z}$ is obtained from the difference in station half-breadths above and below the area to be integrated, whereas the longitudinal waterline slope is obtained as the mean of differences between waterline half-breadths forward of and abaft the station in question.

The calculation routine avoids the necessity for measuring half girths but makes maximum use of the molded surface offsets. As such, it is easily adapted to programming on a digital computer.

7.5 Wetted Surface Coefficients. In order to compare the wetted surface of different ships it is useful to calculate a dimensionless coefficient which relates the wetted surface to the basic characteristics of the hull. Such a coefficient is C_{WS}, where,

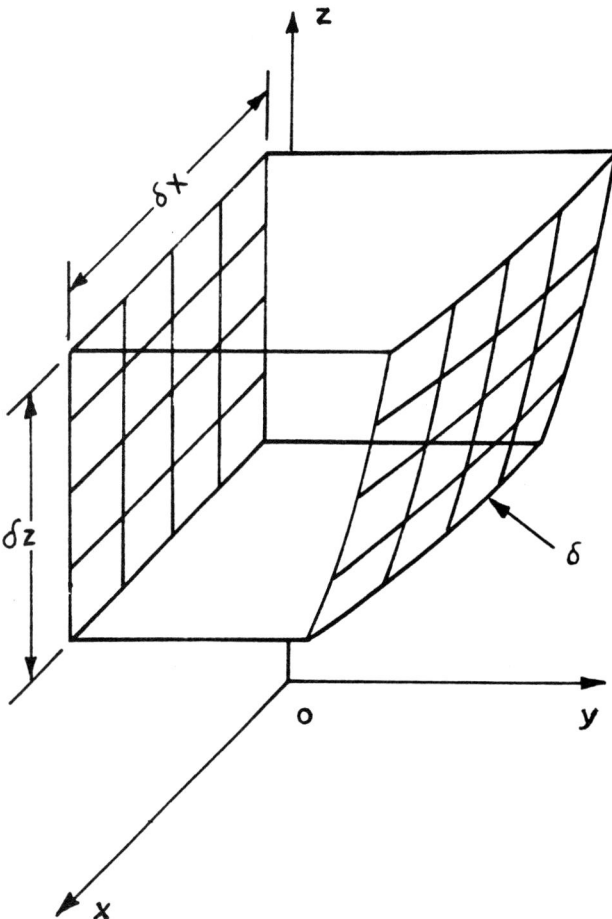

Fig. 37 Differential surface area

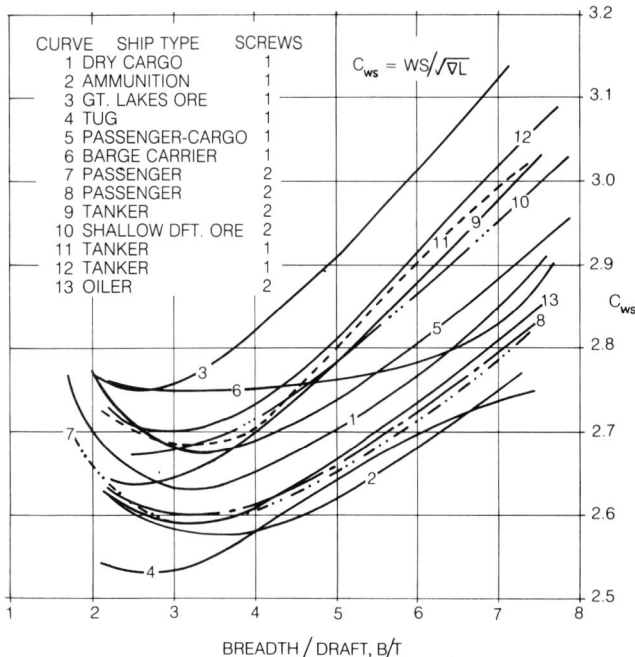

Fig. 38 Wetted surface coefficients

Table 13—Calculation of Wetted Surface of Portion of Tanker by Differential Surface Area Method
(Between Stations 15 and 19 and between the 3 m and 6 m waterlines)

Sta	HB 6 m	HB 3 m	$\left(\dfrac{\delta y}{\delta z}\right)$	$\left(\dfrac{\delta y}{\delta z}\right)^2$	Next sta ford	HB 6 m	HB 3 m	Next sta aft	HB 6 m	HB 3 m	Mean $\left(\dfrac{\delta y}{\delta x}\right)$	$\left(\dfrac{\delta y}{\delta x}\right)^2$	Sum	[Sum]^{½}	SM	Product
15	19.66	18.41	0.42	0.17	14	20.12	19.84	16	17.56	15.56	−0.12	0.01	1.18	1.09	1	1.09
16	17.56	15.97	0.70	0.49	15	19.66	18.41	17	13.38	11.16	−0.24	0.06	1.55	1.24	3	3.72
17	13.38	11.16	0.74	0.55	16	17.56	15.47	18	8.14	6.64	−0.33	0.11	1.66	1.29	3	3.87
18	8.14	6.64	0.50	0.25	17	13.38	11.16	19	2.62	2.16	−0.35	0.13	1.38	1.17	1.444	1.69
18½	5.43	4.39	0.35	0.12	18	8.14	6.64	19	2.62	2.16	−0.36	0.13	1.25	1.12	1.778	1.99
19	2.62	2.16	0.15	0.02	18½	5.43	4.39	19½	−0.22*	−0.28*	−0.37	0.14	1.16	1.08	0.444	0.48
															Σ =	12.84

HB's are waterline halfbreadths

$\delta z = (6 - 3) = 3$ m

$\delta x =$ station spacing $= 13.94$ m.

* Negative halfbreadths from extrapolating waterline to station.

Integrating factor, IF $= 2 \cdot \dfrac{3}{8} \cdot \delta x \cdot \delta z = 2 \cdot \dfrac{3}{8} \cdot 13.94 \cdot 3 = 31.365$. HB$_A$ is halfbreadth aft

Wetted surface, both sides $=$ IF $\cdot$ Σ $= 31.365 \cdot 12.84 = 402.7$ m^2. HB$_F$ is halfbreadth forward

$$\text{Mean}\left(\dfrac{\delta y}{\delta x}\right) = \left(\dfrac{\text{HB}_{A6} - \text{HB}_{F6} + \text{HB}_{A3} - \text{HB}_{F3}}{4 \cdot \delta x}\right),$$

where $4 \cdot \delta x = 4 \times 13.94 = 55.76$ m in way of stations,
$4 \cdot \delta x = 2 \times 13.94 = 27.88$ m in way of half stations.

$$\text{Sum} = 1 + \left(\dfrac{\delta y}{\delta z}\right)^2 + \left(\dfrac{\delta y}{\delta x}\right)^2$$

$$C_{WS} = WS/\sqrt{\nabla \cdot L} \qquad (49)$$

Here WS is wetted surface up to any waterline,
∇ is volume of displacement at that waterline,
L is length of vessel.

Values of C_{WS} range between about 2.6 and 2.9 for usual ships of normal form according to plots in Saunders (1957). There is a noticeable dependence of C_{WX} on B/T, beam to draft ratio, and on C_M, midship coefficient. According to Saunders' plots, minimum C_{WS} occurs approximately with $B/T = 2.9$ and $C_M = 0.92$.

Fig. 38 plots wetted surface coefficient against B/T for a number of unrelated ships, with the ship type and number of propellers noted. The coefficients have been developed from wetted surface curves on the curves of form sheet for each vessel. In all cases C_{WS} reaches a minimum in the range of B/T from 2.5 to 3.75, with C_{WS} increasing for B/T beyond the minimum point. The increase of C_{WS} for low B/T is believed to reflect the greater submergence of the stern overhang which tends to accompany draft increases beyond the design draft.

Wetted surface may also be estimated by reference to data on published hull form series such as Series 60 (Todd, et al, 1957).

Unusual ships may be expected to have wetted surface coefficients substantially different from those of Fig. 38, or those shown by Saunders. In general it may be expected that C_{WS} will increase because of hard sections, chines, knuckles and unusually large appendages. The inclusion of the wetted surface of stern transoms, internal wells or the mating surfaces of an integrated tug-barge in estimating frictional resistance may not be appropriate, as discussed in Chapter V.

Section 8
Capacity

8.1 General. A basic characteristic of any ship is the size of the load that it is able to carry. Thus, two fundamental questions arise: (a) What is the volume of space available for cargo—or cargo *capacity*? (b) What is the weight of cargo that can be carried at full load draft—or cargo *deadweight*?

Under considerations of capacity are included the volume of all cargo spaces, store rooms and tanks and the location, vertically, longitudinally, and transversely of the centroid of each such space to allow finding the weight (and center of gravity) of the variable weights, or deadweight of the ship. This information is needed to check the adequacy of the vessel's size, and to determine its trim and stability characteristics. The calculations are called capacity calculations and lead to capacity curves and plans.

The total deadweight of a merchant ship is the difference between the full-load displacement weight and the light ship weight—the latter consisting of the weight of hull steel, machinery (wet), and outfit. The actual "payload" or cargo deadweight is obtained by deducting the typical maximum values of the variable weights of fuel, stores, fresh water, water (or other removable) ballast, crew and their effects from the total deadweight.

In this book we must use the term weight loosely, for when using SI units of kilograms or metric tons we are really speaking about mass. However, if inch-pound units are retained—lbs and long tons—we are then speaking correctly of weight. Similarly, deadweight and displacement (as explained in section 2.2) can be in either SI mass units or in inch-pound weight units.

8.2 Capacity Plan. Fig. 39 shows, in abbreviated form, the capacity plan for a multipurpose dry cargo ship. Such a plan is prepared for the use of the ship owner, and summarizes in convenient form the amount of cargo, fuel, fresh water and stores which the ship may carry, and the spaces into which these will go. The amount of elaboration on the actual plan varies at different shipyards, and depends upon owners' requirements. There is always an outline inboard profile showing the location of tanks, store and cargo spaces and frequently there are also deck plans showing the arrangement of these spaces, as well as sectional views at various frame locations along the ship. The plan includes the principal dimensions of the ship, and shows, usually in tabular form, the name, location and volume of each cargo space, tank, consumable stores space, etc., as well as its longitudinal and vertical centroid when filled, and its transverse centroid, if the space is unsymmetrically disposed about the vessel's centerline. If large units of cargo are to be carried, such as shipping containers or cargo barges, the location of the various vertical tiers and longitudinal rows are shown, as well as how the units are positioned transversely. In the event that the ship is intended to carry specific amounts of deck cargo, such as deck-stowed containers, these also are shown.

The plan includes a displacement scale alongside a scale of drafts, and draft markings as they appear on the side at the ship at amidships, for all drafts from the light condition to full load. Also usually shown next to the draft scale are *TPcm* and *MTcm*. Freeboard

(Continued on page 54)

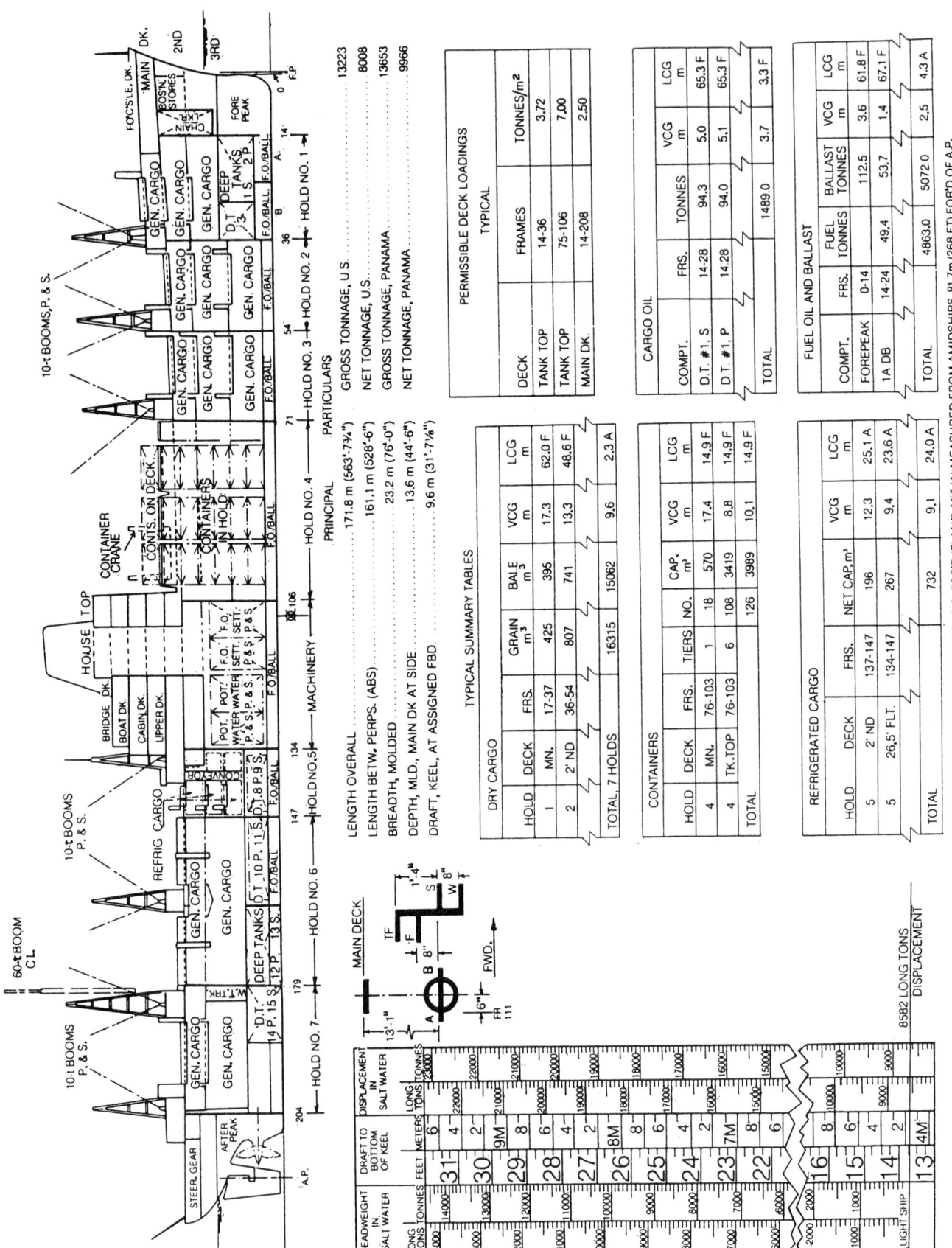

Fig. 39 Capacity plan—multi-purpose dry cargo ship

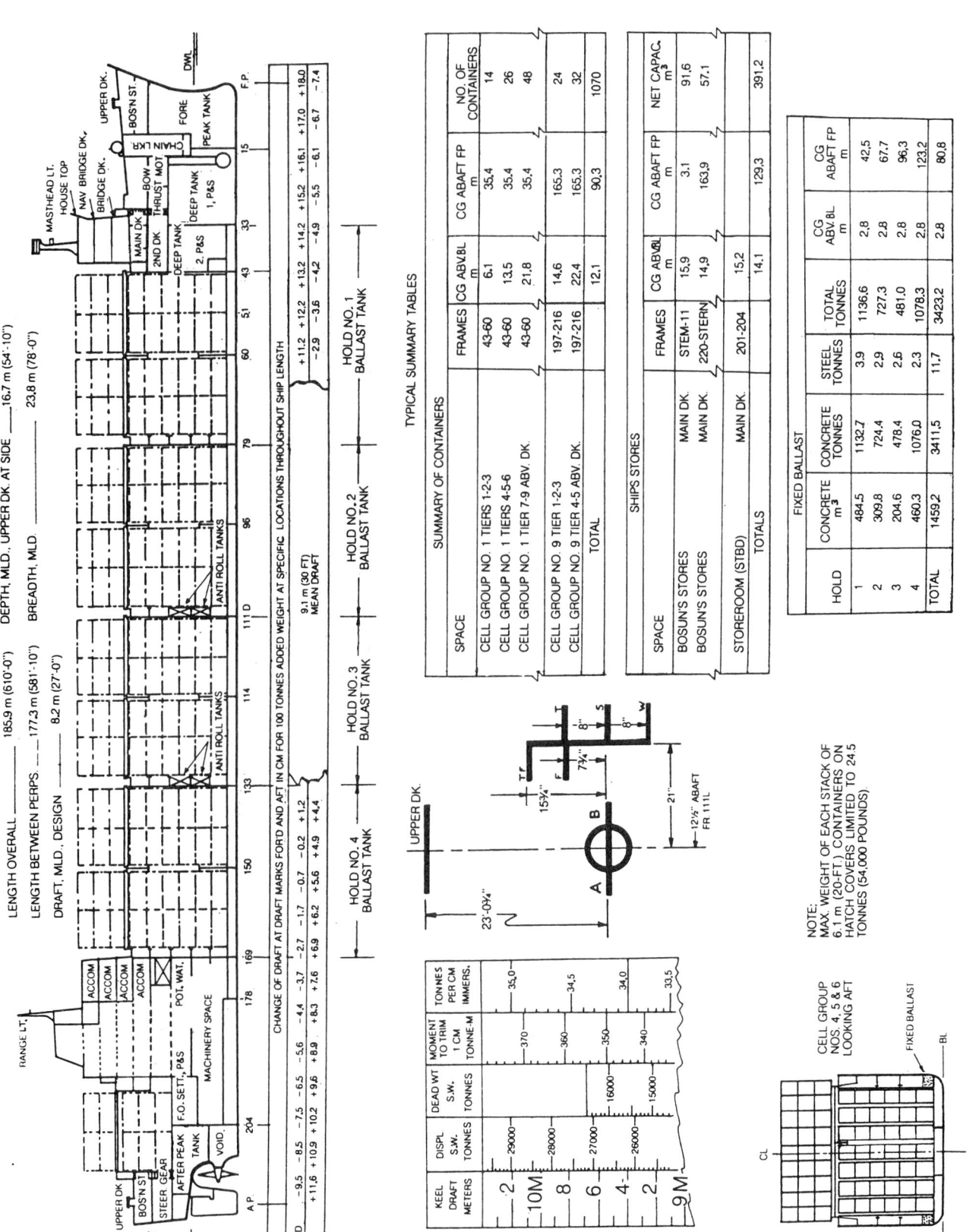

Fig. 40 Capacity plan—containership

(Continued from page 51)
markings and the Plimsoll mark are shown at the top of the draft scale. (The assignment of freeboard is covered in *Ship Design and Construction*, Taggart, 1980). A scale of deadweight is also shown. A trim table is often included to permit the estimation of changes in trim resulting from the addition or removal of weights at various locations along the ship.

Included on the capacity plan may be a list of permissible deck loadings, in tons per m^2, and the outreach and lifting capacity of cargo handling cranes or booms with which the ship is fitted.

The ship shown in Fig. 39 carries general cargo in five holds, containers in guides in one hold, refrigerated cargo in one hold, and cargo oil in deep tanks under four holds and outboard of one hold. Deck-stowed containers, loaded by a shipboard gantry, are regularly carried on the hatch covers over one hold.

The capacity plan for a typical all-container ship is shown in Fig. 40.

8.3 Cargo Capacities. Detail calculations are undertaken to determine the volumes of individual spaces. In the preliminary design phase, approximations to capacities may be adequate. When the design is finalized, exact methods should be used. Where spaces are composed of simple geometrical forms, the standard geometrical formulae may be used. However, on most ships there are numerous spaces bounded on at least one side by the curved hull surface, which are more amenable to calculation by one of the rules of integration.

Two types of capacity are customarily listed on the capacity plan for general cargo holds:

(a) Bale capacity represents the volume below deck beams and inboard of cargo battens which is available for stowing the typical commodities found in general cargo, usually in the form of bales, barrels, bags, crates and boxes.

(b) Grain capacity represents the net molded volume of the space, after deductions for the volume of structure and of ceiling, that is available for carrying granular cargoes in bulk.

In order to calculate bale capacity, a number of frames are selected between the bounding bulkheads. Sectional views are then drawn which show the inside of cargo battens, bottom of deck beams above, and hold ceiling at each frame, as shown in Fig. 41. The cross sectional area available for stowage in way of each frame is then found; these allow a sectional area curve to be drawn, which may be integrated using Simpson's Rule to find the volume, and transverse and longitudinal centroids. Horizontal sections through the space may be taken at a series of levels, and integrated vertically to find the vertical moment, and centroid for partially full and 100 percent full condition.

For grain capacities, the transverse areas are taken to the molded lines, except that deduction is made for ceiling on the inner bottom. A deduction for shifting

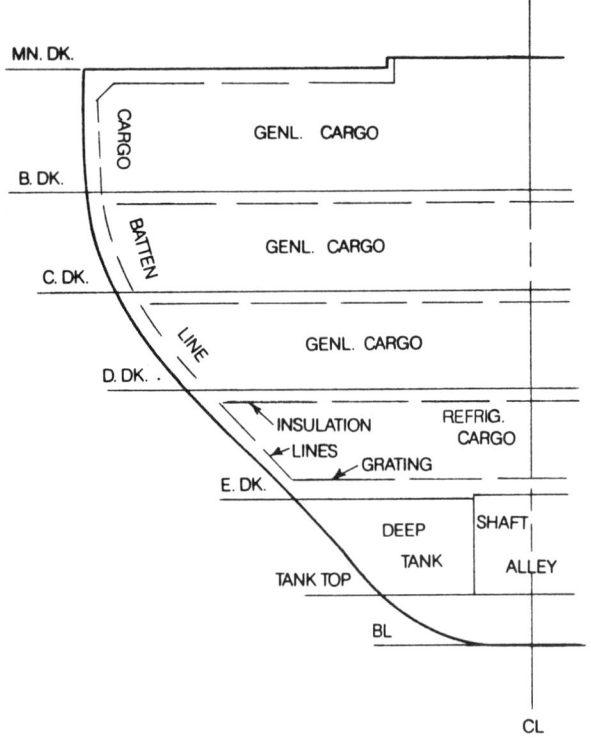

Fig. 41 Section for capacity calculaitons

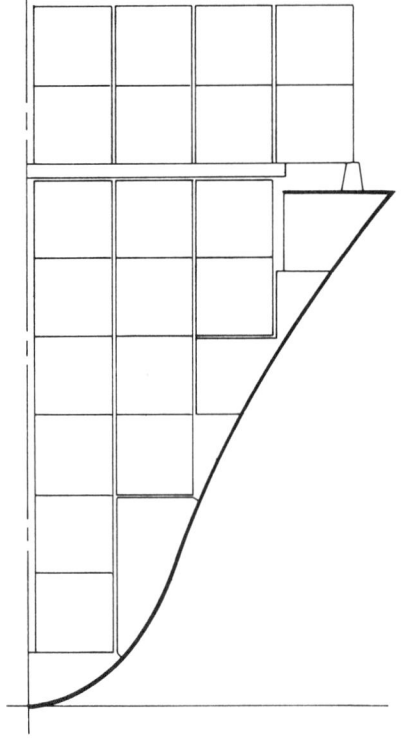

Fig. 42 Section showing container clearances

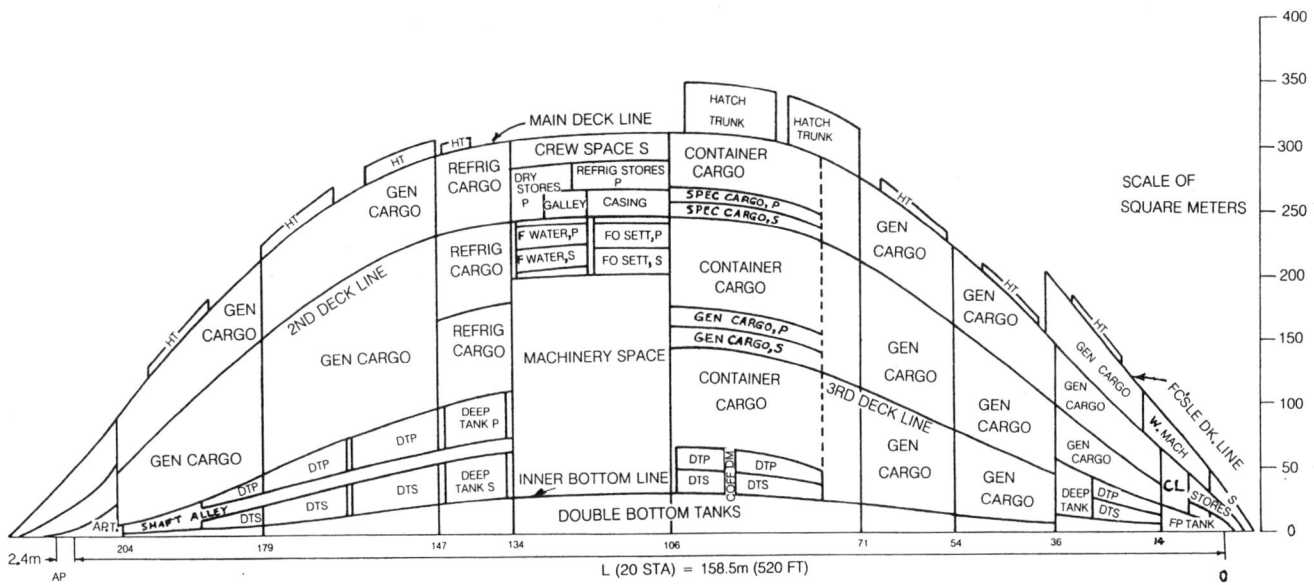

Fig. 43 Underdeck area curve—dry cargo ship

boards may also be necessary. For both bale and grain capacities, deductions must be made for stanchions, pipe covering, deck gratings and other such interferences with stowage.

Refrigerated cargo compartments are figured for total capacity inside of insulation, with deductions for grating on decks, for stanchions, batten protection of refrigerating coils, for air ducts and fan rooms. This volume is usually from 60 to 80 percent of the molded volume. When refrigeration is for fruit carrying, and bins are provided for stowing the fruit, a separate bin capacity is figured, this being the net inside capacity of the bins. The capacity plan, Fig. 39, shows refrigerated cargo forward of hold No. 6. The space taken up by insulation is evident.

Thus far it has been assumed that all cargo is homogeneous; that is, that each cubic meter of it weighs the same. This is by no means always the case. In figuring the centers of gravity of weights in mail rooms, baggage rooms and special storerooms, it may be desirable to estimate the centers of gravity of the weights as they are actually expected to be placed. For some services, it may be desirable to figure that cargo of different weights per cubic meter may be carried in different parts of the vessel. If cargo is hung from the overhead, as for example meat in refrigerated compartments, the center of gravity is effectively at the hook.

When the cargo consists of containers, barges or vehicles, the problem is different, inasmuch as each unit occupies a finite and predictable space. Cross sectional drawings through the container or barge holds, or vehicle decks, showing how the units are to be stowed, are more useful than volumetric calculations. A capacity plan for such vessels shows the number, size and sometimes limiting weight of units, and where they are to be stowed on the ship, rather than the volume of the holds. It is important in the design phase to demonstrate that the lower outboard corner of such specific units and their supporting structure, can be fitted in the space available inside the molded line of the hull, as shown by Fig. 42, and that the number of tiers of units to be stowed in the hold can be accommodated under the hatch covers. In the case of roll-on/roll-off vessels, outline drawings of the vehicles as stowed are sometimes shown to demonstrate available clearance between pairs of lanes and between the tops of vehicles and overhead structure.

A useful drawing sometimes prepared in studies of capacity is a curve of underdeck areas. This is quite similar to a sectional area curve, such as shown in Fig. 24, and has the same ordinate and abscissa units. However, the curve plots sectional areas below the main deck. It may be constructed by the use of Bonjean Curves, modified as necessary for volumes in way of trunks or hatches. The space below the curve is subdivided by lines representing decks and bulkheads, so that all internal spaces are accounted for. Fig. 43 shows an example of such a curve for the dry cargo ship shown in Fig. 39 simplified to illustrate the principles. Many spaces across the ship are made up of readily calculable rectangles, and the cross section of the wing tanks with curved boundaries can sometimes be found by subtracting the rectangular areas from areas bounded by the curve ordinates. An underdeck area curve may be used to check available space in the early stages of design, not only from the point of view of capacity and payload, but also to check tonnage volumes. Tonnage is covered in *Ship Design and Construction* (Taggart, 1980). In the case of certain recent

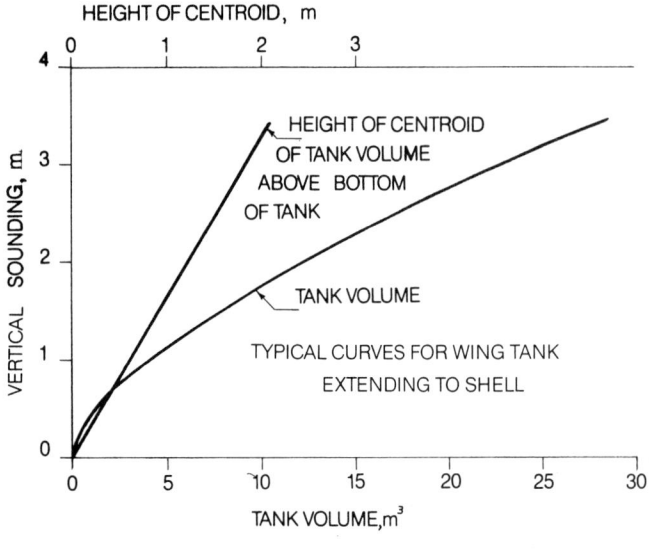

Fig. 44 Tank capacity curve

naval ship designs, internal volumes are at a premium to meet requirements for habitability and electronics. An underdeck area curve is a useful means of determining whether such requirements can be met.

8.4 Tank Capacities. The volumes of rectangular or cylindrical tanks, and their centroids, for any percentage of filling, are readily determined using standard formulas. In the case of irregularly shaped tanks like wing tanks in parts of the ship where the vessel's shape is changing rapidly, the volumes may be found using the method suggested in the preceding section, with the exception that the cross sectional area should be taken from longitudinal boundaries outboard right to the molded surface, rather than to the cargo batten line, as well as between the molded lines of the decks and flats which form the upper and lower boundaries of the tank. Deductions for the structure and other items are then made according to the data in Section 8.5.

Tank capacity tables generally give the tank volumes at a series of closely spaced depths, allowing a curve of volume vs. depth of sounding (depth below surface) to be drawn, as in Fig. 44. The calculations are repetitive and so are adaptable to computer programming. The basic inputs needed are the ordinates from the inboard boundary out to the molded shell line at a series of elevations and frames.

8.5 Deductions from Tank Volumes; Allowance for Expansion. Only molded tank volumes have been discussed thus far. In all tanks there are various internals, such as the frames of the vessel projecting into the tanks, longitudinals and floors in double bottoms, stiffeners on bulkheads and swash plates. There will also be various local deductions and additions which must be calculated separately. For the miscellaneous structural internals, a percent deduction is usually made. Typical data are given in Table 14.

The table gives also the usual allowance for the expansion of petroleum products. The practical operating capacity of an oil tank is not its total net capacity, but from 2 to 5 percent less, since, if the tank could be completely filled with cold oil, the oil would expand and overflow the tank when it becomes warm. The tank capacity is, therefore, calculated as being 95 to 98 percent full, depending upon the usual practice of the owner. Common standards are 95 percent for U.S. Navy practice, and 98 percent for U.S. merchant ma-

Table 14—Typical Corrections to Calculated Tank Capacities

Type of Tank	Allowance For Expansion	Directly Calculated Deductions	Directly Calculated Additions	Average Percent Deductions
Fuel Oil	2 to 3 percent U.S. Navy uses 5 percent	Pockets, sea chests, bulkhead corrugations, heating coils, structure in tank	High coaming hatches, expansion trunks	Double bottom tank without heating coils, 2-1/4 to 2-1/2 percent of molded capacity; add 1/4 percent if with heating coils
Fresh water; salt water ballast	None	Trunks, pockets, sea chests, bulkhead corrugations, structure in tank	High coaming hatches	Double bottom tank, 2-1/4 to 2-1/2 percent of molded capacity
Cargo oil	2 percent	Trunks, pockets, bulkhead corrugations, cargo oil piping, heating coils, structure in tank*	High coaming hatches, expansion trunks	1 percent of molded capacity

* Special considerations may require separate calculations for various levels in tank.

SHIP GEOMETRY

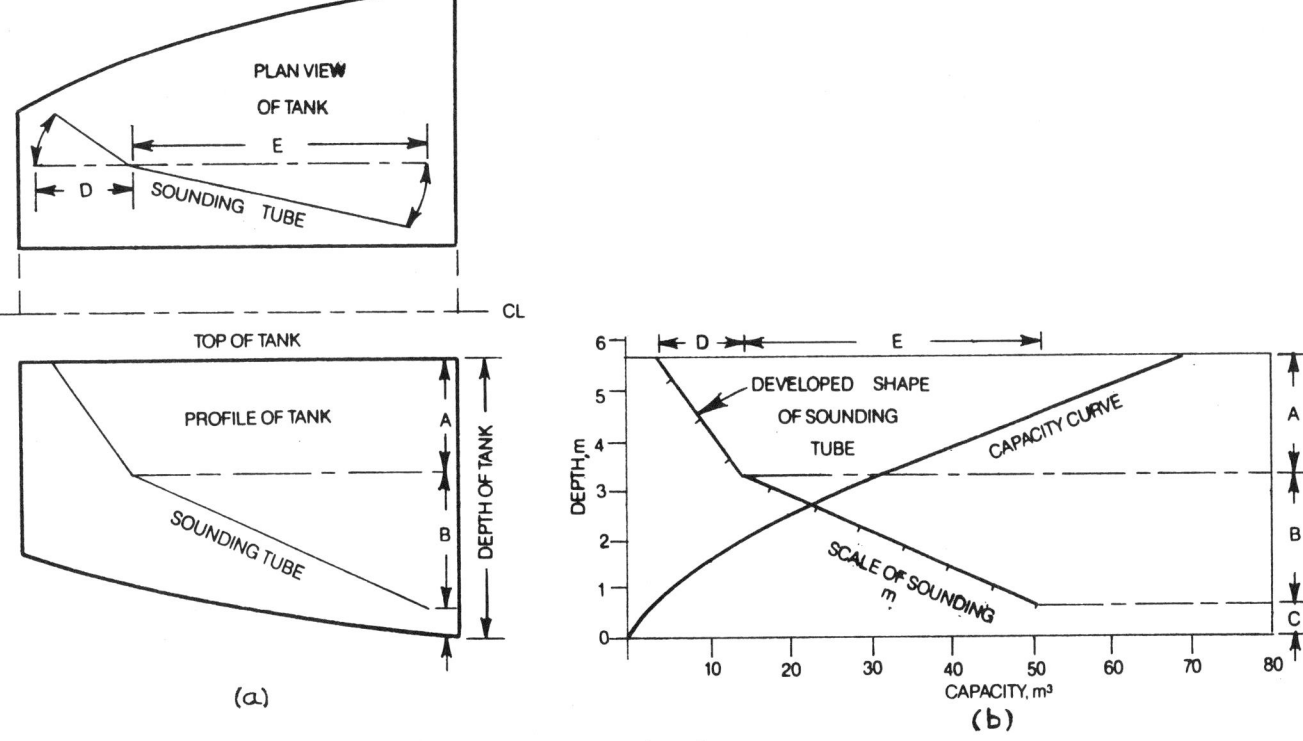

Fig. 45 Sounding tube geometry

rine practice. But more than one percentage is sometimes used on capacity plans. The actual values should, of course, always be stated.

8.6 Capacity Curve and Centroid. The vertical height to the centroid of the contents of any space above the bottom of the space may be readily found provided there is available a capacity curve extending vertically from the bottom of the space. This allows use of the method for VCB of Sec. 5.9.

In Fig. 44 the height of the centroid above the baseline of the curve is obtained by taking the area to any level between the capacity curve and its horizontal base line, divided by the area of the circumscribing rectangle to that level, and multiplying the result by the height from the base line of the curve to that level. The areas may be found by planimeter, or by numerical integration.

8.7 Soundings and Sounding Tables. When the amount of liquid in a tank is determined by lowering a rod or weighted tape measure through a sounding tube, or by any device which otherwise senses the level of the liquid free surface along a line extending into the tank, the sounding is interpreted by use of a table or curve of net tank volume vs. depth, as noted in Section 8.4 and shown in Fig. 44.

It frequently happens that only certain locations are feasible for the upper end of a sounding tube and that a straight vertical pipe from that location will not reach

Table 15—Weights and Conversion Factors

Quantity	Water		Oil				
	Salt	Fresh	Fuel	Diesel	Lube	Gasoline	DFM*
m^3 per t (a)	0.975	1.000	1.059	1.156	1.198	1.393	1.199 − 1.178
Barrels per t (b)	6.139	6.296	6.663	7.277	7.541	8.768	7.548 − 7.417
Gallons per t (c)	261.8	268.5	279.6	305.4	316.4	367.9	316.8 − 311.2
Cubic feet per Long Ton (d)	35	35.9	38	41.5	43	50	43 − 42.3
Pounds per cubic foot	64	62.4	58.95	53.98	52.09	44.80	52.04 − 59.97
Barrels per Long Ton	6.24	6.40	6.768	7.391	7.658	8.905	7.673 − 7.538
Pounds per gallon	8.556	8.342	7.881	7.216	6.964	5.989	6.958 − 7.082

* Diesel Fuel, Marine or Distillate Fuel, Marine—U. S. Navy multi-use fuel.
(a) 1 tonne = 1000 kg; weight of 1 metric ton = 2204 pounds = 1t.
(b) 1 barrel = 5.61 cubic feet.
(c) 1 cubic foot = 7.48 gallons.
(d) 1 Long Ton = 2240 pounds.

Table 16—Stowage Factors, m³ per metric ton

Item	Packing	m³/t	Item	Packing	m³/t
Apples	Boxes	2.23	Lead, Pig	Neat Stowage	0.22
Autos	Assembled and Uncrated	7.52	Lard	Boxes	1.25
			Machinery	Crated	1.39
Barbed Wire	Rolls	1.53	Meat	Cold Storage	2.65
Bauxite	Bulk	1.07	Molasses	Bulk	0.74
Beans	Bags	1.67	Newspaper	Bales	3.34
Beer	Bottled in Cases	2.23	Nitrate	Bags	0.72
Butter	Cases	1.67	Oil	Drums	1.25
Canned Goods	Cases	1.34	Oranges	Boxes	2.17
Carpets	Bales	3.90	Oysters	Barrels	1.67
Cement	Bags	0.97	Paint	Cans	1.00
Cement	Bulk	0.72	Palm Oil	Bulk	1.09
Cheese	Crates	1.81	Paper	Rolls	2.51
Citrus Fruits	Boxes	2.62	Potatoes	Bags	1.67
Coal, Average	Bulk	1.32	Poultry	Boxes	2.65
Cocoanuts	Bulk	3.90	Railroad Rails	Neat Stowage	0.42
Coffee	Bags	1.62	Rice	Bags	1.62
Condensed Milk	Cases of Cans	1.23	Rope	Coils	2.51
Copper Ore	Bulk	0.47	Rubber	Bundles	3.90
Copra	Bags	2.09 to 2.37	Rye	Bulk	1.62
Corn	Bulk	1.41	Salt	Bulk	1.03
Cotton	Bales, Average	1.45	Silk	Bales	3.06
Currants	Crates	1.81	Steel Bolts	Kegs	0.58
Dried Fruit	Boxes	1.25	Steel Sheets	Crated	0.42
Dry-Goods	Boxes	2.79	Sugar	Bags	1.31
Fish	Barrels, Iced	1.39	Tar	Barrels	1.50
Flour	Bags	1.34	Tea	Cases	2.79
Furniture	Crated	4.35	Tile	Boxes	1.39
Glass	Crated	3.62	Timber	Oak	1.09
Gypsum	Bags	1.24	Timber	Fir	1.81
Hardware	Boxes	1.39	Tung Oil	Bulk	1.07
Hides	Bales, Compressed	2.23	Turpentine	Drums	1.59
Iron Ore	Bulk	0.30 to 0.53	Wheat	Bulk	1.31
Iron Ore Pellets	Bulk	0.25 to 0.53	Wheat	Bags	1.45
Iron, Pig	Neat Stowage	0.28	Whiskey	Cases	1.74
Jute	Bales	1.84	Woodchips	Bulk	3.07

the lowest part of the tank. In such cases, it is usual to make the tube sloping and sometimes curved with a large radius. If the tube is sloping or curved, 1 m measured along the tube will not indicate a difference in level of 1 m vertically. In making a sounding table for such a tank, it is, therefore, necessary to allow for this difference. This is done as illustrated in Fig. 45(a). The sounding-tube installation is checked on the ship, as the drawing is usually only diagrammatic. The line of the tube as actually fitted is laid out on the same graph paper as the capacity curve. Any slope or curve in the line of the tube is developed into the plane of the graph paper so that the true shape and length of the tube are shown in the plane of the capacity curve, as illustrated in Fig. 45(b). Even by sloping or curving the sounding tube, it is not always possible to reach the very bottom of the tank. Consequently, the zero sounding as given in the sounding table often shows a considerable number of metric tons or barrels or liters in the tank.

In cases where it may be permissible for tanks to hold alternatively fuel oil or ballast, there should be capacity tables in metric tons of salt water and metric tons, barrels and liters of oil. Table 15 provides conversions for the various liquids and systems of measurement.

8.8 Effects of Heel and Trim. In the event a ship experiences heel or trim, the shift of free surface of the liquid in a tank in order to remain horizontal takes place in such a way that the location of the centroid of free surface area remains fixed with respect to the tank. Thus, a sounding rod experiences no change in reading due to heel and trim only if it passes through the centroid of the tank horizontal cross section. Inasmuch as physical constraints generally lead to a sounding tube location removed from the line of centroids, a trim and/or heel correction in tank sounding reading is generally called for.

It may be shown that the correction, in tank volume for trim is, in any consistent units,

$$A \cdot d \cdot \frac{t}{L}, \qquad (52)$$

where A is tank horizontal cross section area, d is longitudinal distance from centroid to sounding tube, t is trim and L is ship length between draft marks.

The correction is minus if ship trims by the stern

(and sounding tube is abaft centroid), in which case the correction must be deducted from the tank volume.

The correction in volume for heel is,

$$A \cdot d \cdot \tan\phi, \qquad (53)$$

where d is distance of sounding tube to port or starboard of tank centroid and ϕ is angle of heel. The correction is minus if ship heels to starboard and the sounding tube is to starboard of the centroid.

In order to accomplish these corrections completely, the location both of the sounding tube and of the tank free surface centroid must be known at all elevations.

8.9 Ullage. The traditional way of determining the amount of liquid in the tank of a tank vessel is to lower a weighted chain with a scale on it until it touches the surface of the liquid and so measure the distance from the top of hatch to the free surface. This is called *ullage*, and the associated tables are called ullage tables. Capacities should be given in barrels or cubic meters, and weights in metric tons. Ullage tables differ from tables of tank capacity, in that tank capacities are given for varying depths of the liquid in the tank or soundings, while ullage tables give the tank capacity for differing amounts of ullage, or distance from top of tank to liquid surface.

Ships with inert gas systems to protect against explosion should have sealed and remotely operated means of reading liquid levels, inasmuch as the atmosphere within the tank is normally at a slight positive pressure.

8.10 Cargo Stowage Factors. The average specific volume or *stowage factor* of the cargo to be carried may exert a strong effect on the design of a ship. Low stowage factor cargoes, such as ores and finished steel, lead to ships which are weight limited, and tend to have large fullness to obtain a large displacement on fixed dimensions, and have low freeboard. High stowage factor cargoes, such as crated furniture, automobiles and containers tend to result in volume limited ships with high freeboard and relatively low fullness, in order to achieve adequate propeller immersion and draft on fixed dimensions. The volume requirements of containerized cargo are met on many containerships by carrying some of the containers on deck in several tiers, thereby providing the capability to carry enough cargo to reach design displacement when fully loaded.

Table 16 shows approximate stowage factors for a number of typical kinds of cargo. Thomas (1957) gives stowage factors for a wide variety of commodities.

8.11 Consumables. This term includes fuel oil, lubricating oil, fresh water for culinary and drinking purposes, fresh water for washing purposes, fresh water for boiler feed, and stores that are expended on the voyage, such as supplies and food provisions of all kinds. The capacity required for consumables depends upon the main propulsion power, the length of the voyage, the number of passengers and crew, and the quality of the accommodations provided. Owners usually have definite ideas as to the amount of provisions, stores and fresh water necessary for their service. When better information is not available, Table 17 gives reasonable assumptions for approximate weight per day per passenger and member of crew for stores, and liters of fresh water consumed per day per person.

Fresh water capacity will be dependent upon whether water is to be obtained ashore and carried for the length of the voyage, or the ship's distilling plant can meet all requirements at sea. In the latter case, requisite capacity for fresh water will be much reduced.

Many ships provide separate tankage for potable water, distilled water, and boiler feed water (if steam propelled). A survey of post World War II designs shows average fresh water tank capacities of 1.7 metric tons per person for potable water, and 0.0032 and 0.007 metric tons per shaft horsepower of propulsion machinery (maximum rating) for distilled water and boiler feed water, respectively.

Table 17—Weight Allowances for Stores and Water

Item	Kg per person per day Passengers	Crew
Fresh water, moderate ships	40	20
Fresh water, luxury ships	100	45
Stores	10	5
Provisions	4.5	4.5

Baggage for a long voyage may amount to an average of 0.1 t per passenger and on luxury ships to as much as 0.17 t per passenger. For tourist class passengers, it may be as little as 0.07 t per passenger. For short voyages and excursions it will average about 0.08 t per passenger.

Table 18—Fuel Oil Consumption in Port, metric tons per day

Type of Propelling Machinery	Steam Turbine	Diesel
Source of Auxiliary Power	Steam	Diesel
Minimum Fuel Consumption	3.5	2
Add for each 100 t of cargo moved	0.8	0.5
Add for each 100 "tons" of refrigeration	1.2	0.9
Add for each 100 persons complement	0.6	0.4

For a steam tanker with steam-turbine driven cargo pumps about 0.5 t of fuel will be required for each 1000 t of cargo oil pumped. If cargo pumps are diesel-driven, about 0.3 t of fuel will be required. Direct calculations using specific fuel rate of pump prime mover are recommended for precise values.

Table 19—Sewage Flow Rates for Calculating Capacity of Marine Sanitation Devices for Passenger Ships

Type of Vessel Service		Flow Rates (Liters per Person Per Day)		
		Sewage		Gray Water
		Full Flush	Reduced Flush	
Long Trips (1)	Crew	110–170	11–57	114
	Passengers	110–170	11–57	114
Medium Trips (2)	Crew	110–170	11–57	114
	Passengers	38–57	4–19	57
Short Trips (3)	Crew	57–87	5.7–28	57
	Passengers	28–38	2.8–14	5.5

Notes:
(1) 24-hour day for crew; passengers sleep aboard
(2) 24-hour day for crew; passengers aboard for about a four-hour trip
(3) 24-hour day for crew; passengers aboard for about two hours, six trips per day.

8.12 Fuel and Ballast Requirements. The necessary capacity of fuel oil tanks is a matter for specific estimating for the particular machinery installation. Approximate curves are given in *Ship Design and Construction* (Taggart, 1980) whereby the fuel rate in kg per SHP per hr may be estimated in the preliminary design phase, once one has a reliable estimate of the shaft horsepower to be installed, and the type of machinery to be used. If the desired endurance, or the distance the loaded ship can travel at rated horsepower without refueling is known, the weight of fuel needed may be calculated. Hence, after allowing some margin, the needed fuel oil tank capacity can be calculated for any assumed density of the oil.

In addition to the fuel burned while underway, some fuel will be required to maintain ship services while in port, such as electricity for lighting, power for cargo winches or pumps, ventilation and sanitary system, steam for heating accommodations, steam for cargo oil heating etc. Tankage for the fuel and space for the dry stores consumed while in port must be provided when determining the overall capacities for the ship. Fresh water requirements will be about the same as when at sea for the same number of people. Table 18 may be used as a guide in estimating the additional tank capacity for fuel for port services.

The required capacity of ballast tanks will be dependent upon trim and stability considerations, as discussed in Chapters II and III.

8.13 Service Requirements. The foregoing sections have discussed specific and predictable consumables, which will be required by the service contemplated. However, changing conditions often lead to a ship being utilized for a service other than that originally envisioned. Limited capacities—for example, the size of fuel oil tanks—may restrict such a reassignment. For this reason, and because of possible voyage delays caused by adverse weather and port congestion, generous margins are usually allowed in establishing capacities.

Additional tankage called for by present-day U.S. operating procedures are an oily residue or slop tank for contaminated bilge water *(Code of Federal Regulations)*, and the requirement that tank vessels have slop tanks for contaminated ballast and tank washings; i.e., one tank of 2 percent of total cargo tank capacity for vessels under 70,000 DWT, and 2 tanks of 2 percent capacity if 70,000 DWT or over. All new tankers must have a segregated ballast system except product tankers of less than 30,000 DWT (See Coast Guard, 1981).

As a further anti-pollution measure, many ships are now fitted with sewage holding tanks, which are pumped out at sea, or pumped ashore to treatment plants when in port. As a guide in selecting tank capacities, the fresh water daily consumption in Table 18 may be used to estimate gray water. Flushing water (black water) may range between 11 and 170 liters per person per day, depending upon whether reduced flow or full flow flush facilities are fitted. Suggested water requirements are also partly dependent on length of voyage. Table 19 (from Canadian Dept. of Environment, 1976) may be used as a guide in selecting holding tank size. A detailed estimate is needed to determine design values for a specific ship. Tank capacity requirements for a large passenger ship with a prolonged retention period can be substantial.

References

American Bureau of Shipping (Annual), *Rules for Building and Classing Steel Vessels.*

Australian Department of Transport (1974), Notice

to Shipbuilders, Shipowners, Naval Architects, Port Authorities, etc. "Draught and Other Marks on British Ships," Notice No. 8/1974.

Coast Guard, U.S. (1981), "Guidance for Enforcement of the Requirements of the Port and Safety Act of 1978", NVI Circular No. 1-81.

Code of Federal Regulations (U.S.), "Pollution Prevention Regulations", 33CFR 155.330.

Department of Environment, Canada (1976). "Development of Design Guidelines for Shipboard Sewage Holding Tanks", Report EPS-3-WP-76-3.

Fuller, Arthur L., Aughey, Michael E., and Billingsley, Daniel W. (1977), "Computer Aided Ship Hull Definition at the Naval Ship Engineering Center", First International Symposium on Computer-Aided Ship Hull Definition (SCHAD), SNAME, September.

IIT Research Institute (1980), "Hull Definition Fairing Program (HULDEF) Overview", REAPS (Research and Engineering for Automation and Productivity in Shipbuilding) Technical Memorandum 6449-7-013, February.

Kilgore, Ullman (1967), "Developable Hull Surfaces", Fishing Boats of the World; 3—Published by the F.A.O. of the United Nations Organization.

Kuiper, G. (1970), "Preliminary Design of Ship Lines by Mathematical Methods", *Journal of Ship Research*, SNAME, Vol. 14.

Lackenby, H. (1950), "On the Systematic Geometrical Variation of Ship Forms", *Transactions*, RINA.

Lasky, Marc P. and Daidola, John C. (1977), "Design Experience with Hull Form Definition During Pre-Detail Design", SCHAD, SNAME, September.

MARAD (1987). *The MARAD Systematic Series of Full-Form Ship Models*, SNAME.

Miller, N.S. (1963), "The Accuracy of Numerical Integration in Ship Calculations", The Institute of Engineers and Shipbuilders in Scotland, *Transactions*.

Morrish, S. W.F., (1892), "Approximate Rule for the Vertical Position of the Center of Buoyancy", *Transactions*, INA, now RINA.

NAVSEA (Naval Ship Engineering Center) (1976), "Ship Hull Characteristics Program—SHCP, Users Manual", Dep't. of the Navy, Washington, D.C., January.

Posdunine, Valentine. (1925). "Some Approximate Formulae Useful in Ship Design," *The Shipbuilder*, April.

Rawson, J., and Tupper, E. C. (1976), *Basic Ship Theory*, Vol. I, Longman, Inc.

Ridgely-Nevitt (1963), "The Development of Parent Hulls for a High Displacement-Length Series of Trawler Forms", *Transactions*, SNAME.

Saunders, H. E. (1957), *Hydrodynamics in Ship Design*, Vol. II, SNAME.

SNAME, Hydrodynamics Committee (1966), *Model Resistance Data Sheets*, D-1 through D-10, D-13, and Bulletins 1-13, 1-14, 1-23.

Söding, Heinrich and Rabien, Uwe (1977), "Hull Surface Design Modifying an Existing Hull", SCHAD, SNAME, September.

Taggart, Robert (Editor) (1980), *Ship Design and Construction*; SNAME.

Taylor, David W. (1915), "Calculations for Ship's Forms and the Light Thrown by Model Experiments upon Resistance, Propulsion and Rolling of Ships", International Engineering Conference, San Francisco.

Taylor, David W. (1943), *The Speed and Power of Ships*, U.S. Government Printing Office.

Thomas and Thomas (1957), *Stowage—The Properties and Stowage of Cargoes*, Brown, Son and Ferguson, Ltd.

Todd, F. H., Stuntz, G. P., and Pien, P.C. (1957), "Series 60, The Effect Upon Resistance and Power of Variation in Ship Proportions", *Transactions*, SNAME.

Townsend, Harry S. (1967), "A Series of Cargo Vessel Hull Forms", T. and R. Symposium on Some Effects of Hull Performance in a Seaway, SNAME, Oct. 10.

CHAPTER II

Lawrence L. Goldberg | **Intact Stability**

Section 1
Elementary Principles

1.1 Gravitational Stability. Not only must the designer provide adequate buoyancy to give support for the ship and its contents, as discussed in Chapter I, but it must be assured that it will float in the proper attitude, or trim, and remain upright when loaded with passengers and cargo. This involves the problems of gravitational stability and trim, which will be discussed in detail in this chapter, primarily with reference to static conditions in calm water. Consideration will also be given to criteria for judging the adequacy of a ship's stability, considering both internal loading and external hazards. (See Section 7).

It is important to recognize, however, that a ship in its natural sea environment is subject to dynamic forces resulting from accelerations caused primarily by wave action. These are treated separately in Chapter VII, Motions in Waves. Nevertheless, it is possible to consider some dynamic effects, such as the forces related to wind and high-speed turning, while dealing with static stability, for these forces can be treated as static rather than dynamic forces.

Another external hazard affecting a ship's stability is that of damage to the hull by collision, grounding or other accident that results in flow of water into the hull. The stability and trim of the damaged ship will be considered in Chapter III, Subdivision and Damage Stability.

(a) Equilibrium. In general, a rigid body is considered to be in a state of equilibrium when the resultants of all forces and moments acting on the body are zero. In dealing with static floating body stability, we are interested in that state of equilibrium associated with the floating body upright and at rest in a still liquid. In this case the resultant of all gravity forces (weights) acting downward, and the resultant of the buoyancy forces, acting upward on the body, are of equal magnitude and are applied in the same vertical line.

(b) Stable equilibrium. If a floating body, initially at equilibrium, is disturbed by an external moment, there will be a change in its angular attitude. If upon removal of the external moment, the body returns to its original position, it is said to have been in *stable equilibrium* and to have positive stability.

(c) Neutral equilibrium. If, on the other hand, a floating body that assumes a displaced inclination because of an external moment remains in that displaced position when the external moment is removed, the body is said to have been in *neutral equilibrium* and has neutral stability. A floating cylindrical homogeneous log would be in neutral equilibrium.

(d) Unstable equilibrium. If a floating body, displaced from its original angular attitude by an external force, continues to move in the same direction after the force is removed, it is said to have been in *unstable equilibrium* and was initially unstable.

A ship may be inclined in any direction. Any inclination may be considered as made up of an inclination in the athwartship plane and an inclination in the longitudinal plane. In ship calculations the athwartship inclination, called heel or list, and the longitudinal inclination, called trim, are usually dealt with separately. This chapter deals with both athwartship or transverse stability and longitudinal stability.

1.2 Weight and Center of Gravity. This chapter deals with the forces and moments acting on a ship afloat in calm water, which consist primarily of gravity forces (weights) and buoyancy forces. Therefore, equations are usually developed using displacement weight, W, and component weights, w. In the "English" system, displacement, weights and buoyant forces are thus expressed in the familiar units of long tons (or lb). But using SI, as explained in Chapter I, requires that displacement, Δ, be considered as mass, in metric tons, t (or kg), where $\Delta = W/g$, and individual mass components are w/g. Hence, for convenience righting and heeling moments may be considered in mass units—metric ton-meters, t-m. Since a metric ton mass is numerically almost the same as a long ton of weight, results are essentially the same regardless of which units are used.

The total weight, or displacement, of a ship can be determined from the draft marks and Curves of Form, as discussed in Chapter I. The position of the center of gravity may be either calculated or determined experimentally. Both methods are used when dealing with ships. The weight and center of gravity of a ship that has not yet been launched can be established only by a weight estimate, which is a summation of the estimated weights and moments of all the various items that make up the ship. Weight estimating is discussed in Section 2.

After the ship is afloat, the weight and center of gravity can be accurately established by an inclining experiment, as described in detail in Section 9.

To calculate the position of the center of gravity of any object, it is assumed to be divided into infinitesimal particles, the moment of each particle calculated by multiplying its weight by its distance from a reference plane, the weights and moments of all the particles added, and the total moment divided by the total weight. The result is the distance of the center of gravity from the reference plane. The location of the center of gravity of a system of weights, such as a ship, may be calculated by multiplying the weight of each component by the distance of its center of gravity from a reference plane, and dividing the total moment of the components by the total weight. The location of the center of gravity is completely determined when its distance from each of three planes has been established. In ship calculations, the three reference planes generally used are a horizontal plane through the baseline, for the vertical location of the center of gravity (VCG), a vertical transverse plane either through amidships or through the forward perpendicular for the longitudinal location (LCG), and a vertical plane through the centerline for the transverse position (TCG). (The TCG is usually very nearly in the centerline plane and is often assumed to be in that plane.)

1.3 Displacement and Center of Buoyancy. In Chapter I it has been shown that the force of buoyancy is equal to the weight of the displaced liquid, and that the resultant of this force acts vertically upward through a point called the center of buoyancy, which is the center of gravity of the displaced liquid.

Application of these principles to a ship or submarine makes it possible to evaluate the effect of the hydrostatic pressure acting on the hull and appendages by determining the volume of the ship below the waterline and the centroid of this volume. The submerged volume, when converted to weight or mass of displaced liquid, is called the displacement, W or Δ, respectively.

1.4 Interaction of Weight and Buoyancy. The attitude of a floating object is determined by the interaction of the forces of weight and buoyancy. If no other forces are acting, it will settle until the force of buoyancy equals the weight, and will rotate until two conditions are satisfied:

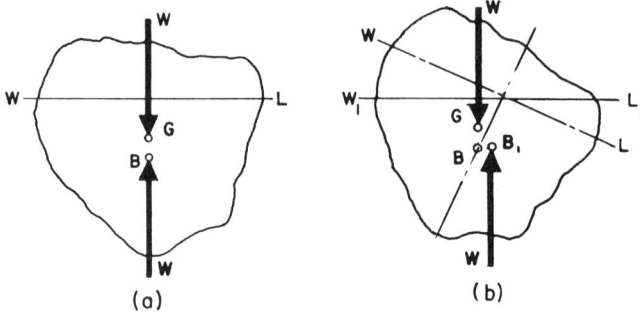

Fig. 1 Stable equilibrium of floating body

(a) The centers of buoyancy B and gravity G are in the same vertical line, as in Fig. 1(a), and

(b) Any slight rotation from this position, as from WL to W_1L_1 in Fig. 1(b) will cause the equal forces of weight and buoyancy to generate a couple tending to move the object back to float on WL *(stable equilibrium)*.

For every object, with the exception noted later, at least one position must exist for which these conditions are satisfied, since otherwise the object would continue to rotate indefinitely. There may be several such positions. The center of gravity may be either above or below the center of buoyancy.

An exception to the second condition exists when the object is a body of revolution with its center of gravity exactly on the axis of revolution as illustrated in Figs. 2(a) and 2(b). When such an object is rotated to any angle, no moment is produced, since the center of buoyancy is always directly below the center of gravity. It will remain at any angle at which it is placed *(neutral equilibrium)*.

A submerged object that is clear of the bottom can come to rest in only one position. It will rotate until the center of gravity is directly below the center of buoyancy. If its center of gravity coincides with its center of buoyancy, as in the case of a solid body of homogeneous material, the object would remain in any position in which it is placed.

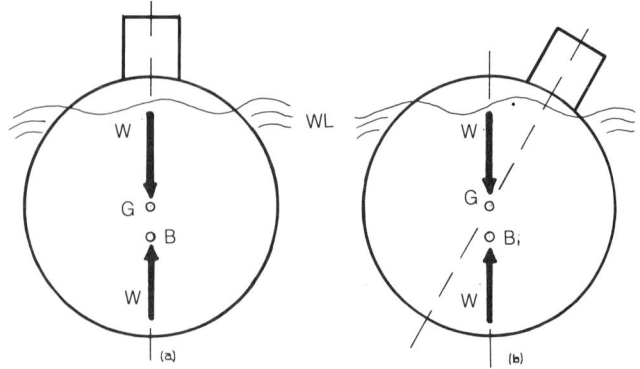

Fig. 2 Neutral equilibrium of floating body

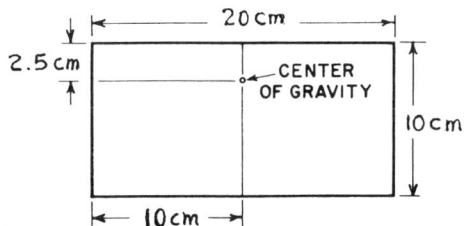

Fig. 3 Example of stability of watertight rectangular body

The difference in the action of floating and submerged objects is explained by the fact that the center of buoyancy of the submerged object is fixed, while the center of buoyancy of a floating object will generally shift when the object is rotated.

As an example, consider a watertight body having a rectangular section with dimensions and center of gravity as illustrated in Fig. 3. Assume that it will float with half its volume submerged, as in Fig. 4. It can come to rest in either of two positions, (a) or (c), 180° apart. In either of these positions, the centers of buoyancy and gravity are in the same vertical line. Also, as the body is inclined from (a) to (b) or from (c) to (d), a moment is developed which tends to rotate the body back to its original position, and the same situation would exist if it were inclined in the opposite direction.

If the 20-cm (8-in.) dimension were reduced, with the center of gravity still on the centerline and 2.5 cm (1 in.) below the top, a situation would be reached where the center of buoyancy would no longer move far enough to be to the right of the center of gravity as the body is inclined from (a) to (b). Then the body could come to rest only in position (c).

As an illustration of a body in the submerged condition, assume that the weight of the body shown in Fig. 3 is increased so that the body is submerged, as in Fig. 5. In positions (a) and (c) the centers of buoyancy and gravity are in the same vertical line. An inclination from (a) in either direction would produce a moment tending to rotate the body away from position (a), as illustrated in Fig. 5(b). An inclination from (c) would produce a moment tending to restore the body to position (c). Therefore, the body can come to rest only in position (c).

A ship or submarine is designed to float in the upright position. This fact permits the definition of two classes of hydrostatic moments, illustrated in Fig. 6, as follows:

Righting moments. A righting moment exists at any angle of inclination where the forces of weight and buoyancy act to move the ship toward the upright position.

Heeling moments. A heeling moment exists at any angle of inclination where the forces of weight and buoyancy act to move the ship away from the upright position.

The center of buoyancy of a ship or a surfaced submarine moves with respect to the ship, as the ship is inclined, in a manner that depends upon the shape of the ship in the vicinity of the waterline. The center of buoyancy of a submerged submarine, on the contrary, does not move with respect to the ship, regardless of the inclination or the shape of the hull, since it is stationary at the center of gravity of the entire submerged volume. This constitutes an important difference between floating and submerged ships. The moment acting on a surface ship can change from a righting moment to a heeling moment, or vice versa, as the ship is inclined, but this cannot occur on a submerged submarine unless there is a shift of the ship's center of gravity.

It can be seen from Fig. 6 that lowering of the center of gravity along the ship's centerline increases stability. When a righting moment exists, lowering the cen-

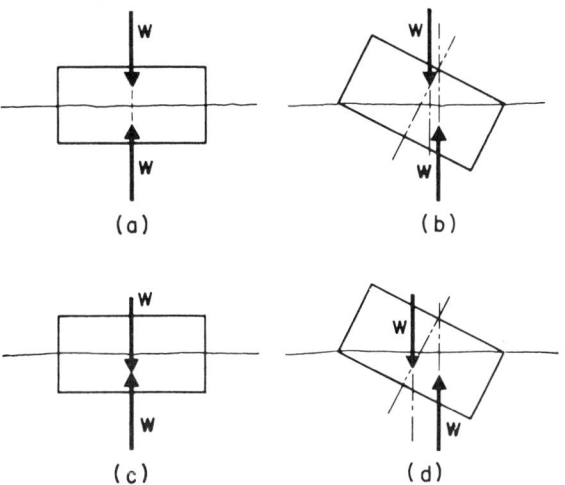

Fig. 4 Alternate conditions of stable equilibrium for floating body

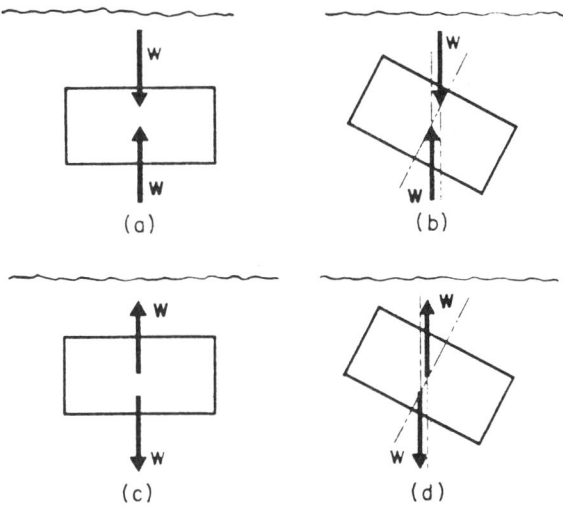

Fig. 5 Single condition of stable equilibrium for submerged body

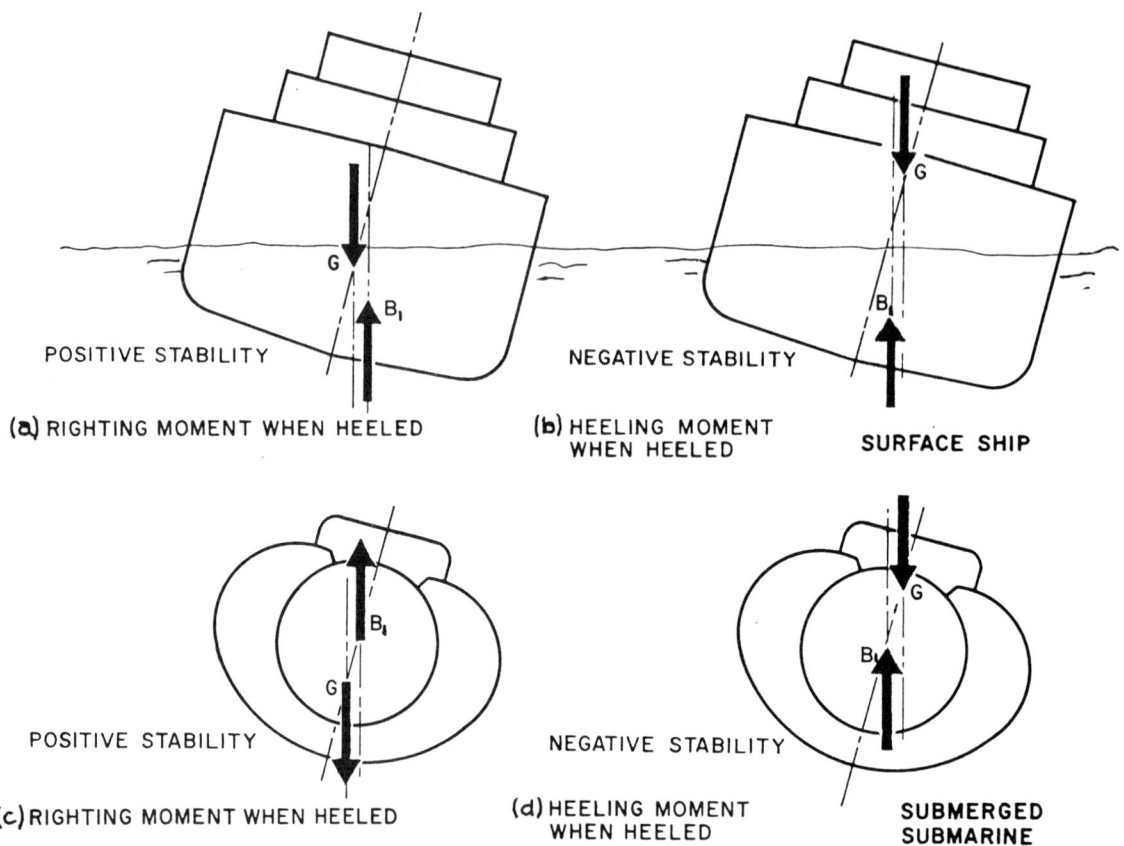

Fig. 6 Effect of height of center of gravity on stability

ter of gravity along the centerline increases the separation of the forces of weight and buoyancy and increases the righting moment. When a heeling moment exists, sufficient lowering of the center of gravity along the centerline would change the heeling moment to a righting moment. Similarly, sufficient lowering of the center of gravity along the centerline could change the initial stability in the upright position from negative to positive.

In problems involving longitudinal stability of undamaged surface ships, we are concerned primarily with determining the ship's draft and trim under the influence of various upsetting moments, rather than evaluating the possibility of the ship capsizing in the longitudinal direction. If the longitudinal centers of gravity and buoyancy are not in the same vertical line, the ship will change trim as discussed in Section 8 and will come to rest as illustrated in Fig. 7, with the centers of gravity and buoyancy in the same vertical line. A small longitudinal inclination will cause the center of buoyancy to move so far in a fore-and-aft direction that the moment of weight and buoyancy would be many times greater than that produced by the same inclination in the transverse direction. The longitudinal shift in B creates such a large longitudinal righting moment that longitudinal stability is usually very great compared to transverse stability.

Thus, if the ship's center of gravity were to rise along the centerline, the ship would capsize transversely long before there would be any danger of capsizing longitudinally. However, a surface ship could, theoretically, be made to founder by a downward external force applied toward one end, at a point near the centerline and at a height near or below the center of buoyancy, without capsizing. It is unlikely, however, that an intact ship would encounter a force of the required magnitude.

Surface ships can, and do, founder after extensive flooding as a result of damage at one end. The loss of buoyancy at the damaged end causes the center of buoyancy to move so far toward the opposite end of the ship that subsequent submergence of the damaged end is not adequate to move the center of buoyancy back to a position in line with the center of gravity, and the ship founders, or capsizes longitudinally. This is also discussed in Chapter III.

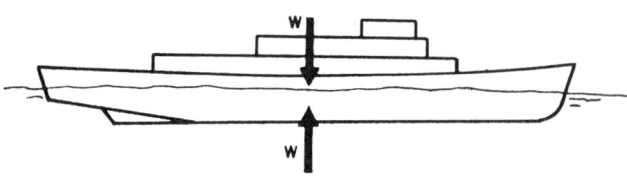

Fig. 7 Longitudinal equilibrium

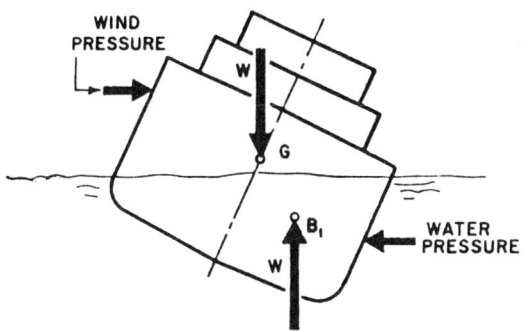

Fig. 8 Effect of a beam wind

In the case of a submerged submarine, the center of buoyancy does not move as the submarine is inclined in a fore-and-aft direction. Therefore, capsizing of an intact submerged submarine in the longitudinal direction is possible, and would require very nearly the same moment as would be required to capsize it transversely. If the center of gravity of a submerged submarine were to rise to a position above the center of buoyancy, the direction, longitudinal or transverse, in which it would capsize would depend upon the movement of liquids or loose objects within the ship. The foregoing discussion of submerged submarines does not take into account the stabilizing effect of the bow and stern planes, which have an important effect on longitudinal stability while the ship is underway with the planes operating.

1.5 Upsetting Forces. The magnitude of the upsetting forces, or heeling moments, that may act on a ship determines the magnitude of moment that must be generated by the forces of weight and buoyancy in order to prevent capsizing or excessive heel.

External upsetting forces affecting transverse stability may be caused by:
(a) Beam winds, with or without rolling.
(b) Lifting of heavy weights over the side.
(c) High-speed turns.
(d) Grounding.
(e) Strain on mooring lines.
(f) Towline pull of tugs.

Internal upsetting forces include:
(g) Shifting of on-board weights athwartship.
(h) Entrapped water on deck.

Section 7 of this chapter discusses evaluation of stability with regard to the upsetting forces listed above. The discussion below is general in nature and illustrates the stability principles involved when a ship is subjected to upsetting forces.

When a ship is exposed to a beam wind, the wind pressure acts on the portion of the ship above the waterline, and the resistance of the water to the ship's lateral motion exerts a force on the opposite side below the waterline. The situation is illustrated in Fig. 8. Equilibrium with respect to angle of heel will be reached when:

(a) The ship is moving to leeward with a speed such that the water resistance equals the wind pressure, and
(b) The ship has heeled to an angle such that the moment produced by the forces of weight and buoyancy equals the moment developed by the wind pressure and the water pressure.

As the ship heels from the vertical, the wind pressure, water pressure, and their vertical separation remain substantially constant. The ship's weight is constant and acts at a fixed point. The force of buoyancy also is constant, but the point at which it acts varies with the angle of heel. Equilibrium will be reached when sufficient horizontal separation of the centers of gravity and buoyancy has been produced to cause a balance between heeling and righting moments.

When a weight if lifted over the side, as illustrated in Fig. 9, the force exerted by the weight acts through the outboard end of the boom, regardless of the angle of heel or the height to which the load has been lifted. Therefore, the weight of the sidelift may be considered to be added to the ship at the end of the boom. If the ship's center of gravity is initially on the ship's cen-

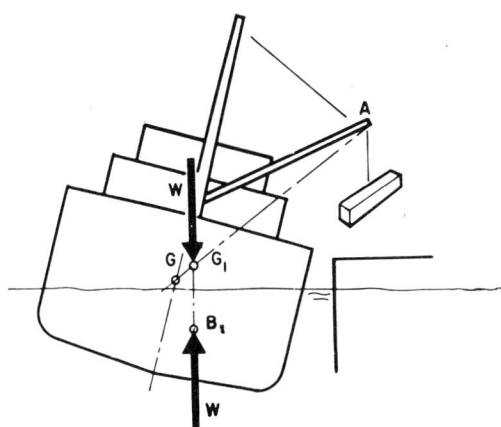

Fig. 9 Lifting a weight overside

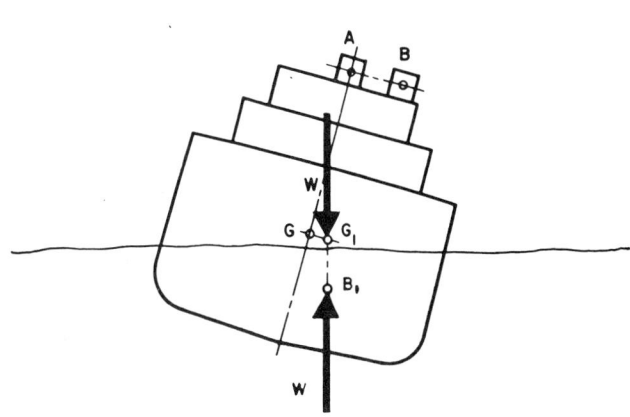

Fig. 10 Effect of offside weight

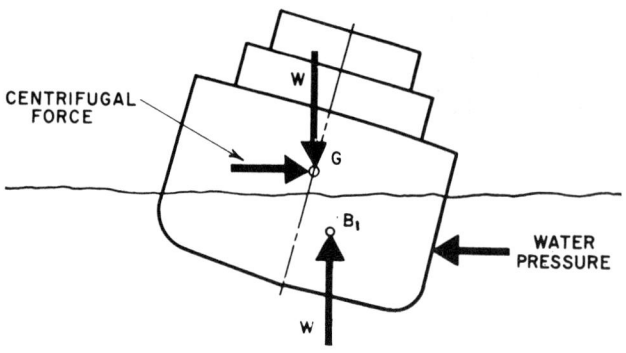

Fig. 11 (a) Effect of a turn   (b) Effect of grounding

terline, as at G in Fig. 9, the center of gravity of the combined weight of the ship and the sidelift will be located along the line GA, and will move to a final position, G_1, when the load has been lifted clear of the pier. Point G_1 will be off the ship's centerline and somewhat higher than G. The ship will heel until the center of buoyancy has moved off the ship's centerline to a position directly below point G_1.

Movement of weights already aboard the ship, such as passengers, liquids or cargo, will cause the ship's center of gravity to move. If a weight is moved from A to B in Fig. 10, the ship's center of gravity will move from G to G_1 in a direction parallel to the direction of movement of the shifted weight. The ship will heel until the center of buoyancy is directly below point G_1.

When a ship is executing a turn, a centrifugal force is generated, which acts horizontally through the ship's center of gravity. This force is balanced by a horizontal water pressure on the side of the ship, as illustrated in Fig. 11(a). Except for the point of application of the heeling force, the situation is similar to that in which the ship is acted upon by a beam wind, and the ship will heel until the moment of the ship's weight and buoyancy equals that of the centrifugal force and water pressure.

If a ship runs aground in such a manner that the bottom offers little restraint to heeling, as illustrated in Fig. 11(b), the reaction of the bottom may produce a heeling moment. As the ship grounds, part of the energy due to its forward motion may be absorbed in lifting the ship, in which case a reaction, R, between the bottom and the ship would develop. This reaction may be increased later as the tide ebbs. Under these conditions, the force of buoyancy would be less than the weight of the ship, since the ship would be supported by the combination of buoyancy and the reaction of the bottom. The ship would heel until the moment of buoyancy about the point of contact with the bottom became equal to the moment of the ship's weight about the same point, when $(W - R) \times a$ equals $W \times b$.

There are numerous other situations in which external forces can produce heel. A moored ship may be heeled by the combination of strain on the mooring

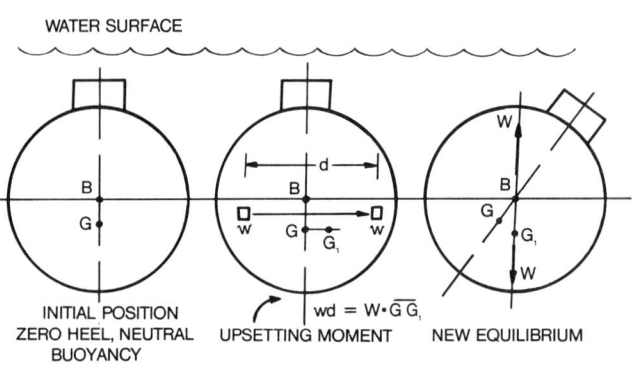
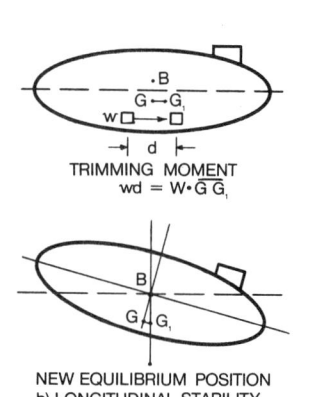

Fig. 12 Effect of weight shift on transverse and longitudinal stability of a submerged submarine

INTACT STABILITY

lines and pressure produced by wind or current. Towline strain may produce heeling moments in either the towed or towing ship. In each case, equilibrium would be reached when the center of buoyancy has moved to a point where heeling and righting moments are balanced.

In any of the foregoing examples, it is quite possible that equilibrium would not be reached before the ship capsized. It is also possible that equilibrium would not be reached until the angle of heel became so large that water would be shipped through topside openings, and that the weight of this water, running to the low side of the ship, would contribute to capsizing which otherwise would not have occurred.

Upsetting forces act to incline a ship in the longitudinal as well as the transverse direction. Since a surface ship is much stiffer, however, in the longitudinal direction, many forces, such as wind pressure or towline strain would not have any significant effect in inclining the ship longitudinally. Shifting of weights aboard in a longitudinal direction can cause large changes in the attitude of the ship because the weights can be moved much farther than in the transverse direction. When very heavy lifts are to be attempted, as in salvage work, they are usually made over the bow or stern, rather than over the side, and large longitudinal inclinations may be involved in these operations. Stranding at the bow or stern can produce substantial changes in trim. In each case, the principles are the same as previously discussed for transverse inclinations. When a weight is shifted longitudinally, or lifted over the bow or stern, the center of gravity of the ship will move, and the ship will trim until the center of buoyancy is directly below the new position of the center of gravity. If a ship is grounded at the bow or stern, it will assume an attitude such that the moments of weight and buoyancy about the point of contact are equal.

In the case of a submerged submarine, the center of buoyancy is fixed, and a given upsetting moment produces very nearly the same inclination in the longitudinal direction as it does in the transverse direction. (Fig. 12) The only difference, which is trivial, is due to the effect of liquids aboard which may move to a different extent in the two directions. A submerged submarine, however, is comparatively free from large upsetting forces. Shifting of the center of gravity as the result of weight changes is carefully avoided. For example, when a torpedo is fired, its weight is immediately replaced by an equal weight of water at the same location.

1.6 Submerged Equilibrium. Before a submarine is submerged, considerable effort has been expended, both in design and operation, to ensure that:

(a) The weight of the submarine, with its loads and ballast, will be very nearly equal to the weight of the water it will displace when submerged.

(b) The center of gravity of these weights will be very nearly in the same logitudinal position as the center of buoyancy of the submerged submarine.

(c) The center of gravity of these weights will be lower than the center of buoyancy of the submerged submarine.

These precautions produce favorable conditions which are described, respectively, as *neutral buoyancy, zero trim*, and *positive stability*. A submarine on the surface, with weights adjusted so that the first two conditions will be satisfied upon filling the main ballast tanks, is said to be in *diving trim*.

The effect of this situation is that the submarine, insofar as transverse and longitudinal stability are concerned, acts in the same manner as a pendulum. This imaginary pendulum is supported at the center of buoyancy, has a length equal to the separation of the centers of buoyancy and gravity, and a weight equal to the weight of the submarine.

It is not practical to achieve an exact balance of weight and buoyancy, or to bring the center of gravity precisely to the same longitudinal position as the center of buoyancy. It is also not necessary, since minor deviations can be counteracted by the effect of the bow and stern planes when underway submerged.

Section 2
The Weight Estimate

2.1 Weight and Location of Center of Gravity. It is important that the weight and the location of the center of gravity be estimated at an early stage in the design of a ship. The weight and height of the center of gravity are major factors in determining the adequacy of the ship's stability. The weight and longitudinal position of the center of gravity determine the drafts at which the ship will float. The distance of the center of gravity from the ship's centerline plane de-

termines whether the ship will have an unacceptable list. It will be clear that this calculation of weight and center of gravity, although laborious and tedious, is one of the most important steps in the successful design of ships.

During the early stages of design, the weight and the height of center of gravity for the ship in light condition are estimated by comparison with ships of similar type or from coefficients derived from existing

Table 1—Typical Summaries of Light Ship Weights and Centers (Computed)

	Dry Cargo Ship			Containership		
	Weight[4] t	VCG m	LCG[2] m	Weight[4] t	VCG m	LCG[2] m
Steel	4968	8.0	82.2	4557	9.0	94.6
Outfit	2099	13.8	81.5	1739	13.9	94.3
Machinery	1004	6.0	96.4	827	9.0	151.5
	8071	9.3	83.8	7123		
Weight Margin[3]	511			76		
$\overline{KG}$ Margin[3]		0.5	+ 1.0		0.1	
Total w/o Ballast	8582	9.8	84.8	7199		
Fixed Ballast	—	—		3329	2.8	80.9
Total	8582	9.8	84.8	10,528	7.9	94.6

Notes:
1. Commercial ships are assumed to be transversely symmetrical, and therefore transverse centers and moments are not included.
2. *LCG* measured from F.P.
3. Margins adjusted to agree with inclining experiment results.
4. "Weight" is mass in S.I. units.

ships. At later stages of design detailed estimates of weights and centers of gravity are required. It is often necessary to modify ship dimensions or the distribution of weights to achieve the desired optimum combination of a ship's drafts, trim and stabiilty, as well as to meet other design requirements such as motions in waves and powering. A summary of only the three major components of light ship weights for two typical ships is given in Table 1.

2.2 Detailed Estimates of Weights and Position of Center of Gravity. The reader is referred to Chapter 1, by R. K. Kiss, of *Ship Design and Construction* (Taggart, 1980)[1] for a detailed discussion of the methodology of weight estimating for each design stage, starting with concept design and ending with detail design.

Ordinarily in design the horizontal plane of reference is taken through the molded baseline of the ship, which is described in Chapter I. The height of the center of gravity above this base is referred to as $\overline{KG}$ and its position as VCG (vertical center of gravity). Sometimes, after a ship's completion, the reference plane is taken through the bottom of the keel.

The plane of reference for the longitudinal position of the center of gravity may be the transverse plane at the midship section, which is midway between the forward and after perpendiculars. In this case the longitudinal position of the center of gravity, called the LCG (longitudinal center of gravity), is measured forward or abaft the midship section. This practice involves the possibility of inadvertently applying the measurements aft instead of forward, or vice versa, and a more desirable plane of reference is one through the forward perpendicular.

The plane of reference for the transverse position of the center of gravity is the vertical centerline plane of the ship, the transverse position of the center of gravity being measured to port or starboard of this plane.

In weight estimates it is essential that an orderly and systematic classification of weights be followed. Two such classifications are in general use in this country, that of the U.S. Maritime Administration (MarAd, 1962), and that of the U.S. Navy (NAVSEA, 1978). Both of these use the three broad classifications of hull steel, outfit, and propelling machinery. A detailed description of either classification may be obtained from the respective organizations. Some design offices may use systems differing in detail from either of these, but the general classification will be similar.

2.3 Weight and Center of Gravity Margins. The weight estimate will of necessity contain many approximations and, it may be presumed, some errors. The errors will generally be errors of omission. The steel as received from the mills is usually heavier, within the *mill tolerance*, than the ordered nominal weight. It is impossible, in the design stages, to calculate in accurate detail the weight of many groups, such as piping, wiring, auxiliary machinery, and many others.

For these and similar reasons, it is essential that margins for error be included in the weight estimate. The amount of these margins is derived from the experience of the estimator, and varies with the accuracy and extent of the available information.

Table 2 is a composite of the usual practice of several design offices, which is acceptable to the Maritime Administration. In each instance, the smaller values apply to conventional ships that do not involve unusual features and for which there is a reliable basis for the estimate. If the estimate is reviewed by several independent interested agencies, there is less chance of substantial error and smaller margins are in order.

---

[1] Complete references are listed at end of chapter.

Table 2—Margins

Margin of weight (in percent of light-ship weight)
Cargo ships.....................................1.5 to 2.5
Tankers..1.5 to 2.5
Cargo-passenger ships..........................2.0 to 3.0
Large passenger liners.........................2.5 to 3.5
Small naval vessels............................6.0 to 7.0
Large naval vessels............................3.5 to 7.0
Margin in *VCG* Meters
Cargo ships...................................0.15–0.23
Tankers.......................................0.15
Cargo-passenger ships.........................0.15–0.23
Large passenger liners........................0.23–0.30
Small naval vessels...........................0.15–0.23
Large naval vessels...........................0.15–0.23

The larger values apply to vessels with unusual features or in which there is considerable uncertainty as to the ultimate development of the design.

The amount of margin will also depend on the seriousness of mis-estimating weight or center of gravity. For example, until the advent of the double bottom for tankers there was no real need for any margin at all in the VCG of a conventional tanker, because such ships generally have considerably more stability than is needed. On the other hand, if there were a substantial penalty in the contract for overweight or for a high VCG, a correspondingly substantial margin in the estimate would be indicated.

The above margins apply to estimates made in the *contract-design* stage, where the calculations are based primarily on a midship section, arrangement drawings, and the specifications. In a final, detailed *finished-weight* calculation, made mostly from working drawings, a much smaller margin, of 1 or 2 percent, or even, if extremely detailed information is available, no margin at all may be appropriate.

For more detailed information on U.S. Navy practice with regard to margin, the reader is referred to *Weight Control of Naval Ships* (NAVSEA, 1978).

Monitoring of weights (including load items) and the coordinates of the center of gravity is continued during ship construction, in order to ensure that all applicable weight and stability requirements will be satisfied in the completed ship.

2.4 Variation in Displacement and Position of Center of Gravity With Loading of Ship. The total weight (displacement) and position of the center of gravity of any ship in service will depend greatly on the amount and location of the deadweight items discussed in Chapter I—cargo, fuel, fresh water, stores, etc. Hence, the position of the center of gravity is determined for various operating conditions of the ship, the conditions depending upon the class of ship. (See Section 3.8).

Section 3
Metacentric Height

3.1 The Transverse Metacenter and Transverse Metacentric Height. Consider a symmetric ship heeled to a very small angle, $\delta\phi$, shown, with the angle exaggerated, in Fig. 13. The center of buoyancy has moved off the ship's centerline as the result of the inclination, and the lines along which the resultants of weight and buoyancy act are separated by a distance, $\overline{GZ}$, the righting arm. A vertical line through the center of buoyancy will intersect the original vertical through the center of buoyancy, which is in the ship's centerline plane, at a point M, called the *transverse metacenter*, when $\delta\phi \to 0$. The location of this point will vary with the ship's displacement and trim, but, for any given drafts, it will always be in the same place.

Unless there is an abrupt change in the shape of the ship in the vicinity of the waterline, point M will remain practically stationary with respect to the ship as the ship is inclined to small angles, up to about 7 or, sometimes, 10 deg.

As can be seen from Fig. 13, if the locations of G and M are known, the righting arm for small angles of heel can be calculated readily, with sufficient accuracy for all practical purposes, by the formula

$$\overline{GZ} \approx \overline{GM} \sin \delta\phi \qquad (1)$$

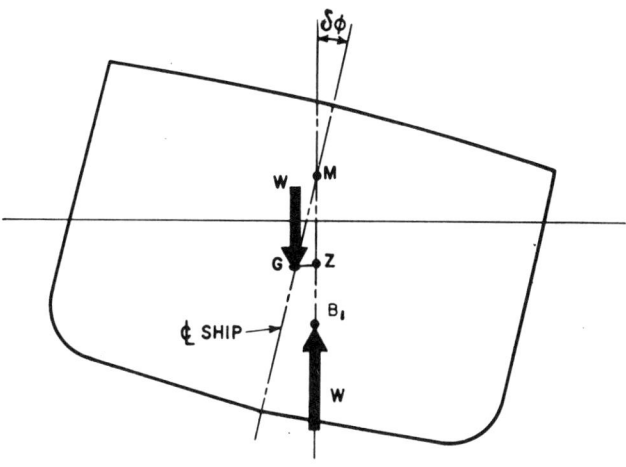

Fig. 13 Metacenter and righting arm

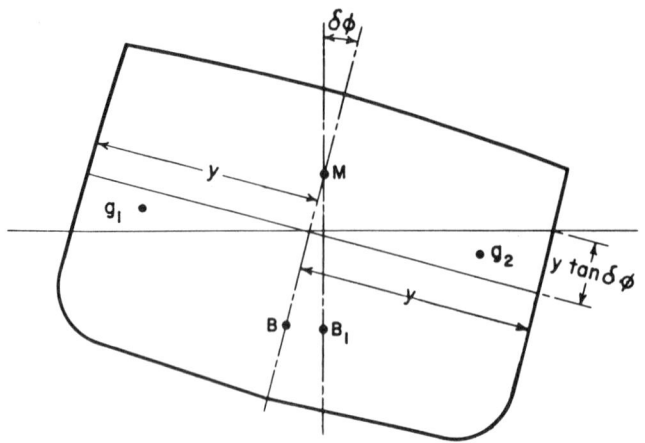

Fig. 14 Locating the transverse metacenter

The distance $\overline{GM}$ is therefore important as an index of transverse stability at small angles of heel, and is called the *transverse metacentric height*. Since $\overline{GZ}$ is considered positive when the moment of weight and buoyancy tends to rotate the ship toward the upright position, $\overline{GM}$ is positive when M is above G, and negative when M is below G.

Metacentric Height ($\overline{GM}$) is often used as an index of stability when preparation of stability curves for large angles (Section 4) has not been made. Its use is based on the assumption that adequate $\overline{GM}$, in conjunction with adequate freeboard, will assure that adequate righting moments will exist at both small and large angles of heel.

3.2 Location of the Transverse Metacenter. When a ship is inclined to a small angle, as in Fig. 14, the new waterline will intersect the original waterline at the ship's centerline plane if the ship is wall-sided in the vicinity of the waterline, since the volumes of the two wedges between the two waterlines will then be equal, and there will be no change in displacement. If v is the volume of each wedge, ∇ the volume of displacement, and the centers of gravity of the wedges are at g_1 and g_2, the ship's center of buoyancy will move:

(a) In a direction parallel to a line connecting g_1 and g_2.

(b) A distance, $\overline{BB_1}$, equal to $(v \cdot \overline{g_1 g_2})/\nabla$.

As the angle of heel approaches zero, the line $\overline{g_1 g_2}$, and therefore $\overline{BB_1}$, become perpendicular to the ship's centerline. Also, any variation from wall-sidedness becomes negligible, and we may say

$$\overline{BM} = \frac{\overline{BB_1}}{\tan \delta \phi} = \frac{v \cdot \overline{g_1 g_2}}{\nabla \tan \delta \phi}$$

If y is the half-breadth of the waterline at any point of the ship's length at a distance x from one end, and if the ship's length is designated as L, then, since the area of a section through the wedge is $\frac{1}{2}(y)(y \tan \delta \phi)$ and its centroid is at a distance of $2 \times \frac{2}{3}y$ from the centroid of the corresponding section on the other side

$$v \cdot \overline{g_1 g_2} = \int_0^L \tfrac{1}{2}(y)(y \tan \delta \phi)(2 \cdot \tfrac{2}{3}y) dx$$

or

$$\frac{v \cdot \overline{g_1 g_2}}{\tan \delta \phi} = \frac{2}{3} \int_0^L y^3 dx$$

Equation (13) of Chapter I shows that the expression

$$\frac{1}{3} \int_0^L y^3 dx$$

is the moment of inertia of a figure bounded by a curve and a straight line with the straight line as the axis. If we consider the straight line to be the ship's centerline, then the moment of inertia of the entire waterplane about the ship's centerline (both sides) designated as I_T, is

$$I_T = \frac{2}{3} \int_0^L y^3 dx = \frac{v \cdot \overline{g_1 g_2}}{\tan \delta \phi}$$

and, therefore, when $\delta \phi \to 0$,

$$\overline{BM} = \frac{I_T}{\nabla} \qquad (2)[2]$$

The calculation of the height of the transverse metacenter above the keel, usually called $\overline{KM}$, is explained in Chapter 1. This distance is the sum of $\overline{BM}$, or I_T/∇, and $\overline{KB}$, the height of the center of buoyancy above the keel. The height of the center of gravity above the keel, $\overline{KG}$, is found from the weight estimate or inclining experiment. Then,

$$\begin{aligned}\overline{GM} &= \overline{KM} - \overline{KG} \\ &= \overline{KB} + \overline{BM} - \overline{KG}\end{aligned} \qquad (3)$$

3.3 The Longitudinal Metacenter and Longitudinal Metacentric Height. The longitudinal metacenter is

---

[2] This theorem was derived by the French hydrographer Pierre Bouguer while on an expedition to Peru to measure a degree of the meridian near the equator. It appeared in his *Traité du Navire* published in Paris in 1746.

similar to the transverse metacenter except that it involves longitudinal inclinations. Since ships are usually not symmetrical forward and aft, the center of buoyancy at various even-keel waterlines does not always lie in a fixed transverse plane, but may move forward and aft with changes in draft. For a given even-keel waterline, the longitudinal metacenter is defined as the intersection of a vertical line through the center of buoyancy in the even-keel attitude with a vertical line through the new position of the center of buoyancy after the ship has been inclined longitudinally through a small angle.

The longitudinal metacenter, like the transverse metacenter, is substantially fixed with respect to the ship for moderate angles of inclination if there is no abrupt change in the shape of the ship in the vicinity of the waterline, and its distance above the ship's center of gravity, or the longitudinal metacentric height, is an index of the ship's resistance to changes in trim. For a normal surface ship, the longitudinal metacenter is always far above the center of gravity, and the longitudinal metacentric height is always positive.

3.4 Location of the Longitudinal Metacenter. Locating the longitudinal metacenter is similar to, but somewhat more complicated than locating the transverse metacenter. Since the hull form is usually not symmetrical in the fore-and-aft direction, the immersed wedge and the emerged wedge usually do not have the same shape. To maintain the same displacement, however, they must have the same volume. Fig. 15 shows a ship inclined longitudinally from an even-keel waterline WL, through a small angle, $\delta\theta$, to waterline W_1L_1. Using the intersection of these two waterlines, point F, as the reference for fore-and-aft distances, and letting:

L = length of waterplane
Q = distance from F to the forward end of waterplane
y = breadth of waterline WL at any distance x from F

the volume of the forward wedge is

$$v = \int_0^Q (y)(x \tan \delta\theta) dx$$

and the volume of the after wedge is

$$v = \int_0^{L-Q} (y)(x \tan \delta\theta) dx$$

Equating the volumes

$$\tan d\theta \int_0^Q xy\, dx = \tan \delta\theta \int_0^{L-Q} xy\, dx$$

$$\int_0^Q xy\, dx = \int_0^{L-Q} xy\, dx$$

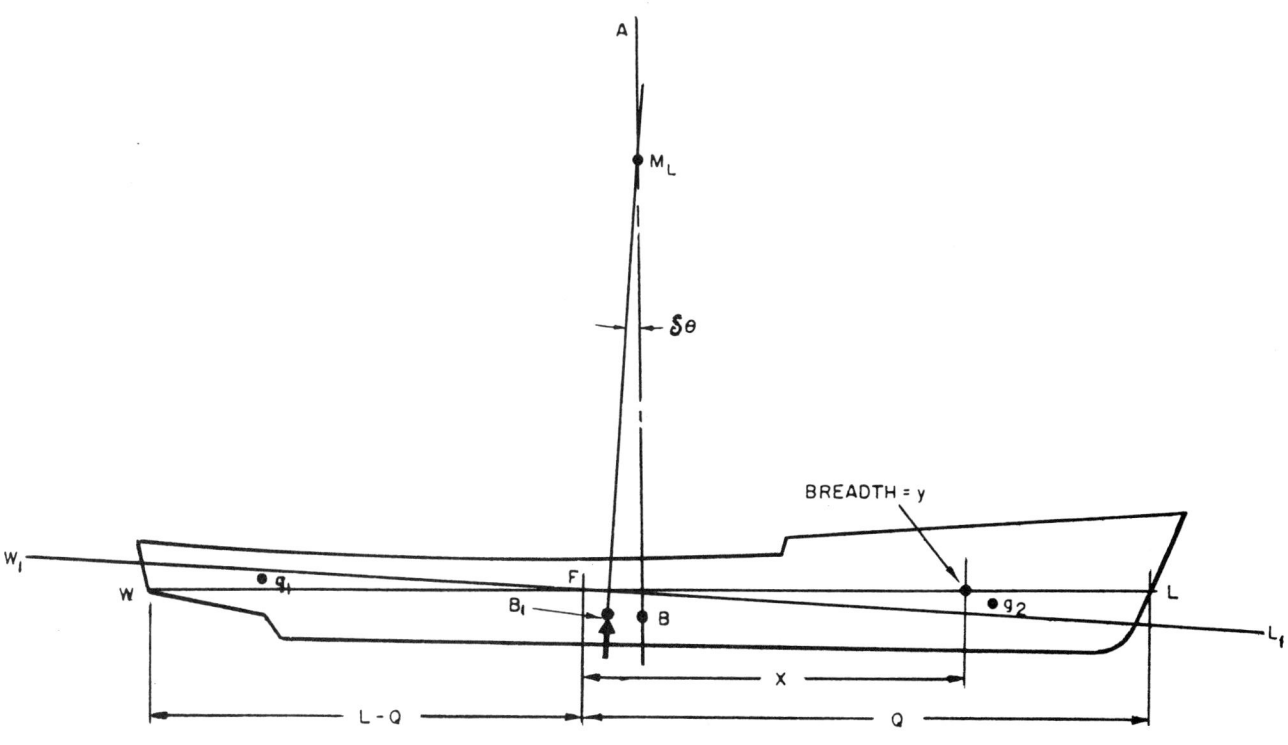

Fig. 15 Longitudinal metacenter

Equation (7) of Chapter I shows that these expressions are, respectively, the moment of the area of the waterplane foward of F and the moment of the area aft of F, both moments being about a transverse line through point F. Since these moments are equal and opposite, the moment of the entire waterplane about a transverse axis through F is zero, and therefore F lies on the transverse axis through the centroid of the waterplane, called the *center of flotation* in Chapter I.

In Fig. 15 AB is a transverse vertical plane through the initial position of the center of buoyancy, B, when the ship was floating on the even-keel waterline, WL. With longitudinal inclination, B will move parallel to $\overline{g_1 g_2}$, or as the inclination approaches zero, perpendicular to plane AB, to a point B_1. The height of the metacenter above B will be

$$\overline{BM}_L = \frac{\overline{BB_1}}{\tan \delta\theta} = \frac{v \cdot \overline{g_1 g_2}}{\nabla \tan \delta\theta}$$

The distance of g_1, the centroid of the after wedge, from F is equal to the moment of the after wedge about F divided by the volume of the wedge, and a similar formula applies to the forward wedge. If the moments of the after and forward wedges are designated as m_1 and m_2, respectively, then the distance

$$\overline{g_1 g_2} = \frac{m_1}{v} + \frac{m_2}{v} = \frac{m_1 + m_2}{v}$$

or

$$v \cdot \overline{g_1 g_2} = m_1 + m_2$$

The moments of the volumes are obtained by integrating, forward and aft, the product of the section area at a distance x from F and the distance x, or

$$m_1 = \int_0^Q (y)(x \tan \delta\theta)(x) dx = \tan \delta\theta \int_0^Q x^2 y \, dx$$

$$m_2 = \tan \delta\theta \int_0^{L-Q} x^2 y \, dx$$

The integrals in the expressions for m_1 and m_2 correspond to equation (11) in Chapter I, which is the formula for the moment of inertia of an area about the axis corresponding to $x = 0$, or, in this case, a transverse axis through F, the centroid of the waterplane. Therefore, the sum of the two integrals is the longitudinal moment of inertia, I_L, of the entire waterplane, and

$$m_1 + m_2 = v \cdot \overline{g_1 g_2} = I_L \tan \delta\theta$$

or

$$I_L = \frac{v \cdot \overline{g_1 g_2}}{\tan \delta\theta}$$

Then, when $\delta\theta \to 0$

$$\overline{BM}_L = \frac{v \cdot \overline{g_1 g_2}}{\nabla \tan \delta\theta} = \frac{I_L}{\nabla} \quad (4)$$

where I_L is the moment of inertia of the entire waterplane about a transverse axis through its centroid, or center of flotation.

The details of calculating the height of the longitudinal metacenter above the keel are explained in Chapter I.

3.5 Metacenter for Submerged Submarines. When a submarine is submerged, as noted in Section 1, the center of buoyancy is stationary with respect to the ship at any inclination. It follows that the vertical through the center of buoyancy in the upright position will intersect the vertical through the center of buoyancy in any inclined position at the center of buoyancy, and the center of buoyancy is, therefore, both the transverse and longitudinal metacenter.

To look at the situation from a different viewpoint, the $\overline{KM}$ of a surfaced submarine is equal to $\overline{KB}$ plus $\overline{BM}$, or $\overline{KB}$ plus I/∇. As the ship submerges, the waterplane disappears, and the value of I, and hence $\overline{BM}$, is reduced to zero. The value of $\overline{KM}$ becomes $\overline{KB}$ plus zero, and B and M coincide.

The metacentric height of a submerged submarine is usually called $\overline{GB}$ rather than $\overline{GM}$.

3.6 Effects of Trim on the Metacenter. The discussion and formulas for $\overline{BM}$, $\overline{KM}$ and $\overline{GM}$ all assumed that the waterline at each station was the same, namely no trim existed. In cases where substantial trim exists, values for $\overline{BM}$, $\overline{KM}$ and $\overline{GM}$ will be substantially different from those calculated for the zero trim situation. It is important to calculate metacentric values for trim for many ship types. The discussion on computer programs for calculating cross curves (Section 4.15) includes the effects of trim. Chapter I in describing the calculation of $\overline{KM}$ also discusses the effects of trim.

3.7 Applications of Metacentric Height. A convenient and frequently used concept is the *moment to heel one degree*. This is the moment of weight and buoyancy, or $W\overline{GZ}$, when the ship is heeled to 1 deg, and is equivalent to the moment of external forces required to produce a 1-deg heel. Using mass units, for convenience with SI notation, moment = $\Delta \overline{GZ}$, and if $\overline{GM} \sin \phi$ is substituted for $\overline{GZ}$, we have:

(a) Moment to heel one degree = $\Delta \overline{GM} \sin 1$ deg

Within the range of inclinations where the metacenter is stationary, the change in the angle of heel produced by a given external moment can be found by dividing the moment by the moment to heel one degree.

The same theory and formula apply to inclinations in the longitudinal direction, and we may say

(b) Moment to trim one degree = $\Delta \overline{GM}_L \sin 1 \deg$, where $\overline{GM}_L$ is the longitudinal metacentric height. We are more interested, however, in the changes in draft produced by a longitudinal moment than in the angle of trim. The expression is converted to *moment to trim one cm* by substituting one cm divided by the length of the ship in cm for sin 1 deg. The formula becomes, with mass units,

$$MTcm = \frac{\Delta \overline{GM}_L}{100 L} \text{ t-m} \quad (5)$$

where L is ship length in meters. As a practical matter, $\overline{GM}_L$ is usually so large compared to $\overline{GB}$ that only a negligible error would be introduced if $\overline{BM}_L$ were substituted for $\overline{GM}_L$. Then $\dfrac{I_L}{\nabla}$ may be substituted for $\overline{BM}_L$, where I_L is the moment of inertia of the waterplane about a transverse axis through its centroid, and $\Delta = \rho \nabla$, where ρ is density. Then, moment to trim one cm:

$$MTcm \approx \rho \nabla \times \frac{I_L}{\nabla} \times \frac{1}{100 L} = \frac{\rho I_L}{100 L} \quad (6)$$

For fresh water $\rho = 1.0$; for salt water $\rho = 1.025$ (t/m³). Since the value of this function depends only on the size and shape of the waterplane, it is usually calculated together with the displacement and other curves, before the location of G is known. Although approximate, this expression may be used for calculations involving moderate trim with satisfactory accuracy.

When using "English" notation it is customary to retain weight (force) units, with displacement, W, in long tons of weight. Then moment to trim one inch,

$$MTI = \frac{W \overline{GM}_L}{12 L} \text{ ton-ft} \quad (5a)$$

where GM_L and L are in ft. Approximately,

$$MTI \approx \frac{\rho g I_L}{12 L} \text{ ton-ft} \quad (6a)$$

where $\rho g = 1/35.0$ in SW and $1/35.9$ in FW.

(c) The period of roll in still water, if not influenced by damping effects, is shown in Chapter VII to be:

$$\text{Period} = \frac{\text{constant} \times k}{\sqrt{\overline{GM}}} = \frac{C \times B}{\sqrt{\overline{GM}}} \quad (7)$$

where k is the radius of gyration of the ship about a fore-and-aft axis through its center of gravity.

The factor "constant $\times$ k" is often replaced by $C \times B$, where C is a constant obtained from observed data for different types of ships.

This formula may be used to estimate the period of roll when data for ships of the same type are available, if it is assumed that the radius of gyration is the same percentage of the ship's beam in each case. For example, if a ship with a beam of 15.24m and a $\overline{GM}$ of 1.22m has a period of roll of 10.5 seconds, then

$$C = \frac{10.5 \times \sqrt{1.22}}{15.24} = 0.76 \text{ sec}/\text{m}^{1/2}$$

If another ship of the same type has a beam of 13.72m and a $\overline{GM}$ of 1.52m, the estimated period of roll would be:

$$T = \frac{0.76 \times 13.72}{\sqrt{1.52}} = 8.5 \text{ sec}$$

The variation of the value of C for ships of different types is not large; a reasonably close estimate can be made if 0.80 is used for surface types and 0.67 is used for submarines. In almost all cases, values of C for conventional, homogeneously loaded surface ships are between 0.72 and 0.91. This formula is useful also for estimating $\overline{GM}$ when the period of roll has been observed.

A snappy, short-period roll may be interpreted as indicating that a ship has moderate to high stability, while a sluggish, slow roll (long period) may be interpreted as an indication of lesser stability, or that other factors such as free surface or liquids in systems may be influencing the roll period. However, the external rolling forces due to waves and wind tend to distort the relationship of $T = \dfrac{CB}{\sqrt{\overline{GM}}}$. Hence, caution must be exercised in calculating $\overline{GM}$ values from periods of roll observed at sea, particularly for small craft.

The case of the ore carrier is an interesting illustration of the effect of weight distribution on the radius of gyration and therefore on the value of C. The weight of the ore, which is several times that of the light ship, is concentrated fairly close to the center of gravity, both vertically and transversely. When the ship is in ballast, the ballast water is carried in wing tanks at a considerable distance outboard of the center of gravity, and the radius of gyration is greater than that for the loaded condition. This can result in a variation in the value of C from 0.69 for a particular ship in the loaded condition to 0.94 when the ship is in ballast. For most ships, however, there is only a minor change in the radius of gyration with the usual changes in loading.

If no other information is available, the metacentric height, in conjunction with freeboard, is a reasonably good measure of a ship's stability, although it must be used with judgment and caution. On ships with ample freeboard, the moment required to heel the ship to 20 deg may be larger than 20 times the moment to heel one degree, but on ships with but little freeboard it may be considerably less. Little effort may be required to capsize a ship with large $\overline{GM}$ but with small-freeboard. When the metacentric height is zero or negative, certain types of ship would capsize, while other types might develop fairly large righting moments at the larger angles of heel. The metacentric height may be used, however, as an approximate index of stability for an undamaged ship with reasonable confidence if the ship can be compared to another with similar lines and freeboard for which the stability characteristics are known.

For approximate estimates of $\overline{BM}$ see Chapter I.

3.8 Conditions of Loading. A ship's stability, and hence $\overline{GM}$, may vary considerably during the course of a voyage, or from one voyage to the next, and it is necessary to determine which probable condition of loading is the least favorable, and will therefore govern the required stability. (The general effect of variations in cargo and liquid load is further discussed in Section 6). It is customary to study, for each design, a number of loaded conditions with various quantities, locations, and densities of cargo and with various liquid loadings. When a ship is completed the builder usually provides such information for the guidance of the operator in the form of a *Trim and Stability Booklet*. A typical booklet contains a general arrangement of the ship, curves of form, capacities and centers, in addition to calculations of $\overline{GM}$ and trim for a number of representative conditions and blank forms for calculating new conditions. Such a booklet is required for all general cargo ships, tankers and passenger ships by international conventions and must be approved in the U.S. by the Coast Guard, American Bureau of Shipping and (if government financing is involved) by the Maritime Administration. See Garzke, Johnson and Landsburg (1974). Similar information is furnished for naval ships.

The range of loading conditions which a ship might experience varies with its type and the service in which it is engaged. Typical conditions usually included in the ship's Trim and Stability Booklet are:

(a) Full-load departure condition, with full allowance of cargo and variable loads. All the ship's spaces are filled to normal capacity with load items intended to be carried in these spaces, which usually implies minimum density homogeneous cargoes, whether general, dry bulk, liquid or containerized. A typical example is given in Table 3.

Naval combatant ships do not carry cargo in the usual sense. Instead, cargo on such ships would be ammunition or jet fuel.

Additional conditions may be included for other heavier cargo densities, involving partially filled or

Table 3—Typical Full Load Departure Conditions Computed Weights and Centers

	Dry Cargo Ship[1]			Containership[2]		
	Weight[5] t	VCG m	LCG m	Weight[5] t	VCG m	LCG m
Light Ship[3] (incl. margin & solid ballast)	8582	9.8	84.8	10,528	7.9	94.6
Crew and Stores	74	12.8	90.1	120	15.8	135.9
Fuel Oil	4785	2.5	83.6	2265	3.6	80.9
Fresh Water	297	6.4	88.0	220	8.0	141.5
Salt Water Ballast	—	—	—	755	0.8	97.1
Cargo, General	7459	9.8	81.7	10,960	11.9	90.2
Refrig.	258	9.2	103.3	—	—	—
Container	1178	7.9	69.2	—	—	—
Deadweight, Total	14,051	7.1	81.9	14,320	10.0	90.3
Displacement	22,633	8.1	83.0	24,848	9.2	92.8
Mean Draft in S.W.	9.63m	(31.59 ft.)		9.11m	(29.88 ft.)	
$\overline{GM} = \overline{KM}-\overline{KG}$	10.4-8.1=	2.3m		10.01-9.21=	0.80m	
Free Surface Cor.[4]		0.6			0.28	
Net $\overline{GM}$		1.7			0.52	

Notes:
1. Similar to ship in Fig. 39.
2. Similar to ship in Fig. 40.
3. Light ship figures from Table 1.
4. The free surface correction to $\overline{GM}$ is discussed in Section 5.
5. "Weight" is really mass in S.I. units.

empty holds or tanks. For ships that carry deck cargoes, conditions with cargo on deck should be included, since they may be critical for stability.

(b) Partial-load departure conditions, such as half-cargo or no-cargo. When no cargo is carried, solid or liquid ballast may be required, located so as to provide sufficient draft and satisfactory trim and stability.

(c) Arrival or minimum operating conditions. These describe the ship after an extended period at sea and are usually the lowest stability conditions consistent with the liquid loading instructions (Section 6.8). Certain cargo ships might be engaged in point-to-point service, while others might make many stops before returning to home port. The amount of cargo and consumables would vary, depending on the service. Conditions for naval ships would reflect the most adverse distribution of ammunition, along with reduced amounts of other consumables.

In all of the above conditions of loading it is necessary to make appropriate allowances for the effects on stability of the free surface of liquids in tanks, as explained in Section 5.2.

U.S. Coast Guard stability requirements are given in the Code of Federal Regulations (46 CFR 170-174) obtainable from the Government Printing Office.

3.9 Suitable Metacentric Height. The stability of a ship design, as evidenced approximately by its metacentric height ($\overline{GM}$), should meet at least the following requirements in all conditions of loading anticipated:

(a) It should be large enough in passenger ships to prevent capsizing or an excessive list in case of flooding a portion of the ship during an accident. The effect of flooding is described in Chapter III.

(b) It should be large enough to prevent listing to unpleasant or dangerous angles in case all passengers crowd to one side. This may require considerable $\overline{GM}$ in light-displacement ships, such as excursion steamers carrying large numbers of passengers.

(c) It should be large enough to minimize the possibility of a serious list under pressure from strong beam winds.

For passenger ships Item *(a)*, as discussed fully in Chapter III, is often the controlling consideration. The International Convention requirements for stability after damage, or other criteria for sufficient stability, may result in a metacentric height which is larger than that desirable from the standpoint of rolling at sea. Since the period of roll in still water varies inversely as the square root of the metacentric height, larger metacentric heights produce shorter periods of roll, resulting in greater acceleration forces which can become objectionable. The period of roll may also be a factor in determining the amplitude of roll, since the amplitude tends to increase as the period of roll approaches the period of encounter of the waves, as explained in Chapter VII.

Of these two conflicting considerations, that of safety outweighs the possibility of uncomfortable rolling, and adequate stability for safety after damage must be provided for passenger ships, and is desirable for cargo ships. However, the metacentric height should not be permitted to exceed that required for adequate stability by more than a reasonable margin.

Numerous international and national maritime organizations have established stability criteria which cover to some degree almost all types of ships, be they commercial or military. The metacentric height has been used as a stability measure for about fifty years. It has been used as a US regulatory standard since the late 1930's when it was introduced as a measure of minimum stability for the large ocean going passenger vessels in conjunction with an evaluation of a storm-wind heeling moment. Hence, for United States commercial ship designs, $\overline{GM}$ criteria established by the U.S. Coast Guard are now applicable, based primarily on limiting wind heel (Code of Federal Regulations, U.S. 46 CFR 170.) All U.S. ships constructed under government subsidy must meet Maritime Administration requirements, which for the intact ship are the same as U.S. Coast Guard requirements. For passenger ships, damage stability requirements are applicable (as discussed in Chapter III), since flooding resulting from hull damage may be the governing hazard.

Since the required stability will vary with displacement, it is convenient to express the required stability as a curve of required $\overline{GM}$ plotted against displacement or draft. Actual $\overline{GM}$ s for various loading conditions are corrected for free surface of liquids in tanks (Section 5) and plotted on the same graph as required $\overline{GM}$, as illustrated in Chapter III. This indicates immediately which conditions are satisfactory and which are not. Conditions of loading which are unsatisfactory must be made satisfactory by increasing the ship's inherent stability (during the design stage) or by issuing loading instructions which will prevent a ship from loading to an unsatisfactory stability condition.

Initial metacentric height is also an integral part of the international recommendation on stability for ships under 100 meters in length, which has been adopted into regulations in many nations.

Required $\overline{GM}$ curves must be used with caution since analysis of the righting arm curve, which defines the stability at large angles, is the only rigorous method of evaluating adequacy of stability. The righting arm curve takes into account freeboard, range of stability and the other features discussed in Section 4.8. Hence, stability criteria are often based on righting arm curves, rather than on $\overline{GM}$ alone, especially for small craft and naval ships. See Section 7.

U.S. Navy ships must meet all the stability requirements of commercial ships, including the ability to operate safely in severe weather. In addition, they must have the capability of withstanding considerable underwater hull damage as a result of enemy action. For these reasons, U.S. Navy ships may have larger initial $\overline{GM}$ than similar sized commercial ships. However, Navy stability requirements are usually based on righting arm curves rather than $\overline{GM}$, as discussed in Section 7.

An alternative approach is to make use of the "allowable $\overline{KG}$" curve, derived from the righting arm curves, which is usually more meaningful than the required $\overline{GM}$ curve, unless the latter has been adjusted to take into account the aforementioned righting arm features. See NAVSEA (1975). A useful tabulation is prepared for ships to permit a quick judgment as to whether a proposed weight change will generally be acceptable or unacceptable with regard to the limits on draft and stability. This tabulation is titled *Ship Status for Proposed Weight Changes* and takes on the following format:

Ship	Status	Allowable $\overline{KG}$ for governing loading conditions
A	1	7.6 m
B	3	8.0 m
C	2	4.2 m
D	4	8.5 m
E	1	

- *Status 1* means that the ship has adequate weight and stability margins with respect to these limits. Thus, a reasonable weight change at any height is generally acceptable.
- *Status 2* means that a ship is very close to both the limiting drafts and the stability ($\overline{KG}$) limits. Thus, a weight increase or rise in the center of gravity is unacceptable.
- *Status 3* means that a ship is very close to the stability limit but has adequate weight margin. If a weight change is above the allowable $\overline{KG}$ value and would thus cause a rise in the ship's center of gravity, the addition of solid ballast low in the ship may be a reasonable form of compensation.
- *Status 4* means that adequate stability margin exists but that the ship is operating at departure very close to its limiting drafts. Tankers and beach landing ships usually fall into this category. A weight addition is at the expense of cargo deadweight, or may adversely affect the ability of a landing ship to land at a designated beach site.

To reduce any necessary compromise between the requirements of a large amount of initial stability to withstand underwater hull damage and the desire to reduce $\overline{GM}$ to obtain more comfortable rolling characteristics, many large ships have anti-rolling tanks or fin stabilizers which operate to reduce roll amplitude. See Chapter VII. Anti-roll tanks operate on the principle of automatic shifting of liquids from side to side out of phase with the ship's rolling. The liquids present a free-surface effect problem (discussed in Section 5) which must be taken into account when evaluating a ship's stability.

Section 4
Curves of Stability

4.1 Description of Cross Curves of Stability. To determine the moment of weight and buoyancy tending to restore the ship to the upright position at large angles of heel, it is necessary to know the distance from the center of gravity, through which the weight force, W, acts downward, to the vertical line through the center of buoyancy—shown as AD in Fig. 16(a)—through which the equal upward force of buoyancy acts. This distance is called the *righting arm* and is usually referred to as $\overline{GZ}$. These *Cross Curves of Stability* provide a means of presenting this distance for any probable value of displacement and for several angles of heel, and are usually calculated before the curves of statical stability discussed in Section 4.8 of this chapter can be drawn.

It is not practical in general to calculate the actual righting arm for all conditions of loading, since the location of the center of gravity varies with the loading of the ship. To overcome this difficulty, it is common practice to assume a location for the center of gravity, such as point O in Fig. 16(a), on the ship's centerline

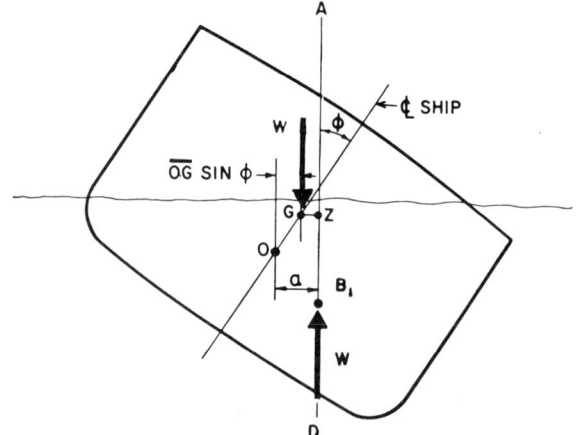

Fig. 16(a) Transverse righting arms

plane. This point or *pole* may be either in the vicinity of the actual location of the center of gravity, or, for

INTACT STABILITY

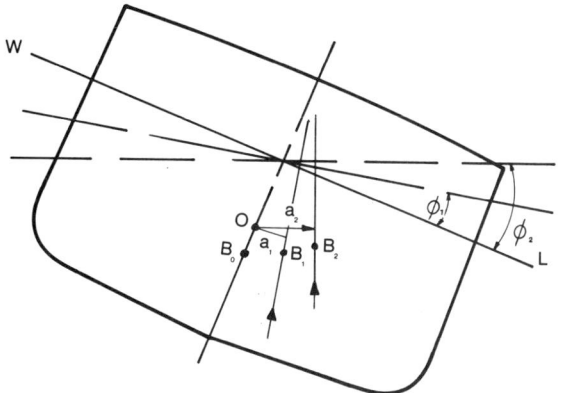

Fig. 16(b) Transverse righting arms for two water lines

convenience, at the baseline. The latter has certain advantages which will be mentioned later. The distance a in Fig. 16(a) from this point to the line AD is calculated for a number of waterlines at various drafts and angles of heel.

An illustration of this with the ship drawn upright and the waterlines inclined at angles ϕ_1 and ϕ_2 is shown in Fig. 16(b). The figure shows the initial center of buoyancy B_0 and new centers of buoyancy B_1 and B_2, corresponding to ϕ_1 and ϕ_2, respectively, with a fixed reference point O, or pole. The two distances a_1 and a_2 correspond to ϕ_1 and ϕ_2, and represent the righting arms if the center of gravity were at the point O. The displacements for the two inclined waterlines are not necessarily the same.

Thus at each inclined waterline, a displacement (W) and a lever ($\overline{GZ}$) between the buoyant force and the weight force is determined. A similar illustration could be made for waterlines inclined from any other initial waterline.

For convenience when using S.I. units it is customary to take displacement to be mass in metric tons and moments in t-m. (In English units, moments would be in ton-ft)

The sample cross curves shown in Fig. 17 consist of a plot of the distance a for several waterlines and angles of heel against displacement. The two circled points correspond to the conditions shown in Fig. 16(b).

For any particular condition of loading of the ship, for which the displacement and vertical location of the center of gravity are known, the values of a for the various angles of heel can be read from the cross curves. If the center of gravity, G in Fig. 16(a) is above O, the actual values of the righting arms can be obtained from $\overline{GZ} = a - \overline{OG} \sin \phi$, where ϕ is the angle of inclination. If G is below O, the value $\overline{OG} \sin \phi$ is added to a. If O is taken at the baseline, point K, the value $\overline{KG} \sin \phi$ is always subtracted from a, a distinct advantage since there is no possibility of interchanging the signs of the correction. In any case, raising G always reduces $\overline{GZ}$.

In some cases, the center of gravity of the ship will be off the ship's centerline, as shown in Fig. 18. As G moves off the centerline a distance b, measured perpendicular to the centerline, to G_1, the righting arms will be decreased by the value $b \cos \phi$ when the ship heels in one direction and increased by the same amount when the ship heels in the opposite direction.

4.2 Methods for Obtaining Cross Curves. The method most commonly used for preparing the cross curves today involves the use of a digital computer in conjunction with numerical offsets. However, it is best to discuss manual methods first in order that the basic principles can be made clear. Most manual methods involve the use of transverse sections of the ship extending up to the weather deck, usually those appearing on the lines drawing, which are also used in preparing the displacement and other curves as described in Chapter I.

Cross curves and righting arm curves are usually prepared on the basis of the assumption that there is a complete watertight envelope consisting of bottom, side shell and weather (or bulkhead) deck. Actually, superstructures and deckhouses having watertight sides, ends and overhead (including poops and forecastles) would contribute to intact stability, but in most cases they are not taken into account on the grounds that there will usually be openings not properly closed even if nominally watertight. On the other hand, if there are openings not properly closed in the weather (or bulkhead) decks, or side shell below the heeled waterline, there is a likelihood of appreciable downflooding which would put the ship in a damaged condition and might either reduce or improve stability. The superstructure or deckhouses are usually not treated as part of the watertight envelope in developing cross curves unless the ship's stability is poor and advantage must be taken of the watertight su-

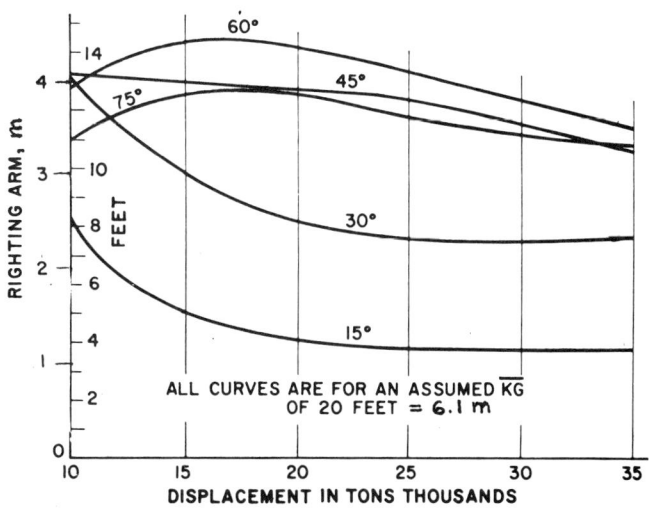

Fig. 17 Cross curves of stability

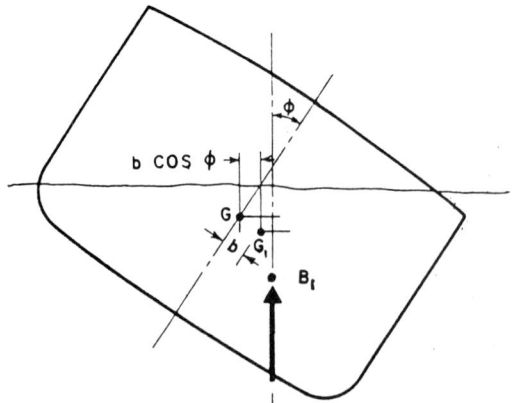

Fig. 18 Effect of off-center location of CG

perstructure. In this case, for a given displacement, the angle up to which a benefit may be taken would be the angle at which water might enter the superstructure through a door or other opening. In such case it is desirable to indicate on each stability curve the angle at which superstructure entry occurs, as well as the angle at which downflooding may occur.

A number of waterlines are selected at various drafts and angles of heel. The angles selected are usually multiples of 10 or 15 deg., the latter being typical for commercial and naval ships. For smaller ship types with low freeboard, particularly those having raised forecastles and flat decks astern, multiples of 5 or 10 degrees are usually selected rather than 15 degree multiples. Offshore supply ships, tugs, and some classes of fishing ships are examples of ship types where the smaller multiples are used in preparing cross curves.

For each angle, four or five drafts are used, so that four or five points will be available for plotting each of the cross curves. The shallowest and deepest drafts used for each angle are selected to give an appropriate range of displacement.

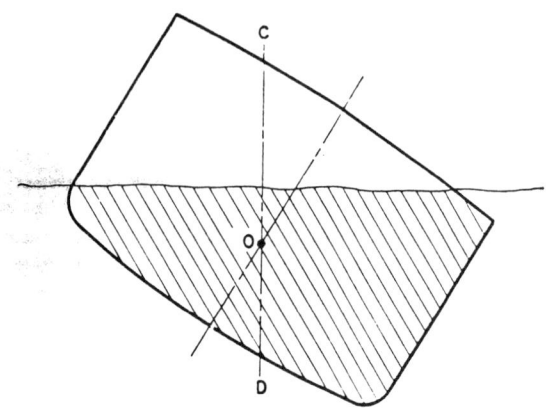

Fig. 19 Typical inclined section

A typical section, with one inclined waterline, is shown in Fig. 19. A fore-and-aft plane, CD, perpendicular to the waterline and passing through the assumed center of gravity, O, is used for reference.

For each inclined waterline, the area of each section below the waterline (the shaded area in Fig. 19), is determined. In addition, the moment of the shaded area about plane CD is determined. This provides the information necessary to establish:

(a) A curve of section areas plotted against the ship's length.

(b) A curve of moments of section areas about the plane CD, plotted against the ship's length.

The areas under these curves are the volume of the ship below the inclined waterline and the moment of this volume with respect to the plane CD. Dividing the moment by the volume gives the distance of the center of buoyancy from the plane CD, which is the righting arm corresponding to the assumed center of gravity O.

Dividing the volume (in cubic meters) by 0.975 cubic meters per metric ton (or multiplying by the specific gravity 1.025) gives the displacement in sea water at the inclined waterline in metric tons. The displacement and its corresponding righting arm provide data for plotting one point on the cross curves.

In practice, the curves of area and moment are not necessarily drawn, but the areas under these curves are obtained arithmetically by one of the integration rules, such as Simpson's rule, which are described in Chapter I. As a practical check, however, there is some advantage in plotting these curves, since any errors or unusual features may become apparent.

If Simpson's first rule is used for integration, a typical calculation of displacement and righting arm, for one inclined waterline, would appear as in Table 4.

Since the purpose of these calculations is to determine righting arms, moments which tend to right the ship are considered to be positive, and those in the opposite direction as negative. In the example used in Table 4, the sections 1 and 2 are quite narrow, and, at this inclination, their centroids would be to the left of plane CD in Fig. 19. This indicates that, if the center of gravity were actually at its assumed position, the buoyancy of the volume of the ship forward of a point somewhere between stations 2 and 3 would tend to upset the ship.

If O is taken at the base line, as is the practice in some offices, all moments are positive. This tends to minimize the likelihood of error.

4.3 Use of Integrator in Preparing Cross Curves. Offices which do not have ready access to electronic computers may use a mechanical integrator to obtain the areas and moments for each station section at various inclined waterlines. A description of the operation of the mechanical integrator is not given here, but Figure 20 illustrates the set-up that is used. Note that each station section is drawn as a complete section and the functions of areas and corresponding

Table 4—Displacement and Righting Arm Calculation

Station	½SM	Area, m²	Product	Moment, m³	Product
0	½	0	0	0	0
1	2	9.4	18.8	− 2.9	− 5.8
2	1	19.8	19.8	− 2.3	− 2.3
3	2	32.3	64.6	+ 5.2	+10.4
4	1	43.1	43.1	+13.1	+13.1
5	2	41.1	95.4	+16.3	+32.6
6	1	44.6	44.6	+18.2	+18.2
7	2	36.4	72.8	+18.4	+36.8
8	1	24.8	24.8	+14.1	−14.1
9	2	11.8	23.6	+ 6.4	−12.8
10	½	0	0	0	0
Sum of products			407.5		+129.9

functions of moments are read from the wheels marked *area* and *moment*. (The *inertia* wheel is not often used). Each instrument has its own conversion factors for areas and moments and an appropriate drawing scale factor must be used. Note that in Fig. 20 the assumed center of gravity is shown to be at the keel.

4.4 Wedge Method for Calculating Cross Curves. This method is similar to the integrator method except that the inclined W_1L_1 displacement and the moment of the inclined volume about a vertical reference line are calculated by integrating numerically in a fore and aft direction the areas and moments of the immersed and emerged wedges of each section, and algebraically adding these to the upright WL displacement and transverse moment (usually zero in the upright position).

Reference to Figure 21 illustrates the general nature of the relationships. In Figure 21 the assumption is made that the immersed and emerged wedges are of

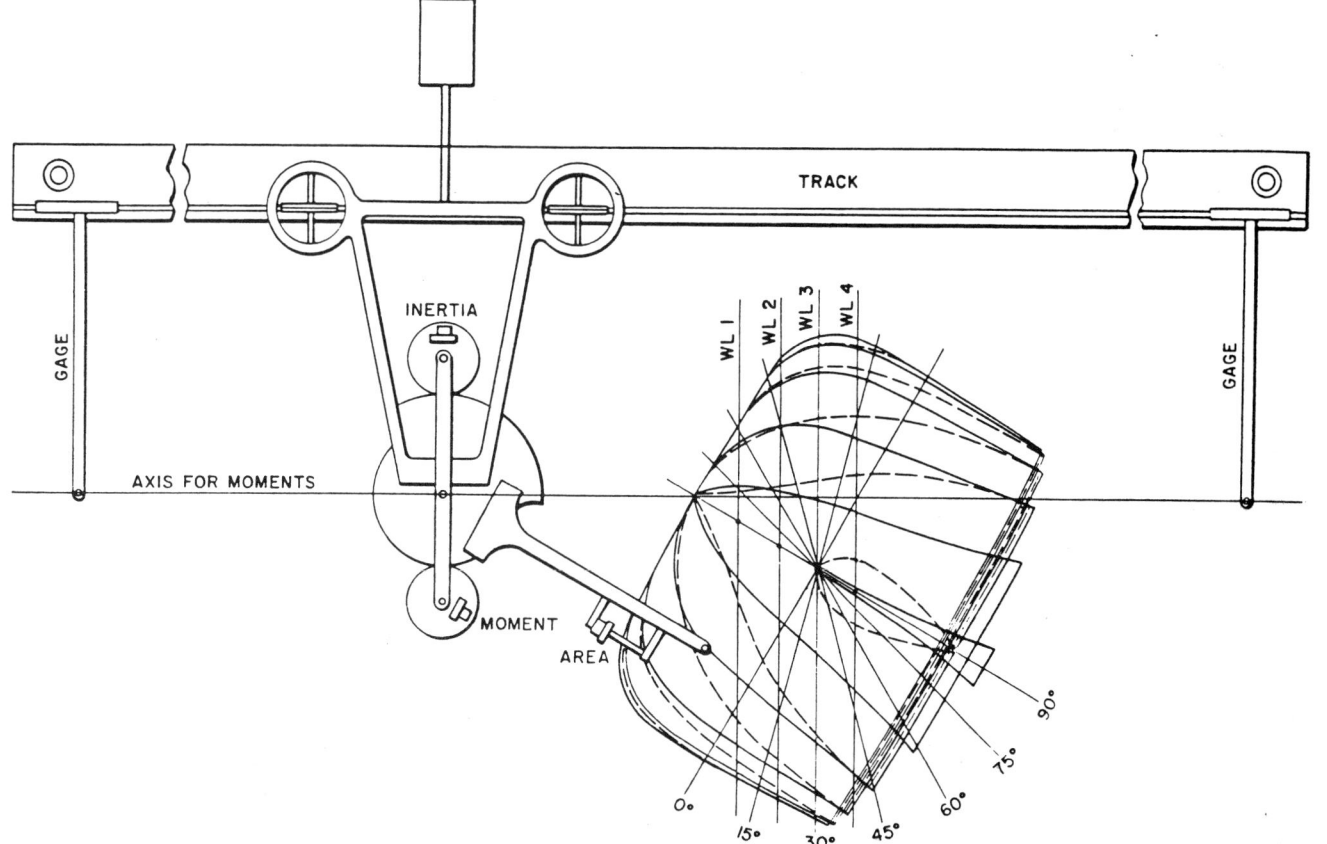

Fig. 20 Use of integrator

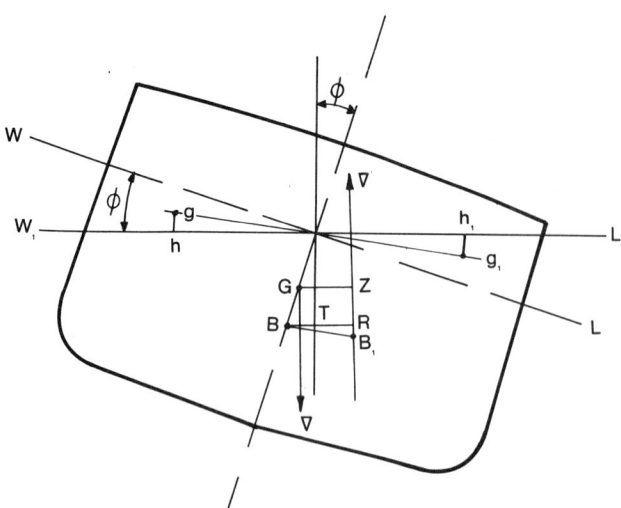

Fig. 21 Wedge method of calculating stability

equal volume.

Let v = volume of wedge
∇ = ship volume.
$\overline{gg_1}$ = distance between centroids of wedges

Moment of righting couple

$$= \nabla \, \overline{GZ}$$

Also, by principle of moments, the shift in buoyancy,

$$\overline{BB} \, \nabla = \overline{gg_1} \, v \qquad (8)$$
$$\therefore \frac{\overline{BB_1}}{\overline{gg_1}} = \frac{v}{\nabla}$$

($\overline{BB_1}$ is parallel to $\overline{gg_1}$)

Similarly, $\dfrac{\overline{BR}}{\overline{hh_1}} = \dfrac{v}{\nabla}$

($\overline{BR}$ is parallel to $\overline{hh_1}$ and $W_1 L_1$)

$\overline{GZ}$ and $\overline{BR}$ are both parallel to the inclined WL and are perpendicular to the vertical weight and buoyant forces. From triangle BGT,

$$\overline{BT} = \overline{BG} \sin \phi$$
$$\overline{GZ} = \overline{BR} - \overline{BT} \qquad (9)$$

Hence,

$$\overline{GZ} = \frac{v \, \overline{hh_1}}{\nabla} - \overline{BG} \sin \phi$$

If the volumes of the two wedges are not equal, as would be the case at high and low drafts, the ship ∇ to the inclined waterline $W_1 L_1$ would be greater or less, especially with ships having large curvature in their sections. Unless troublesome corrections are made, the method must be considered as approximate.

4.5 Computer Methods for Calculating Cross Curves. As previously noted, the most common practice today is to use a high-speed electronic computer to perform calculations necessary to produce cross curves, since the amount of work required to integrate ship sections manually to obtain areas and moments (including the the effect of trim) is very time consuming. Furthermore, the computer makes it possible to arrive at exact solutions by iteration to satisfy equilibrium trim conditions without arbitrary assumptions.

Although a number of programs are available, the use of the Navy's stability sub-routine developed as part of their Ship Hull Characteristics Program (SHCP), described in Chapter I, will be explained in some detail in Section 4.15.

Stability cross curves can also be calculated by using a hand-held programmable calculator, as explained by Cromer (1981).

4.6 Techniques. When the integrator is used, it is necessary to prepare a body plan showing both the forward and after sections on both sides of the ship's centerline. To avoid the confusion resulting from drawing numerous waterlines on the body plan, several parallel waterlines may be drawn on a sheet of clear plastic, or a piece of tracing paper, which is placed over the body plan with the waterlines at the desired drafts and perpendicular to the moment axis of the integrator, as shown in Fig. 20. This sheet should be large enough so that the area and moment wheels do not run over the edge of the sheet during the integration process.

When the moment axis of the integrator has been set to coincide with the vertical plane through the assumed center of gravity for a particular inclination, each section may be integrated up to each waterline without moving the integrator track or the body plan.

If the method of wedges is used, the required dimensions may be scaled directly from the lines drawing. A sheet of clear plastic or tracing paper with a set of radial lines may be placed over the drawing while the dimensions are lifted.

If an electronic computer is used, a one-sided body plan showing each station is prepared, and inputs to define the hull form for each station are made in conformance with the program being used. Section 4.15 describes the specific inputs required to obtain cross curves.

The location of the waterlines used in preparing the cross curves should be kept for future reference. This may be done by recording the angle of inclination and the intersection of the waterline with the ship's centerline or base line. This information will be needed if cross curves for the damaged ship are prepared later by deducting the buoyancy and moment of the flooded volumes, up to the same waterlines, from the buoyancy and moment of the intact ship.

4.7 Appendages. The foregoing discussion of cross curves deals only with the fair main body of the ship. In many cases it will be necessary to make adjustments to the calculations to take account of other

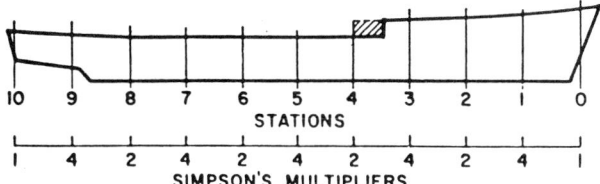

Fig. 22 Example ship with long forecastle

items which are considered to be appendages to the main body of the ship. Examples of these items are as follows:

(a) Fittings such as propellers, rudders, shafts, shaft bossings and sonar domes, which protrude from the main body. In most cases, these items are symmetrical with respect to the ship's centerline and their center of buoyancy falls on the ship's centerline. Their volume and the vertical location of their center of buoyancy may have been calculated in connection with the displacement and other curves. If they remain submerged until large angles of heel are reached, they may be treated as a single item for the purpose of making an adjustment to the cross-curve calculations.

(b) Portions of the ship forward or aft of the perpendiculars that may be submerged as the ship heels. If the effect of these appendages is significant, a separate estimate of their volume and center of buoyancy is made for each waterline.

(c) Free-flooding volumes within the fair main body of the ship. Significant free-flooding spaces are unusual, but they may have an important effect on the cross curves if they exist. Examples are unusually large sea chests, wells, and free-flooding spaces within the fair lines on submarines. These are treated as negative appendages, and their volume and center calculated for each waterline. This treatment must be consistent with the weight estimate; if they are considered as negative appendages in making the cross curves, the weight of water they contain should not be included in the weight estimate.

(d) Where the deck forming the watertight envelope of the ship is not continuous, as in the case of a ship with a forecastle deck illustrated in Fig. 22, there is an abrupt change in the section-area curve where the break occurs for angles of heel where the lower portion of the deck is submerged. The area and moment curves can be integrated accurately by Simpson's first rule if the break occurs at one of the even-numbered stations, where Simpson's multiplier is 2 as shown in Fig. 22, by entering the mean of the values of area and moment immediately forward and immediately aft of the even-numbered station, or by entering the sum of the two values in the "products" column. If the break occurs between Stations 3 and 4 as Fig. 22, it could be assumed, in the process of longitudinal integration, that the forecastle deck extends aft to Station 4 by an imaginary extension of the lines of the deck and shell, and the volume and moment of volume of the submerged portion of the imaginary extension calculated and deducted as a negative appendage. If the break occurs between Station 4 and 5, it could be assumed that the forecastle deck ends at Station 4, and the real submerged volume aft of Station 4 treated as a positive appendage.

The method of adjusting the calculations for the main body to take account of appendages is illustrated in Fig. 23. The 1500-t item represents the buoyancy of the main body which has been found to act at a distance of 0.3 m from the vertical plane CD through the assumed center of gravity O. The shaded area is the submerged portion of the negative appendage shown in Fig. 22; its negative buoyancy for this waterline is 20 t acting at a distance of 4.6 m (15 ft) from plane CD. The 5-ton item is the buoyancy of the rudder, propeller and shaft bossing, centered at a point on the ship's centerline, 2.1 m (7 ft) below point O, or 2.1 × sin 30° = 1.1 m (3.5 ft) from plane CD. The adjusted values of displacement and righting arm are found by adding, algebraically, the forces and moments, as follows:

	Buoyancy	Arm	Moment
Main body................	1500	0.3	450
Negative appendage.......	−20	4.6	−92
Rudder propeller, etc.	5	−1.1	−6
	1485	0.24	352

The adjusted displacement is 1485 t; the righting arm is 352/1485 = 0.24 m.

The shell plating, which is often treated as an appendage in the calculations for displacement and other curves, cannot be handled efficiently in this manner in the calculations for cross curves, since its submerged volume and moment are different for each waterline. If the shell is ignored, as it often is, the values of righting arm will be somewhat smaller than the actual values, and the evaluation of stability will be on the

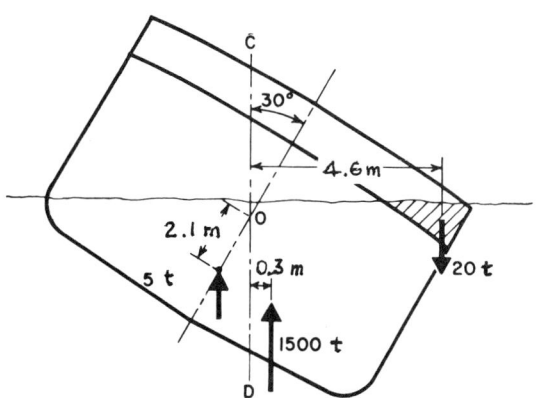

Fig. 23 Example of correction for negative buoyancy

Table 5—Calculations for Righting Arms

	Angle of Inclination, ϕ, deg.					
	0	15	30	45	60	75
Righting Arms, m, from Cross Curves	0	1.17	2.44	3.57	3.79	3.41
Adjustment for actual KG, (6.1-8.8) sin ϕ	0	−0.70	−1.35	−1.91	−2.34	−2.61
Adjustment for off center CG, $-0.30 \times \cos \phi$	−0.30	−0.29	−0.26	−0.21	−0.15	−0.08
Free Surface Correction	0	−0.05	−0.06	−0.05	−0.05	−0.04
Righting Arms, m	−0.30	0.13	0.77	1.40	1.25	0.68

pessimistic, or safe, side. The effect of the shell may be worked into the calculations for the main body of the ship by running the tracing point of the integrator slightly on the outside of the molded lines, or by adding the shell thickness to the measurements taken from the ship sections.

Deckhouses are generally not considered as appendages. Because of the access and ventilation fittings usually found in the house sides, which may be open, and the fact that the deckhouses would contribute righting moment only at rather large inclinations, it is customary to omit the buoyancy of the deckhouses in the cross curve calculations. This does not preclude including the effect of a deckhouse if there is assurance that it will not become flooded if submerged.

4.8 Curves of Statical Stability. The statical stability curve is a plot of righting arm or righting moment against angle of heel for a given condition of loading. For any ship, the shape of this curve will vary with the displacement, the vertical and transverse position of the center of gravity, the trim, and the effect of free liquids. Section 4.1 describes, and Fig. 17 illustrates, the cross curves of stability which show the righting arms, for an assumed height of the center of gravity, for several angles of heel and throughout the range of operating displacements. Section 4.1 also describes the method for adjusting the righting arms read from the cross curves to correspond to the actual height of the ship's center of gravity and for any movement of the center of gravity from the ship's centerline. Section 5 discusses the methods for determining the effect of free liquid on the righting arm.

Table 5 illustrates the method used for obtaining the actual righting arms from the cross curves shown in Fig. 17. The following data are required, in addition to the cross curves and the height of the center of gravity assumed thereon:

Displacement, mt	30,000
CG above keel, $\overline{KG}$, m	8.8
CG from CL, m	0.30
Assumed $\overline{KG}$ (Cross Curves)m	6.1

These values of the righting arm, plotted against angle of heel, form the statical stability curve, shown in Fig. 24. This figure illustrates the general case, in which the center of gravity is not on the ship's centerline (creating an initial list), rather than the usual case, in which the center of gravity is on, or very near, the centerline. Fig. 25 is the statical stability curve for the same ship, at the same displacement, with the same $\overline{KG}$ and free liquids, but with the center of gravity on the ship's center line. It is obtained by omitting the item 0.3 cos ϕ from Table 5.

For ships on which the effect of free liquid is relatively small, and the moment of transference is therefore not calculated, as discussed in Section 5, the righting arms are calculated by considering the effect of free liquid as a virtual rise of the center of gravity of the ship. In the usual case, where the center of gravity is on the ship's centerline, only the first two lines of Table 5 would be required. The value used for $\overline{KG}$ would include the virtual rise due to free liquid.

The curve of righting arms may be converted to a curve of righting moments by multiplying the ordinates by the ship's displacement. The righting-arm curve may therefore be used as a curve of righting moment by adding a scale of moments.

The displacement and center of buoyancy of a submerged submarine are fixed. The effect of free surface is small, so that it may be treated as a virtual rise of the center of gravity. The distance from the center of buoyancy (which is the metacenter) to the center of gravity is fixed at all angles of inclination and the righting arm is equal to $\overline{BG} \sin \phi$, where $\overline{BG}$ is the distance of the center of buoyancy above the virtual position of the center of gravity.

Waves may have a significant effect on static stability, particularly following or overtaking waves of approximately the ship's length. Righting curves can be drawn by superimposing offsets from the wave profile on the body plan used for the calculation of cross curves. In a computer calculation, the wave profile is used for input instead of a straight waterline. Dynamic effects of rolling are excluded.

Fig. 26 shows typical righting arm curves for a ship in a regular wave of the same length as the ship and height equal to $L_w/20$, with either wave crest or wave trough amidships.

Although the dynamic effects of rolling in waves were not included above nor in any published regulations, the designer should consider their possible influence on a new design, as discussed in Chapter VII.

INTACT STABILITY

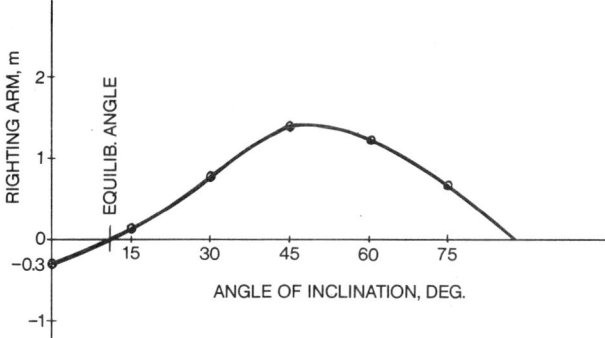

Fig. 24 Typical static stability curve, CG off centerline

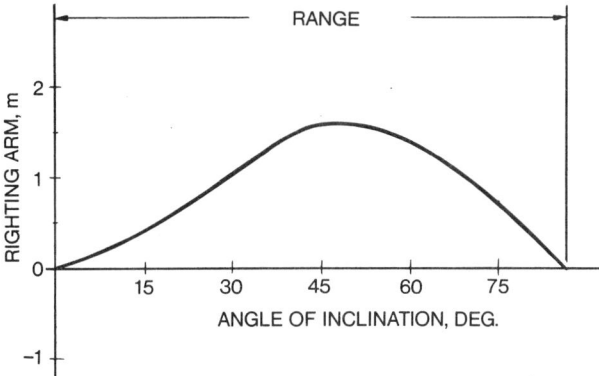

Fig. 25 Typical static stability curve, CG on centerline

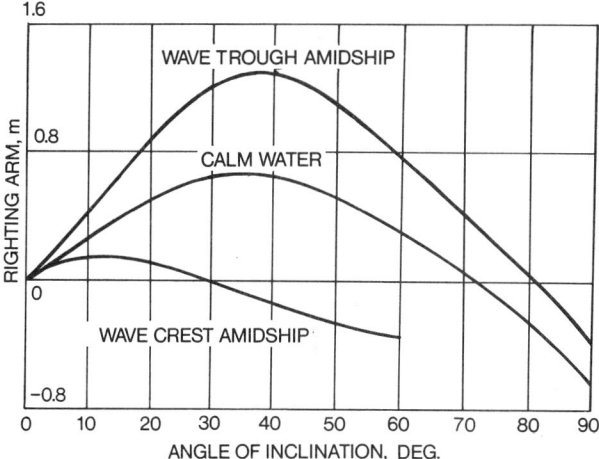

Fig. 26 Typical static righting arms in a wave of ship length and height L/20

Reference is also made to Paulling (1974) for a description of his experiments and a hypothesis explaining the capsizing mechanism for ships in following waves.

4.9 Effect of Beam on Statical Stability. Fig. 27 shows, in solid lines, a typical ship section, and another section, in broken lines, which differs from the other by an increase in beam.

Increasing the beam as illustrated in Fig. 27 will, at any angle of inclination, cause the ship to rise so that the lost displacement of the shaded volume between the two waterlines is equal to the added displacement of the shaded volumes port and starboard between the two shell lines, less the additon of weight involved in increasing the beam. If w_1 is the lost displacement between the waterlines, w_2 the added displacement on the high side, and w_3 the added displacement on the low side, with their centroids located at distances a, b, and c from a vertical through the original inclined center of buoyancy, B_1, the horizontal shift of the center of buoyancy toward the low side, d, would be found as shown in Table 6.

The righting arm is increased by the amount d due to the shifting of the center of buoyancy. The righting arm would be affected also by any movement of the ship's center of gravity caused by the added weight. As a practical matter, the weight change would probably be small and not far from the ship's center of gravity, and the effect of the weight change on righting arm would be small, if not negligible.

Metacentric height, and therefore the righting moments at small angles of inclination, are increased by increasing beam because of the large increase in the moment of inertia of the waterplane, along with small changes in volume of displacement and vertical positions of the centers of buoyancy and gravity.

4.10 Effect of Depth on Statical Stability. Fig. 28 shows a ship section in solid lines, with another, in broken lines, which differs by an increase in depth.

Weight changes caused by an increase in depth will have a more pronounced effect on stability than those associated with a change in beam. This is due to the fact that increasing depth results not only in adding structure such as shell, framing, and bulkheads between the two positions of the deck in Fig. 28 at some

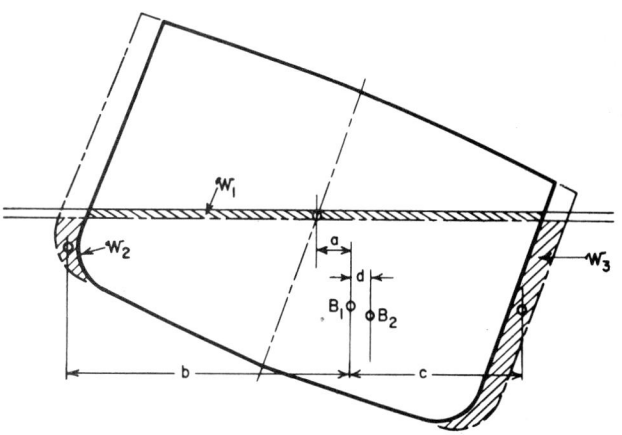

Fig. 27 Effect of increase of beam

Table 6—Effect of Increased Beam

	Weight	Lever	Moment
Original displacement...	W	0	0
Lost displacement.......	$-w_1$	$-a$	w_1a
Added displacement (high side)............	w_2	$-b$	$-w_2b$
Added displacement (low side)................	w_3	c	w_3c
	$W - w_1 + w_2 + w_3$		w_1a $-w_2b$ $+w_3c$

$$d = \frac{w_1a - w_2b + w_3c}{W - w_1 + w_2 + w_3}$$

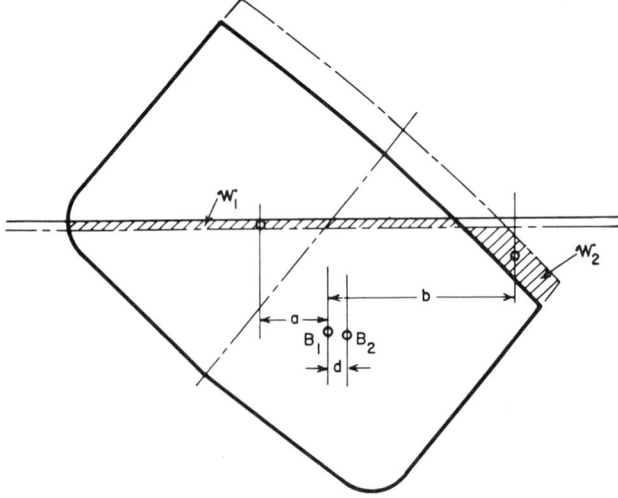

Fig. 28 Effect of increase of depth

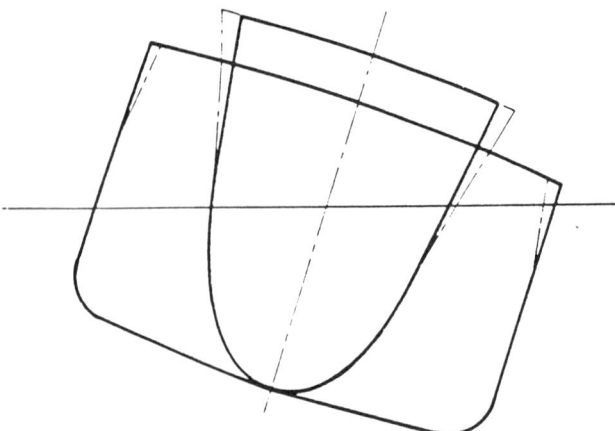

Fig. 29 Effect of tumble-home and flare

distance above the ship's center of gravity, but also in raising the entire superstructure and all items installed above the deck through a distance equal to the increase in depth.

Increasing depth, as illustrated in Fig. 28, causes the center of buoyancy to shift toward the low side, owing to addition of displacement of the shaded volume between the two positions of the deck and the loss of the displacement of the shaded volume between the two waterlines. In this case, the shift of B,

$$d = \frac{w_1a + w_2b}{W + w_4} \quad (10)$$

where w_4 is the weight of the added structure. Note that results would be the same if mass units were used.

Increasing depth will not result in an increase in righting arm due to shifting of the center of buoyancy until an angle is reached at which the original deck edge is submerged. Beyond this angle, substantial increases in righting arm may be obtained, particularly if the original freeboard was relatively low. At all angles, righting arms will be decreased because of the upward shift of the center of gravity caused by the added weight and raising of the superstructure and other topside items. Thus the net effect of increasing depth on right arms is a decrease until the original deck edge is immersed and a significant increase above this angle (assuming the beneficial buoyancy effect is greater than the adverse effect of the rise in G). Therefore, depth should not be increased for the purpose of improving stability unless some decrease in righting arms at the smaller angles of heel can be accepted or the rise of the center of gravity is offset by some measure such as adding low weight or increasing the beam.

When depth is increased, metacentric height is decreased by the amount that the center of gravity is raised and by the amount $\overline{KM}$ may be reduced.

4.11 Effect of Other Changes in Form on Statical Stability. Tumble-home or flare above the waterline in the upright position, as illustrated in Fig. 29, have effects on stability quite similar to a change in beam except that a major effect on righting arm will be delayed until the larger angles of heel are reached.

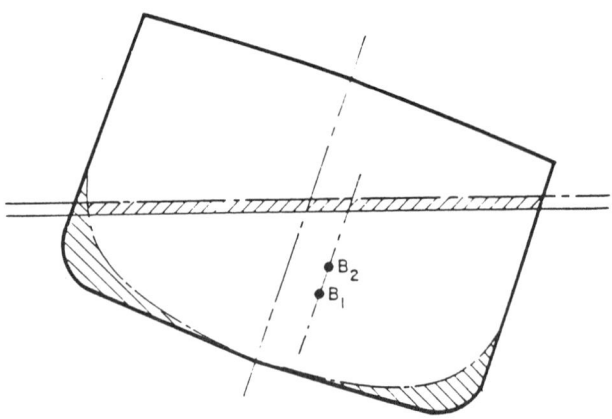

Fig. 30 Effect of fining the bilges

INTACT STABILITY

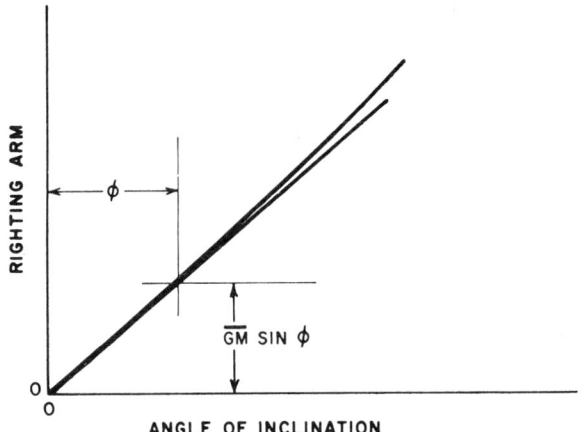

Fig. 31 Slope of stability curve at the origin

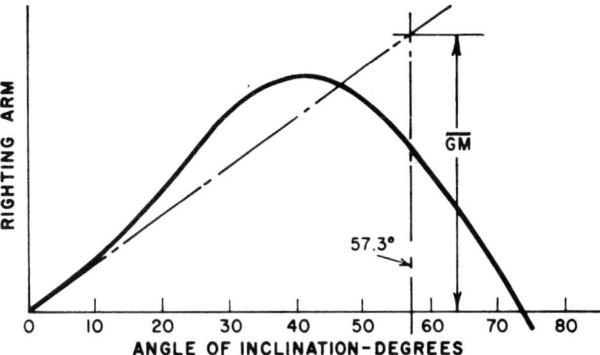

Fig. 32 Slope of stability curve at the origin defined by $\overline{GM}$

Righting arms at the larger angles are increased by flare and decreased by tumble-home.

Flare and tumble-home have no effect on metacentric height except for a small change in the height of the ship's center of gravity caused by differences in structure.

Replacing full sections by finer sections with the change wholly below the waterline in the upright position, as illustrated by Fig. 30, has the effect of raising the center of buoyancy because the shaded volume at the bilges is replaced by the shaded volume between the two waterlines. Before the shaded volume at the bilge on the high side emerges, the effect is to move the center of buoyancy upward, parallel to the ship's centerline, as from B_1 to B_2, increasing the righting arm by the horizontal shift, or $(B_1B_2) \sin \phi$. As the shaded volume at the bilge on the high side emerges, the gain in righting arm tends to decrease, since the centroid of volume of the combination of the two shaded volumes at the bilges moves toward the low side, and, since this volume is negative, the ship's center of buoyancy moves toward the high side. The increase in righting arm due to movement of the center of buoyancy is offset, to some extent, by the rise of the ship's center of gravity caused by the reduction in low weight.

When the lines are made finer in this manner, the metacentric height is increased by the difference in the rise of the center of buoyancy and the rise of the center of gravity. An additional effect on metacentric height may be produced by a change in the moment of inertia of the waterplane as the ship settles to the deeper waterline.

4.12 Significance of the Statical Stability Curve. The statical stability curve has a number of features that are significant in the analysis of the ship's stability.

Where the ship's center of gravity is not on the centerline, as in the case illustrated in Fig. 24, the point at which the curve crosses the horizontal axis corresponds to the static angle of heel at which the ship will come to rest in still water.

The slope of the curve at zero degrees is the metacentric height. As discussed in Section 3.1, the righting arm for small angles of heel may be expressed by the formula

$$\overline{GZ} = \overline{GM} \sin \phi \qquad (11)$$

The slope of the curve at the origin, as shown in Fig. 31, is therefore $\overline{GM} \sin \phi / \phi$ or, since $\sin \phi$ approaches ϕ (in radians) as ϕ approaches zero, the slope is the metacentric height. If the righting arm continued to increase at the same rate as at the origin it would be equal to $\overline{GM}$ at an inclination of 1 radian, or 57.3 deg, as illustrated in Fig. 32. Therefore, if the value of $\overline{GM}$ is plotted as an ordinate at 57.3 deg, a line connecting the plotted point with the origin would be tangent to the statical stability curve at the origin. This is a convenient check for major error in the initial portion of the righting-arm curve.

In cases where there is considerable free surface in wide, shallow tanks, and the moment of transference of the liquid is used to modify the righting arms, the metacentric height is not calculated, and is therefore not available for use in checking the curve. In such

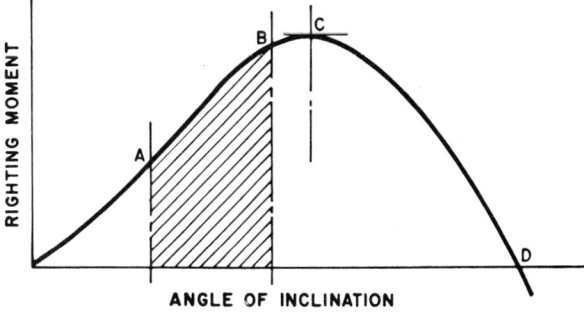

Fig. 33 Work required to heel a ship

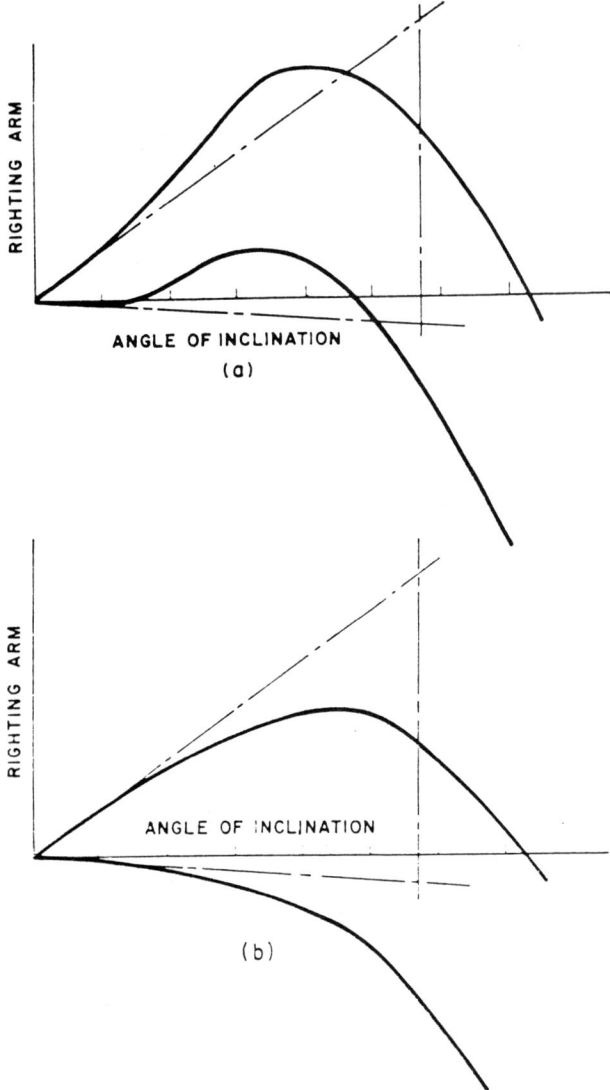

Fig. 34 Typical stability curves for different ships

which the deck edge enters the water (Rawson and Tupper, 1965):

$$\overline{GZ} = \overline{GM} \sin \phi + (\overline{BM}/2) \tan^2 \phi \sin \phi \quad (12)$$

A more general expression, taking account of hull shape can be obtained by substituting a factor F for $(\overline{BM}/2) \tan^2 \phi$ and calculating F for a variety of ship types and hull forms. (Niedermair, 1932.) This is sometimes done in damage stability calculations, as discussed in Chapter III.

The maximum righting moment, which occurs at angle C in Fig. 33, determines the maximum external upsetting moment that the ship can withstand without capsizing. If the ship is forced over to angle C by an external moment that does not thereafter diminish faster than does the righting moment, it will continue to heel until capsizing occurs.

The range of positive stability is indicated by point D in Fig. 33. If the ship heels beyond this angle, the forces of weight and buoyancy will act to capsize, rather than to right, the ship. On a normal ship, the range of positive stability is somewhat indefinite. As discussed in Section 3, the cross curves of stability are usually based on the assumption that the superstructure is not effective, and it was pointed out that, at very large angles of heel, there is a possibility that water may be shipped through topside openings with a consequent reduction in stability. If point D is determined from the cross curves as they are customarily calculated, positive righting arms would probably exist if the ship were to roll beyond angle D for a brief period because of the effect of the superstructure. If, on the other hand, the ship were to roll repeatedly to angles approaching point D, shipping water through topside openings (downflooding) might cause a progressive reduction in stability, which could eventually result in negative righting arms before point D is reached. Shifting of cargo can have a similar result.

The direction of curvature of the statical stability curve near the origin determines whether the ship will develop positive righting arms when the metacentric height is reduced to zero or becomes slightly negative. Two statical stability curves are shown in Fig. 34 for two ships having the same metacentric height but different forms. In Fig. 34(a), which is typical of cargo and passenger types, the curve is concave upward, while the curve in Fig. 34(b) is concave downward. Assume that the center of gravity of each ship is shifted upward the same distance, so that the metacentric height in each case becomes slightly negative. At any angle, the righting arm will be reduced by the same amount in each case, the reduction being the product of the vertical movement of the center of gravity and the sine of the angle of inclination. There is an important difference between the resulting statical stability curves: in case (a) the ship will heel to a small angle beyond which there will be some positive righting arm, while in case (b) the ship will capsize. The

cases, however, the process may be reversed, and the righting arm curve may be used to determine the effective metacentric height by determining the slope of the curve at the origin. This procedure might be useful if there is occasion to estimate the period of roll from the metacentric height, as discussed in Section 3.7. The slope at the origin, or the effective metacentric height, is equal to the righting arm at 10 deg divided by 10 deg in radians or $10°/57.3°$, which is equivalent to multiplying the righting arm at 10 deg. by 5.73.

The reason that the stability curve usually rises above the $\overline{GZ} = \overline{GM} \sin \phi$ line at first, as angle of inclination increases (Fig. 32), is that there is an upward shift of buoyancy as well as a lateral shift, as the ship inclines. At large angles of inclination this shift produces a significant increase in the righting arm, $\overline{GZ}$, until the deck edge is reached. It can be shown that for a *wall-sided* ship, up to the angle at

condition of negative metacentric height shown in case (a) may be recognized by the behavior of the ship, since there will be a list with no apparent heeling moment; the ship comes to rest with a small angle of heel either to port or to starboard but will not remain upright. It is possible for a condition of negative metacentric height to develop gradually in normal operation owing to consuming or unloading low weight, to developing a large free surface or to accumulating topside ice. In such situations, if the righting-arm curve is of the type shown in Fig. 34(a), and if the existence of negative metacentric height is recognized, there will be some warning that a precarious situation is developing. With a curve of the type shown in Fig. 34(b), the only warning prior to capsizing would be a lengthening of the period of roll, which would not be apparent if the ship were in still water.

4.13 Work and Energy Determination from Statical Stability Curve. Although the statical stability curve, as the name implies, is the representation of the righting arm, $\overline{GZ}$, or righting moment ($W\,\overline{GZ}$) of a ship when in a fixed-heel attitude, the curve can be used to determine the work involved in causing the ship to heel from one angle to another against the righting moment.

The area under any portion of a curve of righting moment, such as the shaded area in Fig. 33, represents the work required to heel the ship from angle A to angle B. A moment, multiplied by the angle through which it is exerted, represents work. In the case of a ship, where the moment varies with the angle, if M is the moment at any angle of heel, ϕ, then the work required to rotate the ship against this moment through an angle $\delta\phi$ is $M\,\delta\phi$ (ϕ in radians), and the work required to rotate it from A to B is

$$\text{Work} = \int_A^B M\,d\phi \qquad (13)$$

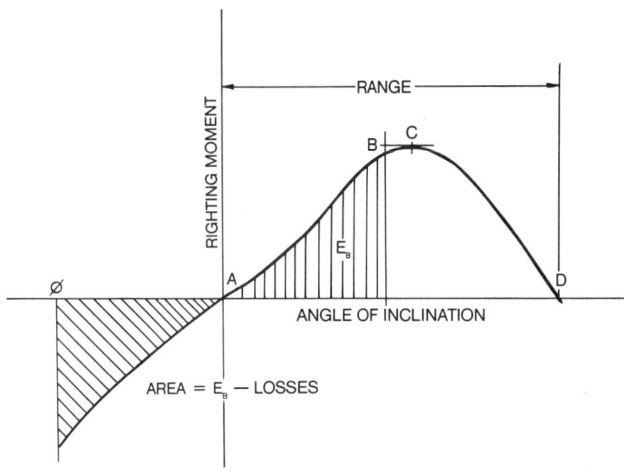

Fig. 35 Effect of rolling on dynamic stability

which is the area under the curve between A and B.

The total area between the righting-moment curve (at zero degrees to angle D) and the horizontal axis represents the total work required to capsize the ship from the upright position. This is often referred to as dynamic stability, although it does not really involve dynamics because wave-induced rolling velocities and accelerations are not considered.

One can also say that the total area under the curve from A to B in Fig. 35, which represents the work done in heeling the ship from 0 to B, is the potential energy, E_B, acquired by the ship at B. If all external heeling moments are then released, this energy will bring the ship back to the upright, zero-heel condition. But at this point the potential energy will have been transformed into kinetic energy equal to E_B minus energy loss (i.e., energy expended in overcoming the

Table 7—Intact Curves of Statical Stability

DISPL	LCG	POLE HT	HEEL	RA	TCB	VCB	DRAFT	TRIM
7739.807	4.828	26.00	15.000	6.892	9.591	16.815	23.138	-.722
			30.000	12.160	17.462	20.076	20.145	-3.003
			45.000	16.130	23.849	24.961	13.670	-0.710
			60.000	17.667	27.956	30.260	1.546	-4.446
			70.000	16.893	29.388	33.281	-16.054	-13.233
			80.000	14.742	30.036	35.674	-68.703	-20.470
			89.000	12.096	30.219	37.570	-1009.560	-121.799
11298.148	3.217	28.60	15.000	5.794	8.229	20.277	29.462	-.564
			30.000	10.734	15.475	23.265	27.268	-2.015
			45.000	13.422	20.515	27.068	23.066	-4.182
			60.000	13.276	23.204	30.533	16.052	-4.139
			70.000	12.206	24.245	32.765	8.813	-2.327
			80.000	10.617	24.853	34.996	-19.343	4.971
			89.000	8.445	25.012	36.610	-498.674	134.475
15175.417	1.574	31.10	15.000	4.851	7.014	23.665	35.659	-.498
			30.000	9.033	13.259	26.212	34.454	-1.346
			45.000	10.320	16.809	28.865	33.289	.350
			60.000	9.429	18.632	31.230	32.164	5.147
			70.000	8.093	19.315	32.082	31.109	12.190
			80.000	6.431	19.710	34.155	28.808	32.923
			89.000	4.828	19.847	35.582	-3.056	412.462

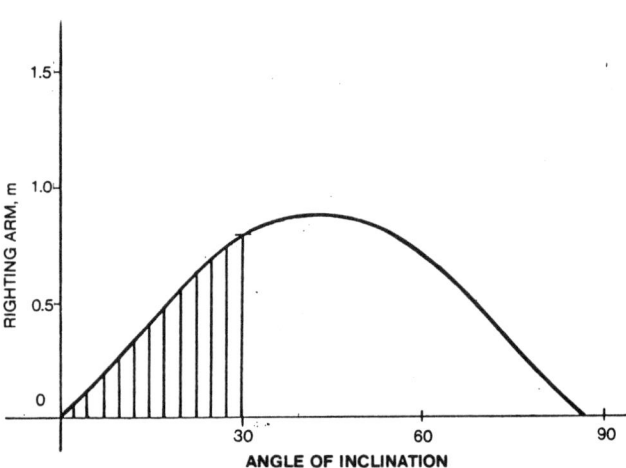

Fig. 36 Example of dynamic stability

resistance of the water to rolling.) This kinetic energy (proportional to the square of angular velocity) will carry it to an angle such that the area under the righting-moment curve, from the upright to that angle, is equal to the ship's kinetic energy at zero inclination minus the energy absorbed by the resistance of the water.

The following numerical example illustrates how a value of dynamical stability at a particular angle of heel, $\phi = 30$ deg, is determined for the case shown in Fig. 36 with $\Delta = 4000$ t. The corresponding weight displacement (force), $W = 4000 \times 9.81 = 39{,}240$ kN.

Righting Arm at 30 deg = 0.8 m
Righting Moment at 30 deg = $0.8 \times 39{,}240$ kN
 = 31,400 kN-m
Angle of 30 deg = $30 \pi / 180$ rad.
 = 0.52 rad.

Assume area under curve to 30 deg is a triangle. Therefore, dynamical stability up to 30 deg;

Area = $\frac{1}{2} \times 31{,}400 \times 0.52$
 = 8164 kN-m

4.14 Representation of Heeling Moments. In Section 1, the nature of certain heeling forces was discussed. If the heeling moments developed by these forces are calculated for several angles of inclination, these moments may be plotted on the same coordinates as the statical stability curve, as illustrated in Fig. 37. Note that both curves are extended to the left to show heel in the opposite direction. If the curve labeled "heeling moment" represents the moment of a beam wind, the moment will vary with the angle of inclination because of changes in the "sail" area, projected on a vertical plane, and in the vertical separation of the centroids of the wind pressure and the water pressure acting on the hull. If the heeling-moment curve represents the effect of high-speed turning, the mo-

ment will decrease at the larger angles, since the vertical separation of the centrifugal force and the water pressure will vary approximately as the cosine of the angle of inclination. A heeling moment due to the crowding of passengers to one side will similarly vary as the cosine of the angle of inclination. In general, heeling moments will vary with inclination because of variations in forces, levers, or both.

At points A and B in Fig. 37 the heeling moment equals the righting moment and the forces are in equilibrium. For example, if the heeling moment is caused by a lateral shift of weight (mass), the heeling moment is $w\,d\,\cos\phi$, where w is the weight (or mass) and d is the distance moved. At equilibrium points A and B,

$$W\,\overline{GZ} = w\,d\,\cos\phi \qquad (14)$$

If the ship is heeled to point A, an inclination in either direction will generate a moment tending to restore the ship to position A. If the ship is heeled to point B, a slight inclination in either direction will produce a moment tending to move the ship away from position B, and the ship will either come to rest in position A or capsize. The range of positive stability is decreased by the effect of the heeling moment to point B.

When a heeling moment exists, as in Fig. 37, the vertical distance between the heeling-moment and righting-moment curves at any angle represents the net moment acting at that angle either to heel or right the ship, depending on the relative magnitude of the righting and heeling moments, i.e., the net righting moment for the case of a weight shift is $W\,\overline{GZ} - w\,d\,\cos\phi$.

Coming now to energy considerations, assume that the ship has rolled to the left to angle C in Fig. 37, has come to rest, and is about to roll in the opposite direction. Between C and the origin, the heeling moment and the ship's righting moment will act in the same angular direction, and the total moment acting on the ship will be represented by the vertical distance between the two curves. To the right of the origin, these moments will act in opposite angular directions, and the moment acting on the ship will still be rep-

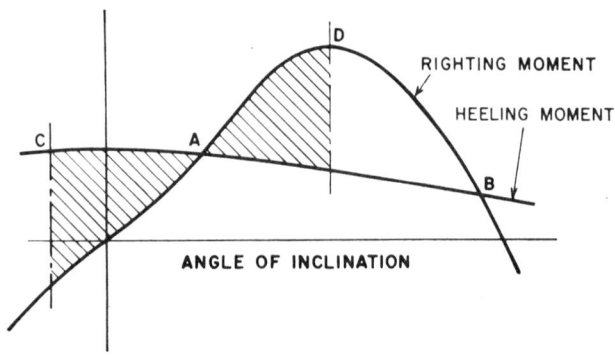

Fig. 37 Heeling and righting moments

Table 8—Intact Cross Curves at Pole Height 26.00 feet above BL

TRIM	HEEL	DISPL	RA	TCB	VCB	LCB	DRAFT
0.000	5.000	4693.896	2.591	3.812	12.155	6.215	18.000
		6179.310	2.526	3.587	13.987	5.474	21.000
		7791.031	2.430	3.332	15.796	4.724	24.000
		9513.395	2.318	3.063	17.585	3.952	27.000
		11331.266	2.210	2.800	19.354	3.144	30.000
		13231.552	2.119	2.555	21.103	2.328	33.000
		15198.822	2.058	2.341	22.860	1.522	36.000
0.000	10.000	4886.383	5.044	7.435	12.883	5.688	18.000
		6355.224	4.911	6.990	14.642	5.067	21.000
		7943.898	4.735	6.504	16.380	4.402	24.000
		9641.080	4.548	6.010	18.103	3.691	27.000
		11436.151	4.370	5.528	19.814	2.928	30.000
		13317.787	4.222	5.077	21.518	2.136	33.000
		15271.757	4.115	4.666	23.231	1.355	36.000
0.000	15.000	5195.171	7.256	10.724	14.016	4.922	18.000
		6636.504	7.058	10.077	15.664	4.458	21.000
		8197.238	6.847	9.418	17.309	3.879	24.000
		9861.163	6.635	8.761	18.943	3.256	27.000
		11622.906	6.441	8.123	20.572	2.565	30.000
		13473.623	6.277	7.516	22.203	1.829	33.000
		15404.332	6.164	6.956	23.858	1.069	36.000
0.000	20.000	5610.333	9.172	13.594	15.467	4.007	18.000
		7025.126	8.956	12.808	16.996	3.697	21.000
		8553.606	8.762	12.040	18.540	3.186	24.000
		10180.904	8.569	11.277	20.073	2.649	27.000
		11904.034	8.405	10.540	21.615	2.039	30.000
		13717.408	8.284	9.841	23.181	1.370	33.000
		15613.634	8.196	9.176	24.751	.660	36.000
0.000	30.000	6759.630	12.322	18.183	19.152	2.039	18.000
		8136.714	12.220	17.296	20.462	1.731	21.000
		9604.458	12.144	16.437	21.818	1.381	24.000
		11160.012	12.094	15.600	23.167	.987	27.000
		12762.044	11.977	14.717	24.464	.749	30.000
		14386.652	11.739	13.750	25.661	.610	33.000
		16009.060	11.392	12.713	26.765	.569	36.000
0.000	40.000	8430.367	15.064	21.499	23.813	-.182	18.000
		9701.415	14.885	20.479	24.751	-.031	21.000
		10989.129	14.647	19.433	25.628	.222	24.000
		12288.363	14.336	18.352	26.433	.539	27.000
		13593.076	13.975	17.249	27.186	.853	30.000
		14890.214	13.562	16.124	27.882	1.181	33.000
		16175.236	13.114	14.992	28.535	1.531	36.000
0.000	60.000	11753.547	15.280	22.636	30.575	1.496	18.000
		12494.304	14.955	21.751	30.711	1.915	21.000
		13233.200	14.649	20.886	30.857	2.360	24.000
		13968.554	14.359	20.038	31.011	2.796	27.000
		14699.867	14.079	19.204	31.169	3.213	30.000
		15426.092	13.809	18.384	31.331	3.611	33.000
		16146.982	13.547	17.577	31.495	3.987	36.000

resented by the distance between the two curves. Between C and A, the shaded area, minus the energy absorbed by water resistance, corresponds to the energy imparted to the ship that will exist, as kinetic energy, when the ship rolls through point A. This energy will carry the ship to some angle D such that the area between the curves and between A and D is equivalent to the kinetic energy at point A, less the energy absorbed by the water between A and D. If there is not sufficient area between the curves and between A and B to absorb this energy, the ship will roll past point B and capsize. To reduce the danger of capsizing under these conditions, the area between the heeling and righting-moment curves and between A and B should be greater, by some margin, than that between C and A. As a practical matter, it may be desirable to establish a limit for rolling which is considerably smaller than angle B because of the unfavorable attitude of the ship and the probability of shipping water through topside openings, i.e. down flooding, at very large angles.

The heeling moment produced by the movement of a weight aboard ship is the product of the weight of the item, the distance perpendicular to the ship's centerline through which it is moved, and the cosine of the angle of inclination. If a weight is added to or removed from a ship, the heeling moment is the product of the weight, its distance from the ship's centerline plane, and the cosine of the angle of inclination. The righting-moment curves on which these heeling moments are plotted should be adjusted, in the first case, for any movement of the item upward or downward parallel to the ship's centerline, which will affect the ship's center of gravity, or in the second case, for the height of the item relative to the ship's center of gravity and the effect of the weight change on the

Table 9—Intact Curves of Statical Stability in Waves

DISPLACEMENT 15175.44 TONS SW   LCG 1.575 FT (+FWD)   POLE HT 26.00 FT ABOVE BL

WAVE HEIGHT	WAVE LENGTH	WAVE CENTER	HEEL	RA	TCB	VCB	LCB	DRAFT	TRIM
19.053	1.000	150.000	0.000	0.000	0.000	23.526	1.568	25.520	.822
			5.000	1.962	2.178	23.614	1.569	25.488	.773
			10.000	3.933	4.368	23.879	1.571	25.385	.615
			15.000	5.907	6.564	24.326	1.572	25.202	.345
			20.000	7.731	8.630	24.894	1.575	25.003	.027
			25.000	9.309	10.469	25.534	1.575	24.817	-.306
			30.000	10.629	12.142	26.227	1.575	24.635	-.621
			35.000	11.694	13.600	26.965	1.572	24.442	-.787
			40.000	12.524	14.680	27.750	1.571	24.226	-.710
			45.000	13.070	15.953	28.531	1.571	24.003	-.468
19.053	1.000	75.000	0.000	0.000	0.000	23.125	1.368	26.949	21.592
			5.000	2.147	2.397	23.231	1.376	26.914	21.520
			10.000	4.282	4.780	23.547	1.401	26.804	21.301
			15.000	6.392	7.134	24.072	1.440	26.601	20.934
			20.000	8.447	9.429	24.794	1.493	26.277	20.446
			25.000	10.397	11.617	25.690	1.554	25.813	19.980
			30.000	11.976	13.463	26.634	1.617	25.329	19.695
			35.000	13.045	14.866	27.514	1.674	24.935	19.610
			40.000	13.726	15.958	28.335	1.728	24.599	19.708
			45.000	14.111	16.837	29.120	1.783	24.284	20.045
19.053	1.000	0.000	0.000	0.000	0.000	23.664	1.601	28.706	-3.363
			5.000	2.241	2.444	23.772	1.600	28.667	-3.390
			10.000	4.453	4.857	24.093	1.597	28.542	-3.495
			15.000	6.603	7.207	24.616	1.592	28.313	-3.674
			20.000	8.589	9.401	25.284	1.584	28.005	-3.601
			25.000	10.375	11.416	26.067	1.574	27.604	-3.205
			30.000	11.332	13.233	26.944	1.567	27.105	-2.433
			35.000	13.206	14.808	27.875	1.567	26.536	-1.239
			40.000	14.044	16.025	28.751	1.577	26.014	.178
			45.000	14.504	16.955	29.557	1.595	25.540	1.716
19.053	1.000	-75.000	0.000	0.000	0.000	23.016	1.818	27.203	-24.414
			5.000	2.029	2.290	23.114	1.810	27.171	-24.431
			10.000	4.060	4.580	23.408	1.786	27.067	-24.476
			15.000	6.081	6.858	23.901	1.746	26.872	-24.547
			20.000	7.984	9.023	24.554	1.693	26.614	-24.467
			25.000	9.663	10.982	25.313	1.630	26.325	-24.059
			30.000	11.082	12.715	26.142	1.564	26.004	-23.296
			35.000	12.230	14.215	27.016	1.500	25.649	-22.212
			40.000	13.061	15.464	27.891	1.445	25.301	-20.701
			45.000	13.621	16.493	28.769	1.399	24.946	-19.027

ship's displacement. A large weight added or removed may change the displacement so that the values of the righting arms read from the cross curves may be appreciably different from those read at the initial displacement.

The concept of righting-arm curves, rather than righting-moment curves, in evaluating stability was used in 4.8 and 4.12. Similarly, the heeling arm is the heeling moment divided by the ship's displacement. When heeling arms are plotted on a curve of righting arms, the resulting diagram has the same appearance as the moment curves; both sets of curves will intersect at the same angles; and the areas, such as the shaded areas between C and A and between A and D in Fig. 37 will have the same ratios.

Static stability of any ship in any condition can be evaluated by superimposing various heeling arms resulting from specific upsetting forces (wind, turning, etc.) on a curve of righting arms (see 7.2).

4.15 The Stability Subroutine of the U.S. Navy's Ship Hull Characteristics Program. The Navy's Ship Hull Characteristics Program (SHCP) is a widely used set of computer programs for naval architectural computations, including stability, as described in Chapter I. The reader should bear in mind that other computer programs are satisfactorily in use, and it is not the intention to imply that SHCP is the only acceptable program for obtaining cross curves and stability curves.

In this section, calculation of both cross curves and righting arm curves for the intact ship, using the SHCP stability sub-routine, will be discussed. Because of the ease of obtaining the righting arm curves for as many conditions as desired by use of a computer, the computations are seldom stopped at the completion of the cross curves. However, the availability of cross curves enables righting arm curves for additional loading conditions to be prepared whenever needed.

Further details will be found in the Naval Sea Systems Command (formerly NAVSEC) SHCP User's Manual (NAVSEA, 1976).

Once the Ship Data Table (SDT) has been established for the ship (Chapter I), both cross curves and righting arm curves (statical stability curves) can be calculated in one computer run. These curves can be calculated at up to 10 heel angles, with the maximum heel angle less than 90 degrees. Up to seven combinations of design conditions, consisting of either displacement and drafts, or displacement and LCG, or mean draft and trim are permitted. The assumed vertical center of gravity or the pole height (reference height for the cross curves) is entered as additional input data. If the vertical center of gravity is not known at the time the program is run, an assumed value is used and an adjustment is made later to the stability curves when an accurate value of the vertical center of gravity is known (the adjustment to $\overline{GZ}$ is equal to the difference between the actual and assumed VCG times $\sin \phi$).

The program can be run in still water or in a specified wave. The program takes the ship at the initial specified displacement and trim and heels the ship to a specified angle, allowing the ship to reach an equilibrium position with regard to forward and after drafts.

Cross curves are usually produced for a ship assumed to be initially at zero trim at each initial displacement or mean draft. They can be prepared for different trimmed waterlines, but usually it is much easier to input the specific loading condition (displacement, VCG, and LCG or trim) and obtain as output the appropriate statical stability curves.

Tables 7-9 have been extracted from NAVSEA (1976), and show typical stability curve and cross curve outputs for a hypothetical ship with various positions for the longitudinal center of gravity, various trims, and the quasi-static effect of waves. The program also includes the effects of any appendages specified.

Section 5
Effect of Free Liquids and Special Cargoes

5.1 Free-Surface Effect. The motion of the liquid in a tank that is partially full reduces a ship's stability because, as the ship is inclined, the center of gravity of the liquid shifts toward the low side. This causes the ship's center of gravity to move toward the low side, reducing the righting arm.

The mathematical processes applying to the motion of the center of gravity of the liquid are identical to those applying to the motion of the center of buoyancy of a ship. At small angles of inclination, the liquid in each tank has a metacenter located at a distance equal to i_r/v above its center of gravity in the upright position, where i_r is the moment of inertia of the surface of the liquid about an axis through its centroid and parallel to the centerline, and v is the volume of the liquid. A set of cross curves can be developed for a tank which shows, for each of the various angles of heel, and for varying quantities of liquid, the distance that the center of gravity moves in a direction parallel to the inclined waterline.

The usual practice in evaluating the effect of free surface in a ship's tanks is to assume the most unfavorable disposition of liquids likely to occur. If a tank is empty or completely full, there is no effect. The maximum effect occurs when a tank is about half full. Therefore, it is customary to assume that the largest tank in each of the systems, or the largest pair of tanks if they are in pairs, is half full. This assumption is made even when a full-load condition is being studied, since a free surface will develop shortly after the ship leaves port. In the fuel-oil system, the settling tanks also are assumed to be half full. Fuel-oil tanks which are nominally full of fuel are considered to be either about 95 percent full in naval practice or about 98 percent full in the case of a merchant ship, to allow for expansion of the oil without overflowing, and will therefore have some free-surface effect. A ballasted fuel-oil tank or a nominally full water tank should be completely full, and have no effect. If "split plant" operation is practiced, which involves dividing the system into two or more independent sections to enhance reliability in the event of damage, the largest tank or pair of tanks in each section is assumed to be half full.

When the stability of a damaged ship is being investigated, it is essential that the assumptions as to tank loading be consistent with the flooding assumed. For example, if one of a pair of tanks is assumed to be a source of off-center flooding, that pair should not be assumed to contain liquid or have a free-surface effect in preparing the weight estimate.

5.2 Evaluation of Effect of Free Surface on Metacentric Height. The theoretical effect of free surface on metacentric height can be assessed by assuming that the weight of the liquid in each tank acts at the metacenter of the tank, because, at any small angle of heel, a vertical line through the actual center of gravity will pass through this point. This is equivalent to assuming that the weight of the liquid in each tank is raised from its centroid in the upright position to its metacenter, a distance of i_r/v. This increases the vertical moment of the mass of the ship by $(w/g)(i_r/v)$, where w is the weight of the liquid. If the specific volume of the liquid, expressed as volume/mass, is designated as δ, then $w/g = v/\delta$ and the increase in vertical mass moment becomes

$$\frac{v}{\delta} \cdot \frac{i_r}{v} = \frac{i_r}{\delta}$$

an expression which is independent of the quantity of liquid in the tank. Therefore, for any condition of loading, free surface may be evaluated for small angles of heel, by adding the values of i_r/δ for all tanks in which a free surface exists. If this summation, which is the increase in vertical moment due to the free surface, is divided by the ship's displacement, the result will be the rise in the ship's center of gravity caused by the free-surface effect. This rise, called the *free-surface correction* is added to $\overline{KG}$, the height of the ship's center of gravity above the keel, resulting in an equivalent reduction in the metacentric height. Hence, with displacement in mass units,

$$\overline{GM}_{\text{COR}} = \overline{KB} + \overline{BM} - \overline{KG} - i_r/\delta \cdot \Delta \quad (15)$$

or with displacement in weight units,

$$\overline{GM}_{\text{COR}} = \overline{KB} + \overline{BM} - \overline{KG} - gi_r/\delta \cdot W \quad (15a)$$

It has been pointed out previously that the metacentric height, when multiplied by the weight of the ship and the sine of the angle of inclination, provides a fairly accurate evaluation of the righting moment up to 7, or perhaps 10, deg. The same is true of the metacentric height when reduced in the manner just described to take account of the effect of free liquid, provided that the surface of the liquid does not reach the top or bottom of the tank during this inclination. In cases where the tank is about half full, this treatment of free surface is accurate except in the case of a wide double-bottom tank. When a tank is 95 or 98 percent full, this treatment would be accurate only in the case of a narrow deep tank where the surface of the liquid would not reach the top of the tank at those inclinations at which the metacentric height provides a reasonably accurate index of righting moment.

In making the loading calculations discussed in Section 3.8, free surface corrections to $\overline{GM}$ must be made on the basis of reasonable assumptions regarding the condition of all tanks. In the full load departure condition, for example, it is customary to modify the condition to allow for free surfaces shortly after departure by assuming that service fuel tanks are half-full. The largest pair of storage fuel tanks may be assumed empty, but with maximum free surface effect acting (for a non-compensating system), and potable and reserve feed water may be reduced to two-thirds full. For naval ships with a compensating system the same assumptions are made except that 100 percent sea water ballast is carried in one pair of storage fuel tanks instead of assuming a pair of empty tanks.

5.3 Evaluation of Effect of Free Surface on Righting Arm. The effect of free liquid in a tank is to cause the center of gravity of the liquid to shift through a certain distance d, parallel to the inclined waterline. If the weight of the liquid is w and the displacement of the ship W, the center of gravity of the ship will move parallel to the inclined waterline through a distance $(d \cdot w)/W$, reducing the righting arm by that amount. Hence, the reduction in $\overline{GZ}$ is $(d \cdot w)/W$, or $d \cdot w/g)/\Delta$ in mass units. The quantity $d \cdot w/g$ is known as the *moment of transference*. When free surface

Table 10—Factors for Moment of Transferrence—Tanks 50 Percent Full

Ratio of depth to breadth	Angle of inclination, deg								
	10	20	30	40	50	60	70	80	90
0.1	0.13	0.14	0.14	0.12	0.11	0.09	0.06	0.04	0.02
0.15	0.17	0.21	0.21	0.19	0.16	0.14	0.10	0.07	0.03
0.2	0.18	0.27	0.27	0.26	0.23	0.20	0.16	0.11	0.06
0.25	0.18	0.31	0.34	0.33	0.30	0.26	0.21	0.16	0.09
0.3	0.18	0.35	0.40	0.40	0.37	0.33	0.27	0.21	0.14
0.4	0.18	0.36	0.50	0.53	0.51	0.47	0.41	0.33	0.24
0.5	0.18	0.36	0.57	0.65	0.66	0.63	0.56	0.47	0.38
0.6	0.18	0.36	0.58	0.74	0.80	0.79	0.74	0.65	0.54
0.7	0.18	0.36	0.58	0.83	0.94	0.96	0.92	0.85	0.74
0.8	0.18	0.36	0.58	0.87	1.06	1.13	1.12	1.06	0.96
0.9	0.18	0.36	0.58	0.87	1.16	1.30	1.34	1.30	1.22
1.0	0.18	0.36	0.58	0.87	1.24	1.47	1.56	1.56	1.50
1.2	0.18	0.36	0.58	0.87	1.31	1.7	2.0	2.1	2.2
1.5	0.18	0.36	0.58	0.87	1.31	2.0	2.7	3.1	3.4
2.0	0.18	0.36	0.58	0.87	1.31	2.2	3.7	5.0	6.0
3.0	0.18	0.36	0.58	0.87	1.31	2.2	4.5	9.3	13.5
4.0	0.18	0.36	0.58	0.87	1.31	2.2	4.5	13.4	24.0
5.0	0.18	0.36	0.58	0.87	1.31	2.2	4.5	16.2	37.0
6.0	0.18	0.36	0.58	0.87	1.31	2.2	4.5	16.8	54.0
7.0	0.18	0.36	0.58	0.87	1.31	2.2	4.5	16.8	73.0
8.0	0.18	0.36	0.58	0.87	1.31	2.2	4.5	16.8	96.0
9.0	0.18	0.36	0.58	0.87	1.31	2.2	4.5	16.8	121.0
10.0	0.18	0.36	0.58	0.87	1.31	2.2	4.5	16.8	150.0

Table 11—Factors for Moment of Transferrence—Tanks 95 Percent Full

Ratio of depth to breadth	Angle of inclination, deg								
	10	20	30	40	50	60	70	80	90
0.1	0.02	0.02	0.02	0.02	0.02	0.01	0.01	0.01	0.00
0.15	0.04	0.04	0.04	0.03	0.03	0.02	0.02	0.01	0.01
0.2	0.05	0.05	0.05	0.04	0.04	0.03	0.03	0.02	0.01
0.25	0.06	0.06	0.06	0.06	0.05	0.04	0.03	0.03	0.02
0.3	0.06	0.07	0.07	0.07	0.06	0.05	0.04	0.04	0.03
0.4	0.08	0.09	0.09	0.09	0.08	0.07	0.06	0.05	0.05
0.5	0.10	0.11	0.11	0.11	0.10	0.09	0.08	0.07	0.07
0.6	0.11	0.13	0.13	0.13	0.12	0.11	0.10	0.09	0.10
0.7	0.12	0.14	0.15	0.15	0.14	0.13	0.12	0.11	0.14
0.8	0.13	0.16	0.17	0.17	0.16	0.14	0.13	0.14	0.18
0.9	0.14	0.18	0.19	0.18	0.18	0.16	0.15	0.16	0.23
1.0	0.15	0.19	0.20	0.20	0.20	0.18	0.17	0.18	0.28
1.2	0.16	0.22	0.24	0.24	0.24	0.23	0.22	0.23	0.41
1.5	0.17	0.25	0.28	0.29	0.29	0.29	0.28	0.31	0.64
2.0	0.18	0.30	0.35	0.38	0.38	0.38	0.39	0.45	1.14
3.0	0.18	0.36	0.46	0.52	0.56	0.58	0.62	0.77	2.6
4.0	0.18	0.36	0.53	0.64	0.71	0.78	0.87	1.12	4.6
5.0	0.18	0.36	0.57	0.74	0.85	0.96	1.12	1.5	7.1
6.0	0.18	0.36	0.58	0.80	0.97	1.14	1.36	1.9	10.3
7.0	0.18	0.36	0.58	0.85	1.09	1.30	1.6	2.3	14.0
8.0	0.18	0.36	0.58	0.87	1.16	1.46	1.9	2.7	18.2
9.0	0.18	0.36	0.58	0.87	1.22	1.6	2.1	3.2	23.0
10.0	0.18	0.36	0.58	0.87	1.27	1.7	2.3	3.6	28.5

is present in a number of tanks, the summation of $d \cdot w/g$ for the various tanks, divided by the displacement of the ship, gives the total reduction in righting arm.

If the surface of the liquid has not reached the top or bottom of the tank, the distance d is equal to the distance from the center of gravity of the liquid with the ship in the upright position to the metacenter of the liquid, which is equal to i_r/v, multiplied by the sine of the angle of inclination. Therefore,

$$d \cdot \frac{w}{g} = \frac{w}{g} \cdot \frac{i_r}{v} \sin \phi$$

or since $w/g = v/\delta$

$$d \cdot \frac{w}{g} = \frac{v}{\delta} \cdot \frac{i_r}{v} \sin \phi = \frac{i_r}{\delta} \sin \phi$$

The moment of transference is thus seen to be independent of the quantity of liquid in the tank.

If the surface of the liquid has reached the top or bottom of the tank, the moment of transference will be reduced, and may be expressed by the product of i_r/δ and some factor less than $\sin \phi$, or

$$\frac{d \cdot w}{g} = C \cdot \frac{i_r}{\delta}$$

The value of C depends on the degree of fullness, the ratio of depth to breadth of the tank and the angle of inclination, each of which has some influence on the degree to which the motion of the liquid is suppressed.

Evaluation of the factor C is simplified by the fact that the tanks that contain liquid are assumed to be full, half full, 95 percent full in naval practice or 98 percent full in merchant practice.

Tables 10, 11 and 12 contain the results of a series of calculations, similar to those for the cross curves, which can be used to evaluate the free-surface effect. These tables show, for rectangular tanks 50, 95 and 98 percent full, for various values of the depth/breadth ratio, and for various angles of heel, the factors to be applied to the value of i_r/δ to obtain the moment, i.e., the weight of the liquid times the component of its motion parallel to the inclined waterline, generated by the movement of the liquid in the tank. The summation of these moments, for all slack tanks, divided by the displacement of the ship, is the reduction in righting arm due to the free-surface effect.

To illustrate the method of using the tables to find the effect of free surface on the ship's righting arm, assume the following:

Displacement of ship: 2000 t
Dimensions of tank: $l = 9.15$m (30 ft)
$b = 6.1$m (20 ft)
$d = 1.22$m (4 ft)
Contents of tank: 50 percent full F.W.
Angle of heel: 30 deg

Then

$$i_r = \frac{1}{12}(9.15 \times 6.1^3) = 173 \text{m}^4$$

δ (for fresh water) = 1m^3 per t
Depth/breadth = 1.22/6.1 = 0.20
From Table 10 (tanks 50 percent full) for d/b of 0.2, at 30 deg inclination, C is 0.27

Moment of transference = $0.27 \times \frac{i_r}{\delta} = 0.27 \times 173 = 46.7$ t−m

Reduction in righting arm at 30 deg = $\dfrac{46.7}{2000}$ =

0.023m

Fig. 38 illustrates the method used in determining the factor 0.27. Since the depth of the tank is 1.22m (4 ft), the center of gravity of the liquid in the upright position (as used in the weight estimate) is on the centerline and 0.3m (1 ft) above the bottom. At the 30 deg inclination, the center of gravity of the liquid is 0.54m (1.77 ft) above the bottom and 1.46m (4.79 ft) from the centerline.

Considering the shift of the center of gravity both parallel and normal to the bottom, the horizontal component when inclined is,

$$1.46 \cos 30 \deg + (0.54 - 0.3) \sin 30 \deg = 1.38m$$

The mass of the liquid is,

$$\dfrac{\frac{1}{2} \times 6.1 \times 1.22 \times \ell}{\delta} = 3.72 \dfrac{\ell}{\delta}$$

where ℓ is the length of the tank and δ is the specific volume of the liquid.

The moment of transference is,

$$1.38 \times 3.72 \dfrac{\ell}{\delta} = 5.13 \dfrac{\ell}{\delta}$$

The value $\dfrac{i_T}{\delta} = \dfrac{1 \times \ell \times (6.1)^3}{12 \, \delta} = 18.9 \dfrac{\ell}{\delta}$ where i_T is the transverse moment of inertia.

Thus, the ratio of the moment of transference to $\dfrac{i_{T'}}{\delta}$ which is the factor C in the table, is,

$$\dfrac{5.13 \, \ell/\delta}{18.9 \, \ell/\delta} = 0.27.$$

Thus it is clear that this factor is independent of the length of the tank and the specific volume of the liquid.

An accurate evaluation of the free-surface effect on a normal ship would require an excessive amount of calculation. In many cases, the tanks are not rectangular; often, when formed by the shell and a longitudinal bulkhead, they are not symmetrical. If certain approximations are adopted, however, a practical degree of accuracy can be attained, with results somewhat on the safe side, without undue effort. These approximations follow.

(a) Where the total moment of inertia of all tanks having free liquid in m⁴ is numerically not more than 17 times the displacement in t, assume that the weight of the liquid in each slack tank acts at its metacenter at all angles of inclination. With this assumption, the reduction in righting arm is the summation of i_r/δ for the various tanks, divided by the ship's displacement and multiplied by the sine of the angle of heel. When the ratio of i_r to Δ is 17 or less, the tanks would be predominantly of the deep variety, for which the values of the factor C approach the value of the sine. Since the free surface is not large, this approximation is acceptable.

(b) Where the total moment of inertia is more than 17 times the displacement, calculate the moment of transference instead of assuming that the weight of the liquid in each tank acts at its metacenter.

(c) Where a tank is not rectangular in plan view, the breadth at the narrow end may be used to determine the value of depth/breadth to be used in entering

Table 12—Factors for Moment of Transferrence—Tanks 98 Percent Full

Ratio of depth to breadth	Angle of inclination, deg								
	10	20	30	40	50	60	70	80	90
0.1	0.01	0.01	0.01	0.01	0.01	0.01	0.01	0.00	0.00
0.15	0.02	0.02	0.02	0.01	0.01	0.01	0.01	0.01	0.00
0.2	0.02	0.02	0.02	0.02	0.02	0.01	0.01	0.01	0.01
0.25	0.03	0.03	0.02	0.02	0.02	0.02	0.01	0.01	0.01
0.3	0.03	0.03	0.03	0.03	0.02	0.02	0.02	0.01	0.01
0.4	0.04	0.04	0.04	0.04	0.03	0.03	0.02	0.02	0.02
0.5	0.05	0.05	0.05	0.04	0.04	0.03	0.03	0.02	0.03
0.6	0.05	0.06	0.06	0.05	0.05	0.04	0.03	0.03	0.04
0.7	0.06	0.07	0.07	0.06	0.06	0.05	0.04	0.04	0.06
0.8	0.07	0.07	0.07	0.07	0.06	0.06	0.05	0.04	0.08
0.9	0.08	0.08	0.08	0.08	0.07	0.06	0.05	0.05	0.10
1.0	0.08	0.09	0.09	0.09	0.08	0.07	0.06	0.05	0.12
1.2	0.09	0.11	0.11	0.10	0.10	0.09	0.08	0.07	0.17
1.5	0.11	0.13	0.13	0.13	0.12	0.11	0.10	0.09	0.27
2.0	0.13	0.16	0.17	0.17	0.16	0.15	0.13	0.14	0.47
3.0	0.16	0.22	0.24	0.24	0.24	0.22	0.22	0.23	1.06
4.0	0.17	0.27	0.30	0.31	0.31	0.30	0.30	0.34	1.9
5.0	0.18	0.30	0.35	0.38	0.38	0.38	0.39	0.45	2.9
6.0	0.18	0.33	0.40	0.44	0.46	0.46	0.48	0.58	4.2
7.0	0.18	0.35	0.44	0.49	0.52	0.54	0.58	0.70	5.8
8.0	0.18	0.36	0.48	0.55	0.59	0.62	0.67	0.84	7.5
9.0	0.18	0.36	0.51	0.60	0.65	0.70	0.77	0.98	9.5
10.	0.18	0.36	0.54	0.64	0.71	0.78	0.87	1.12	11.8

INTACT STABILITY

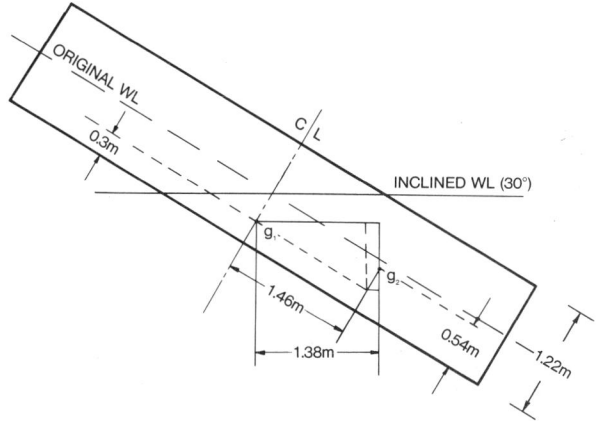

Fig. 38 Free surface effect at large angles

the tables. If the variation in breadth is very large, the breadth of a rectangular tank having the same moment of inertia may be used.

(d) Where a tank is not rectangular in cross section, the greatest depth of the tank may be used in determining the depth/breadth ratio. If the variation in depth is large, or if the tank if not approximately rectangular in section, the depth may be taken as that of a rectangular tank having the same ullage and therefore very nearly the same movement of liquid. For example, if the surface of the liquid is 0.15m (0.5 ft) from the top of the tank, the movement of the liquid at the smaller angles of heel would be very nearly the same as in a rectangular tank, 95 percent full, having the same breadth and a depth of 3.05m (10 ft).

(e) Sufficiently accurate results may usually be obtained without interpolating in the tables. If the value of depth/breadth falls between two in the table, slightly pessimistic results will be obtained if the next higher value of depth/breadth in the table is used. This will not only eliminate interpolating, but may also permit grouping a number of tanks having very nearly the same depth/breadth ratio so that the values of i_r/δ may be added before the coefficients in the table are applied.

While the foregoing approximations will give satisfactory results for the normal ship with tanks of normal size and shape, they should not be applied indiscriminately. For some unusual craft, these approximations may be unsuitable. The only rigorous method of evaluating free-surface effect is by calculating the actual moment of transference for the liquid in each slack tank for several angles of inclination.

5.4 Adjustment of Metacentric Height for Free Surface. Most merchant ships have several wide double-bottom tanks containing fuel oil, of which all but one, or one pair, will be assumed to be 98 percent full. In such cases, if the effect of free liquid on metacentric height were to be evaluated by assuming that the center of gravity of the liquid in each of the tanks were at its metacenter, a gross exaggeration of the loss of righting moment would be obtained at angles beyond 1 or 2 deg. The practice which has been adopted to produce a more reasonable value for the free-surface effect is to determine the effect of free liquid on the righting arm at an arbitrarily selected angle of 5 deg, and translate the reduction in righting arm at 5 deg into a reduction in metacentric height by dividing it by the sine of 5 deg. The effect of this assumption is to produce a value of metacentric height, adjusted for free liquid, which, when multiplied by the weight of the ship and the sine of the angle of inclination, will give, very nearly, the correct value for the righting moment at 5 deg inclination.

To facilitate application of this assumption, Fig. 39 may be entered with the breadth and depth of a tank that is 98 percent full, and a value of moment of inertia read that may be used in place of the actual moment of inertia of the surface of the liquid. If this is done for all tanks that are 98 percent full, and the actual moment of inertia used for those tanks that are half full, the adjusted value of metacentric height will give, very nearly, the correct righting moment at 5 deg inclination. (Note that Fig. 39 is in English units of feet for length and feet[4] for moment of inertia. Conversion is necessary when using SI units).

The value of the moment of inertia read from Fig. 39 is not the moment of inertia of any actual free surface, but is the moment of inertia of the surface in an imaginary tank which would have the same effect on the righting arm at 5 deg as the tank being considered, if the depth of the imaginary tank were great enough so that the surface of the liquid would not reach the top or bottom of the tank at 5 deg heel.

As an example, consider a rectangular tank 6.10m wide (20 ft) and 1.52m deep (5 ft), 98 percent full, and heeled to 5 deg. When the ship is upright, the center of gravity of the liquid would be on the centerline of the tank, 0.75m (2.45 ft) from the bottom. When the ship is heeled to 5 deg. the center of gravity of the liquid is 0.05m (0.16 ft) from the centerline and 0.75+ m (2.45 ft) from the bottom. The shift of the center of gravity is parallel to the waterline at 5 deg. heel and is equivalent to a horizontal plus a vertical shift,

$$0.05 \cos 5 \deg + (0.75+ - 0.75) \sin 5 \deg$$

$$= 0.05 \text{m} (0.16 \text{ ft})$$

If v/δ is the mass of liquid in the tank, the shift of the center of gravity of the liquid parallel to the waterline would move the ship's center of gravity a distance parallel to the water line,

$$0.05 \times \frac{v}{\delta \times \Delta}$$

$$\text{or } \frac{0.05 \times 0.98 \times 1.52 \times 6.10 \times \ell}{\delta \times \Delta} = \frac{0.45 \, \ell}{\delta \times \Delta}$$

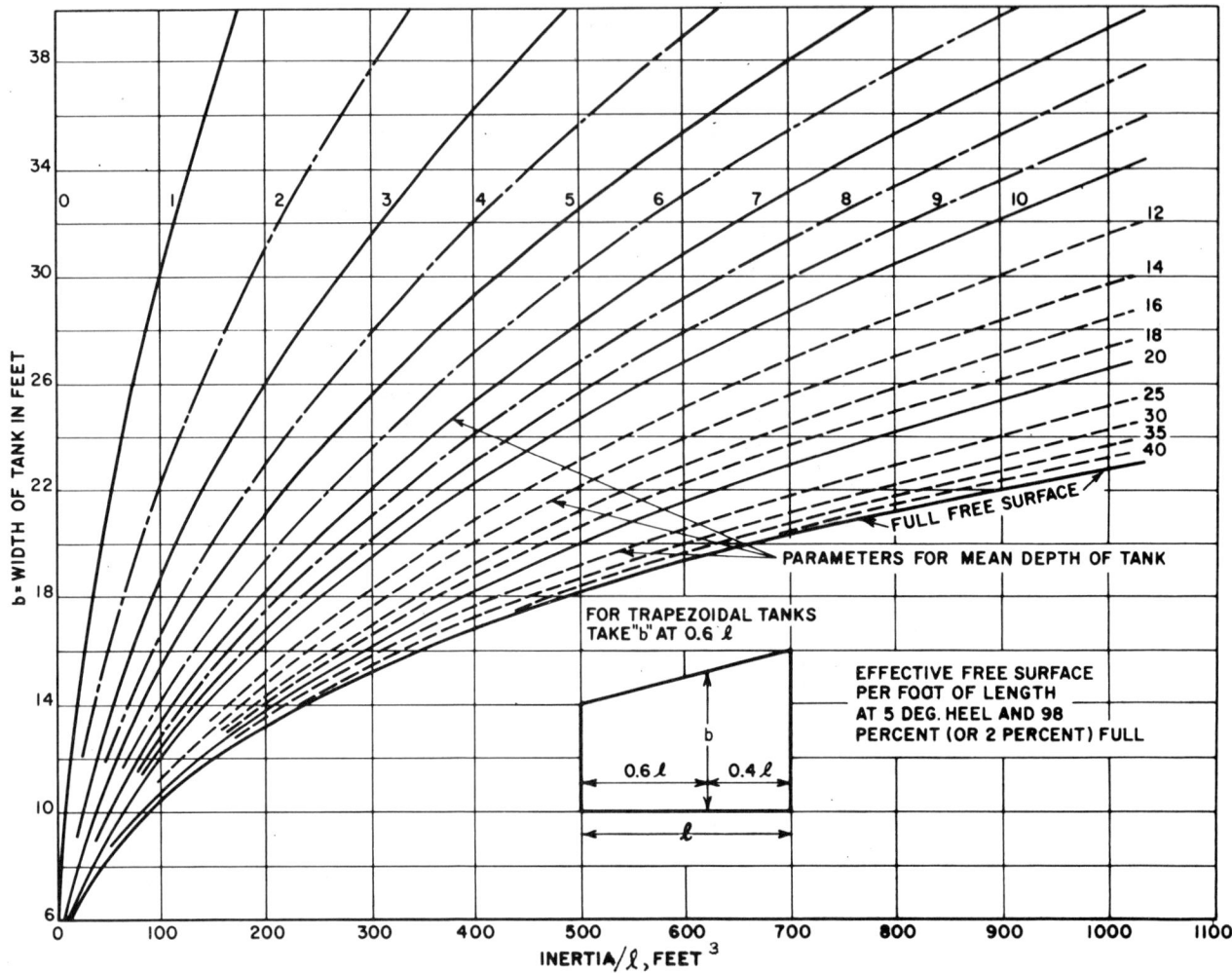

Fig. 39 Effective free surface at 5 deg. heel (English units)

where the length of the tank is ℓ and the specific volume is δ. Thus the shift will reduce the righting arm by that amount. Moving the ship's center of gravity this distance parallel to the waterline at 5 deg inclination has the same effect on righting arm as moving it upward parallel to the ship's centerline a distance of $\dfrac{0.45\ell}{\delta \Delta \sin 5 \deg}$ or $\dfrac{5.16\ell}{\delta \Delta}$ and reducing the metacentric height by that amount. Since the reduction in metacentric height produced by the liquid in the imaginary tank in which the surface does not reach the top or bottom would be $\dfrac{i_r}{\delta \Delta}$, then $\dfrac{i_r}{\delta \Delta} = \dfrac{5.16\ell}{\delta \Delta}$ or $i_r = 5.16\ell$, or 180ℓ in ft. The latter value of moment of inertia per unit of length may be read from Fig. 39 for a tank having a depth of 1.52m (5 ft) and a width of 6.10m (20 ft).

When this adjustment to metacentric height for tanks 98 percent full is used to determine whether stability is adequate, it should be realized that the righting moment obtained from the expression $\Delta \cdot \overline{GM} \sin \phi$ is valid only at 5 deg inclination. For smaller angles, the actual righting moments will be less than indicated, and for greater angles, they will be greater. The fact that this formula produces excessive righting moments at angles less than 5 deg is not important, since the ship is in no danger if the angle of heel is less than 5 deg. When the upsetting moment is sufficient to heel the ship beyond 5 deg, however, it may be profitable to make a more accurate evaluation of righting moment by the use of cross curves with adjustments to righting arm for the effect of free liquid at larger angles of heel, to avoid too greatly underestimating the ship's stability.

The empirical nature of the above procedure is emphasized. It is used when calculations for actual reduction in righting arms due to free surface are too laborious to be practical. Use of computers in determining righting arm and free surface corrections will often obviate the need to use the above approximation.

5.5 Determination of Moment of Inertia of Free Sur-

face. The process of determining the moment of inertia of the surface in a rectangular tank is relatively simple, since the moment of inertia about an axis through the centroid of the surface is given by

$$i_r = \frac{\ell b^3}{12}$$

Many tanks, however, are trapezoidal in plan view, while others may have one or more curved sides; in such cases the process is more complex.

In most cases it will be sufficiently accurate to convert each tank shape to the best equivalent rectangular shape and then approximate the moment of inertia for the rectangular shape. This procedure is particularly practical when other calculations are approximate and the free surface correction is used to provide an indication of the reduction in $\overline{GM}$ due to free surface when making an early determination of adequacy of stability.

5.6 Numerical Example of Free Surface Calculation. The following sample calculation is for a small Navy AM421-class ship, with tank layout and dimensions shown in Fig. 40:

5.7 Effect of Free Surface on Trim. Free liquid on a ship acts in the fore-and-aft direction in the same manner as in the transverse direction. For an intact ship with normal tankage, the effect of free liquid on trim is so small that it may be ignored. Its magnitude is small in comparison to the assumption in the formula for *moment to trim one cm*, described in Section 3, so that the center of gravity is at the same height as the center of buoyancy. On unusual craft, however, free liquids may have an important effect on trim.

For a submerged submarine, the effect of free liquid is significant in comparison to the longitudinal stability, but under normal circumstances the trim is carefully adjusted to zero in the submerged condition.

5.8 Effect of Tank Size on Free Surface. The subdivision of large tanks into two or more smaller tanks may be an effective method of improving stability by suppressing the motion of free liquids.

To illustrate, assume that a fuel tank 1.22m (4 ft) deep, 12.20m (40 ft) long and 12.20m (40 ft) wide has been subdivided at the center by a longitudinal bulkhead. The value of i_T/δ before the division was $\frac{1}{12}\frac{(12.2)^4}{1.06} = 1742$ meter-tons which is approximately equivalent to $\frac{1}{12}\frac{40\,(40)^3}{38.0} = 5600$ ft-tons in English units, allowing for rounding off.

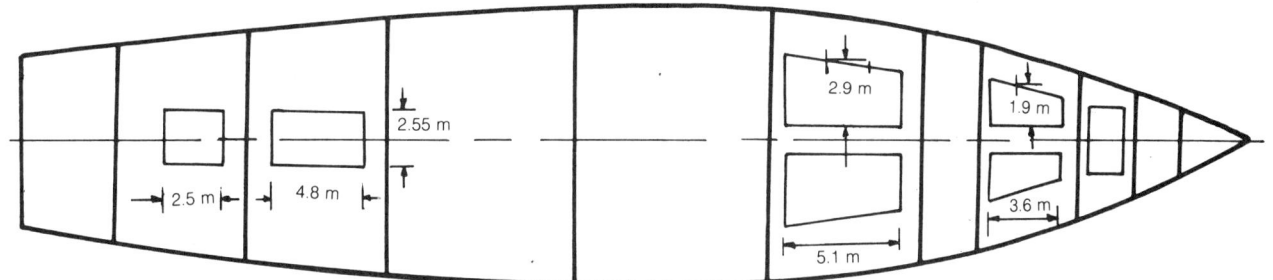

Fig. 40 Tank arrangement for free surface calculations

Diesel Oil tanks
2 tanks × 1/12[5.1 m × (2.9 m)³] = 20.7m⁴
1 tank: 1/12[4.8 × (2.55)³]   = 6.6
1 tank: 1/12[2.5 × (2.55)³]   = 3.5
                                $i = 30.8 m^4$

$$i/\delta = \frac{30.8}{1.17 m^3/t} = 26.3 \text{ m-t}$$

Fresh Water tanks
2 tanks × 1/12[3.6 m × (1.9m)³] $i = 4.15 m^4$

$$i/\delta = \frac{4.15 \text{ m}^4}{1.0 \text{ m}^3/t} = 4.15 \text{ m-t}$$

Free Surface correction:
Total $i/\delta = 26.3 + 4.15$ = 30.5 m-t
Displacement = 741 t
$\frac{i/\delta}{\Delta} = \frac{30.5 \text{ m-t}}{741 \text{ t}}$ = 0.04 m

After the subdivision there would be two tanks, each tank having an i_r/δ of $\dfrac{1}{12} \times \dfrac{12.2\text{m }(6.1)^3}{1.06\text{m}^3/\text{t}} = 217.7$ meter-tons, or 435 meter-tons for both tanks (1400 ft-tons). This means a free surface correction one quarter that of the original single tank. If the tanks were half full, the moments of transference from Table 10 with $\dfrac{d}{b}$ ratios of 0.1 and 0.2 would be as given in Table 13.

Table 13—Effect of Tank Size, Shallow Tank

Angle of heel, deg	Before		After	
	Factor $d/b = 0.1$	Moment $i_r/\delta = 1742$	Factor $d/b = 0.2$	Moment $i_r/\delta = 435$
10	0.13	226	0.18	78
20	0.14	244	0.27	117
30	0.14	244	0.27	117
40	0.12	209	0.26	113
50	0.11	192	0.23	100
60	0.09	157	0.20	87

For a deeper tank, the percentage reduction would be more pronounced at the smaller angles, but about the same at 60 deg. If the depth had been 6.1m (20 ft), the figures would be as given in Table 14.

Table 14—Effect of Tank Size, Deep Tank

Angle of heel, deg	Before		After	
	Factor $d/b = 0.5$	Moment $i_r/\delta = 1742$	Factor $d/b = 1.0$	Moment $i_r/\delta = 435$
10	0.18	314	0.18	78
20	0.36	627	0.36	157
30	0.57	923	0.58	252
40	0.65	1132	0.87	375
50	0.66	1150	1.24	539
60	0.63	1097	1.47	639

If the change in metacentric height due to subdividing the same tanks were to be evaluated, the effect for the shallow tank would be the same as for the deeper tank. In each case the value of i_r/δ would be 1742m-tons before subdivision and 435m-tons after, indicating that the effect of free surface is reduced to 25 percent of its former value. It can be seen from Tables 13 and 14 that this degree of improvement is not attained at the larger angles of heel, after the surface of the liquid has reached the top or bottom of the tank. Since we are interested in the improvement in righting arms at large angles of heel rather than at very small angles, the change in metacentric height produced by subdividing tanks is not an adequate index for judging the effectiveness of this measure.

Subdividing tanks in order to improve stability involves a compromise, since it requires considerable increase in structure, piping, and fittings. If only the largest tank is subdivided, the improvement is not the difference between that tank and the two into which it is divided, but the difference between that tank and the one that was second largest, which may not be much smaller. To obtain an effective improvement, it may be necessary to subdivide nearly all the tanks on the ship, with additional tail pipes, manholes, sounding, overflow and air-escape piping, and additional complication in the operation of the system.

5.9 Effect of Two Liquids. In some cases, a tank will contain two different liquids. For example, some mixed liquid cargos are not true solutions and may separate. Compensating tanks are known to have different liquids in the same tank.

Examples are gasoline tanks in which sea water is introduced at the bottom as gasoline is drawn from the top to avoid the accumulation of explosive mixtures, submarine diesel oil tanks in which the oil is replaced by sea water to preserve submerged equilibrium, and tanks on some diesel-driven surface type ships in which a compensating system is used to improve stability. In the latter case, the stability improvement results from maintaining low weight in the ship, reduction of free surface and reduction in the effect of possible off-center flooding after damage.

Although these tanks are always completely full of liquid, a free-surface effect exists at the interface which will remain parallel to the waterline as the ship inclines. There will be a wedge on the low side in which the lighter liquid will be replaced by the heavier, and a wedge on the high side in which the heavier liquid will be replaced by the lighter. This effect may be evaluated by using, as the value of i_r/δ, the i_r/δ for the heavier liquid minus the i_r/δ for the lighter.

5.10 Effect of Anti-roll Tanks. Chapter VII discusses various anti-roll devices, one of which is the anti-roll tank. Such a tank may have a deck view configuration as shown in Fig. 41. The tank is usually filled to about 50 percent capacity and, thus, has a significant free surface effect when the ship is not rolling. The best calculation method is to treat the shift of liquid from one side to the other as a *moment transference* at each angle of heel and to make the appropriate correction to the righting arm curve. Free surface effect on initial $\overline{GM}$ at zero heel should be calculated as though there are two separate tanks, each filled to 50 percent capacity. (This free surface correction is added to the other free surface corrections for tanks to obtain a $\overline{GM}$ (or $\overline{KG}$) corrected for free surface.

5.11 Bulk Dry Cargo. Bulk dry cargo, such as ore, coal or grain, may redistribute itself if the ship rolls or heels to an inclination greater than the angle of repose of the substance carried (angle of repose is the angle between a horizontal plane and the cone slope obtained when bulk cargo is freely poured onto this plane). Thus a ship may start a voyage with the upper surface of such a cargo horizontal and with the cargo evenly distributed throughout the space. But if the ship rolls sufficiently to cause a cargo shift, a list will result. A ship which has listed due to even a slight shift of cargo is open to the danger that it may later roll to increasing angles on the low side with further shifting of the cargo. Ships have been known to capsize from such progressive shifting of cargo.

Furthermore, all cargoes are directly influenced by

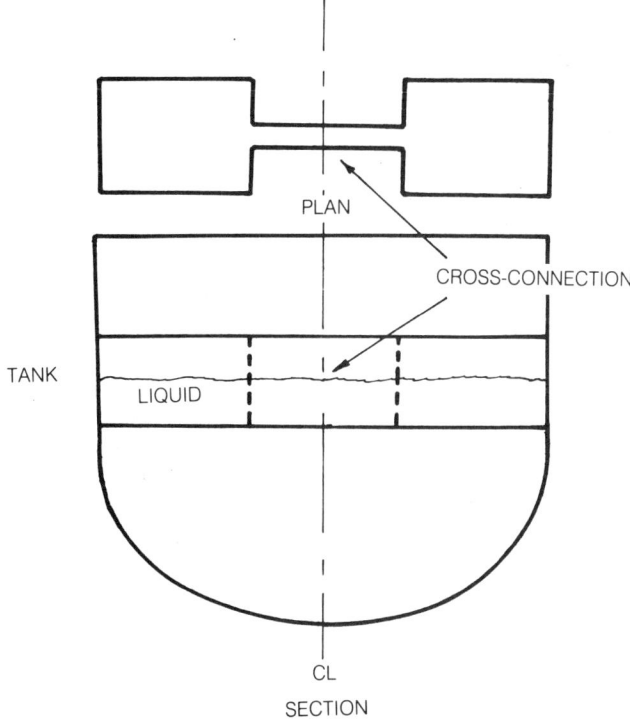

Fig. 41 Anti-roll tanks

the seaway-induced motions of the ship, which produce significant angular and lateral accelerations. In a rapidly rolling ship, such cargoes may shift even when the maximum angle of roll is less than the angle of repose of the cargo, because of the dynamic effects of rolling. Calculations using motion dynamics show that the accelerations involved in rolling produce a greater likelihood of cargo shifting when the cargo is located above the ship's CG (as in the 'tween decks) rather than below (in the hold). See Chapter VII.

One design approach for ships intended to carry dry bulk cargoes is to adjust hold volumes to suit the cargo density, so that holds will normally be full when loaded. Small hatches and sloping sides at the top of the compartment will reduce the danger of shifting cargo. For general cargo ships that may sometimes carry bulk cargo it is essential to provide for fitting one or more longitudinal subdivisions in the holds and 'tween decks to minimize the possibility of shift of cargo in heavy seas. Such temporary subdivision bulkheads are called *shifting boards*. Usually they consist of wooden planks laid edge to edge in steel channels or the equivalent. In all cases it is essential to ascertain that adequate stability can be attained in operation to cope with any anticipated cargo, considering the restraints actually available. Of course, the ship operator is responsible for reviewing such factors prior to every voyage.

Some guidance in the design of ships to carry solid cargoes in bulk is provided by the IMO *Code of Safe Practices for Solid Bulk Cargoes* (IMO, 1980.) The code includes technical information for a list of possible bulk cargoes, including static angles of repose. However, it has been recently determined that some of these cargoes change their physical characteristics with changes in moisture content, temperature and other factors. Some cargoes, especially certain ores, may act much like a semi-liquid slurry when only a small amount of moisture is present and will then shift much more readily. See Green (1980). These phenomena have been the subject of much international discussion, and the IMO is encouraging research.

Grain has long been recognized as a dangerous cargo because of its tendency to flow or shift in the hold of a rolling ship. In the past both national and international regulations (SOLAS, 1948 and 1960) relied heavily on the use of *feeders* from 'tween decks to holds, which were intended to allow grain to flow downward to keep the hold full as the grain settled. Continued reports of grain cargo shifting, with some ship losses, led to a new investigation of the problem, which showed that even with feeders holds could not be assumed to be full and that shifting boards were still of great value in many cases. New grain regulations were developed that changed emphasis from attempting to prevent grain shifting to making sure that the worst possible heeling moments will not exceed acceptable limits for each ship and loading condition. These regulations (IMO, 1973) have been adopted for cargo ships by the U.S. Coast Guard (46 CFR 93.20). The IMO resolution, along with SOLAS (1960) and other useful information, is given in, *General Information on Grain Loading*. (National Cargo Bureau, 1978.) See also SOLAS (1974). In the United States, the National Cargo Bureau represents the Coast Guard for review of grain loading plans and safe bulk cargo stowage on U.S. flag vessels.

5.12 Suspended Cargo or Weight. In the case where meat or similar cargo is suspended from a point above its center of gravity, sometimes it is hooked into eyes under the deck above the hold. This method of stowage calls for special correction in the calculation of $\overline{GM}$. A weight suspended from a boom is a similar case and serves as a convenient explanatory example.

The center of gravity of a weight suspended freely from a boom will remain vertically below the end of the boom, regardless of the list of the ship. The point of suspension, therefore, is the metacenter through which the weight acts. It makes no difference in the stability of the vessel whether the weight hangs high above the deck or not, provided the point of support remains the same. A suspended weight may be treated as though its center of gravity were at the point of support. Obviously, if a full cargo, such as meat, were suspended from several feet above its own center of gravity, the metacentric height of the vessel would be appreciably less than it would have been with an equal weight of unsuspended cargo.

Section 6
Effect of Changes in Weight on Stability

6.1 Effect on Displacement and Center of Gravity.

(a) Principles. The effect of changes in weight on the ship's displacement and center of gravity may be evaluated by the method used in the weight estimate (Section 2) and loading calculations (Section 3). This is a tabulation of items of weight with corresponding levers from the three reference planes, the resulting moments, and a summation of weight and moments. An item which is removed is entered as a negative weight, and, if the lever is positive, the moment will be negative. If the lever is negative, as it may be in the case of port and starboard moments about the centerline plane, the sign of the moment will be opposite to that of the weight. If the present weight and moments of the ship are entered in the tabulation, followed by those of the items to be added and/or removed, the totals will represent the ship after the changes are accomplished.

The use of the tabular form is convenient when a number of items are involved. When only a single item already aboard ship is moved, the ship's center of gravity will move in a direction parallel to the motion of the center of gravity of the item moved. The shift $\overline{G_1 G_2}$ parallel to any axis can be determined from the principle that

$$W \overline{G_1 G_2} = wd \text{ or} \qquad (16)$$
$$\overline{G_1 G_2} = wd/W$$

where w is the weight (mass w/g) of the item moved and d is the distance moved.

A problem which frequently arises is to find how much weight must be added in a given position to move the ship's center of gravity a given distance.

(b) Numerical Example. Assume that just enough ballast is to be added at a level 0.5m above the keel to lower the center of gravity of a 10,000-t ship from 7m (17 ft) to 6.5m (15.7 ft) above the keel. This problem could be solved in either of two ways, both giving the same result. Let the added ballast = w/g, t.

Method 1

	Mass, t	VCG	Vertical moment, m-t
Original condition	10,000	7.0	70,000
Add ballast	w/g	0.5	$0.5w/g$
New condition	$10,000 + w/g$		$70,000 + 0.5w/g$

Since required $VCG = 6.5$m, the required vertical moment will be:

$(10,000 + w) 6.5 = 70,000 + 0.5w$
and $w/g = 833$ t

Method 2

Original $\Delta = 10,000$ t @ $VCG = 7.0$m.
Add ballast, w/g @ VCG; then lower to 0.5 m.
New $\Delta = 10,000 + w/g$ @ $VCG = 6.5$m.

To lower VCG from 7m to 6.5m, a reduction in vertical moment of $(10000 + w/g) 0.5$m is required. This is obtained by lowering the ballast from 7m to 0.5m (the height of the installation.) Therefore, $w/g(6.5m) = (10000 + w/g) 0.5$m and $w/g = 833$ t.

6.2 Effects on Stability.

The effect of weight changes on a ship's stability can be evaluated by recalculating the righting arms and the metacentric height for the revised values of displacement and the vertical and transverse positions of the center of gravity. The effect of any changes in free liquid should be included. A large change in the ship's displacement will have some effect on the change in righting arm and metacentric height. The free surface effect will also change, even though there is no change in tankage, since this effect is the moment of the free liquid divided by the displacement.

In general, weights added above the ship's VCG will reduce stability, while weights added below the ship's VCG will improve stability. Weights removed will have the opposite effects. An exception to the above generalization would be those cases which involve large weight changes. The changes in righting arms (as read from the cross-curves) may have effects opposite to the effects of the vertical location of the changes. Off-center weight changes that result in the center of gravity being shifted from the centerline plane to a position off-center, port or starboard, will result in a heel either to port or starboard. Section 3.7 illustrates the determination of small heel angles caused by moderate athwartship weight movements (moment to heel 1 deg).

Movement of a weight that is already aboard ship has no effect on displacement, and, at a given angle of heel, no effect on the center of buoyancy. The effect of such a weight movement on the righting moment at this angle is to move the ship's center of gravity in a direction parallel to the movement of the center of gravity of the weight. The distance through which the ship's center of gravity moves depends only on the magnitude of the weight, the distance through which its center of gravity moves and the weight of the ship.

Thus, the effect on transverse stability when shifting an *on-board* weight depends on whether the shift results in a rise or lowering of the ship's VCG and whether or not the shift causes the upright ship's center of gravity to move from the centerline plane to a position to either side of the centerline plane. A large weight shifted a moderate fore and aft distance, or a smaller weight shifted a large fore and aft distance, would be necessary to affect longitudinal stability in the form of a change in trim. Longitudinal moment changes affect trim (produce a change in longitudinal inclination) in the same manner that athwartship moment changes affect heel.

6.3 Compensation for Initial Heel. Since the angle of heel of an undamaged ship may be caused either by an external moment, such as is produced by a beam wind, or by an off-center location of the center of gravity, the measures involving changes in weight that would increase stability in these two cases would be different and as follows:

Case 1. The center of gravity is on the ship's centerline and the buoyant volume is symmetrical with respect to the ship's centerline; that is, the ship is undamaged. Since an external moment may act in either direction, any measure used should increase the righting moment in both directions and should be effective at small angles. The measures available, therefore, consist of adding weight with its center of gravity on or near the ship's centerline below the waterline, removing weight with its center of gravity on or near the ship's centerline above the waterline, or moving weight downward in a direction parallel to the ship's centerline. Theoretically, a very large weight may increase stability if it is added slightly above the waterline, but practically, only a small increase could be so obtained. The same situation exists with respect to very large weight removals which could, theoretically, be slightly below the waterline.

Case 2. The ship's center of gravity has moved off the centerline as a result of a transverse shift of weight aboard ship, or the addition or removal of an off-center weight, or the rise in the ship's VCG due to weight additions or weight removals that result in negative metacentric height. Corrective action should be taken after a determination is made as to the cause of the initial heel. For example, in the case of negative $\overline{GM}$, the recommended action is to lower the ship's VCG rather than effecting a moment change in a direction opposite to the heel, since in such a case the ship would heel to a larger angle to the other side and might even capsize. If the initial heel is caused by an off-center weight, then a change in transverse moment should be effected to offset the heel by means of off-center weight additions or removals or athwartship shifts of on-board weights. If empty tankage were available on the opposite side from the weight, adding liquids to an empty tank or tanks on the opposite side might provide an off-setting athwartship moment about equal to the heeling moment. However, if adding liquids in empty low tanks is selected as the means to lower the ship's center of gravity (and, thus, improve stability), the adverse effect of free surface in the process of filling the tank must be taken into consideration before a final decision is made. The large free surface effect, reducing stability, often offsets the stability gain in filling low, wide double bottom tanks in the early stages of filling.

6.4 Large Trim Changes. If a change in weight results in a very large change in trim, the shape of the underwater body may be quite different from that in the even-keel attitude. The trim may be so large as to make the conventional displacement and other curves and the cross curves inapplicable, since they are based on an even-keel condition. It may be, therefore, that a satisfactory analysis of stability requires a recalculation of the data usually taken from these curves.

Current U.S. Navy practice when using electronic computers to produce displacement and other curves, and cross curves of stability, is to incorporate the effects of trim. Hence, the appropriate stability characteristics can readily be obtained for a specific trim by interpolation.

6.5 Weight Changes in Submarines. Changes in weight on submarines are limited by the requirement that the weight and buoyancy must be very nearly equal in the submerged condition. Compensation for changes in the ship's fixed weight is usually obtained by an equal change in the lead ballast. This process may increase or decrease righting moment, in both the surfaced and submerged conditions, depending on the relative heights of the items added and removed. If it is not feasible to remove high weight and add ballast low in the ship, stability can be increased by adding submerged buoyancy and adding an equivalent weight of solid ballast at a lower level than the added buoyancy.

6.6 Effect of Cargo on Stability. In the typical cargo ship, the weight of the cargo may be twice the weight of the light ship; in a tanker it may be more than four times the light ship weight. If the cargo is not homogeneous or does not fill the available space, a large variation in the center of gravity of the loaded ship can result from varying the distribution of cargo. If a ship with a full-load displacement of 10,000 tons has a metacentric height of 0.6m (2 ft) when loaded with 5000 tons of cargo at the centroid of the holds, the metacentric height would be reduced to zero if the same weight of cargo were loaded with its center of gravity 1.2m (4 ft) above the centroid of the holds. On the other hand, the ship might be quite uncomfortable if the cargo were stowed with the heavy items in the lower holds.

The optimum height of the center of gravity of the cargo may vary with the displacement and therefore with the total weight of the cargo to be loaded. A light cargo might be stowed with safety higher than a heavier load, since the lighter cargo would result in a greater freeboard.

In many ships, there is little option in the disposition of cargo, in which case the stability characteristics can be predicted quite accurately.

A factor which should be considered, particularly in stowing ore cargoes, is the effect on radius of gyration and, thus, on rolling characteristics. (Chapter VII). If most of the heavy cargo were stowed topsides, or very low in the ship, rather than homogeneously, the effect on the radius of gyration could be sufficient to affect rolling characteristics of the ship. Another factor affecting the stability of a large containership with several layers of topside stowage is the additional sail area to beam winds.

6.7 Consumption of Liquids and Stores. During the course of a voyage, there may be a considerable variation in displacement and in the position of the ship's center of gravity resulting from the reduction in the consumable load. The major factor in this variation is usually the consumption of fuel and water, which are usually carried low in the ship and constitute a significant percentage of the ship's full-load displacement. The consumable stores are usually a smaller item and are usually stowed fairly close to the vertical position of the ship's center of gravity. In any case, the effect of consumables requires consideration in the evaluation of stability. (See 3.8).

The low weight represented by the fuel can be replaced by taking aboard sea-water ballast, either in the empty fuel tanks or in tanks provided for this purpose, usually designated as *clean* or *segregated* ballast tanks. The practice of ballasting fuel-oil tanks is objectionable, because of the possibility of a temporary loss of power if sea water is inadvertently introduced into the fuel-oil service system and because of the difficulty in disposing of the ballast, which has been contaminated by the residual oil in the fuel tanks, before the ship is refueled. A secondary disadvantage of ballasting fuel-oil tanks is that the percentage of water in the fuel reaching the boilers is increased with minor adverse effects on the boilers.

Furthermore, the discharge of oily ballast water, even at sea, has become such a pollution hazard that current international convention standards require most new ships to be designed with segregated ballast tanks, thus obviating the need to add sea-water ballast to empty fuel-oil tanks.

Oil tanker operators usually reserve a pair of deep tanks (identified as *slop tanks*) on their ships for storing and chemically treating washings from the cargo oil tanks so that the effluent can be safely discharged overboard without polluting the waters, and the recovered oil can be reused as desired.

For a specific ship design, the designer should check with the governing regulatory organizations in order to make sure that all required criteria have been included in the calculations. U. S. Coast Guard requirements are published regularly in the Code of Federal Regulations (46 CFR 170-174).

The Navy has always recognized that in diesel-driven ships, ballasting of empty fuel tanks is less objectionable because of the greater difference in density between the fuel and the sea water than in the case of Bunker C oil. Hence, a *compensating* system was developed for early naval diesel ships in which each bunker tank is always kept completely full of oil or water. This is accomplished by piping a group of tanks in series with a sea-water ballast connection in the tank farthest from the machinery space. As oil is drawn from the top of the tank closest to the machinery space, sea water enters the tank farthest away at the bottom. When this tank is full of sea water, the ballast passes through a pipe from the top of the ballasted tank to the bottom of the tank next in the series, and this process is continued through the successive tanks. With this system, there is an increase in the weight of liquid in the fuel system as the fuel is consumed. The rate of increase may exceed the rate of consumption of stores and water, resulting in a slight increase in the ship's displacement during a voyage, along with a lowering of the ship's center of gravity.

Consumption of liquids and stores on diesel submarines has very little effect on stability. Nearly all fuel tanks are of the compensating type. Continuous compensation is obtained for variation in weight of the variable load by adjusting the quantity of sea water in tanks provided for this purpose in order to maintain a balance of weight and buoyancy when the ship is submerged. This process involves only minor changes in the vertical position of the center of gravity.

However, because of the previously mentioned pollution problems and resulting anti-pollution regulations, the tendency in new and future naval surface ship designs is toward the adoption of clean ballast tanks to achieve the necessary stability, immersion and trim. Meanwhile, for existing surface ships and submarines with compensating systems, better warning devices, controls and other equipment have been installed to prevent spillage of oily ballast or fuel overflows during refueling. There is not much of a problem on nuclear submarines, which carry little fuel oil. While the fuel oil tanks are of compensating type, usually refueling is done in port where *doughnuts* are available to receive the dirty ballast water.

6.8 Liquid Loading Diagrams. *Liquid Loading Diagrams* are provided to U.S. Navy ships and are in the form of Fig. 42. The diagram shows:

(a) The tankage of the ship and the weight of liquid normally carried in each tank.

(b) The liquid loading instruction *(Notes)* which the ship should follow in order to enable it to maintain adequate stability to withstand the governing hazard to stability (either in the intact condition or the damaged condition).

(c) In addition, there is shown for each tank the change in trim and heel when filling or emptying the tank. The liquid loading diagram in Fig. 42 is for an

INTACT STABILITY

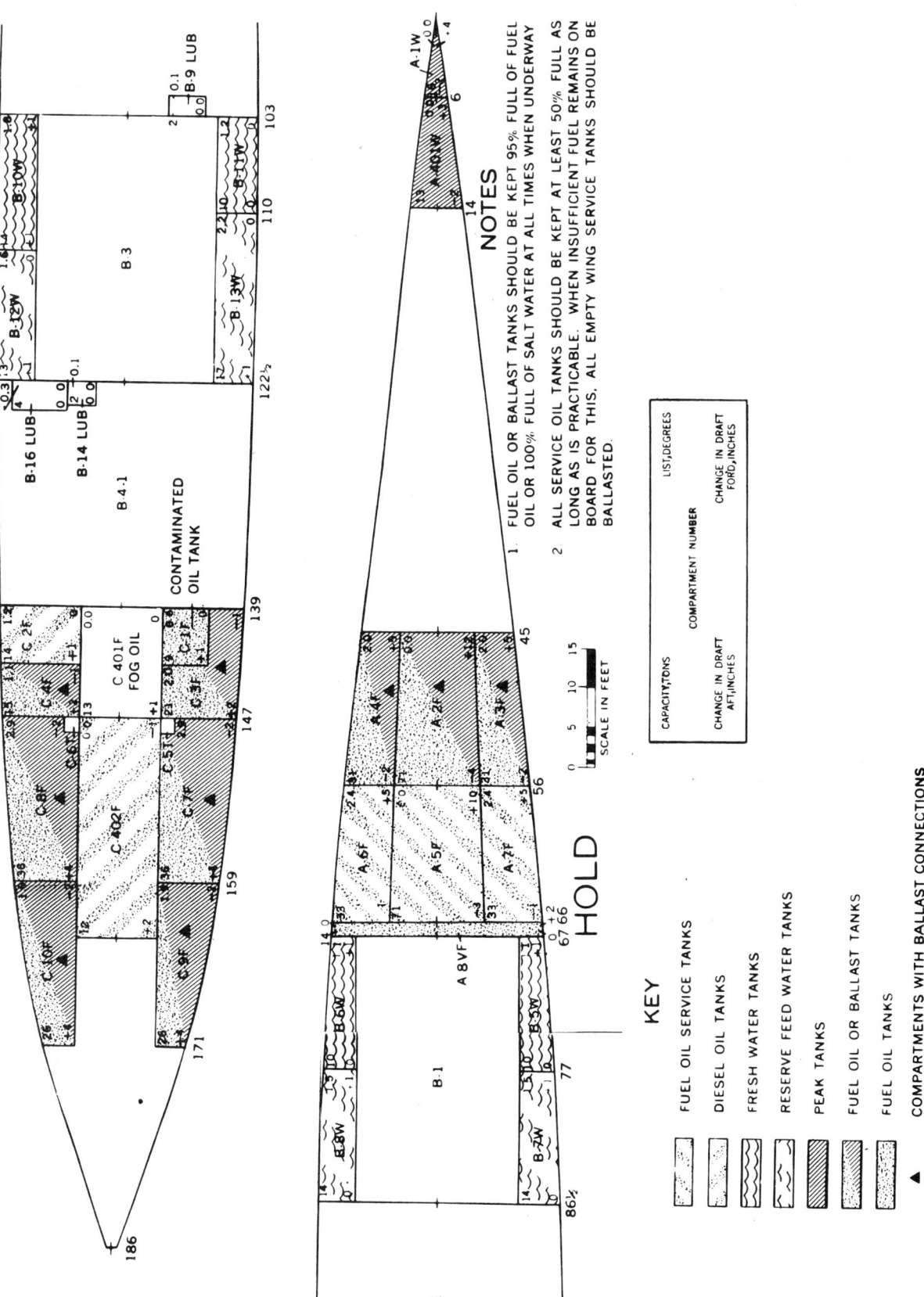

Fig. 42 Liquid loading diagram for destroyer

old Navy design without clean ballast tanks; it is unclassified. (Note that the values are in English units and are presented here for illustration only).

Chapter III discusses a companion diagram issued to certain Navy ships called the *Flooding Effect Diagram*, which illustrates the effects of flooding water on trim and stability, and indicates appropriate measures which should be taken to compensate for them.

6.9 Trim and bending moment computers. On ships where the longitudinal distribution of weights has a significant effect on a ship's bending stresses, a trim and bending moment computer may be used by the ship master to achieve a safe loading from the standpoint of trim and bending moment (or stress) at several locations along the length. Tankers often have such instruments. Specific information on the instruments in use may be obtained by requesting such information from the manufacturers. Each such instrument must be tailored to the individual ship on which it is to be placed. For U.S. ships all instruments must be checked and approved by the Coast Guard as part of the approved stability information. There have been instances of unapproved instruments leading the master to erroneous conclusions.

Section 7
Evaluation of Stability

7.1 Stability Criteria in General. As explained in Section 3.9, stability is often evaluated approximately on the basis of metacentric height alone, without the benefit of a complete righting-arm curve. This is equivalent to assuming that the righting-arm curve has the form of a sine curve, since it is assumed that the righting arm is equal to $\overline{GM} \sin \phi$. For ships with large freeboard and the type of form that produces a righting-arm curve concave upward near the origin, this practice is usually safe, but may result in underestimation of the ship's stability. For ships having little freeboard and the type of form that results in a righting-arm curve concave downward near the origin, this practice may not be acceptable, since it does not assure an adequate range of positive stability or adequate *residual dynamic stability* (defined below.)

This section will deal mainly with criteria based on consideration of actual shape and other characteristics of the curves of righting and heeling moment (or arm) for an undamaged ship through large angles of heel. One of the most important criteria in the evaluation of stability, a ship's ability to survive flooding due to damage, is discussed in Chapter III. So far as the intact ship is concerned, the righting-arm curve for the least favorable condition of loading and the heeling-arm curves for the various upsetting forces provide useful data for judging the adequacy of the ship's stability. Features of the the curves that warrant consideration from a purely static viewpoint are:

(a) The angle of steady heel under the influence of a static heeling moment, as indicated by point A in Fig. 37.

(b) The range of positive stability, point B in Fig. 37.

(c) The relative magnitudes of the heeling arm and the maximum righting arm.

The angle of steady heel is important from two standpoints: First, its absolute value determines its adverse effect on personnel and the operation of the ship. Also, its value with respect to the angle of heel at which the deck edge will be submerged is a measure of the ship's resistance to capsizing, since the righting arm increases at a lesser rate after the deck edge is awash.

The range of residual positive stability is important since it is the limit of the angle to which the ship can heel without capsizing.

The excess of maximum righting arm over the heeling arm, in addition to providing a margin for the upsetting forces of wind and wave, is essential as an allowance for inaccuracies in calculating the heeling and righting arm.

Some stability criteria have been based only on such static considerations (as Wendel, 1960), but most also include the work and energy considerations (dynamic stability) discussed in 4.13 and 4.14. In particular, comparisons are made of the areas under the two curves up to certain angles, and sometimes of the residual dynamic stability, which represents the work required, in addition to the effect of the heeling moment being considered, to capsize the ship. Such additional energy might be supplied by wave action, ship rolling or by wind pressure in cases where upsetting forces other than wind are being considered. The *residual dynamic stability* is represented by the area below the righting-arm curve and above the heeling-arm curve.

The optimum criterion would allow full flexibility of operation while maintaining enough stability to avoid capsize in the roughest seas. To date, there is no such criterion applicable to all sizes of ship, in all of the many varying types of marine operations, with differently shaped hulls.

In addition to the different methods of statical evaluation outlined in previous sections, it must always be borne in mind that a floating object leads a dynamic life, reacting continuously to the sea, Chapter VII. In addition to meeting published regulations the ship designer should review the available criteria, and use those which fit each particular design best. It may be necessary to combine the best facets of several criteria for a specific design.

Various international and national organizations have established criteria for determining adequacy of intact stability to withstand specific upsetting forces. These organizations are:

International Maritime Organization (IMO)
U.S. Coast Guard (USCG)
U.S. Maritime Administration (MARAD)
American Bureau of Shipping (ABS)
(when acting on behalf of USCG)
U.S. Navy

7.2 Merchant Ship Stability Criteria. The development of intact stability criteria for merchant ships has been slow due to the great variety of geometric forms, range of loadings, regional and national differences in type of ship used for various commercial enterprises, rapid development of new marine transport systems, and the naturally slow process of international agreement.

Since 1966, stability evaluation satisfactory to each nation has been required by the International Convention Load Lines. Thus, while almost every nation has adopted stability criteria, each is formed upon the type of ships normally in use. Some nations use various forms of evaluation of the stability curve as in Section 4, and others use the metacentric height (Section 3) because of its simplicity in operational usage. Still others use combinations of the two methods in their criteria.

The International Maritime Organization (IMO) has recently sent out to member nations a proposed severe wind and rolling criterion for ships over 24 m (78 ft) in length. Previously IMO had published recommendations on intact stability for ships under 100 m (328 ft) long and codes for dynamically supported craft, mobile offshore drilling units, offshore supply ships, and fishing vessels, all of which refer to intact stability or include intact stability standards. In each code and recommendation, both $\overline{GM}$ and stability curve evaluations are utilized.

Current National Stability Standards for U.S. flag merchant ships are almost all now contained in a new portion of the Code of Federal Regulations (1983). These standards include intact stability criteria for:

46 CFR 170—General.
　Weather (steady wind heeling)—$\overline{GM}$
46 CFR 171—Passenger ships.
　Passenger heel—$\overline{GM}$
　Wind heel—$\overline{GM}$
　Sailing ships (monohull)—RA curve.
　Sailing catamaran—Moment limit.
46 CFR 172—Bulk Cargoes.
　Barges with hazardous cargo—RA curve, transverse and longitudinal $\overline{GM}$.
46 CFR 173—Special use ships.
　Lifting—Counter ballast—RA curve.
　Towing $\overline{GM}$ & RA curve.
46 CFR 174—Specific vessel types.
　Barges with deck cargo—RA curve.
　Offshore drilling units—RA curve.
　Towboats—RA curve.

Small ships, particularly those used for fishing, towing and offshore supply, have posed special problems, not only because of their size but because of the unusual duties they are called upon to perform, often under severe weather conditions. Hence, some attention of researchers and IMO has been concentrated on these smaller ships.

Among various criteria developed in the past, mainly for small craft, the one by Rahola (1939) is of particular interest because it formed the basis for subsequent work by IMO. On the basis of data on 30 small ship capsizings, he presented an empirical criterion that specified required righting arms at several angles, a minimum value of the maximum $\overline{GZ}$ angle and a minimum area under the curve of righting arms (dynamical lever) up to a specified maximum angle. This empirical approach did not attempt to specify the heeling moments that might be expected from the various hazards encountered in service.

The IMO recommendation on intact stability for passenger and cargo ships under 100 m. (328 ft.) (U.S. Coast Guard, 1973) and a similar recommendation for fishing ships (Coast Guard, 1976) are similar in principle, but IMO adopted slightly different numerical values. They are summarized as follows:

(a) The area under the righting level curve should be not less than 0.055 m-radians up to $\phi = 30$ deg, where ϕ is the heeling or inclining angle (degrees).

(b) The area under the righting lever curve should be not less than 0.09 m-radians up to $\phi = 40$ deg or up to an angle where the non-weather tight openings come under water (whichever is less).

(c) The area under the righting lever curve should be not less than 0.03 m-radians between the angles of heel ϕ-30 to ϕ-40 deg or such lesser angle mentioned under Standard (b).

(d) The righting lever should be at least 0.2 m at an angle of heel equal to or greater than 30 deg.

(e) The maximum righting lever should occur at an angle of heel exceeding 30 deg.

(f) The initial metacentric height $\overline{GM}$ should be not less than 0.15 m (0.35 m for fishing ships.)

Additionally, for passenger ships:

Angle of heel < 10 deg due to movement of passengers.

Angle of heel in turning < 10 deg with moment,

$$M = 0.02 \frac{V^2}{L} \Delta (\overline{KG} - T/2),$$

where V is service speed and T is draft.

The International Conference for Safety of Fishing Vessels, 1977, adopted the recommended Fishing Vessel Criteria but noted the need to examine the physical nature of the various heeling moments expected in service, such as weights overside, severe wind and waves, and water on deck.

The U.S. Coast Guard has set up a criterion for offshore supply ships, which is similar to ISO Res. 167 but does not require the maximum righting arm to occur between 30 and 40 deg heel angles. IMO is investigating the problem with a view to setting up its own recommended Code for Safety of Offshore Supply Vessels.

Criteria associated with stowage of bulk dry cargoes are discussed in Subsection 5.11.

U.S. Navy criteria for all sizes of ships, discussed in the next sub-section, are based on efforts to make a physical assessment of the heeling moments expected due to various specific hazards, as well as considering the curve of righting moments.

Research also continues on true dymanic approaches to the problem of stability criteria, taking account of the rolling motion of ships under different sea conditions, making use of both ship motion theory and model experiments. (Kuo 1975; Kuo and Welaya, 1981; Amy, et al 1976). This includes the special case of capsizing in following seas (Paulling, 1974). New developments can be expected in the future, with IMO playing an important role in the evolution of stability criteria toward greater safety for both large and small craft.

Meanwhile, it is incumbent on the designer to ensure that the ship as designed meets all the applicable national and international stability requirements of the country in which the ship is to be certified. In addition, the designer should make sure that all anticipated operations of the ship are investigated for stability, even though not required for certification.

7.3 U.S. Navy Criteria. *(a). General.* U.S. Navy criteria are intended to ensure the adequacy of stability of all types and sizes of naval ships, as evidenced by sufficient righting energy to withstand various types of upsetting or heeling moments. The fundamental energy relationships have been discussed in Section 4.13, and this Subsection deals with specific criteria and factors used by the U.S. Navy. See NAVSEA (1975), Sarchin & Goldberg (1962), Goldberg & Tucker (1975). The Navy criteria are of value for designers of commercial, as well as naval ships, for although the limits were established through the particular experience of the Navy, the basic principles apply to all ships.

It is first necessary to establish the loading conditions for which the ship will be expected to withstand the upsetting forces (Section 3.8 discusses conditions of loading). It is important to note at this point that stability righting arm curves should be prepared for all conditions and corrected for free surface, as discussed in 5.3. The curve having the least area is then selected for use. Each heeling arm curve is superimposed on this graph so that the resulting plot has the appearance of Fig. 37.

The various types of upsetting moments considered by the Navy will now be discussed, beginning with the special case of beam winds.

(b) Beam Winds Combined with Rolling.

1. Beam winds and rolling are considered simultaneously, since a fairly rough sea is to be expected when winds of high velocity exist. If the water were still, the ship would require only sufficient righting moment to overcome the heeling moment produced by the action of the wind on the ship's "sail area." When the probability of wave action is taken into account, an additional allowance of dynamic stability is required to absorb the energy imparted to the ship by the rolling motion.

2. Wind velocities. The wind velocity which an intact ship is expected to withstand depends upon its service. Specific wind velocities to be assumed should be obtained from the appropriate regulatory body that governs the ship design, if available. If not, U.S. Navy values may be used, as given in Table 15.

Table 15—Wind Velocities Assumed by U.S. Navy

Service	Minimum wind velocity for design purposes (knots)
1. Ocean	
(a) Ships which must be expected to weather full force of tropical cyclones. This includes all ships which will move with the amphibious and striking forces	100
(b) Ships which will be expected to avoid centers of tropical disturbances	80
2. Coastwise	
(a) Ships which will be expected to weather full force to tropical cyclones	100
(b) Ships which will be expected to avoid centers of tropical disturbances, but to stay at sea under all other circumstances of weather	60
(c) Ships which will be recalled to protected anchorages if winds over Force 8 are expected	60
3. Harbor	60

3. Wind heeling moment. A general formula that is used to describe the unit pressure on a ship due to beam winds is as follows:

$$P = C\rho_a \frac{V_w^2}{2g} \qquad (17)$$

INTACT STABILITY

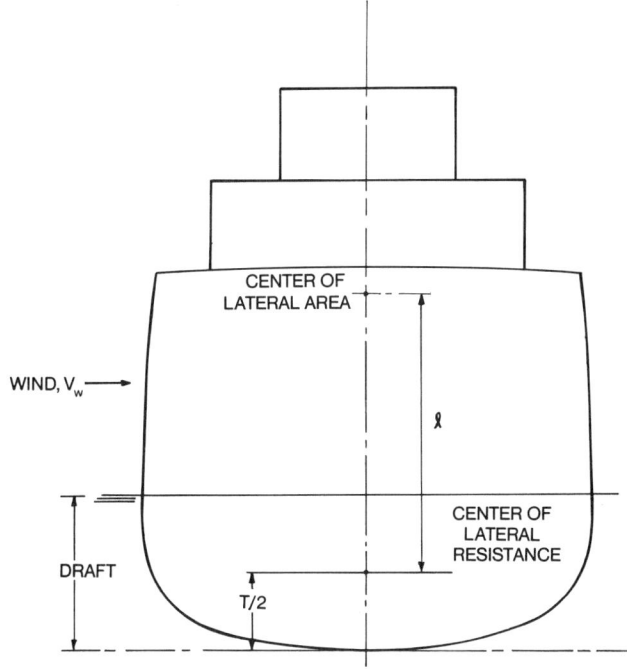

Fig. 43 Heeling effect of wind

where

C = dimensionless coefficient for ship type
ρ_a = air density (mass per unit volume)
g = gravity acceleration
V_w = wind velocity

There is considerable uncertainty regarding the value of C. Similarly, the variation of wind velocity at different heights above the waterline is not universally agreed upon.

The most widely used expression for P, in English unit, (lb per sq ft) is $P = 0.004 V_w^2$ (where V is in knots). Heeling arm $H.A.$ due to wind,

$$H.A. = \frac{0.004 V^2 A \ell \cos^2 \phi}{2240 \times W} \text{ ft} \qquad (18)$$

where (see Fig. 43):

A = projected sail area, sq ft
ℓ = lever arm (vertical distance) from center of lateral resistance of underwater hull (usually assumed at half draft) to centroid of hull and superstructure lateral area, ft
V_w = wind velocity, nominal (knots)
ϕ = angle of inclination
W = displacement, long tons.

The formula using S.I. units (except V in knots) is

$$H.A. = \frac{0.0195 \, V^2 A \ell \cos^2 \phi}{1000 \, \Delta} \qquad (19)$$

where A is in m², ℓ in meters (m) and Δ in metric tons (t).

It is recognized that as the ship heels to large angles, the use of $A\ell \cos^2 \phi$ is not rigorous since the exposed area varies with heel and is not a cosine function. However, other effects are also ignored and the above formula may be used to obtain gross effects. Recent wind tunnel tests at the David W. Taylor Naval Ship Research and Development Center on models representing different ship types and superstructure forms have indicated that an average coefficient of 0.0035

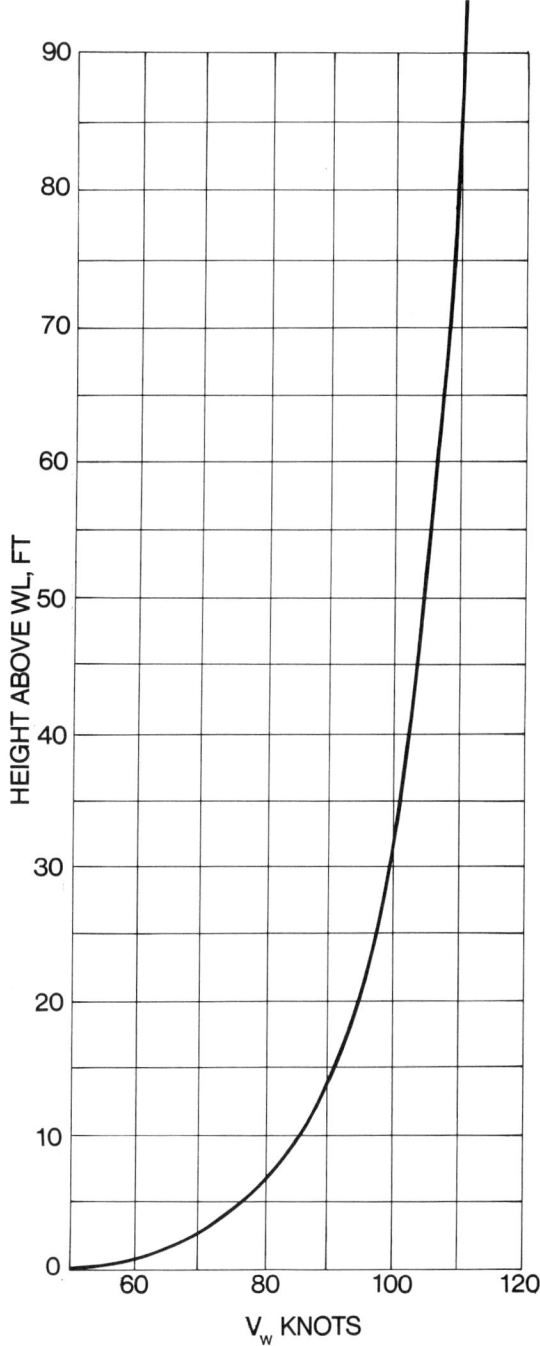

Fig. 44 Wind velocities at various heights above W.L. for a nominal 100-knot wind at 33 ft above W.L. (English units)

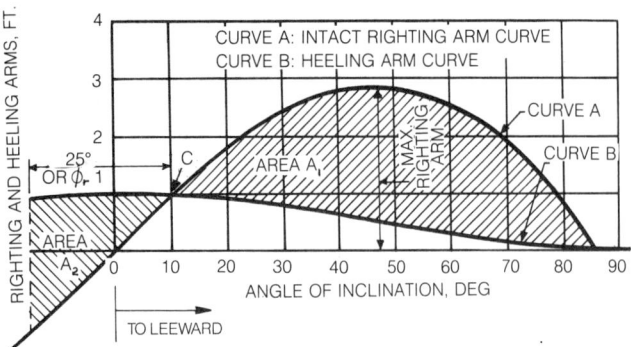

Fig. 45 U.S. Navy criterion of stability in wind and waves

rather than 0.004 (0.017 vice 0.0195 for metric units) should be used in the foregoing formula, which assumes a constant wind gradient. In order to account for actual full-scale velocity gradient effects, an average coefficient value of 0.004 in conjunction with the wind velocity-gradient curve, Fig. 44, is used by the U.S. Navy.

Fig. 44 is a composite of various values described in the literature. The nominal velocity is assumed to occur at about 10 m (33 ft) above the waterline. Table 16 has been prepared for a nominal 100-knot wind as an aid in determining wind heeling moments (from NAVSEA, 1975).

4. Adequate stability. The criteria for adequate stability when encountering adverse wind and wave conditions are based on a comparison of the righting arm and heeling arm curves, Fig. 45. (See NAVSEA, 1975.)

Stability is considered satisfactory if:

- The heeling arm at the intersection of the righting arm and heeling-arm curves (Point C) is not greater than six-tenths of the maximum righting arm; and
- Area A_1 is not less than 1.4 A_2 where area A_2 extends 25 deg or ϕ_r (if roll angle is determined from model tests) to windward from Point C.

The foregoing criteria for adequate stability with respect to adverse wind and sea conditions are based on the following considerations:

- A wind heeling arm in excess of the ship's righting arm would cause the ship to capsize in calm water. The requirement that the heeling arm be not greater than six tenths of the maximum righting arm is intended to provide a margin for gusts, and for inaccuracies resulting from the approximate nature of the heeling-arm calculations.
- In the second criterion, the ship is assumed to be heeled over by the wind to Point C and rolling 25 deg or ϕ_r from this point to windward, the 25 deg being an arbitrary but reasonable roll amplitude for heavy wind and sea conditions. Area A_2 is a measure of the energy imparted to the ship by the wind and the ship's

righting arm in returning to point C. The margin of 40 percent in A_1 is intended to take account of gusts and for calculation inaccuracies. Energy losses mentioned in 4.13 are ignored, and it is assumed that no downflooding occurs.

Upsetting moments caused by lifting weights overside, personnel crowding and high-speed turning will now be discussed, followed by U.S. Navy combined criteria for all three.

(c) Lifting of Heavy Weights Over the Side.

1. Effect of lifting weights. Lifting of weights will be a governing factor in required stability only on small ships which are used to lift heavy items over the side. Lifting of weights has a double effect upon transverse stability. First, the added weight, which acts at the upper end of the boom, will raise the ship's center of gravity and thereby reduce the righting arm. The second effect will be the heel caused by the transverse moment when lifting over the side.

2. Heeling Arms. For the purpose of applying the criteria, the ship's righting-arm curve is modified by correcting VCG and displacement to show the effect of the added weight, assumed to be at the end of the boom. The heeling arm curve is calculated by the formula:

$$\text{Heeling arm} = \frac{wa \cos \phi}{W} \qquad (20)$$

where

w = weight of lift
a = transverse distance from centerline to end of boom
W = displacement, including weight of lift
ϕ = angle of inclination, deg.

(d) Crowding of Personnel to One Side.

1. Effect of crowding of personnel. The movement of personnel will have an important effect only on smaller ships that carry a large number of personnel. The concentration of personnel on one side of a small ship can produce a heeling moment which results in a significant reduction in residual dynamic stability.

2. Heeling arms. The heeling arm produced by the transverse movement of personnel is calculated by (English or metric):

$$\text{Heeling arm} = \frac{wa}{W} \cos \phi \qquad (21)$$

where

w = weight of personnel
a = distance from centerline of ship to center of gravity of personnel
W = displacement
ϕ = angle of inclination, deg.

In determining the heeling moment produced by the personnel, it is assumed that all personnel have moved

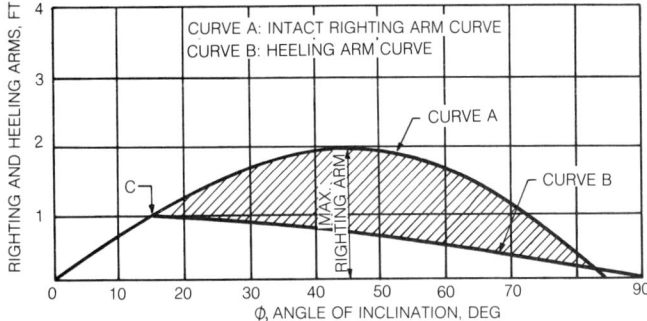

Fig. 46 U.S. Navy criteria of stability for weights overside, personnel crowding

to one side as far as possible. Each person occupies 0.2 m² (2 sq ft) of deck space.

(e) *High-Speed Turning.*
1. *Heeling arms produced by turning.* The centrifugal force acting on a ship during a turn may be expressed by the formula (English or metric):

$$\text{Centrifugal force} = \frac{WV^2}{gR}$$

where

W = displacement of ship (weight)
V = linear velocity of ship in the turn
g = acceleration due to gravity
R = radius of turning circle

The lever arm in conjunction with this force to obtain the heeling moment is the vertical distance between the ship's center of gravity and the center of lateral resistance of the underwater body. This lever will vary as the cosine of the angle of inclination. The center of lateral resistance is taken vertically at the half draft.

If the centrifugal force is multiplied by the lever arm and divided by the ship's displacement, an expression for heeling arm is obtained.

$$\text{Heeling arm} = \frac{V^2 a}{gR} \cos \phi \qquad (22)$$

where

a = distance between ship's center of gravity and center of lateral resistance (half draft) with ship upright
ϕ = angle of inclination, deg

For all practical purposes R may be assumed to be one half of the tactical diameter. If the tactical diameter is not available from model or full-scale data, an estimate is made.

2. *Criteria for adequate stability.* The U.S. Navy criteria for adequate stability for lifting weights, personnel crowding and high-speed turning are based on a comparison of the righting arm and heeling arm curves, Fig. 46.

Stability is considered satisfactory if:

- The angle of heel, as indicated by point C, does not exceed 15 deg.
- The heeling arm at the intersection of the righting arm and heeling arm curves (point C) is not more than six tenths of the maximum righting arm; and
- The reserve of dynamic stability (shaded area) is not less than four tenths of the total area under the righting-arm curve.

The criteria for adequate stability are based on the following considerations:

- Angles of heel in excess of 15 deg will interfere with operations aboard the ship and adversely affect safety and comfort of personnel.
- The requirements that the heeling arm be not more than six-tenths of the maximum righting arm and that the reserve of dynamic stability be not less than four tenths of the total area under the righting-arm curve are intended to provide a margin against capsizing. This margin allows for possible overloading and for possible inaccuracies resulting from the empirical nature of the heeling-arm calculations.

(f) *Topside Icing.*
1. *Effects.* The criterion for topside icing is not as definitive as the other criteria. The reason for this is the inability to estimate an upper limit for accumulation of ice. Once ice has started to form, it will continue to accumulate under unfavorable conditions and the only recourse is to institute ice-removal measures or leave the area. High winds are likely to occur during periods of icing and it is appropriate to consider combined icing and wind effects. A new ship of destroyer size, which is capable of withstanding a 100-knot beam wind without ice, can withstand a beam wind of only 80 knots with an ice accumulation of 200 tons. A cruiser type in service, which can withstand a 90-knot beam wind without ice, can withstand a beam wind of only 78 knots with an accumulation of 600 tons of ice. The foregoing ice weights correspond roughly to a 15-cm (6-in) coating on horizontal and vertical surfaces where ice would build up. An actual build-up of ice would of course be nonuniform, but the ice weights determined on the basis a uniform 15 cm coating may be used in estimating maximum beam-wind velocity for which the stability criterion will be met. For destroyer sizes and above, the criteria will be met for a 70-knot wind in combination with topside icing. For smaller ships, topside icing results in a more significant reduction in righting arms and the allowable beam-wind velocity is accordingly less. For example, a 59-m (194-ft) patrol ship, which can meet the wind criterion for a 75-knot beam wind without ice, will have to avoid beam winds in excess of 50 knots if there has been substantial ice accumulation. In the case of a smaller mine sweeper of 46 m (151 ft) 50 tons of topside ice reduces the maximum righting arm from 0.4 m (1.3 ft) to about

Table 16—Heeling Moments (ft-Tons) per Sq Ft for a Nominal 100-Knot Wind

Height above WL, ft	Center of lateral resistance below waterline, ft																	
	2	3	4	5	6	7	8	9	10	11	12	13	14	15	16	17	18	19
0–5	0.04	0.05	0.06	0.07	0.07	0.08	0.09	0.10	0.11	0.11	0.12	0.13	0.14	0.15	0.16	0.17	0.18	0.19
5–10	0.11	0.12	0.14	0.15	0.16	0.18	0.19	0.20	0.20	0.22	0.23	0.24	0.26	0.27	0.28	0.29	0.31	0.32
10–15	0.20	0.21	0.23	0.24	0.26	0.27	0.29	0.30	0.32	0.33	0.34	0.35	0.37	0.38	0.40	0.41	0.43	0.44
15–20	0.30	0.32	0.33	0.34	0.36	0.37	0.39	0.41	0.42	0.44	0.45	0.46	0.48	0.49	0.51	0.53	0.54	0.56
20–25	0.40	0.41	0.43	0.45	0.46	0.47	0.49	0.51	0.53	0.54	0.56	0.58	0.60	0.60	0.62	0.64	0.66	0.67
25–30	0.50	0.52	0.54	0.55	0.57	0.59	0.60	0.62	0.64	0.65	0.67	0.69	0.71	0.73	0.74	0.75	0.77	0.79
30–35	0.61	0.62	0.64	0.66	0.68	0.70	0.72	0.75	0.75	0.77	0.79	0.80	0.82	0.84	0.86	0.87	0.89	0.91
35–40	0.72	0.73	0.75	0.77	0.79	0.81	0.83	0.85	0.86	0.88	0.90	0.92	0.94	0.96	0.98	1.00	1.01	1.03
40–45	0.83	0.85	0.86	0.88	0.90	0.92	0.94	0.96	0.98	0.99	1.01	1.03	1.05	1.07	1.09	1.11	1.13	1.15
45–50	0.95	0.97	0.98	1.00	1.02	1.04	1.06	1.08	1.10	1.12	1.13	1.15	1.18	1.20	1.22	1.24	1.26	1.27
50–55	1.06	1.08	1.10	1.12	1.14	1.16	1.18	1.20	1.22	1.24	1.26	1.27	1.30	1.31	1.34	1.36	1.38	1.40
55–60	1.18	1.20	1.22	1.24	1.26	1.27	1.30	1.32	1.34	1.36	1.38	1.39	1.41	1.43	1.46	1.48	1.50	152
60–65	1.30	1.32	1.34	1.36	1.38	1.39	1.41	1.44	1.46	1.48	1.50	1.52	1.53	1.56	1.58	1.60	1.62	1.64
65–70	1.41	1.44	1.46	1.48	1.50	1.52	1.54	1.56	1.58	1.60	1.62	1.65	1.66	1.68	1.70	1.72	1.75	1.77
70–75	1.54	1.56	1.58	1.60	1.62	1.65	1.66	1.68	1.70	1.73	1.75	1.77	1.79	1.80	1.83	1.85	1.87	1.89
75–80	1.66	1.67	1.70	1.72	1.74	1.76	1.79	1.80	1.82	1.84	1.87	1.89	1.91	1.93	1.95	1.97	1.99	2.01
80–85	1.79	1.80	1.82	1.84	1.87	1.89	1.91	1.93	1.95	1.97	1.99	2.02	2.04	2.06	2.07	2.10	2.12	2.14
85–90	1.91	1.92	1.94	1.96	1.99	2.01	2.03	2.06	2.07	2.09	2.11	2.14	2.16	2.18	2.20	2.22	2.24	2.26
90–95	2.02	2.05	2.06	2.08	2.11	2.13	2.15	2.18	2.19	2.21	2.23	2,26	2.28	2.30	2.32	2.34	2.36	2.39
95–100	2.14	2.17	2.18	2.20	2.23	2.25	2.27	2.29	2.32	2.33	2.35	2.38	2.40	2.42	2.45	2.46	2.48	2.51

NOTE: To obtain the total heeling moment from this table, Navy procedure is as follows:
(a) Divide sail area into 5-ft layers, starting from waterline.
(b) Determine number of square feet in each layer.
(c) Multiply area of each layer by appropriate figure from table and add products. This is heeling moment for a 100-knot wind.
(d) For wind velocities other than 100 knots, multiply moment by $(V/100)^3$.

0.2 m (0.7 ft) with a reduction in range from 90 to 55 deg. The maximum allowable wind is reduced from 85 to about 40 knots.

2. Design criteria. The design approach to topside icing is to determine the maximum allowable beam winds combined with icing for a ship whose stability has been established from other governing criteria. The design would be considered satisfactory if the allowable wind at time of icing was in excess of winds that are likely to be encountered in the intended service.

The publication *Climatological and Oceanographic Atlas for Mariners* Volume I, North Atlantic Ocean (Commerce Dept., 1959) provides a guide for expected winds in combination with icing. Winds up to Beaufort 9 (41-47 knots) are very likely to occur off the west coast of Greenland. Heavy to severe icing is expected to occur from 5 to 15 percent of the time in February based on simultaneous occurrence of winds equal to or greater than 34 knots and air temperatures equal to or less than 28°F. The Coast Guard recently reported that one of its ships on station in the same area experienced 70-knot winds with severe ice accumulation.

A guide for estimating weight of ice accumulation is shown in Table 17, prepared for *Wind*-class icebreakers (from NAVSEA, 1975). The Torremolinos Convention for the Safety of Fishing Vessels (IMO, 1977) contains recommendations of minimum requirements for icing of fishing vessels, with specific guidelines as to amounts of ice accumulation to be assumed. This is reproduced in the circular, Coast Guard (1976).

The University of Alaska has published comparative icing charts developed from experience in the Gulf of Alaska and the Aleutian Islands which show ice accumulation as a time dependent phenomenon. (Alaska, 1980; Wise and Comiskey, 1980).

7.4 Stability Criteria for Certain Ship Types.

(a) Special hazards. While the stability curve is a good indication of a ship's stability and its resistance to capsizing, certain ship types have been known to capsize under certain wave and following sea conditions even though their stability curves indicated good stability. Considerable experimental work has been done in this field, as reported by Paulling (1974) and by Amy, et al (1976). Such work provides plausible explanations of the mechanisms that cause unexpected capsizings.

Aside from the above sea-motion related capsizings, fishing ship types have peculiar operating conditions which contribute to stability degradation. Among these are collection of water on the deck, free surface in the fish tanks and the effects of deck loads and nets suspended from a boom. It is also a common, although an unwise practice, to take on a very heavy load that reduces freeboard to dangerously low levels. IMO has published recommended criteria regarding fishing vessels (reproduced in Coast Guard, 1976).

(b) Stability of towboats. 1. Towboats may also be prone to sea-motion related capsizings. In addition, these ship types are characteristically designed with

low freeboard, which enhances the danger of taking on sea water through topside openings. Other hazards frequently experienced by tugs are the towline forces generated by the tug's own engines, called *self-tripping* and by the movement of the ship being towed, called *tow-tripping*. These special stability problems are discussed in the paper by Amy, et al(1976).

2. *Heeling arm.* The formula for calculating the transverse heeling arm curve for tow-line pull, used by the U.S. Coast Guard, is as follows (English units):

$$\text{Heeling Arm} = \frac{2N(SHP \times D)^{2/3} \times s \times h \times \cos\phi}{38W} \quad (23)$$

where:

N = number of propellers
SHP = shaft horsepower per shaft
D = propeller diameter, ft
s = effective decimal fraction of propeller slip stream deflected by the rudder—assumed to be equal to that fraction of the propeller circle cylinder that would be intercepted by the rudder if turned to 45 degrees. Use $s = 0.55$ if no other value has been determined.
h = vertical distance from propeller shaft centerline at rudder to towing bitts, ft
W = displacement, long tons
ϕ = angle of heel, degrees

3. *Criteria for adequate stability.* The U.S. Navy criteria for adequate stability are based on the angle of heel, and a comparison of the ship's righting-arm and the heeling-arm curve, Fig. 46.

Stability is considered satisfactory if:

- The angle of heel, as indicated by point C, does not exceed the angle at which unrestricted down flooding may occur, or 40 degrees, whichever is less. The limit on range is to provide a margin of safety in the event a watertight door or vent duct is open and could be a pathway for serious down flooding due to wave and heel action.
- The heeling arm at the interception of the righting arm and heeling arm curves (point C) is not more than six tenths of the maximum righting arm, and
- The reserve of dynamic stability (shaded area) is not less than four tenths of the total area under the righting arm curve.

7.5 Evaluation of Stability of Submarines. The foregoing principles apply to a surfaced submarine as well as to surface ships. There are some peculiarities of submarines, however, which should be mentioned.

The form of the hull of a submarine is such that the righting arms in the surfaced condition are positive at angles well beyond 90 deg, a condition that is seldom, if ever, found in surface ships. The only significant heeling moment to which a surfaced submarine is subjected results from wind and wave action. Unlike surface ships, all topside openings can be closed to prevent shipping of water during heavy rolling except for the

Table 17—Icing-up Chart for *Wind*-Class Icebreakers

Iced area of ship	Thickness of ice, in.									Area, in sq ft	$\overline{kg}$, ft	Dist. cent. ice to ship's CL, ft
	1	2	3	4	5	6	8	12	24			
	Tons of Ice											
Main dk fwd of bkwtr*	1.1	2.2	3.4	4.5	5.6	6.7	9.0	13.4	27.0	525	43.7	0
Main dk bkwtr to fr 68*	1.6	3.2	4.7	6.3	8.0	9.5	12.6	19.0	38.0	738	43.1	12
Main dk fr 66 to 127*	2.4	4.3	6.4	8.6	10.7	12.8	17.0	25.6	51.0	1000	41.1	20
Main dk fr 127 to aft*	3.6	7.3	10.9	14.6	18.2	21.8	29.1	43.6	87.0	1700	41.2	18
Lifelines, etc., fr 66 fwd*	0.6	1.3	2.0	2.6	3.2	3.8	5.1	7.6	15.3	300	44.0	15
Lifelines, etc., fr 66 aft*	1.0	2.1	3.1	4.1	5.2	6.2	8.2	12.4	24.7	480	44.0	20
Breakwater	0.6	1.3	1.9	2.6	3.2	3.8	5.1	7.6	15.3	300	44.0	0
5 in. mount (all sides)	0.9	1.7	2.6	3.4	4.3	5.1	6.4	10.2	20.4	400	48.0	0
Bhd (main dk) fr 66 fwd*	0.7	1.5	2.2	2.9	3.7	4.4	5.8	8.8	17.6	340	46.0	15
Bhd (main dk) fr 66 aft*	2.4	4.7	7.1	9.2	11.6	14.1	18.4	28.2	56.0	1100	44.0	18
Flight dk (plus netting)	6.4	12.8	19.2	25.6	32.2	38.4	51.2	77.0	144	3000	50.1	0
20MM (fwd & lifelines)*	0.5	1.0	1.5	2.0	2.6	3.7	4.1	6.1	12.0	240	50.2	9
01 deck*	1.3	2.6	3.8	5.1	6.4	7.7	10.2	15.4	31.0	600	50.1	20
Bhd (01 dk) fr 61 to 92*	1.0	1.2	2.9	3.9	4.8	5.8	7.7	11.5	23.0	450	53.0	10
Sky lookout (dk & bhd)*	0.8	1.5	2.3	3.1	3.9	4.6	6.2	9.2	18.0	360	57.0	15
Open bridge (40mm's)*	1.0	1.9	2.8	3.8	4.5	5.8	7.7	11.5	23.0	450	61.6	23
Bhd (open bridge)*	0.5	1.0	1.5	2.0	2.6	3.1	4.1	6.1	12.0	240	64.0	9
Top of pilot house	1.4	2.7	4.0	5.4	6.7	8.1	10.8	16.2	29.0	630	68.0	0
Main bat director	1.4	2.7	4.0	5.4	6.7	8.1	10.8	16.2	29.0	630	72.0	0
02 deck*	0.3	0.7	1.0	1.4	1.7	2.0	2.7	4.0	8.0	160	57.1	11
Misc fr 66 fwd*	0.2	0.4	0.6	0.8	1.0	1.2	1.5	2.0	4.0	80	43.0	10
Misc fr 66 aft*	0.4	0.9	1.3	1.7	2.2	2.6	3.5	5.2	11.0	200	46.0	15
Bhd (01 dk) fr 62 fwd*	0.6	1.2	1.8	2.4	3.0	3.6	4.8	7.2	14.0	280	53.0	15

* Indicates identical areas port and starboard.

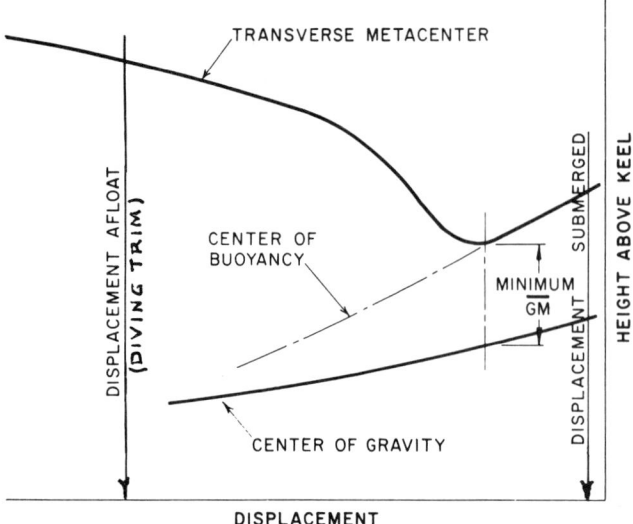

Fig. 47 Submarine stability while submerging

old diesel-powered submarines which must operate with open diesel engine manifolds. As a result, capsizing of an intact submarine is extremely unlikely. The major stability problem is rolling to very large angles with adverse effects on personnel and the operation of the ship.

The righting-arm curve for a submerged submarine is equal to the metacentric height, $\overline{GB}$, multiplied by the sine of the angle of inclination. Its maximum value, therefore, occurs at 90 deg. Except for the minor effect of shifting of liquids and loose items in the ship, the range of positive stability would be 180 deg. A submarine is subjected to only minor heeling moments when submerged. Therefore, there is no danger of capsizing an intact submerged submarine, provided the metacentric height has at least a small positive value.

During the period while a submarine is submerging or surfacing, its transverse stability is less than when either surfaced or submerged, because of the free liquid in the main ballast tanks. On the surface, there is only a small free-surface effect in the main ballast tanks, caused by the small quantity of *residual water* above the tops of the flood openings that cannot be blown. When the ship is submerged, there is no free surface in the main ballast tanks, since they are completely full.

An approximate evaluation of stability during submerging and surfacing can be made by a series of calculations of displacement, height of the center of gravity of the ship and the free-surface effect, assuming that the main ballast tanks are filled to successively greater depths. The only variables in these calculations are the weight, vertical center of gravity, and vertical moment of free surface of the water in the main ballast tanks. The effect of the water in the main ballast tanks at each assumed level is added to the weight, vertical moment and vertical moment of free surface of the ship in the surfaced condition, after the vertical moment of free surface of the residual water has been deducted.

The results of these calculations, consisting of the displacement and height of the center of gravity of the ship, adjusted for free-surface effect, are plotted as shown in Fig. 47, together with the height of the metacenter, and the minimum metacentric height is determined as the smallest vertical distance between the two curves. Stability is satisfactory if the metacentric height has a small positive value, since the nature of the righting-arm curve during submergence is such that positive values will be developed at small angles of heel when the metacentric height is zero.

The height of the metacenter drops, as displacement is increasing, from its value in the surfaced condition, shown to the left of Fig. 47, until it meets the curve of the height of the center of buoyancy. The vertical separation of these two curves, $\overline{BM}$, is equal to I_T/∇, which has been reduced to zero as the ship submerges, owing to the disappearance of the waterplane when the hull is submerged.

The assumption in these calculations that all main ballast tanks are filled to the same waterline is somewhat unrealistic because the actual levels in the various tanks depend on the area of the flood openings, the shape of the individual tanks, and the depths to which the openings are submerged. The flood openings are sized to flood the forward tanks faster than the after tanks, to produce a down angle on the ship and expedite submerging. In addition, any rolling of the ship will increase the depth to which the tanks on the low side are submerged, causing them to fill faster than those on the high side.

When the main ballast tanks are arranged in pairs, the moment of inertia of the individual port and starboard tanks is used, rather than the moment of inertia of the pair considered as a single tank, since there is no flow from one side to the other.

Section 8
Drafts, Trim and Displacement

8.1 Trim. Trim, as used in this section, defines the longitudinal inclination of the ship. Trim may be expressed as the angle between the baseline of the ship and the waterplane, but it is usually expressed as the difference in drafts at the bow and at the stern.

8.2 Center of Flotation. The center of flotation is the point in the ship's waterplane through which the axis of rotation passes when the ship is inclined, either transversely, longitudinally, or both. It is shown in Section 4 that, for longitudinal inclinations, this point is the centroid of the waterplane, and similar reasoning would apply to inclinations in any direction.

The center of flotation is useful in the determination of drafts for two reasons. When the ship is trimmed with no change in displacement, as when a weight is moved forward or aft, there is no change in draft at the center of flotation, if the change in trim is moderate, and, if the original waterline and the change in trim are known, the new waterline can be established. Also, if a small weight is added to the ship at the center of flotation, there is no change in trim because the increment of weight is added at the same distance from the initial position of the centers of gravity and buoyancy as the increment of buoyancy, and, since the two increments are equal, the two centers will move the same distance, maintaining equilibrium.

The longitudinal position of the center of flotation is plotted on the curves of form, as described in Chapter I. The curve may be labeled *Center of Flotation* or *Center of Gravity of Waterplanes*.

8.3 Moment to Trim One Cm. The formulas for moment to trim one cm, $\Delta \overline{GM_L}/100L$, and approximate moment to trim one cm, $\rho I_L/100L$ are derived in Section 3.7.

The trim produced by a moderate longitudinal moment can be calculated by dividing the moment by the approximate moment to trim one cm with sufficient accuracy for normal ships. There may be some unusual craft for which the difference between $\overline{GM_L}$ and $\overline{BM_L}$ is large enough to make the use of the approximate moment to trim unacceptable, particularly if there is a large longitudinal free-surface effect which affects the position of the center of gravity substantially.

The moment to trim one cm is usually plotted on the displacement and other curves. (See Chapter I.)

8.4 Tons per cm Immersion. The tons per cm immersion, which is the displacement of a layer of water one cm thick at the waterplane, is used to calculate the change in draft at the center of flotation caused by a moderate change in displacement. The increase or decrease in draft is equal to the change in displacement divided by the tons per cm immersion.

This function, $TPcm$, is plotted on the displacement and other curves, or curves of form. (Chapter I.)

8.5 Determination of Drafts from Weight and Location of Center of Gravity. There are two methods by which the drafts forward and aft may be obtained when the displacement and longitudinal position of the ship's center of gravity are known. The first involves the displacement and other curves, and is used when the trim is moderate. The second is based on the Bonjean curves, and is used when the trim is so large that the approximations used in the first method are not acceptable.

The steps in determining the forward and after drafts from the curves of form (Chapt. I, Fig. 23) are as follows:

(a) The even-keel draft, or draft at *LCF*, is read from the displacement curve at the value indicated by the weight estimate.

(b) At this draft, the longitudinal location of the center of buoyancy, the longitudinal location of the center of flotation and the approximate moment to trim one cm are read from the appropriate curves of form.

(c) Assume, for the moment, that the ship's center of gravity is at the longitudinal position of the center of buoyancy as read from the displacement and other curves. If this were the case, the ship would be floating at even keel, and the drafts forward and aft would be equal to the draft read from the displacement curve. If, as is usually the case, the center of gravity obtained from the weight estimate is not at the longitudinal position of the center of buoyancy, there is a trimming moment, M_T equal to the weight of the ship multiplied by the distance, parallel to the keel, from the center of buoyancy to the center of gravity.

(d) The trim, in cm, produced by the moment, M_T, is calculated by dividing M_T by the approximate moment to trim one cm. This is the difference between the forward and after drafts, assuming the drafts are equidistant from ⊗. The trim will be by the bow if the center of gravity is forward of the center of buoyancy in the even-keel attitude; otherwise the trim will be by the stern.

(e) The slope of the trimmed waterline with respect to the even-keel waterline is determined by dividing the trim by the length between perpendiculars.

(f) The draft at the center of flotation will be equal to the even-keel draft read from the displacement and other curves in step (a), since the trimmed waterline will intersect the even-keel waterline at this point. The draft at either perpendicular will be equal to the draft at the center of flotation plus or minus the product of the slope of the trimmed waterline and the distance from the center of flotation to that perpendicular.

As an illustration of this process, the following example is presented. Given:

Length between perpendiculars, $L = 161$ m
Displacement, S.W. $= 19,000$ t
Even keel draft $= 8.32$ m
LCG from amidships $= -3.66$ m
At 8.32m draft:
LCB from amidships $= -2.42$ m
LCF from amidships $= -5.73$ m
$M_T = 214$ t-m

Trimming moment,

$$19,000\,(-3.66 + 2.42) = -23,560 \text{ t-m}$$

$$\text{Trim} = \frac{-23,560 \text{ t-m}}{214 \text{ t-m}} = -110 \text{ cm}$$

Slope of trimmed $WL =$

$$\frac{\text{Trim}}{L} = \frac{-110}{161 \times 100} = -0.0068$$

Center of Flotation to after perpendicular,

$$(80.5 - 5.73) = 74.77 \text{ m}$$

Draft aft $= 8.32 + 74.77 \times 0.0068$
$\quad\quad\quad\; = 8.32 + 0.51 \quad\quad = 8.83$ m

Draft fwd. $= 8.83 - \dfrac{110}{100} = 7.73$ m

The displacement and other curves (Fig. 23 of Chapter I) indicate that these drafts are drafts to the bottom of the keel.

In some cases, the moment to trim one cm found on the displacement and other curves may be based on the length between draft marks rather than the length between perpendiculars, in which case the slope of the trimmed waterline is equal to the trim divided by the length between the draft marks.

The foregoing method for determining drafts involves two approximations which are sufficiently accurate for a small trim but become less accurate as the trim increases. One is the assumption that the trimmed and even-keel waterlines intersect at the center of flotation of the even-keel waterline. The other is that there is no change in the moment to trim one cm as the ship is trimmed. If the drafts must be determined very accurately, or if the trim is more than about 1/150 of the ship's length, or if the trim produces a marked change in the shape of the waterplane, the more rigorous method described subsequently is appropriate.

This method involves finding, by trial and error, using the Bonjean curves, the forward and after drafts that correspond to the given displacement and produce a center of buoyancy in the same longitudinal position as the given center of gravity. The Bonjean curves and the method of applying them to determine the displacement and center of buoyancy are described in Chapter I. Computer programs are available to perform these calculations at no significant cost once the hull form configuration has been entered into the computer for other calculations.

The initial trial may be based on the foregoing calculation, in which the drafts are estimated from the displacement and other curves. The displacement and center of buoyancy found from the Bonjean curves on the first trial, when compared to the given values, will indicate the direction in which the draft at the center of flotation should be varied and whether the trim should be increased or decreased for the second trial. The tons per cm immersion, when divided into the difference between the calculated and given displacements, will indicate the amount by which the draft at the center of flotation should be changed. The moment to trim one cm divided into the product of the displacement and the distance from the calculated to the given position of the center of buoyancy, will indicate the appropriate change in trim for the second trial. Satisfactory results are usually obtained on the third trial, if not before. This method is as accurate, for a trimmed waterline, as the displacement and other curves are for an even-keel waterline, since the methods are essentially the same.

8.6 Determining Displacement and Center of Gravity from Drafts.

(a) Methods. The displacement and the longitudinal location of the ship's center of gravity can be determined if the drafts forward and aft are known. If the trim is large, use of the Bonjean curves, as described in Chapter I, is appropriate. For moderate trim, with no abrupt changes in the waterplane between the trimmed and corresponding even-keel waterlines, the displacement and other curves may be used with less effort.

When the displacement and other curves are used, the displacement may be obtained by either of two methods, one of which depends upon the availability of the function Increase (or Decrease) in Displacement per cm of Trim Aft, as shown in Fig. 23 of Chapt. I. The derivation of this function is illustrated in Fig. 48, which shows a ship with a trim by the stern, the midship perpendicular, and the center of flotation for an even-keel waterline. For moderate trim, the displacement is equal to that under an even-keel waterline passing through point F and can be read from the displacement curve at draft T_1. If the draft readings are taken at the perpendiculars or at two sets of draft marks nearly equidistant from amidships, it is convenient to enter the displacement curve with the mean of the draft readings, T_2, read the displacement, and add (or subtract) the precalculated correction representing the difference in displacement at the even-keel drafts T_1 and T_2. The value of this correction is simply the difference in drafts T_1 and T_2 (in cm) multiplied by the TPcm. If T_1 and T_2 are in meters, this is

$$100\,(T_1 - T_2)\,TP\text{cm}.$$

The value of $(T_1 - T_2)$, or h in Fig. 48, may be obtained conveniently from similar triangles,

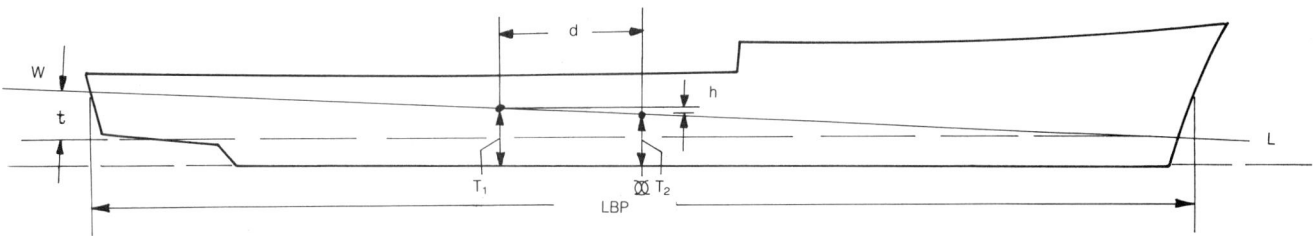

Fig. 48 Trimmed waterline and center of flotation

$h/d = t/L$, or $h = d\,t/L$

Hence, the change in displacement becomes

$$100\,d\,(t/L)\,TPcm$$

For $t = 1$ cm $= 1/100$ m, the change in displacement per cm of trim is $(d/L)\,TP$cm.

If this function does not appear on the displacement and other curves, the procedure is to enter the displacement and other curves at the mean of the forward and after drafts and find the location of the center of flotation. Then, the draft at the center of flotation is found from the draft reading on one set of marks, the slope of the trimmed waterline, and the distance of the center of flotation from the draft reading. This procedure is advantageous if the draft readings are not substantially equidistant from amidships, even though the value of the increase (or decrease) in displacement per cm trim aft is available, since the same effort is involved in finding the draft amidships as in finding the draft at the center of flotation.

In cases where the center of flotation is at a considerable distance from amidships, resulting in an appreciable difference between T_1 and T_2, and where the values of the other functions, such as transverse metacenter above base line, longitudinal center of buoyancy or moment to trim one cm are changing rapidly with change in draft, more accurate values of these functions will be obtained if they are read at T_1 rather than at T_2, since they will correspond to the draft at which the displacement is equal to the actual displacement.

After the displacement has been found, the center of gravity may be located as follows:

1. The location of the center of buoyancy and the moment to trim one cm at the even-keel draft T_1, or, as an approximation, T_2, are read from the displacement and other curves.

2. The trim between the perpendiculars is found by taking the difference between the drafts at the forward and after perpendiculars, or, if the draft marks are not located at the perpendiculars, by multiplying the difference in the readings by the ratio of the length between perpendiculars to the distance between the marks.

3. The ship is assumed to be floating in equilibrium on an even keel at the displacement just found, in which case the center of gravity would be in the same longitudinal position as the center of buoyancy. If we assume that the given trim was produced by movement of a weight already aboard, the moment generated by the weight movement would be equal to the trim multiplied by the moment to trim one cm. The movement of the center of gravity would be equal to the moment produced by the weight movement divided by the ship's displacement.

4. The location of the center of gravity in the trimmed condition is found from its known position with the ship on even keel, at the center of buoyancy, and the distance it moved. If the trim is by the bow, the center of gravity would have moved forward; otherwise it would have moved aft.

(b) *Numerical examples.* Illustrative examples of the above procedures are shown below.

1. To determine displacement.

Given: $LBP = 161$m

Draft at fwd perp. $= 7.45$m

Draft at aft perp. $= 8.38$m

Mean Draft $= \dfrac{7.45 + 8.38}{2} = 7.92$m

Trim $= 8.38 - 7.45 = -0.93$m

At the 7.92m WL (Curves of Form):

Displacement $= 17,960$t

Increase in displacement per cm trim aft
$= 0.886$t

Increase in displacement

$$0.93 \times 100 \times 0.886 = 82.4\text{t (say 82)}$$

Corrected displacement $= 18,042$t

OR

Given: Mean draft $= 7.92$m

LCF at 7.92m draft from ⊗ $= -5.3$m

Trim by stern $= -0.93$m

Then slope of trimmed $WL = \dfrac{-0.93}{161} = -.0058$

Draft at center of flotation $= 7.92 + 5.3$m $\times$ $.0058 = 7.95$m

Displacement at 7.95m $WL = 18,042$t

2. To Find The Longitudinal Center of Gravity.

At the 7.95m WL:

LCB from ⊗ $= -2.26$m

Moment to trim one cm $= 207$t-m

Trimming moment
$= -0.93\text{m} \times 207 \text{ t-m} \times 100 = -19{,}251 \text{ t-m}$

Shift of center of gravity

$$\frac{-19{,}251}{18{,}042} = -1.05\text{m}$$

Location of CG from ⊗ $=$
$-2.26 - 1.05 = -3.31$m

8.7 Determining Drafts after Change in Loading. When a change in loading is contemplated and the existing drafts are known, there are two basic methods for determining the drafts after the change is made. The first, which is appropriate when the changes in draft and trim are small and no large initial trim is present, involves determining the differences in drafts before and after the change. The second, which is suitable when the changes or the trim, or both, are large, involves finding the displacement and the location of the center of gravity after the change is made.

When the first method is used, the change in loading is reduced to a single equivalent item by adding, algebraically, the weights and the longitudinal moments of the individual items, and dividing the total moment by the total weight to find the distance of the center of gravity of the equivalent item from the reference plane being used. Then making use of the fact that if a weight is added at the center of flotation, there is no change in trim, it is assumed that the equivalent item is added at the longitudinal position of the center of flotation. This would change the draft an amount equivalent to the added weight divided by the tons per cm immersion, at each end of the ship, which is generally called the *parallel sinkage*. It is next assumed that the equivalent item is moved from the center of flotation to its actual position. This would generate a trimming moment equal to the weight of the equivalent item times the distance, parallel to the keel, from the center of flotation to its actual location. The change in trim would be equal to this trimming moment divided by the moment to trim one cm. If it is assumed, for example, that the weight is added forward of the center of flotation so that the trim will be by the bow, the increase in draft forward will be equal to the parallel sinkage plus the change in trim multiplied by the ratio of the distance of the center of flotation abaft the forward perpendicular to the length between perpendiculars. The change in draft aft, which may be negative, will be equal to the parallel sinkage minus the change in trim multiplied by the ratio of the distance of the center of flotation forward of the after perpendicular to the length between perpendiculars.

The following example illustrates the procedure. Start with the ship given in Section 8.6. Assume 300 tons are to be added 61m forward of amidships. From the example in 8.6, at the 7.95m WL:

Tons per cm $= 26.5$t

LCF from ⊗ $= -5.3$m

Moment to trim 1 cm $= 207$ t-m

Parallel Sinkage $=$

$$\frac{300\text{t}}{26.5 \text{ t/cm}} = 11.32 \text{ cm}$$

Center of gravity of added weight, from the center of flotation, $(61 + 5.3) = +66.3$m

Trimming moment $=$
$300 \times 66.3 = +19{,}890$ t-m

Change in trim $=$

$$\frac{+19{,}890 \text{ t-m}}{207 \text{ t-m/cm} \times 100} = +0.96\text{m}$$

Center of flotation:

From fwd perpendicular $(80.5 + 5.3) = 85.8$m

From aft perpendicular $(80.5 - 5.3) = 75.2$m

Change in draft due to trim:

$$\text{Forward} \left(0.96\text{m} \times \frac{85.8\text{m}}{161\text{m}}\right) = 0.51\text{m}$$

$$\text{Aft} \left(0.96\text{m} \times \frac{75.2\text{m}}{161\text{m}}\right) = 0.45\text{m}$$

Draft forward $= (7.45 + 0.11 + 0.51) = 8.07$m
Draft aft $= (8.38 + 0.11 - 0.45) = 8.04$m

By the second method, the initial displacement and location of the center of gravity of the ship are determined from the drafts as described in 8.6, the weights and moments of the changes added algebraically to those of the ship in the initial condition, and the total longitudinal moment divided by the total weight to find the center of gravity of the ship after the changes are made. From this information, the drafts forward and aft in the final condition may be calculated by one of the methods described in Section 8.5.

8.8 Navigational Drafts. The navigational draft of a ship in any condition is the draft to the bottom of the greatest projection below the waterline of the hull or any appendage. This draft determines the depth of water required for safe navigation. The navigational

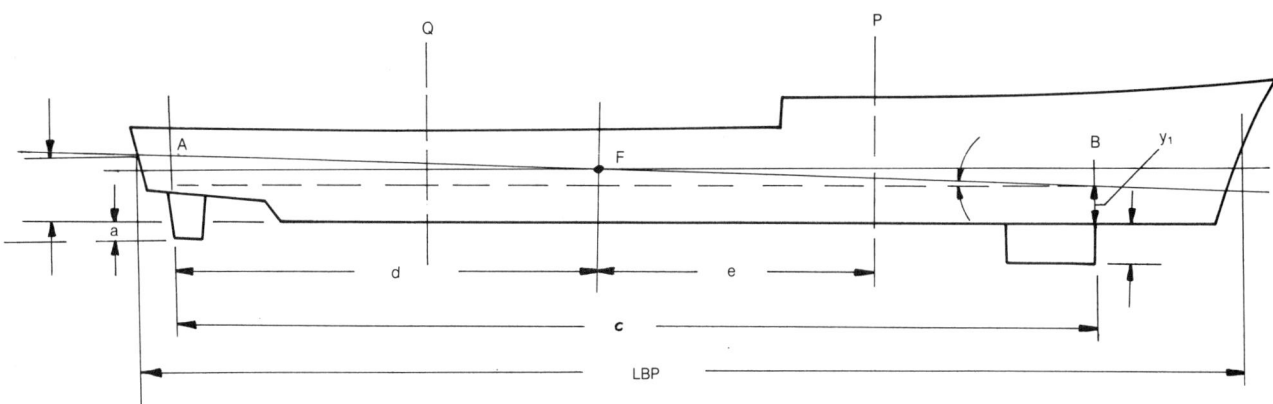

Fig. 49 Projections determine navigational drafts

draft may be increased somewhat over its static value due to settling of the ship caused by forward motion.

It is sometimes necessary for a ship to pass through a channel or enter a drydock in which the depth of water is less than the ship's normal navigational draft. This situation requires a study of the possible adjustments of the load to minimize the navigational draft. In extreme cases, it may be desirable to remove, or, in the case of an uncompleted ship, to omit some items of light ship weight. The problem may be complicated by the projection of appendages below the keel.

Fig. 49 shows a ship on which the rudder projects below the line of the keel a distance a, and an appendage forward projects a greater distance b. The navigational draft will occur either at A or B, depending upon the direction and amount of trim. For any given displacement, the navigational draft will be minimized when the drafts at A and B are equal. The slope of the trimmed waterline which will produce this condition is equal to $(b - a)/c$ with the ship trimmed by the stern, which is derived as follows:

Let the fwd. navigational draft $= y_1 + b$

Let the aft navigational draft $= y_2 + a$

For equal drafts at A and B,
$$y_2 + a = y_1 + b, \text{ or}$$
$$y_2 - y_1 = b - a$$

Then
$$\tan \theta = \frac{y_2 - y_1}{c} = \frac{b - a}{c}$$

The trim for minimum navigational draft will be the slope of the waterline multiplied by the length between perpendiculars (L), or $\dfrac{(b - a) L}{c}$.

Assume that the drafts of the ship in some condition of loading are known, and the problem is to reduce the navigational draft to a certain figure, based on the channel depth. Assume that the draft at the rudder in Fig. 49 exceeds that at the forward appendage. In this case, there is a plane, P, located in the forward portion of the ship, at which an addition or removal of weight will have no effect on the draft at the rudder. If a weight is added in plane P, the effect on the after draft may be considered, as discussed in Section 8.7, to consist of parallel sinkage as if the weight were added at the center of flotation, and the effect of the trim produced by moving the weight from the center of flotation to plane P. If these two effects are equal and in opposite directions, there will be no change in draft at the rudder. This may be expressed as follows.

$$\frac{w}{T P \text{cm}} = \frac{w\, e}{MT \text{cm}} \frac{d}{L}$$

where: $w =$ added weight

Then the plane P can be located by solving for e,

$$e = \frac{MT\text{cm}\, L}{d\, TP\text{cm}}. \qquad (24)$$

This equation applies not only to the rudder, but to any point on the ship, if d is the distance from the point at which no change of draft is desired to the center of flotation.

If the weight were removed rather than added, these equations would still apply, since the signs of both sides would be reversed. If the moment to trim one cm is based on the length between draft marks, rather than the length between perpendiculars, the length between draft marks would be substituted for L in the equations.

Having determined the optimum trim and the location at which weight changes have no effect on navigational draft, we can reduce the navigational draft by removing weight aft of plane P, adding weight forward of plane P, or both, until the optimum trim is attained. If the draft is still excessive, further reduction can be obtained only by changes which are equivalent to the removal of weight from the portion of the ship between plane P and the corresponding plane Q

in the after part of the ship in which changes have no effect on the draft at the forward appendage. This does not preclude the use of weight additions as a means for further reduction of navigational draft, but any addition must be made in conjunction with a removal of greater magnitude, such that the center of gravity of the equivalent removal will fall between planes P and Q.

In selecting items to be removed or the location at which an item is to be added, it should be realized that the effect of changes in weight fairly close to plane P or Q, although favorable, may be too small to justify the effort involved.

8.9 Hog and Sag. Deviation of the keel from a straight line, which may be of a permanent or temporary nature, or a combination of both, is known as *hog* when the keel is concave downward, or *sag* when the keel is concave upward. Permanent deflection may be caused by shrinkage associated with welding, while temporary deflections are caused by flexure due to the ship's loading or by thermal expansion. If hog or sag is appreciable, an adjustment should be made when determining displacement from draft readings.

Hog or sag is detected by calculating the draft amidships from the readings of the forward and after draft marks under the assumption that the keel is straight and comparing the calculated value with the actual reading of the midship draft marks. Hog or sag is measured by the difference in the calculated and actual drafts amidships. If the actual reading is greater than the calculated value, the ship is sagging; if the actual reading is less, the ship is hogging.

Assume that a ship with a straight keel line is floating at certain drafts forward and aft, that the waterline corresponding to these drafts is drawn on the ship's side, and that the displacement under this waterline is determined by one of the methods discussed in Section 8.6. Assume next that the ship is deflected in sag, at the same forward and after drafts. The line drawn on the ship will be curved and will be submerged by the amount of the sag amidships, and to lesser amounts toward the ends. The displacement calculated by the methods of Section 8.6 will not include the volume of the ship between the waterline drawn on the ship's side and the actual waterline. If the ship were deflected in hog, the calculations would include a volume, between these two waterlines, that is not submerged.

To find the displacement in a hogged or sagged condition, when the Bonjean curves are used, it is customary to enter the curves at the actual drafts forward, amidships and aft, and to determine the drafts at the intermediate stations by the assumption that the keel deflection is parabolic. When the displacement and other curves are used, the practice is to increase the draft at the center of flotation or at amidships, calculated as described in Section 8.5, by 75 percent of the sag, or to decrease it by 75 percent of the hog.

For a ship with a rectangular waterplane, the percentage would be 67, since the area under a parabola is two thirds of the area of the circumscribing rectangle, and this percentage would tend to increase as the waterline becomes finer at the ends. The 75 percent figure is an approximation which has been found to give good agreement with the calculations from the Bonjean curves, for the normal ship form.

8.10 Drag. Some ships, particularly tugs and small high-speed craft that have considerable power for their size, are designed to float deeper at the stern than at the bow in order to submerge their relatively large propellers. Craft designed to come ashore on beaches are also designed with a permanent stern drag so that on landing the forward part of the vessel will be aground while the after part with the propeller will be submerged. Drawings for these ships show the keel sloping downward toward the stern. The designed drag is the linear vertical dimension corresponding to the distance the keel line at the after perpendicular is below the keel line at the forward perpendicular.

When a ship has a designed drag, the waterlines used in preparing the displacement and other curves are not parallel to the keel, but are parallel to the waterline at which the ship is designed to float. The ship is considered to have zero trim when floating with the keel sloped to suit the designed drag, as discussed in Chapter I.

Great care should be used in relating drafts to displacement and center of buoyancy when the ship has a designed drag. There is no fixed convention for locating the baseline from which drafts are measured in drawing the displacement and other curves. It may be a horizontal line at the intersection of the bottom of the keel with the after perpendicular or the midship perpendicular, or the intersection of the molded base line with one of the perpendiculars. It may be completely below the ship. The draft marks installed on the ship may indicate drafts above any of these base lines, or above the bottom of the keel at the location of the draft marks. Errors of several feet of draft or several hundred tons of displacement may be made if the proper baselines are not used.

8.11 Reference Planes. Calculations of draft, displacement and longitudinal position of the center of gravity involve the use of transverse and horizontal reference planes which are a potential source of serious error. If, for example, the weight estimate and the displacement and other curves do not use the same transverse reference plane, it is necessary to take into account the separation of the two reference planes in finding the distance from the center of gravity to the center of buoyancy. A similar adjustment may be necessary in vertical measurements, as, for example, when the waterlines on the displacement and other curves are measured from the molded base line while the drafts are measured on draft marks which use the bottom of the keel as a reference. On wooden ships,

INTACT STABILITY

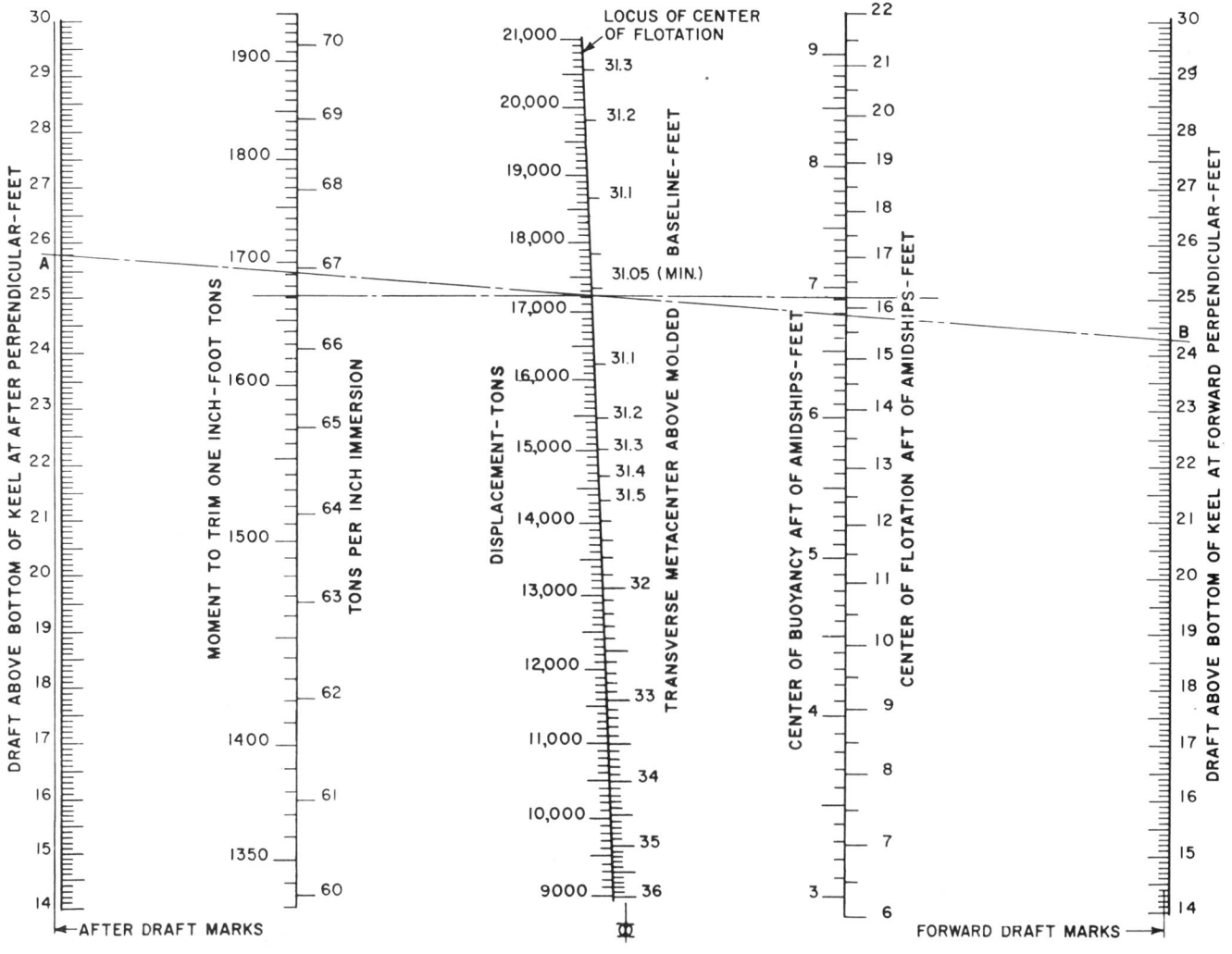

Fig. 50 Draft diagram (English units)

these reference planes may be as much as 0.3 m (1 ft) apart. Each calculation should be checked to ensure that the reference planes are consistent or that the proper adjustment has been made.

8.12 Draft Diagram. The processes of dealing with displacement, draft, and trim can be simplified by the preparation of a draft diagram, illustrated in Fig. 50, which has been developed from curves of form, such as Fig. 23 of Chapter I. The locus of the center of flotation for even-keel waterlines is plotted to a convenient scale, in proper relation to the longitudinal positions of the forward and after draft marks and the forward and after perpendiculars. If a scale of even-keel displacement is plotted along the locus of the center of flotation, the displacement corresponding to any combination of forward and after drafts, representing a moderate trim, may be found directly by connecting the draft readings, either at the draft marks or the perpendiculars, by a straight line such as AB, and reading the displacement where this line intersects the displacement scale. If the displacement and trim are known, the drafts at any point of the ship's length can be found by drawing a line at the proper slope, through the appropriate displacement on the displacement scale.

If scales for the transverse metacenter, moment to trim one cm, tons per cm immersion and longitudinal centers of buoyancy and flotation are added, either along the locus of the center of flotation or any vertical line, the draft diagram provides a more convenient presentation of the functions of form than the displacement and other curves. Functions plotted along vertical lines are read at the point where they are intersected by a horizontal line at the appropriate displacement. For ships having appendages that project below the keel, a scale of draft above the bottom of the appendage may be added at the proper fore-and-aft location, which will permit reading nagivational drafts directly when the displacement and trim, or the forward and after drafts, are known.

Section 9
The Inclining Experiment

9.1 Basic Principles. Toward the end of the construction period, an inclining experiment is conducted to establish, experimentally, the weight of the ship and the vertical and longitudinal coordinates of its center of gravity. The results of the inclining experiment customarily supersede the corresponding figures in the weight estimate.

The International Convention on Safety of Life at Sea requires that "every passenger or cargo vessel shall be inclined on completion."

An inclining experiment is also conducted after a ship has been converted if the conversion is extensive enough to preclude a reliable estimate of the effect of the conversion on weight and center of gravity or after extended service if it is felt that the weight or center of gravity may have been affected significantly by the accumulation of many minor changes.

The inclining experiment consists of heeling the ship to a small angle by moving a known weight, already aboard, perpendicular to the ship's centerline plane through a measured distance, allowing the ship to settle until the righting moment adjusts itself to equal the heeling moment, and measuring the angle of inclination. The process is repeated at several angles in both directions. Draft readings also are taken.

The displacement and the longitudinal position of the ship's center of gravity, in the condition in which it is inclined, are found from the observed drafts as discussed in Section 8. The metacentric height, $\overline{GM}$, is determined in the following manner:

(a) As discussed in Section 3, the righting arm, $\overline{GZ}$, at a small angle of inclination, ϕ, is

$$\overline{GZ} = \overline{GM} \sin \phi \qquad (1)$$

from which it follows that the righting moment is

$$W \cdot \overline{GZ} = W \cdot \overline{GM} \sin \phi \qquad (25)$$

where W is the displacement of the ship determined from the drafts.

(b) The heeling moment, M, produced by moving a weight, w, aboard ship perpendicular to the ship's centerline plane through a distance, d, is

$$M = w \cdot d \cos \phi \qquad (26)$$

(c) Since the righting moment and heeling moment are equal at the time the inclination is measured

$$W \cdot \overline{GM} \sin \phi = w \cdot d \cos \phi$$

or

$$\begin{aligned}\overline{GM} &= \frac{w \cdot d \cos \phi}{W \sin \phi} \\ &= \frac{w \cdot d}{W \tan \phi}\end{aligned} \qquad (27)$$

The height of the ship's center of gravity is found by subtracting $\overline{GM}$, the metacentric height, from $\overline{KM}$, the height of the metacenter above the keel.

When the ship as inclined has a considerable trim, it is usually necessary to calculate displacement and $\overline{KM}$ by the method described in Section 6, Chapter I, using Bonjean curves, for the actual trimmed condition. However, if the ship is inclined in a nearly zero-trim condition, the height of the metacenter may be read from the displacement and other curves at the draft at the center of flotation, after the draft at the center of flotation has been obtained from the observed drafts.

It is desirable to perform the inclining experiment when the ship is as nearly complete as is practicable, and at this late stage of construction there are usually some tanks containing fuel oil or fresh water. It is prudent to adjust these liquids, before the experiment, to avoid having any tank nearly full or nearly empty, so that there will be no appreciable change in the moment of inertia of the surface during the expected inclination. When this is done, the value of $\overline{GM}$ "as inclined" is the virtual $\overline{GM}$, and the height of the center of gravity is the virtual height, which includes the free-surface effect. The value of the free-surface effect, in meters, is equal to the summation of i_r/δ for the various tanks divided by the ship's displacement, where i_r is the moment of inertia of the free surface and δ the specific volume of the liquid in the tank in cubic meters per ton, as discussed in Section 5. To find the real center of gravity of the ship, the free-surface effect, as well as the virtual $\overline{GM}$, must be subtracted from the height of the metacenter, or

$$\overline{KG} = \overline{KM} - \overline{GM} - \frac{1}{W} \sum \frac{i_r}{\delta} \qquad (28)$$

Occasionally, there may be an unavoidable free surface that is not constant throughout the range of inclination, and therefore cannot be treated as a virtual rise of the center of gravity. Its effect may be taken into account by considering the shifted liquid as part of the inclining weight by adding the weight of the liquid, multiplied by the distance its center of gravity moves in the direction perpendicular to the ship's centerline plane, to the moment of the inclining weight, $w \cdot d$, in the foregoing formula for $\overline{GM}$.

In almost all cases, a new ship is inclined in a condition that is of no particular interest, in that the construction or conversion work has not yet been completed, there is a considerable amount of foreign material, such as staging, aboard, and the ship may be partially loaded as with fuel oil and feedwater. However, it is desirable to delay the inclining until most of the foreign material is no longer needed and then to conduct a thorough cleaning of the ship to get most of it off the ship. Then as an immediate preparation for the test, it is necessary to make an estimate of the weight and the vertical and longitudinal moments of all items which are part of the light ship and have not yet been put aboard, all foreign items which will be removed, and all items of load which are aboard, and to apply these estimates to the weight and the vertical and longitudinal moments of the ship as determined from the inclining experiment to produce the *lightship* condition. If there are items of light-ship weight aboard but not in their final positions, the moments that will result from shifting these items must be included.

9.2 Preparation for Inclining. The following points require attention prior to the inclining experiment:

Schedule. If the inclining test is to be officially witnessed by the U.S. Coast Guard for purposes of stability approval, it is necessary to submit a schedule of the major tasks and procedure to a Coast Guard inspection or technical office in advance.

Drafts and trim. While it is possible, and sometimes necessary, to incline a ship with a large trim, there are advantages in reducing trim nearly to zero, as previously noted. Excessive trim will also make it necessary to correct readings of tank capacities and centroids given in the tank-capacity tables, and make it more difficult to adjust the levels in partially filled tanks so that the liquid level will not reach the tops or bottoms of the tanks. Drafts at which abrupt changes in the waterplane will occur as the ship is inclined should be avoided.

List. While a small initial heel is not objectionable, it should be small enough so that the list, plus the expected inclination, will not exceed the angle at which the relationship $\overline{GZ} = \overline{GM} \sin \phi$ no longer applies.

Metacentric height. The ship must have positive metacentric height at the time of the experiment, after allowances are made for free surface and the effect of the inclining weights and gear.

Free surface in tanks. Free liquids can be dealt with, provided the surface does not reach the top or bottom of the tank as the result of the combination of list and trim, and provided that the moment of inertia of the free surface can be determined accurately. However, consideration of free surface can be eliminated entirely if the tanks are either completely full or completely empty, and if possible these conditions should be obtained. A tank cannot be assumed to be empty unless it is known that the liquid below the suction has been substantially removed, nor assumed full unless the sounding is well above the top of the tank and it is known that no large air pockets exist. To accomplish this, an air escape must be available at the highest point of the tank, but even this will not eliminate the numerous small air pockets between the structural members. It may be possible to heel the ship while the tank is filling to assist the escape of air. The best procedure is to have all tanks that must contain liquid about half full, provided the resulting free-surface effect can be accurately calculated and will not produce negative metacentric height.

Personnel aboard. Arrangements should be made to reduce the personnel aboard to a minimum necessary for the test. Those permitted to remain should not be allowed to conduct any work involving movement of the people, their equipment, or items in or attached to the ship.

Transfer of liquids. Arrangements should be made to prevent changes in the liquid load during the experiment. Valves next to the tanks in all systems should be closed. Precautions should be taken to prevent both deliberate and accidental transfer. The latter could occur as the ship is inclined if port and starboard tanks are inadvertently cross-connected through a piping system.

Swinging weights. Items such as boats and booms, which are normally fixed in a stowed position, should be secured to prevent swinging during the experiment.

Forces affecting heel. During the experiment, the inclination of the ship should not be influenced appreciably by any forces other than the effect of the inclining weights. Gangways should be lifted clear of the ship during the experiment. The effect of floats, fenders and submerged objects should be eliminated. The effect of wind, current, pier, mooring lines, cable and hose should be reduced to a minimum. If possible, the experiment should be performed at slack tide, or, if feasible, in a drydock. Consideration should be given to the possibility of heading the ship into the wind or current. Lines, cable and hose from ship to shore should be well slacked while inclination readings are taken.

Selection of inclining weights. Inclining weights should be selected that will produce an angle of heel sufficient to ensure accurate results, but inclinations should not be carried to an angle at which $\overline{GZ}$ no longer equals $\overline{GM} \sin \phi$. In practice, it is customary to estimate in advance the probable $\overline{GM}$ at the time the inclining experiments is to be performed, and a weight is then selected which will give an angle of heel of about 1 deg on each side of the upright for large vessels, 1½ deg for 30m (100 ft) vessels of normal form, and 2 to 3 deg for very small craft of normal form. This practice assures that no appreciable change in $\overline{KM}$ will occur during the experiment. In the case of ships whose sides flare appreciably at the waterline amidships, the angle of inclination should not exceed 1 deg from the upright. The following equation from

Equation (27) gives the weight required:

$$w = \frac{\overline{GM} \cdot W \tan \phi}{d} \quad (29)$$

It is advisable to incline the ship, by means of the weights selected, some time prior to the experiment to ensure that a suitable angle can be attained. A car carrying the weights and rolling on transverse rails gives excellent results since little rolling of the ship is induced and the movement of the weights can be measured accurately. Handling weights by a crane is a practical method, although not as satisfactory. The weight of the inclining weights and car, if used, must be accurately determined and recorded.

Measurement of inclination. Provision should be made for measuring the angle of inclination independently at three stations. This will permit rejecting a reading which is obviously inconsistent with the others. Numerous devices have been used for this purpose, some of which give a direct reading of the tangent of the angle. A pendulum of string or fine wire, 4 to 6m (to 18 ft) long, with a heavy bob damped in a bucket of oil, will give excellent results and is required for inclinings witnessed by the Coast Guard, unless prior approval is obtained for another device. To obtain a pendulum of this length in a location protected from the wind, it is usually necessary to run the line through one or more hatches. A check should be made to ensure that the pendulum is free to swing to the expected angle without interference. A horizontal transverse batten, fixed to the ship's structure, is provided at the lower end, above the bob, for recording pendulum deflections. The length of each pendulum, from the point of suspension to the batten, is recorded so that tangents of angles of inclination may be calculated.

Reading draft marks. Provision should be made for reading the draft marks. A glass tube with a small hole in the bottom and a scale inside, or a similar device, is recommended to damp out minor wave action.

Measurement of water density. Provision should be made for obtaining samples of the water in which the ship is floating at the time of the experiment. A weighted bottle which can be opened while submerged is useful in obtaining samples from various depths. The latest U.S. Navy instructions on preparing a ship for inclining may be obtained from NAVSEA.

9.3 Conducting the Inclining Experiment. The operations involved in conducting the inclining experiment are as follows:

Inventory. A comprehensive definition of the lightship condition is necessary as the basis for the inventory. The inventory consists of three summations, first, the weight and the vertical and longitudinal moments of those items which would be removed to bring the ship to the light condition, then similar figures to be added, and last, the moments produced by moving to their final positions those items of lightship weight which are aboard but not in their proper locations. The weights to be removed include all items of load and all material aboard which are foreign to the ship, determined by a thorough survey of all spaces. Each tank and void should be sounded, preferably both before and after the experiment, and the specific gravity of the contents recorded. Spaces which are presumably empty should be investigated. The inclining weights and gear should not be overlooked.

Draft readings. Draft readings should be taken simultaneously on the port and starboard sides, at the

Table 18—Condensed Inclining Experiment

TEST DATA

Drafts — Feet

Location of draft marks:
- Forward, abaft FP — 1.3
- Aft, for'd. of AP — 15.0

Observed drafts:
- Forward — 16'-3 7/8" — 16.32
- Aft — 18'-4 3/4" — 18.39
- *Midship, avg. P & S — 17.63

Drafts at perpendiculars:
- FP — 16.31
- AP — 18.52
- Mean — 17.42

Sag: 17.63 − 17.42 — 0.21

Trim: 18.52 − 16.31 — −2.21

Displacement

Longitudinal CF, from midship — −23.61#

Draft at CF:
- Uncorrected — 17.61
- Corrected for sag (factor 0.7) — 17.76

Specific volume of water — 35.165 cuft/T

Total displacement at 17.76-ft draft:
- In salt water — 3720 T#
- Corrected for spec. vol. — 3702 T

Inclining Weights

No.	Weight, T	VCG, ft	LCG, ft
1	4.93	39.7	67.5 aft
2	4.89	39.5	71.5 aft
	9.82	39.6	69.5 aft

Weight Shifts (See plot)

Moments, P, ft-T	Tan φ	Moments, S, ft-T	Tan φ
72	0.0068	216	0.0190
216	0.0212	145	0.0139
145	0.0131	72	0.0076
0	0	216	0.0197

* Midship is 134.5 aft abaft FP.
Length between perps. is 269.0 ft.
\# From hydrostatic curves.

Table 18—Condensed Inclining Experiment (Continued)

Report, Survey Ship *Discoverer* (English units)

SHIP IN CONDITION 0:
AT TIME OF STABILITY TEST

Corrected displacement, W	3702 T
$\overline{GM}$ from plot, Moment/W tan ϕ	2.87 ft
Correction for free surface	0.25
Corrected $\overline{GM}$	3.12
$\overline{KM}_T$ corrected for trim	#24.00
$\overline{KG} = \overline{KM} - \overline{GM}$ (cor.)	20.88
$\overline{GM}_L = \overline{KM}_L^{\#} - \overline{KG}$	322.
Mt. to change trim 1 ft	#4434 ft-T
Trim by stern	2.21 ft
Trim lever = Trim × mt. to trim 1 ft/W	2.65 ft
LCB abaft midship	#7.75
LCG abaft midship	10.40

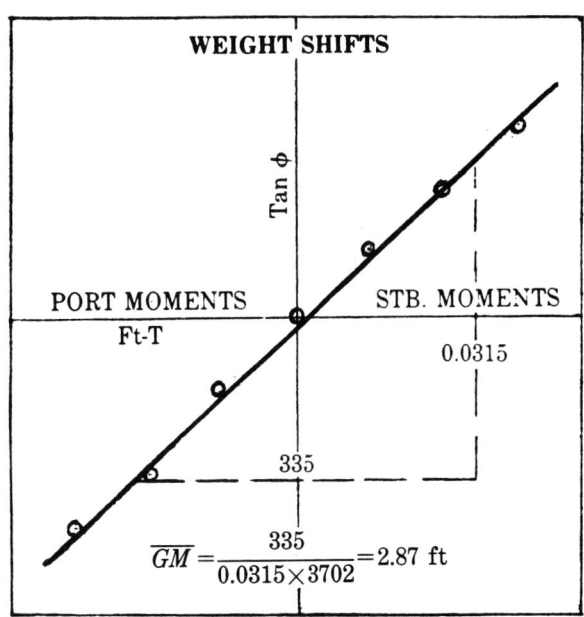

ITEMS TO DEDUCT FOR LIGHT SHIP

		CG abv. BL		LCG from midship			
	Weights, Tons	Lev., ft	VM ft-T	Aft lev., ft	Aft Mt ft-T	Fwd. Lev., ft	Fwd. Mt ft-T
Liquids in tanks	774.82	8.14	6309	19.02	14,737		
Misc. gear, equip., stores	104.93	21.03	2207			28.17	2956
Personnel	1.86	33.33	62	33.46	62		
Inclining weights	9.82	39.6	389	69.5	682		
Total	891.43	10.06	8967	14.05	15,481		2956
					12,525		

SHIP IN LIGHT CONDITION I**

Ship in Condition 0	3702.	20.88	77,300	10.4	38,500	
Weights to complete	1.	44.69	45	9.7	10	
	3703		77,345		38,510	
Weights to be deducted	891	10.06	8967	14.05	12,525	
Ship in Condition I	2812	24.32	68,378	9.24	25,985	

**Ship complete in every respect, with water in boilers at steaming level and liquids in machinery and piping, but with all tanks and bunkers empty and no passengers, crew, cargo, stores or baggage aboard.

---

forward, amidships and after draft marks, at the time of inclining. If no midship draft marks are available, the midship drafts should be obtained by measurement from the deck amidships.

Determination of water density. Hydrometer readings should be taken for several samples from various locations along the ship's length and at various depths. The hydrometer readings should be converted to indicate the density of the water in air.

Movement of personnel. Movement of personnel during the experiment should be restricted.

Weight movements. Inclining weights are customarily moved to produce at least two inclinations to port and two to starboard, the intermediate inclinations being about half the maximum. Weights should be moved slowly, or set down easily, to avoid inducing a roll. The transverse displacement of each weight from its initial position is measured and recorded after each movement.

Measurement of inclination. An initial mark is made simultaneously on each batten, if pendulums are used, or any other device is set to zero, while the inclining weights are in their initial position. Thereafter, readings of the inclination are taken, simultaneously, at each station, after each weight movement. The signal to read the inclination should be given after allowing sufficient time for the ship to come to a position of equilibrium after the weight movement. The ship should be clear of the pier and all lines should be well slacked. If the ship is not absolutely steady, the reading of inclination should be taken at the midpoint of the residual motion.

It is essential to incline to both sides of upright, and it is highly desirable to use either multiple weights or multiple levers so as to heel the ship to at least two different angles on either side of upright. Thus, if anything occurs to cause an erroneous reading it can be identified as an error more easily.

Plot of tangents. During the inclinations, the tangents of the angles of inclination should be plotted against the moments of the inclining weights. Variation of the resulting plot from a straight line indicates that conditions are not favorable or that an error has been made, in which case a check should be made to determine the cause. Trials should be repeated until a satisfactory set of readings has been obtained. The plot of tangents will indicate only certain types of error. For example, an error in measuring the length of one of the pendulums would be apparent since the tangent would be consistently larger or smaller than those from the other pendulums. Or, if a weight movement were measured incorrectly, the corresponding point on the plot would be out of line with the others, but this lack of alignment might be caused by an external force that was acting only at this particular inclination. On the other hand, if a single inclining weight is used and its weight is recorded incorrectly, this would not be apparent from the plot.

9.4 Inclining Experiment Report. This report consists of a recording of the observed data, the calculations necessary to determine the displacement and center of gravity at the time the ship was inclined, and the calculations made to arrive at the light-ship condition by modifications to the condition of the ship at the time of inclining. It is advisable to record the basic data, such as the weight of each inclining weight and the distance it was moved and the lengths and deflections of each pendulum, rather than only the moments and tangents, in order to permit further checking in case any data appear later to be questionable. Actual tank soundings should be recorded, and determination of liquid weights from sounding tables shown—including trim corrections, if any.

Table 18 shows a condensed summary of a typical USCG report. Note that the units are in the English system. More detailed reporting forms are available from the USCG, and similar forms are used by the U.S. Navy.

The displacement and the longitudinal position of the ship's center of gravity are calculated by one of the methods described in Section 8. If the midship draft readings indicate that the ship has hog or sag, correction for this should be made. The displacement thus obtained is multiplied by the ratio of the specific gravity of the water in which the ship was floating to 1.025 (the specific gravity of salt water) to obtain the displacement as inclined.

The plot of tangents is prepared showing the tangents measured at each of the three stations, plotted at the appropriate inclining moment for each inclination. Under ideal conditions, a straight line can be drawn through the plotted points. In most cases, however, some judgment must be applied in drawing the straight line which best represents the information plotted on the plot of tangents. If at any particular trial one measurement of inclination does not agree well with the other two, it may be appropriate to disregard it. Or, it may be possible that one of the recorded moments does not represent the actual moment acting at that time, because of the effect of wind or current. It is usually necessary to draw the line that best represents the slope corresponding to the plotted points. This line does not necessarily pass through the origin, since this point carries no more weight than the others, and it is possible that some upsetting force was acting at the time that the initial reading was made. After this line has been established, its slope $w \cdot d / \tan \phi$ is divided by the displacement to find the metacentric height, which, from subsection 9.1, is

$$\overline{\text{GM}} = \frac{w \cdot d}{W \tan \phi} \quad (27)$$

After the metacentric height has been obtained, the vertical position of the center of gravity in the inclined condition can be found and the characteristics of the light-ship condition developed as described in Section 9.1. See Table 18.

It is desirable to record any major features of the light ship at the time of inclining, such as the weight and center of gravity of any permanent ballast, for future reference.

9.5 Inclining in Air. When a small boat is to be inclined, it is preferable, and may be more convenient, to perform the experiment in air rather than in water.

The boat is suspended by slings forward and aft which pass over a knife edge, and the slings adjusted so that the base line used for calculations is parallel to the knife edge. The knife edge is supported by two scales, one forward and one aft. Inclining weights and pendulums are provided and used in the same manner as for inclining in water. See Fig. 51(b), where ϕ is the inclination due to the movement of the weight w through distance A, G is the center of gravity of the boat with inclining weight, and B is the height of knife edges above the keel.

The weight of the boat W_B is obtained by adding the two scale readings, $W_1 + W_2$. The distance x of the center of gravity of the boat from the forward perpendicular is obtained by taking moments. (Fig. 51a),

$$(W_1 + W_2) x = (A + B) W_1 + A W_2.$$

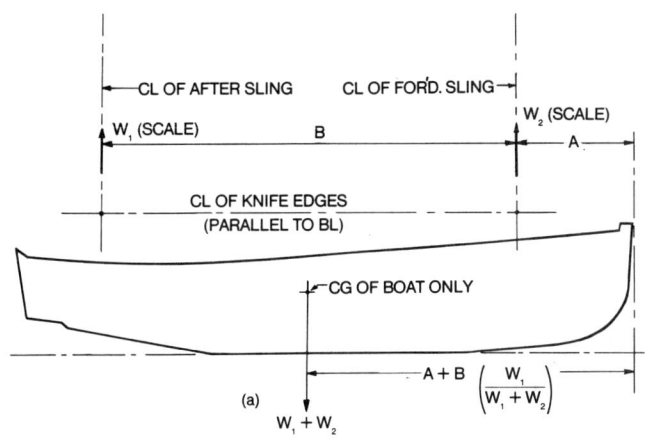
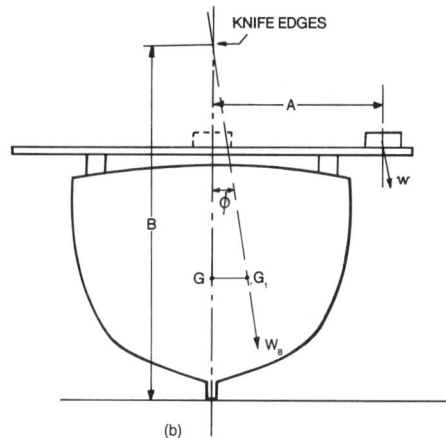

Fig. 51 Inclining experiment in air

Hence,

$$x = A + BW_1 (W_1 + W_2)$$

The accuracy of the scales may be checked by reweighing the boat with the scales interchanged.

The vertical location of the center of gravity of the boat is obtained by the same process used in inclining in water, except that the "metacenter" is located at the knife edge. Hence, the center of gravity above the keel is,

$$\overline{KG} = B - \overline{GG_1} \cot \phi$$

$$= B - \frac{wA}{W_B} \cot \phi$$

9.6 Accuracy. Consideration of the subject of accuracy will not only tend to improve the reliability of experiments, whether in water or air, but may also result in avoiding laborious refinements which do not have an appreciable effect on the results.

An error in measurement of pendulum length, pendulum deflection, inclining weights or weight movement will result in a proportional error in the metacentric height.

The effect of inaccuracy in draft readings on displacement can be evaluated by considering the ship's tons per inch immersion. If the height of the metacenter is changing rapidly with a change in draft, as it often does at light displacements, an error in the height of the center of gravity may result from inaccuracy in draft readings.

Errors in inventory appear as equal errors in displacement. The degree of accuracy used in locating centers of gravity for items of the inventory depend on the weight of the item. Exact location of centers of light items is not necessary.

The contribution of different tanks to the total free-surface effect varies widely. If a tank has a width equal to about half the ship's beam, the shape of the surface should be determined accurately, and precise methods used in finding the moment of inertia. For smaller tanks, less accurate methods may be used. There may be small tanks, or tanks containing small quantities of liquid, for which the movement of transference at the expected angle of heel is negligible compared to the moment of the inclining weights.

The proper degree of importance to be attached to any item can be evaluated by an approximate calculation of its effect on the ship's displacement and the height of the center of gravity.

9.7 Induced Rolling (Sallying). Rolling may be induced, for small ships, by *sallying*, in which a group of people moves across the deck in synchronism with the ship's natural period, or a vertical force may be applied on one side and suddenly released. For larger ships, a weight may be landed on one side and then lifted and lowered several times, again in synchronism with the natural period of roll.

The period of roll may be found quite accurately by measuring the total elapsed time of a number of rolls. Then, from Section 3.7, the period of roll,

$$T_\phi = \frac{CB}{\sqrt{GM}} \tag{7}$$

where C is the sally constant
and B is the beam of ship

Since the sally experiment gives the period of roll

for a full cycle, using the *as inclined* $\overline{GM}$ yields the sally constant C. At subsequent times, sallying the ship in calm waters again yields a period of roll, and an approximate $\overline{GM}$ can be determined by using T_ϕ, C and B in the above formula.

If at the time of inclining, the sally constant, C, which is determined from the data, is out of line with constants found for previous similar ships, this would indicate a significant difference in construction of the new ship or a possible error in the experiment. The construction drawings and calculations should be closely checked.

Section 10
Submerged Equilibrium

10.1 Definition. A submerged submarine is in equilibrium when its weight is equal to the buoyancy of the total hull and when its center of gravity is in the same longitudinal position as its center of buoyancy so that it has zero trim. Actually there may be very small differences between the weight and buoyancy and between the longitudinal positions of the centers of gravity and buoyancy. These differences are overcome by use of submarine's planes when underway in a submerged condition. When a submarine must be *dead* underwater to maintain silence, it is important that the weight equal the submerged buoyancy exactly.

On the surface, the weight of the submarine is lighter than its submerged displacement and it is necessary to take on sea water in the ballast tanks to enable the submarine to submerge and be in a condition of equilibrium with zero trim while submerged. There are certain loading conditions on the surface that tend to make the submarine "heavy" or "light," which result in correspondingly small or large amounts of *variable ballast water* that must be carried on the surface, so that upon complete flooding of the main ballast tanks, a submerged condition of equilibrium and zero trim will result. A submarine that is properly designed with respect to weight, buoyancy and variable ballast tank capacity, will always be in *diving trim* on the surface regardless of the actual load variations, and will be able to successfully dive and be in equilibrium with *zero trim* in a submerged condition with the main ballast tanks full of sea water.

Diving trim, diving ballast and variable ballast are discussed in sections 10.2, 10.3, and 10.4.

10.2 Items of Weight. The items of weight that are considered in studies of submerged equilibrium are illustrated in Figs. 52 and 53, and are defined as follows:

(a) Submerged displacement. The displacement of the entire envelope of the ship minus any free-flooding spaces. The submerged displacement is fixed by geometry rather than by weight. For a given configuration of the ship, the submerged displacement will vary only with the density of the sea. Weight must be adjusted to conform to the submerged displacement.

(b) Light ship. The sum of the weights of the components making up the ship. This weight is fixed unless some alteration to the ship is made.

(c) Lead. Solid ballast. This item is the margin in the weight estimate. In submarine design, the volume of the submerged displacement is made larger than the anticipated weight in the submerged condition by a generous margin. Some of this margin is usually needed to compensate for inaccuracies in the weight and displacement calculations or for unexpected installations. When the ship is completed, any unexpended margin must be installed as solid ballast to achieve submerged equilibrium. Solid ballast is part of light ship. Lead is most often used as solid ballast material because of its high density, and because it causes few corrosion problems.

(d) Load to submerge. The weight that must be added to the *light ship—with lead* to bring the ship to a condition of submerged equilibrium. Assuming no changes are made that affect the geometry or weight of the ship, the load to submerge will vary only when there is a change in the density of the sea water.

(e) Normal fuel-oil tanks. Fuel-oil storage tanks that are fitted with a sea water-compensating system so that they are always full of oil or sea water.

(f) Main ballast tanks. Tanks that are flooded to submerge and blown to surface. They are fitted with vents at the top, which are opened to flood the tanks, flood openings at the bottom and air connections for blowing.

(g) Fuel ballast tanks. Tanks that may be rigged either as normal fuel oil tanks or as main ballast tanks. When used for fuel, they are compensating tanks and are handled in the same manner as the normal fuel tanks. After the oil in the fuel ballast tanks has been burned, they may be converted to serve as main ballast tanks, and thereafter flooded upon submerging and blown upon surfacing. Conversion to main ballast tanks reduces the surface displacement and increases the reserve buoyancy until the ship is refueled. Fuel ballast tanks are not fitted on nuclear-powered submarines because of the small amount of fuel oil.

(h) Residual water. The water in main ballast tanks and fuel ballast tanks, located below the top of

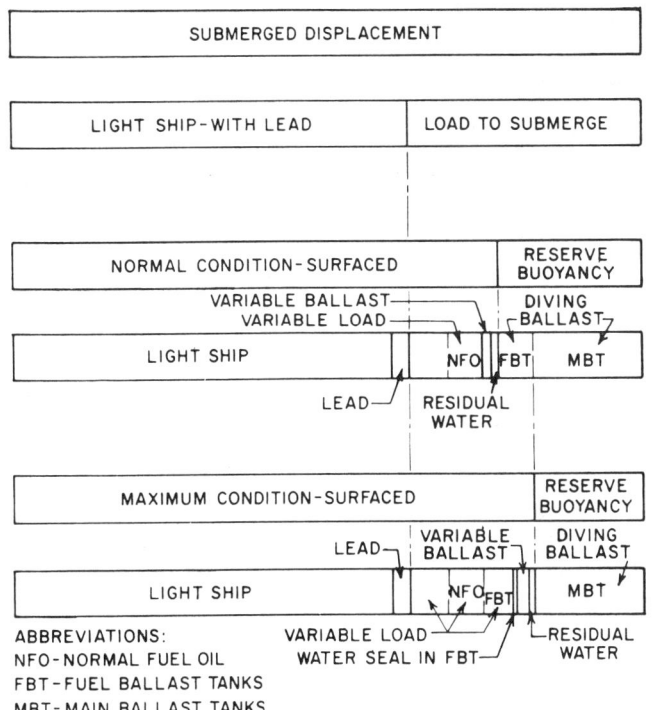

Fig. 52 Weights, diesel-powered submarine

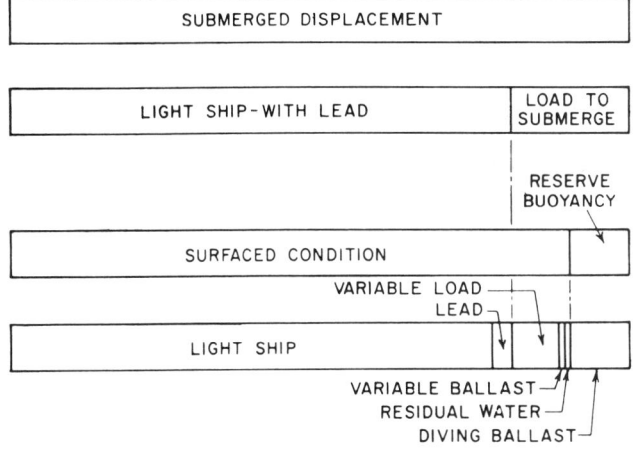

Fig. 53 Weights, nuclear-powered submarine

the flood opening, which cannot be blown upon surfacing.

(i) Water seal in fuel ballast tanks. The layer of water above the top of the flood opening and below the bottom of the compensating water pipe, which is maintained when the fuel ballast tanks are serving as normal fuel-oil tanks to prevent spilling of oil through the flood openings as the ship rolls. When the tank is nominally full of fuel, its contents, starting from the top, consist of fuel, water seal, and residual water.

(j) Diving ballast. The term applied to the water that is admitted to the ship upon diving and blown to bring the ship to the surface. When the fuel ballast tanks are used for fuel, the diving ballast is equal to the capacity of the main ballast tanks above the residual water. When fuel ballast tanks are rigged for main ballast, the diving ballast is equal to the capacity of the main ballast tanks and the fuel ballast tanks above the residual water. Some submarines have a tank near admidships designated as *safety tank*, which is blown upon surfacing and is considered to be a part of the diving ballast. The safety tank may be only partly filled in the submerged condition when the ship is heavily loaded. The safety tank, when installed, is intended to be equal in volume and with about the same location as the topside conning area and may be blown while the submarine is submerged to regain buoyancy in case of topside damage.

(k) Normal condition—surfaced. The displacement of the ship on the surface when fuel ballast tanks are rigged as main ballast tanks.

(l) Maximum condition—surfaced. The displacement of the ship on the surface when fuel ballast tanks are rigged as normal fuel oil tanks.

(m) Reserve buoyancy. The displacement of the volume of the envelope of the ship above the waterline in the surfaced condition, minus any free-flooding spaces.

(n) Variable load. Such items as personnel and their effects, missiles, torpedoes, provisions, stores, cargo, passengers, potable water, reserve feedwater, battery water, reserve reactor coolant, lubricating oil, oxygen, reserve hydraulic oil, contents of sanitary tanks and hovering tanks, and fuel oil. The variable load in the normal and maximum conditions is identical except that the oil in the fuel ballast tanks is included in the maximum condition.

(o) Variable ballast. Sea-water ballast that is adjusted continuously at sea to compensate for changes in variable load or in sea-water density. Variable water ballast is carried in forward and after tanks called *trim tanks* and in midship tanks called *auxiliary tanks* to permit adjustment of the longitudinal moment as well as the weight. Some diesel-powered ships have *variable fuel oil tanks* that are non-compensating tanks sized so that the weight of oil that they carry is approximately equal to the increase in weight that occurs when the oil in the compensating tanks is replaced by sea water. Burning oil from the variable fuel-oil tanks so that the percentage remaining in these tanks is the same as the percentage remaining in the compensating tanks will tend to keep the weight of the contents of all oil tanks nearly constant. Variable fuel-oil tanks are piped so that oil may be transferred between them and the compensating tanks. This transfer involves an increase or decrease in the ship's weight as the compensating water is drawn or expelled from the normal fuel-oil tanks. Because of this capability variable fuel-oil tanks, when fitted, are consid-

ered as part of the variable ballast, rather than part of the variable load.

10.3 Relationships Between Items of Weight. As conditions for submerged equilibrium, each of the bars in Fig. 52 and each in Fig. 53, must represent the same weight, and each must have its center of gravity in the same longitudinal position. Two other equations are indicated by the vertical lines in Figs. 52 and 53. First, the reserve buoyancy is equal to the diving ballast and the longitudinal positions of their centroids must coincide. Also, the load to submerge (the difference between the submerged displacement and the light ship with lead) is equal to and has the same longitudinal center of gravity as the sum of the variable load, variable ballast, residual water and diving ballast, and, in the maximum condition, the water seal in the fuel ballast tanks.

These necessary equalities are achieved by variations in weight and longitudinal position of the center of gravity of the lead ballast and the variable ballast. To conserve space, the variable ballast tanks are sized to accommodate only the probable variation in the variable load plus the variation in the submerged displacement caused by changes in sea-water density. Lead ballast, since it occupies less space per ton, is used for the required adjustment beyond the capacity of the variable ballast tanks. In general, changes in lead are made in the shipyard, to compensate for changes in the light ship weight or in the volume of the submerged displacement, while the variable ballast is used to compensate for changes that occur at sea.

10.4 Diving Trim. A submarine on the surface at sea is normally kept in *diving trim*, which means that the weights aboard are adjusted so that completion of flooding of the main ballast tanks, and any fuel ballast tanks rigged as main ballast tanks, will submerge the ship in a condition of equilibrium, with the ship's weight equal to the submerged displacement and the center of gravity in the same longitudinal position as the center of buoyancy.

It can be seen from Fig. 52 that the weight and longitudinal moment of the surfaced ship in diving trim must be equal to the difference between the figures for the submerged displacement and those for the reserve buoyancy, and that the reserve buoyancy corresponds, in weight and center, to the diving ballast. Since the volume and moment of both the submerged displacement and the diving ballast depend only on the configuration of the ship, the surface drafts in diving trim, in either the normal or the maximum condition, are determined by the geometry of the hull.

Diving trim is maintained at sea by adjustment of water in the variable ballast tanks. Where variable fuel tanks are fitted, the oil in these tanks may also be adjusted. While the submarine is submerged at very low speeds, variable ballast may be admitted, discharged or transferred longitudinally until any fore-and-aft inclination is eliminated, and any appreciable tendency to rise or settle disappears. Between such experimental adjustments, the proper quantity and disposition of variable ballast is maintained by recording all changes in weight such as the replacement of fuel by sea water, ejection of trash or blowing of sanitary tanks, and making compensating changes in the variable ballast.

In addition to such gradual or minor changes in weight, there are large and abrupt changes that may occur in the submerged condition, due to firing of weapons, which require immediate compensation. This is accomplished by admitting a quantity of water, as

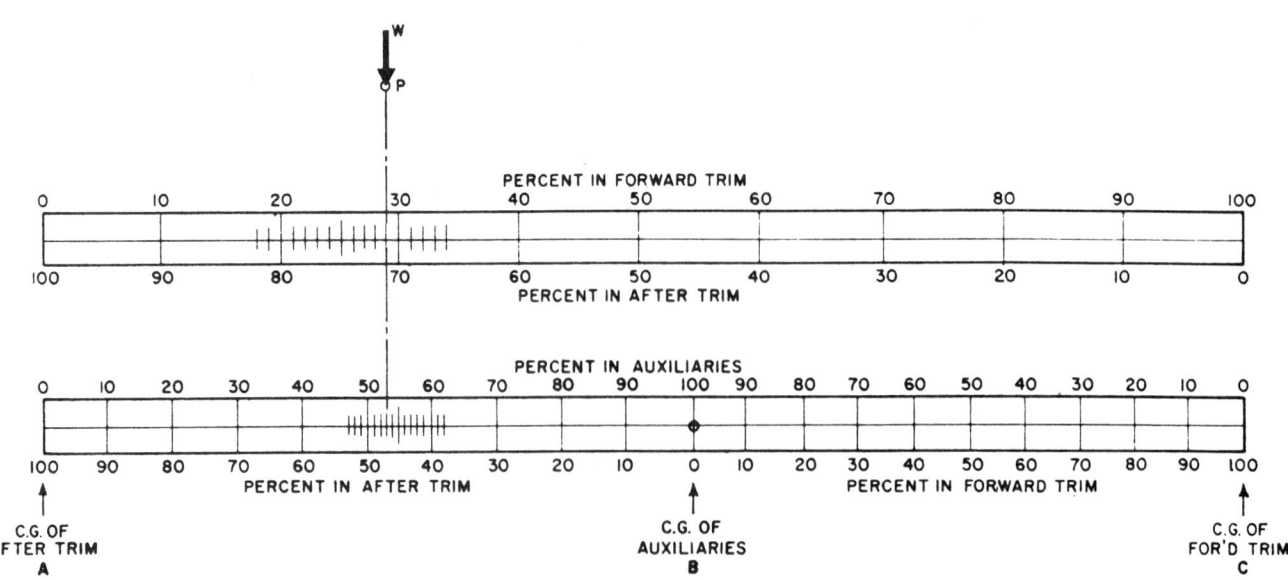

Fig. 54 Moment diagram for submarine

part of the firing operation, equal to the weight of the weapon ejected.

The moment diagram, illustrated in Fig. 54 is a convenience in finding the change in weight that must be made in the variable ballast to compensate for a change in the variable load. If a weight, w, is added at some point, P, along the ship's length, reading the scales directly below point P will indicate the percentage of the weight, w, to be removed from the after trim and the auxiliaries, or from each of the two trim tanks, to compensate for the added weight. In this diagram, which is usually plotted below an inboard profile of the ship to which it applies, points A, B, and C are located, respectively, at the locations of the centroids of the after trim, auxiliaries, and the forward trim tanks. The scales are constructed by dividing the distances between A and B, B and C, and A and C into 100 equal divisions.

As an example of the use of the moment diagram, if 1000 kg is added at point P in Fig. 54 there will be no change in the weight or in the longitudinal position of the center of gravity of the ship if either:

(a) 290 kg of water are blown from the forward trim tank and 710 kg are blown from the after trim tank, or

(b) 530 kg are blown from the auxiliary tanks and 470 kg are blown from the after trim tank.

10.5 Equilibrium Conditions. Since the variable ballast must be adjusted to compensate for changes in sea-water density and for changes in the variable load, it is necessary to evaluate the magnitude of probable changes in both the weight and the longitudinal moment of these two items in order to select the proper size for the variable ballast tanks.

To ensure that diving trim can be achieved with any possible variation in sea-water density and variable load, it would be necessary to develop the lightest possible condition in the heaviest sea water, the heaviest possible condition in the lightest water, and conditions with maximum longitudinal moments in each direction in both heavy and light sea water, and to make the variable ballast tanks large enough to compensate for these changes. This would result in very large variable ballast tanks on a ship having a limited amount of space. By using judgment to eliminate improbable extreme conditions, the variable ballast tanks can be held to a reasonable size.

In displacement calculations for surface ships, the sea water specific volume is assumed to be 0.975 m³ per ton and the normal variations from this figure are negligible, since they would produce only a small change in draft. A small change in the displacement of a submerged submarine, such as 10 tons, would result in an unacceptable unbalance between weight and buoyancy. Since 10 tons is a small percentage of the submerged displacement, only a small change in the density of the sea water is required to produce such an unbalance.

The specific volume of sea water has been found to vary from 0.981 to 0.971 m³ per ton. The extreme variation in variable ballast to compensate for this effect would occur if the ship were to dive in light water, filling the main ballast and fuel ballast tanks, and then pass, submerged, into heavy water. It is customary to assume that this extreme, or the opposite, will not occur, and that the diving ballast is of the same density as that in which the submarine is operating. Under this assumption, the quantity of variable ballast needed for variation in sea water density is equal to the submerged displacement less the diving ballast, in cubic meters, multiplied by the change in density. As an example, if the submerged displacement is 4000 t and the diving ballast 500 t, the quantity of variable ballast needed to counteract this effect would be:

$$(4000 - 500)\text{t} \times 0.975 \times \left(\frac{1.0}{0.971} - \frac{1.0}{0.981}\right)$$
$$= 3500 \times 0.975 \times (1.03 - 1.02)$$
$$= 3500 \times 0.975 \times 0.01 = 34 \text{ t}.$$

For the purpose of studying the additional variation in variable ballast necessary to compensate for changes in the variable load, calculations are made for a series of *equilibrium conditions* representing heavy loads in light water, light loads in heavy water and heavy forward and heavy aft loadings in both light and heavy water. In calculating the loads aboard for any particular condition, judgment and familiarity with operating procedures are necessary in deciding on the quantities of the various items of variable load if very large variable ballast tanks are to be avoided.

The *heavy forward* and *heavy aft* conditions are not necessarily heavy. The term heavy forward, for example, means that loads in the forward end are heavy while those aft are light. In a ship that carries most of the variable load forward, the heavy-aft condition might be quite light. In the heavy-aft condition, the quantities of torpedoes and dry cargo, for example, would be assumed to be zero in the forward portion of the ship while a full load of such items would be assumed aft. As in the case of the heavy and light conditions, it is advisable to investigate two heavy-forward and two heavy-aft conditions, one with a large percentage of fuel aboard in which only the oil in the fuel ballast tanks at the heavy end has been burned, and a condition occurring later when all oil in the fuel ballast tanks and in the normal fuel-oil tanks at the heavy end has been burned. Since the heavy-forward and heavy-aft conditions are not necessarily either heavy or light, calculations should be made for both heavy and light sea water.

The final result of the equilibrium-condition calculations is the weight and longitudinal moment of the *variable ballast to balance,* which is the variable ballast required under the assumed loading and sea-water density to bring the ship to diving trim on the surface and to submerged equilibrium after diving.

As shown in Figs. 52 and 53, the variable ballast to balance can be established by subtracting the summation of weight and longitudinal moment of the variable load from the figures for the load to submerge. Figs. 52 and 53 also show that the load to submerge may be found by deducting the weight and longitudinal moment of the light ship, with lead, from the figures for the submerged displacement. Two sets of values for the load to submerge are found by using figures for the submerged displacement at both 0.981 and 0.971 m^3/t. The load to submerge in light water is used for the heavy conditions, the figures for heavy water used for the light conditions, and both are used for the heavy-forward and heavy-aft conditions. Two summations of variable load are required for the heavy-forward and heavy-aft conditions, so that the density of the diving ballast will correspond, in each case, to that used for the submerged displacement.

10.6 The Equilibrium Polygon. The equilibrium polygon of a typical diesel-powered submarine, illustrated in Fig. 55, is a device for presenting graphically the envelope of variation in weight and longitudinal moment which can be obtained by adjusting the variable ballast. In Fig. 55 the weight of variable ballast is plotted vertically and the longitudinal moment, about the transverse reference plane used for the equilibrium conditions, is plotted horizontally. Each side of the polygon represents the effect of filling one of the variable ballast tanks. The polygon is constructed by adding, algebraically and successively, the weights and moments of each of the variable ballast tanks, starting with the forward, most tank and proceeding aft, then repeating the process starting with the aftermost and proceeding forward. Each summation is plotted as in Fig. 55, where line OA represents the weight and moment developed as the forward trim tank is filled, line AB the effect of filling the forward variable fuel-oil tank after the forward trim tank has been filled, and so forth until point E, representing the weight and moment of all the variable ballast tanks, is reached. The same point E is reached, by a different route, by plotting the various stages of the summation starting with the aftermost tank and proceeding forward.

The weight in each of the variable ballast tanks is taken to be equal to the net capacity of the tank at specific volume of 0.975 m^3 per ton. This volume is applied to the variable fuel tanks, even though they contain oil, because the transfer of 0.975 m^3 of oil from a variable fuel tank to a normal fuel tank would force

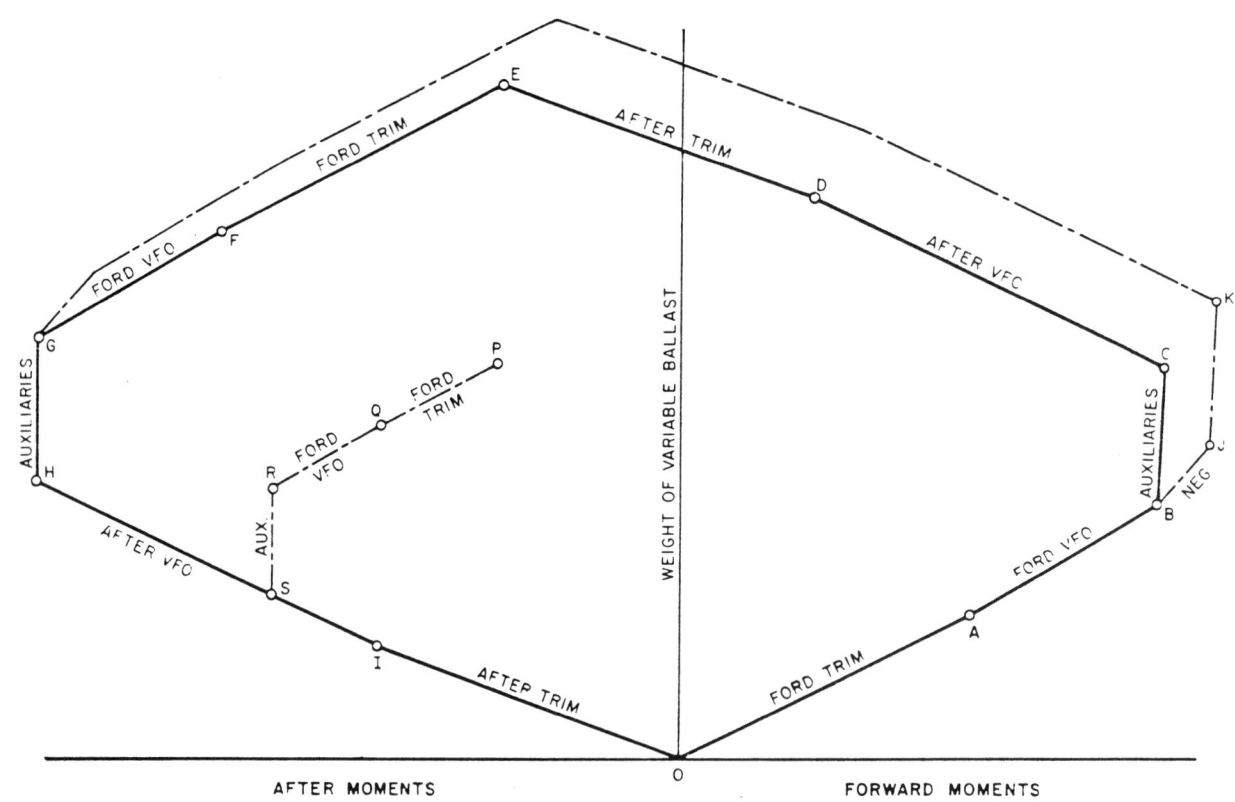

Fig. 55 Equilibrium polygon for submarine

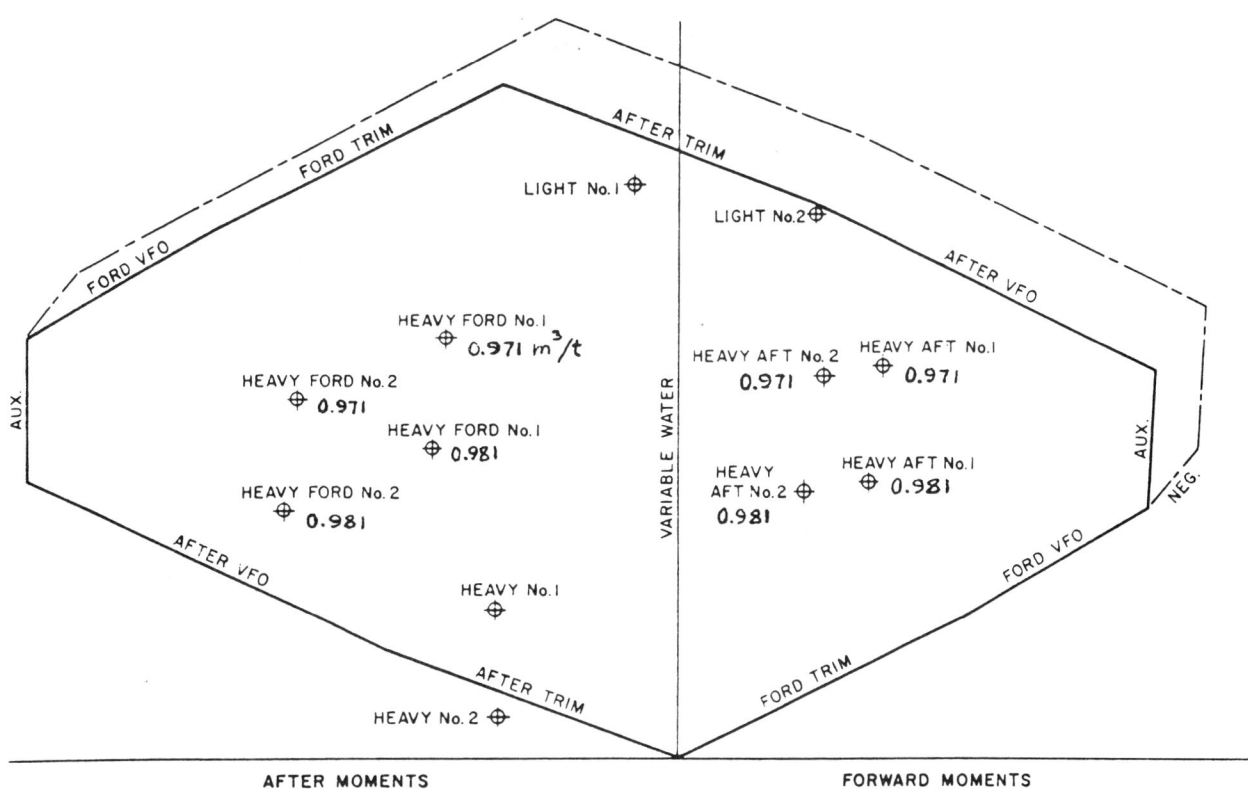

Fig. 56 Equilibrium polygon for Fig. 55, with required weight and moment plotted for several loading conditions

0.975 m³ sea water, or 1 ton overboard. The change in weight is assumed to occur at the location of the variable fuel tank, although actually there will also be a small change at the indeterminate location of the normal fuel-oil tank to which the oil is transferred.

The exterior broken line in Fig. 55 shows the effect of considering the *negative tank* as part of the variable ballast. This tank, located forward of the center of buoyancy and normally empty in the surfaced and submerged conditions, is customarily filled just prior to diving in order to expedite the operation. This causes the weight of the ship to exceed the submerged displacement by a few tons and produces a down angle on the ship, both of which are favorable to rapid submerging. The negative tank is blown when the ship reaches the ordered depth, restoring equilibrium. If necessary to meet very light equilibrium conditions, the negative tank may be treated as part of the variable ballast, if the loss of the advantage of its normal function in the light condition is accepted.

The variable ballast can be adjusted so that its weight and moment correspond to the coordinates of any point within the polygon. Point P in Fig. 55, for example, can be reached by filling the after trim tank, moving from O to I; part of the after variable fuel tank, moving from I to S; then partially filling the auxiliaries, forward variable fuel tank and the forward trim tank. Line SR is parallel to and not longer than HG, RQ is parallel to and not longer than GF, and QP is parallel to and not longer than FE. This is only one of many ways in which point P can be reached.

Fig. 56 is the polygon of Fig. 55, with the weight and moment of the variable ballast to balance for the various equilibrium conditions plotted. It is immediately apparent from Fig. 56 that the ship cannot be brought to submerged equilibrium in Condition Heavy No. 2 in light sea water, and that all other conditions can be met without the use of the negative tank, although there is but little margin in the case of Light no. 2.

10.7 Adjustment of Lead and Variable Ballast Tankage. Under the conditions shown in Fig. 56, it is apparent from Figs. 52 and 53 that, if some lead ballast were removed, the amount of variable ballast required would be increased, and Condition Heavy No. 2 in Fig. 56 would move upward into the polygon. Also, if some lead were shifted aft with no change in the total amount, this point would move horizontally to the right into the polygon. All other points in Fig. 56 representing equilibrium conditions would, in either case, move the same distance and in the same direction. This shows that all points may be moved, as if they were plotted on a separate piece of paper, either fore and aft or up and down, or a combination of both, by adjustment of the lead ballast. There may be, of course, some physical limitation on the adjustment of lead, or its removal may be precluded by considerations of transverse stability.

In the case shown in Fig. 56, it is apparent that the spread of the points representing the equilibrium conditions is beyond the capacity of the variable water and variable fuel tanks, but that the constellation of points can be embraced by the polygon that includes the effect of the negative tank. If the ship were already built, the loss of complete effectiveness of the negative tank in diving in the light condition in heavy water would probably be accepted and Light No. 2 allowed to move into the area representing the effect of the negative tank. Otherwise, it would be prudent to consider an increase in the size of the polygon. In this situation, the polygon is useful in deciding which tank or tanks should be enlarged. It is apparent from Fig. 56 that no improvement would result from increasing the capacity of the after variable fuel tank or the after trim tank, since this would only extend the polygon to the left. The greatest gain, per ton increase in capacity, would be obtained by increasing the size of the auxiliary tanks, but if this were not feasible, increasing either the forward trim or the forward variable fuel tank would be effective.

10.8 Stability in Depth. While the forces of weight and buoyancy can be brought very nearly to equilibrium when a submarine is submerged, most submarines have no inherent stability with respect to depth since, as the ship settles or rises, no force is generated to return the ship to the original level. A situation may exist in which the water at greater depths may be appreciably heavier than that near the surface, because of differences in temperature and salinity, which will enable the submarine to rest on the interface if its weight is greater than its displacement in the lighter water but less than its displacement in the heavier water. Otherwise, unless some force is applied, as by the planes or a hovering system, most submarines would eventually either rise to the surface or settle to the bottom.

On the normal submarine, the pressure of the sea on the hull tends to produce an unstable condition. The loss of buoyancy due to compression of the hull as the ship settles exceeds the gain in buoyancy due to compression of the sea water and the resulting slight increase in its density. The net result is that buoyancy is decreased as the ship settles and increased as the ship rises. The effect of sea pressure would be aggravated if there were a partially filled tank open to the sea, since the air therein would expand and compress readily with changes in depth, expelling water when the ship was rising or admitting water settling.

On some very rigid hulls, the effect is reversed, since the effect of compression of the hull is less than the effect of the compression of the sea water in increasing the water density. This results in a small gain in buoyancy as the ship settles, a small loss when it rises, and hence a minor stabilizing effect.

Section 11
The Trim Dive

11.1 Basic Principles. The trim dive is an experimental determination of the weight and longitudinal moment of the *load to submerge*, as defined in Section 10.

Theoretically, the load to submerge could be obtained, as illustrated in Figs. 52 and 53 of Section 10, by deducting the light ship with lead, as determined from the inclining experiment, from the calculated figures for the submerged displacement. The submerged displacement, however, cannot be calculated accurately because of numerous topside appendages. Also, the load to submerge, determined in this manner, would represent a small difference between two large quantities, and therefore subject to a larger error than if it were determined directly. It is therefore customary to find the load to submerge experimentally, by an inventory of all weights aboard that comprise the load to submerge, taken while the ship is in submerged equilibrium.

The load to submerge is used, as discussed in Section 10, as the basis for calculating the variable ballast to balance in the various equilibrium conditions, which, in turn, determine the optimum weight and disposition of the lead ballast.

11.2 Conducting the Trim Dive. The ship is completely submerged in an area that is free from strong currents and sharp density gradients. The variable ballast is carefully adjusted to bring the ship to submerged equilibrium. The ship is held at rest long enough to ensure that there is no fore-and-aft inclination and no appreciable tendency to rise or settle.

While the ship is in submerged equilibrium, a sample of sea water is taken, preferably from a circulating system in operation, and the density determined.

An inventory is taken of the weight and longitudinal moment of all items aboard (other than lead ballast) that are not part of the light-ship weight. As in the case of the inclining experiment, this inventory must be based on a comprehensive definition of the light-ship condition. The total weight and moment resulting from this inventory are the load to submerge and its longitudinal moment at the sea-water density observed concurrent with the inventory.

11.3 Report of the Trim Dive. The calculations made in the report of the trim dive involve converting the load to submerge at the density of the sea water in which the ship was submerged to its values at specific volumes of 0.981, 0.975 and 0.971 m^3/t. As mentioned in Section 10, these values represent the variation in sea-water specific volume, and are used in the equilibrium conditions, where small variations in specific volume are important. The value of 0.975 m^3/t is used in stability calculations, as in the case of surface ships.

When the inclining experiment and the trim dive have been completed, the weights and longitudinal moments of the light ship with lead and of the load to submerge are known. Theoretically, the sum of these two items should correspond to the volumetric calculations for the submerged displacement, but minor discrepancies are to be expected due to the inaccuracies involved in each of the three items. It is customary to regard the submerged displacement from the inclining experiment and trim dive as being more accurate than that obtained from the volumetric calculations.

The values of the load to submerge at the various sea-water specific volumes are obtained as follows:

(a) The submerged displacement and its longitudinal moment at the time of the trim dive are obtained by adding the weights and moments of the light ship with lead from the inclining experiment and the load to submerge from the trim dive.

(b) The submerged displacement and its longitudinal moment at 0.981, 0.975, or 0.971 m^3/t are found by multiplying the weight and moment obtained in step (1) by the ratios of 0.981, 0.975, and 0.971 m^3/t to the specific volume of the outside sea water.

(c) The load to submerge and its longitudinal moment at sp. vol of 0.981, 0.975, and 0.971 m^3/t are obtained by subtracting the figures for the light ship with lead from the submerged displacement at those specific volumes.

Section 12
Methods of Improving Stability, Drafts and List

12.1 Changes in Form. In the early stages of design, variations in the ship's form, usually changes in beam or depth, may be used effectively to obtain optimum stability and drafts. The effects of changes in form on stability are discussed in detail in Section 4. Draft can be varied by changes in beam or fullness, and trim can be varied by increasing the fullness at one end of the ship and decreasing it at the other, thus moving the center of buoyancy in the desired direction. This must be done with caution, however, as it may have a serious adverse effect on resistance. Moreover, in cargo ships, and especially in tankers and ore carriers, the benefit is partially offset by the resulting movement, in the same direction, of the center of gravity of the total cargo space.

When extensive topside weight is to be added to an existing ship, drafts and stability may be improved by the installation of blisters which have the effect of increasing the ship's beam, and may be used to adjust trim by moving the center of buoyancy longitudinally. There is also at least one case in which unsymmetrical blisters have been used successfully to remove an inherent list, with improvement in drafts and stability.

The full effect of intact stability improvement due to widening the ship's beam by blisters or other means is not a complete measure of the net improvements in the case of damage stability. Where damage stability is governing in the ship's required initial stability, the net stability improvement must be considered rather than the improvement in intact stability alone.

12.2 Adjustment of Load. The effectiveness of adjusting the load, as a means of improving stability, correcting trim and minimizing list, depends upon the ratio of the load to the total weight of the ship and the freedom that the operator has in the distribution of loads within the ship. One extreme is the cargo ship, loading mixed cargo of various densities, which can be stowed to obtain a wide variation in the ship's center of gravity. The other exteme is a ship such as a tug, which carries very little load and each item is loaded in a specific location.

In the former case, it is advisable to calculate the effect of each proposed loading on the stability and drafts, and to ensure that the cargo is stowed symmetrically. The forward draft should be adequate to ensure against pounding in a seaway, and the after draft adequate to provide sufficient propeller immersion. The expected variation in the consumable load during the voyage should be taken into account. Any prescribed draft limitation should be observed.

12.3 Permanent Ballast. Permanent ballast, judiciously located, can often be used to improve stability or trim, or to remove list. When installed primarily to improve one of these characteristics, ballast can often be used to improve the others as well.

Permanent topside or 'tween-deck ballast has also been used to decrease the metacentric height of ships found to be too "stiff" for satisfactory operation.

When the object is to increase stability, the use of a dense material, such as lead, iron, or concrete made with a dense aggregate, is usually appropriate, since this will permit the installation of ballast at a lower center of gravity than a less dense material, and therefore with a greater effect, per ton of ballast, in lowering the ship's center of gravity.

If solid ballast is to be located in fuel tanks, there is a marked advantage in using metallic ballast, preferably lead. The effective weight of such ballast is its weight in oil. A dense material will displace less oil, per ton of effective weight, than a less dense type.

When ballast is used to improve drafts or remove list, and lowering of the ship's center of gravity is not required, ordinary concrete may be effective. In many cases, the ballast may be installed on several levels, in relatively thin layers, and the additional volume occupied by the less dense ballast is not important.

The use of liquid ballast, preferably fresh water with a rust inhibitor, is often effective, and has the advantages of a lower material cost and easy removal. In some cases, however, considerable expense may be involved in reinforcing the structure to contain the ballast, as, for example, when it is located in a cargo hold and the strength of the deck above is not adequate to withstand the head imposed when the ship rolls or heels to a large angle. When permanent liquid ballast is used to improve stability, its location and effectiveness in case of underwater hull damage must be considered. See Chapter III.

The use of permanent ballast in surface ships occurs most often in ships that are being used for some purpose other than that for which they were designed. A common example is the naval auxiliary which has been converted from a merchant type, carries little or no cargo and has had considerable topside weight added.

There is a common feeling that the installation of permanent ballast in new designs is undesirable, and an indication that the design is less than optimum. This is not necessarily the case. The outstanding example of essential ballast installation is the case of the submarine, as previously discussed.

12.4 Weight Removal. Removal of topside weight is the most effective method of improving stability because it will not only lower the ship's center of gravity, but will also, in most cases, cause the center of buoyancy to move farther to the low side as the ship is inclined. Examination of the slope of the cross curves will indicate the increase in righting arm that will be obtained because of the decrease in displacement, to which will be added the increase in righting arm produced by lowering of the ship's center of gravity.

Removals of topside weight are particularly advantageous in improving the stability of a ship that has little freeboard, when the addition of low weight may be completely ineffective in increasing the righting arms at significant angles of heel. In addition, weight removal is generally beneficial in improving the ship's resistance to foundering after underwater damage.

12.5 Loading Instructions. The studies of stability and trim made during the design stage may indicate that certain limitations must be placed on the ship's loading in order to obtain satisfactory characteristics. These should be transmitted to the operating personnel. This is particularly important in the case of stability problems, because unsatisfactory stability, unlike unfavorable drafts or list, is an invisible source of potential danger, and the ship may be operated unknowingly in an unsafe condition.

These limitations, together with information regarding the drafts and stability resulting from a variety of probable or possible loading conditions are usually presented to the owner by the designer in the form of a *Stability Booklet*, which in the United States must be approved by the U.S. Coast Guard for all merchant ships before a certificate of approval is issued. This booklet also usually contains the information regarding *required* $\overline{GM}$ referred to in Section 3 of this chapter and in Chapter III. The U.S. Navy provides similar loading instructions to its operators.

Section 13
Stability When Grounded

The problem of the stability of grounded ships is limited, in general, to the dry docking of ships of relatively small $\overline{GM}$, and to salvage operations.

13.1 Stability During Dry Docking. When a vessel enters dry dock it generally has a trim and hence the keel makes an angle with the keel blocks.[3]

As the water level falls, due to the pumping out of the dry dock, the keel of the ship comes into contact with the keel blocks. Vessels are usually trimmed by the stern, in which case the after part of the keel will ordinarily touch first. The weight that is supported by the keel blocks at any subsequent time is the difference between the displacement when fully water borne and the displacement to the waterline in the aground condition. As the water continues to recede, the slope of the keel gradually approaches the slope of the keel blocks.

---

[3] Frequently the keel blocks of dry docks are given a slight slope to facilitate dock drainage and also to make it easier to dock ships which trim by the stern.

The force exerted by the keel blocks has the same effect on the $\overline{GM}$ of the ship as would the removal of a corresponding weight from a position at the vessel's keel. If the bilge blocks are considered as contributing to the ship's stability, the condition of minimum stability occurs when the keel blocks first make contact throughout the entire length of the keel just before the workmen have had time to haul the bilge blocks into place.

If the weight on the keel blocks materially reduces the effective $\overline{GM}$, a ship with a small fully water-borne $\overline{GM}$ may become unstable and list to an appreciable angle before the bilge blocks can be hauled. To avoid this situation a ship with little $\overline{GM}$ should be trimmed[4] as nearly as practicable to the slope of the keel blocks before it enters the dry dock. The vessel can then be lowered in such a manner that all of the blocks will touch the keel at approximately the same time. The bilge blocks and the shores to the dock sides can be placed in their proper positions before much weight has been placed on the blocks, and hence before appreciable loss in effective $\overline{GM}$ has taken place.

When a ship has just landed on the keel blocks, part of its weight is borne by the blocks and part of it is water borne. Consider the ship in Fig. 57 resting on the keel blocks with the waterline at W_1L_1 which is below the water-borne waterline WL.

P = upward force exerted by keel blocks
M_1 = metacenter at waterline W_1L_1
G = center of gravity of the water-borne ship
W = displacement of the water-borne ship
$W - P$ = water-borne displacement at waterline W_1L_1
$\overline{GM}_1$ = virtual metacentric height of ship at waterline W_1L_1

If the force P is considered as a weight removed from the ship at point K, the center of gravity of the ship rises to point G_v. This point is the virtual center of gravity of the partially water-borne ship. The virtual rise of the center of gravity is

$$GG_v = \frac{\overline{KG} \cdot P}{W - P} \qquad (30)$$

The virtual height of the center of gravity will be

$$\overline{KG}_v = \overline{KG} + \frac{\overline{KG} \cdot P}{W - P}$$

$$\overline{KG}_v = \frac{\overline{KG} \cdot W}{W - P} \qquad (31)$$

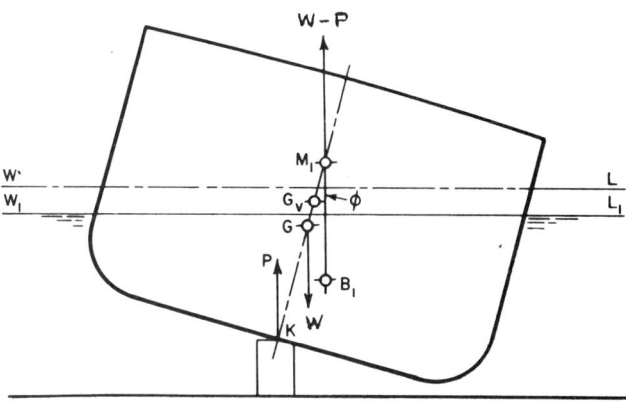

Fig. 57 Stability in dry deck

The virtual metacentric height of the grounded ship is

$$\overline{G_vM_1} = \overline{KM_1} - \frac{\overline{KG} \cdot W}{W - P} \qquad (32)[5]$$

If the value of $\overline{G_vM_1}$ is low or negative, there is danger that the ship will assume a list before the bilge blocks are hauled unless it is otherwise supported.

When a vessel enters dry dock with a trim, generally the most critical stage of docking is at the time when the vessel's keel comes in contact with the keel blocks throughout its length. To investigate this special condition and to determine the force P at this stage the following calculations may be made:

(a) Calculate the displacement and LCB of the fully water-borne ship for the trim at which it enters dry dock.

(b) If the vessel is trimming by the stern and hence the after end of the keel touches the keel blocks first, calculate the moment of weight about the after end of the keel. This will be the product of the water-borne displacement and the horizontal distance from the after keel block to a point vertically below the LCB.

(c) Draw several waterlines at an inclination to the keel equal to the slope of the dry-dock blocks representing different displacements both greater and less than the water-borne displacement. Calculate the displacement and moment of buoyancy about the after end of the keel for each of these sloping waterlines. If the keel blocks should be level, then these waterlines will be parallel to the keel.

---

[4] Often it is not practicable to place the vessel in the condition indicated. In that event is must be handled with special care during the critical interval between the grounding at one end of the ship and the hauling of bilge blocks. If the trim is not excessive, the ship can usually be held upright by judicious use of side shores until the keel bears fore and aft, when the bilge blocks are hauled.

[5] Some authorities use the expression

$$\overline{GM_1} - \frac{P\overline{KM_1}}{W}$$

for the virtual metacentric height. This value is used in conjunction with the full water-borne displacement W to obtain the righting moment.

(d) On graph paper, with displacements as abscissas and the moments as ordinates, plot a curve of moment of buoyancy about the after end of the keel. Also plot the moment of weight on the same graph (this will be a straight horizontal line). The point at which the moment-of-buoyancy curve crosses the moment-of-weight line gives the displacement at the time the keel first bears fore and aft on the keel blocks. The difference between this displacement and the fully water-borne displacement is the weight supported by the keel blocking; i.e., force P in Fig. 57. Having determined P, one can calculate the virtual metacentric height from Equation (32) above.

The problem of stability during docking is often eliminated entirely, especially for large heavy ships, by providing a *cradle* of bilge blocks and cribbing, shaped to conform to the contour of the bottom of the ship. This requires accurate control of the fore-and-aft position of the ship, which is not necessary if sliding bilge blocks are used.

13.2 Stability Stranded. When a ship is stranded on a fairly flat bottom, there is no question of transverse stability.

There is only a remote possibility of a stranded ship capsizing as the result of ebbing tide. For this to occur it would be necessary for the ship to be grounded on a bottom such that there is no restraint to heeling in one or both directions until a very large angle is reached, as, for example, on a peak which was considerably higher than the surrounding bottom. When a ship is aground in this manner, as illustrated in Fig. 11(b), the heel would increase as the tide ebbs. The attitude of the ship would always be such that the moment of buoyancy equals the moment of weight, or $W \cdot b$ in Fig. 11(b) would be equal to $(W - R)a$. Before the ship could capsize, leaving the point of support, the reaction R must be reduced to zero. Since $Wb = (W - R)a$, when $R = 0$ a and b would be equal, or G would be directly above B_1. The situation would be the same as if the ship were heeled to its range of positive stability, and the angle of heel could be determined from the statical stability curve corresponding to the ship's weight and the position of its center of gravity. It can therefore be said that a stranded ship will not capsize, in the absence of other upsetting forces, until it reaches an angle of heel equivalent to its range of positive stability if it were afloat.

The following conclusions may be drawn:

(a) It is unlikely that a stranded ship will capsize unless its range of positive stability is much less than usual. Unless impaled, the ship would slide from the point of support when the tangent of the angle which the bottom of the ship makes with the horizontal exceeds the coefficient of static friction between the bottom of the ship and the support. This angle is generally much smaller than the range of positive stability.

(b) If the angle of inclination approached the range of positive stability, only a relatively small strain would be required to free the ship as the reaction of the support approached zero. The point of application of the tow-line should be low, since only a small heeling moment would be required to capsize the ship.

(c) The likelihood of capsizing with the expected variation in tide can be evaluated by assuming the ship heeled to its range of positive stability, drawing the waterline at the lowest expected level relative to the point of support, and estimating the displacement below this waterline. If this displacement exceeds the weight of the ship, the range of positive stability will not be reached.

Section 14
Intact Stability of Unusual Ship Forms

14.1 Floating Platforms. Semi-submersible drilling platforms obtain static stability from surface-piercing columns that connect their submerged flotation bodies to the above-water platform. Analysis of their ability to withstand the upsetting forces of winds and waves (under varying loading conditions) is similar to the type of analysis made for conventional ship forms. The intact statical stability curve and the appropriate heeling arm curve superimposed on the stability curve are used to assess adequacy of stability. As noted in Section 7.2, the minimum regulatory criterion for intact stability is published by USCG in the Code of Federal Regulations (46 CFR 174).

Recent model experiments have shown (Numata, Michel and McClure, 1976), however, that design criteria which emphasize overturning may be inappropriate, since even with high winds and seas, the likelihood of pure wind/wave-induced overturning is minimal. The above authors state, "The major needs for adequate stability are indicated to be for lessening the possibility of wave impact on the upper structure in heavy weather, and for minimizing motions due to secondary effects..." such as wave-induced heeling. These are problems discussed in Chapter VII.

The American Bureau of Shipping publication, *Rules for Building and Classing Offshore Mobile Drilling Units*, contains specific criteria for drilling units and may be used as one source for such criteria.

14.2 Advanced Marine Vehicles.

(a) *General.* Various types of advanced marine

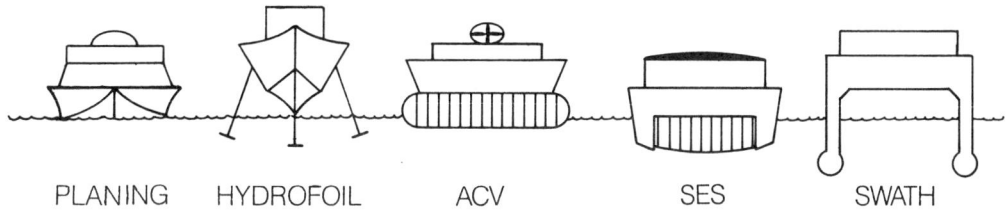

Fig. 58 Types of Advanced Marine Vehicles

vehicles are able to attain unusually high speeds by means of special design features or devices, such as dynamic lift, fan-generated lift or narrow, twin hulls. A number of these are shown diagramatically in Fig. 58. For such craft, separate treatment is required when dealing with intact stability for low-speed-displacement and high-speed modes of operation.

When operating at low speeds in the hullborne or displacement mode, their stability problems are similar to those of conventional displacement ships. In general, high-performance craft are exposed to the same hazards as conventional ships, and thus the requirements for adequate stability and buoyancy to resist the effects of these hazards are similar (Section 7.2). Hence, the same general approach to evaluating stability is used for the advanced vehicles. Additionally, special problems peculiar to some of these vehicles may arise, all of which should be considered in the stability and buoyancy analysis. Examples of such problems are: low freeboard, with the hazard of shipping and trapping sea water, large center of gravity shifts due to extension or retraction of foils, weight constraints that might limit the number of watertight bulkheads, thin shell structure which is susceptible to damage from impact with waves or debris at high speeds or from collisions and, for the low waterplane catamaran types, the potential for large unsymmetrical flooding. Although the submerged foils of hydrofoil craft generally have a small effect on static stability, they do have a favorable effect on damping of rolling motion. Specific intact stability criteria for advanced marine vehicles in the displacement condition are contained in Goldberg & Tucker (1975).

At the high speeds for which advanced marine vehicles are designed, however, their stability characteristics and the means of achieving stability are very different from those of conventional ships and vary considerably from one type to another. In November 1977 IMO published a proposed code of safety for submerged foil-borne craft (IMO, 1977) but only very general guidance on the subject of stability for high-speed operation was included. In 1978, the U.S. Coast Guard initiated a study with the objective of developing intact stability criteria for the high-speed operation of advanced marine vehicles (Band, Lavis, 1979, 1981). Even though preliminary criteria were proposed, a great deal of work still requires to be performed before generalized definitive criteria can be established. Acceptable criteria should encourage the development of safe designs without arbitrarily or unnecessarily restricting the designer.

(b) *Planing Hull.* Compared with other types of advanced marine vehicles, the planing boat is well known and has been in wide use for very many years. For chine boats, the question of transverse stability has, over the years, been given very limited treatment. However, the fact that these craft were always beamy, with very large transverse $\overline{GM}$ when planing, seems to have ensured satisfactory stability in service. For the deep-V, high deadrise planing hulls, it has been found that longitudinal strakes or spray strips must be used to obtain satisfactory transverse stability.

Planing hulls tend to be considerably stiffer in roll when operating at high speeds than when at rest. This is particularly true of low deadrise hull forms. As deadrise is increased, the stability is decreased. Longitudinal instability can result in *porpoising*, which is discussed by Martin (1976).

Existing stability standards do not necessarily recognize the special considerations of the high speeds achievable by modern-day passenger-carrying planing (or semi-displacement) craft. Stability problems of planing craft have, traditionally, been solved empirically and successfully by simple, practical remedies such as the use of ballast to move the center of gravity forward or the use of transom flaps or *wedges* to reduce the running trim angle. In any case, the modes of instability that do occur, during the operation of planing craft at moderate speeds, are normally rather mild and can be avoided by the operator by decreasing trim (by thrustline or transom-flap control), by reducing speed or by moving passengers or crew forward.

(c) *Small Waterplane Area Hull (SWATH).* The SWATH is essentially a displacement catamaran. The small waterplane area featured in its design serves to decrease wave drag and to provide good seakeeping qualities. Intact stability must be provided in the same way as for a conventional ship. Ballast tanks are usually provided to allow trim and draft to be carefully controlled and to provide damage control in the event of loss of buoyancy in any compartments. Some SWATH designs include horizontal control surfaces to provide additional trim control and stabilization at high speed. Since longitudinal metacentric height, $\overline{GM}_L$, will generally be of a similar magnitude to transverse metacentric height, $\overline{GM}_T$, stabilization of heave, pitch and roll are possible. This is unlike a conventional monohull for which only roll stabilization is usually feasible. Stability standards for SWATH forms are adequately covered in Goldberg, et al (1975).

(d) Hydrofoil Craft. In the case of hydrofoil craft, intact stability while foilborne is attained entirely by hydrodynamic means, although there may be some incidental aerodynamic effects. Surface-piercing foils make it possible to attain stability passively, i.e. without automatic controls, since heel to one side will result in increased immersed foil area and hence lift on that side—and less on the opposite side—which produces a righting moment. However, some adjustment of one or more of the fixed foils is usually possible, particularly to control running trim. Sometimes electronic stability augmentation systems are provided, actuating trailing edge flaps, to improve the ride quality.

For hydrofoil craft with fully submerged foils, an automatic control system or autopilot is essential, not only for transverse stability but for attaining proper elevation and trim relative to the water surface. The control system operates either by changing the angles of incidence of the foils individually or by adjusting trailing edge flaps. Such automatic systems can also control to some extent the wave-induced motions of pitch, heave and roll in moderate seas. See Mantle (1980).

Current practice indicates that the nondimensional foilborne longitudinal metacentric height ($\overline{GM}_L/L_T$) for a surface-piercing foil craft should lie between 3.5 and 5.5, where L_T is the longitudinal distance between foils. Also, the yawing moment contribution of the aft foils should be at least 20 percent greater than the moment contribution of the forward foils for adequate directional stability.

One approximate method for determining the required foilborne transverse metacentric height of a surface-piercing V-foil configuration is given in IMO (1977a), where it is shown that adequate transverse stability, for hydrofoil craft having V-foil systems, is obtained when

$$\left(\frac{\overline{GM}_T + H}{H}\right) \leq 1.9$$

where H is height of CG above water surface.

Boats with fully-submerged foils depend almost entirely on active controls for transverse stability. On the other hand, the available righting moment is usually so large at normal foilborne speeds that wind loads and off-center passenger loads have negligible effect. Only at the lowest foilborne speed, or when forced below minimum flight speed, are such disturbances considerable. The mode of escape in such an event is to ditch the craft, which can be done quickly by cutting the throttle and reducing the height command. Potential difficulties at takeoff can be avoided by choice of a suitable course with respect to wind and sea directions.

One important source of roll disturbance is the water velocity due to the orbital motion in waves. In beam seas, the horizontal component of the orbital velocity produces sideslip and resultant side forces on the struts, while the vertical component alters the angle of attack, and hence the lift, on the foils. At any foilborne speed, the righting moment obtainable from deflection of the ailerons (and the rudder, if heel-to-steer control is provided) must be larger than the heeling moment produced by any possible beam-sea condition within the craft's operational envelope.

No widely accepted stability criteria for the foilborne mode have been established. See Goldberg, et al (1975) and bibliography.

(e) Air Cushion Vehicles. When an air-cushion vehicle (ACV) operates in the cushion-borne mode, the craft is supported aerostatically by the air pressure in the cushion system. The transverse (or longitudinal) stability of a simple craft with a single air cushion and no flexible skirts or fixed sidewalls is inherently negative. The reason is that the uniform air pressure acting across the under surface provides an upward buoyancy force that always acts through the center of pressure (CP) of the upper surface of the air chamber (which is on the centerline of a symmetrical hull), while the downward weight force acting through the CG (usually above the CP) will be displaced in the direction of heel (or trim), thus producing an upsetting moment.

One method of improving stability at low speeds is to subdivide the air cushion by means of air jets or flexible skirt *keels*—at least one fore-and-aft and one athwartships. In the hypothetical case of little or no air leakage, when the craft heels (or trims), the air in the heeled-down compartment would be compressed, thus increasing the pressure, while the air in the raised compartment would expand, thus decreasing the pressure. The result would be a righting moment of buoyancy which would tend to counteract the effect of the off-center weight vector.

In actual ACVs air is supplied by a lift system of fans and leaks continuously through the open gap between the water and the craft's seal system. In early designs, the seal system consisted of peripheral jets or skirts that are flexible and designed so that heel or trim will give rise to a skirt deflection that will shift the geometrical center of pressure to cause a righting moment, Fig. 59. In some early designs, these air leakage forces were partially controlled with variable geometry skirts. As skirt designs have progressed the complex peripheral jets and controllable geometry designs have been replaced by much simpler and more effective skirts of the *bag-and-finger* type. If cushion compartmentation as well as flexible skirts are provided, additional stability will result. When heeled the gap and hence air leakage in the down side are reduced

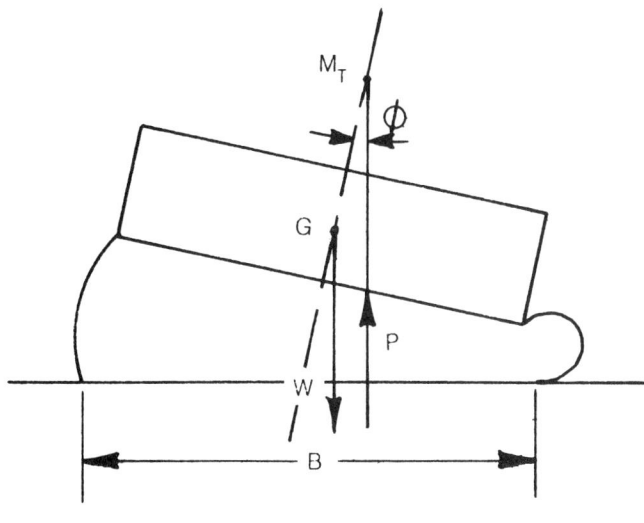

Fig. 59 ACV stability

and the air pressure increases while the leakage increases on the high side, with a corresponding pressure drop. Again a righting moment results.

Under way, particularly at higher speeds, the air cushion vehicle develops aerodynamic and hydrodynamic forces and moments affecting the craft's stability, and these must be considered in design. Incidents of trim and heel instability have occasionally caused some craft to *plow-in* and/or capsize, usually during turning maneuvers under severe weather conditions. Hence, efforts must be made by experiments and full-scale trials to ensure that dynamic effects on stability are favorable.

It is customary among ACV designers to make use of a non-dimensional transverse stiffness measure for evaluating stability, $\overline{GM}_T/B$, where $\overline{GM}_T$ is the transverse metacentric height on cushion and B is the overall beam. In calculating or estimating $\overline{GM}_T$ the principal effect to be taken into account is the shift of the center of cushion pressure with heel relative to the vertical through the CG. For convenience, approximate empirical relations have been devised for the design evaluation of stiffness.

Criteria for minimum acceptable values of on-cushion stiffness have been adopted by designers in individual cases on the basis of specified maximum angles of heel under the influence of various possible heeling moments. More generally, statistical analyses of casualty records have indicated acceptable values of stiffness for different types of craft. However, no widely accepted criteria have yet been established. Information on high-speed cushion-mode criteria is given by Trillo (1971), Mantle (1980) and Wachnik and Livingston (1980), as well as Band, Lavis Assoc. (1979).

(f) Surface Effect Ship (SES). Surface Effect Ships (SES) utilize another method of obtaining transverse stability: immersed sidewall structures containing sufficient volume to provide significant buoyancy and dynamic lift when under way in the cushion-borne mode. The static shift of center of buoyancy when the craft heels can then be calculated and stability determined in a similar manner as for a conventional hull, except that the waterline inside the hulls is lower than on the outside. Longitudinal stability is attained by the use of flexible seals at bow and stern, supplementing the buoyant moments of the sidewalls.

Transverse stability must be carefully evaluated during the design stages. This is particularly true for SES with high ratios of length to beam. The Navy's XR-5, for example, which has a length-to-beam ratio of 6.54, was subjected to a series of rudder-reversal tests, in simulated collision-avoidance maneuvers, with a range of CG heights, to determine acceptable ratios of CG height-to-beam.

As in the case of ACVs, no formal criteria for transverse and longitudinal stability have been established. But acceptable values of $\overline{GM}_T/B$ have been tentatively determined by empirical analysis of experimental and casualty data for different types of SES. See Band, Lavis (1981) and Lavis, et al (1982).

REFERENCES

Alaska, University of (1980), "Superstructure Icing Forecast Guide for Alaskan Waters," Nov.

American Bureau of Shipping (1980), *Rules for Building and Classing Offshore Mobile Drilling Units.*

Amy, J.R., Johnson, R.E., Miller, E.R. (1976), "Development of Intact Stability Criteria for Towing and Fishing Vessels," *Trans.* SNAME.

Band, Lavis & Associates (1979) *Background Study of Intact Stability Standards for Dynamically Supported Craft,* Volumes I through VI, U.S.C.G. Report No. CG-D-75-79.

Band, Lavis & Associates (1979) *Classification of Intact Stability Standards for Dynamically Supported Craft,* U.S.C.G. Report No. CG-D-76-79.

Band, Lavis & Associates (1981) *Development of Intact Stability Standards for Rigid-Sidehull Surface-Effect Ships,* U.S.C.G. Report No. CG-D-53-81.

Civil Aviation Authority (U.K.) (1975) "Report of the ARB Special Committee on Hovercraft Stability and Control," CAA75017, June.

Coast Guard, U.S. (1973), Navigation and Vessel Inspection Circular, NVC 3-73, quoting IMO Res. 167, Recommendations on Intact Stability for Passenger and Cargo Ships under 100m. (328 ft)

Coast Guard, U.S. (1976), Navigation and Vessel Inspection Circular No. 3-76, quoting IMO Res. 168, Recommendations for Safety of Fishing Vessels.

Code of Federal Regulations, Title 46—Shipping, Chapters I and II, containing U.S. Coast Guard stability regulations applicable to commercial ships.

Commerce Dept., U.S. (1959), *Climatological and Oceanographic Atlas for Mariners*, Vol. I, North Atlantic Ocean.

Cromer, P.B. (1981), "Calculating the Cross Curves of Stability Using the Hand-Held Programmable Calculator and Printer," *Marine Technology*, April, p. 188.

Garzke, W.H., Johnson, R.E. and Landsburg, A.C. (1974), "Trim, Stability and Loading Manuals," *Trans.* SNAME, Vol. 82.

Goldberg, L.L., and Tucker (1975), "Current Status of Stability and Buoyancy Criteria Used by the U.S. Navy for Advanced Marine Vehicles," *Naval Engineers Jo.*, Oct.

Green, P.V. (1980), "The Hazard of Flow in Bulk Mineral Cargoes," (Warren Springs Lab., U.K.), *Safety at Sea*, Jan.

IMO (1973), *Grain Regulations*, Resolution A.264(VIII), Amendment to Chapt. VI of the Int. Conv. for SOLAS, 1960. Available in *General Information for Grain Loading*, National Cargo Bureau, Inc., New York, 1978.

IMO (1977), International Convention for Safety of Fishing Vessels, Res. A.168. See Coast Guard, 1976.

IMO (1977a), *Code of Safety for Dynamically Supported Craft*, Resolution A-373(x), November 14.

IMO (1980), *Code of Safe Practice for Solid Bulk Cargoes*, National Cargo Bureau, Inc., New York.

Kiss, R.K. (1980), Chapter I, "Mission Analysis and Basic Design," *Ship Design and Construction*, R. Taggart, Ed., SNAME.

Kuo (1975), "Proceedings of the International Conference on Stability of Ships and Ocean Vehicles," University of Strathclyde, Glasgow (24 papers and discussions.)

Kuo, C., and Welaya, Y. (1981), "A Review of Intact Ship Stability Research and Criteria," *Ocean Engineering*, Vol. 8, pp. 65-84, London.

Lavis, D.R. (1980), "The Development of Stability Standards for Dynamically Supported Craft, A Progress Report," High-Speed Surface Craft Exhibition & Conference, U.K., 24-27 June.

Lavis, D.R., Band, E.G.U., and Hoyt, E.D. (1982), "The Development of Stability Standards for Rigid-Sidehull Surface-Effect Ships," *Trans.* Royal Institution of Naval Architects, Second International Symposium on Small Fast Warships and Security Vessels, U.K., 18-21 May.

Mantle, P.J. (1980), "Air Cushion Craft Development," DWTNSRDC report 80/012, Jan.

MarAd (1962), "Classification of Weights—Standard Grouping for Merchant Ships," U.S. Maritime Administration (Clearing House for Federal Scientific and Technical Information, Springfield, VA 22151.)

Martin, Milton (1976), "Theoretical Determination of Porpoising Instability of High-Speed Planing Boats," DWTNSRDC Report 76-0068, April.

National Cargo Bureau (1978) U.S. Salvage Association, New York.

NAVSEA (1976), Ship Hull Characteristics Program, SHCP, *User's Manual*, CASDAC No. 231072.

NAVSEA (1975), *Stability and Buoyancy of U.S. Naval Surface Ships*, Design Data Sheet DDS 079-1, Dept. of the Navy.

NAVSEA (1978), *Weight Control of Naval Ships*, S9096-AA-WCM-010/(u) WT CNTRL.

NAVSEA (1977), *Naval Ships Technical Manual*, NAVSEA 0901-LP-096-000, Chapt. 096.

Niedermair, J.C. (1932), "Stability of Ships After Damage," *Trans.* SNAME, Vol. 40.

Numata, Michel and McClure, (1976) "Assessment of Stability Requirements for Semisubmersible Drilling Units," *Trans.*, SNAME Vol. 84.

Paulling, J.R. (1974), "Studies of Ship Capsizing in Heavy Seas," Pacific Northwest Local Section SNAME, March.

Rahola, J. (1939), "The Judging of the Stability of Ships and the Determination of the Minimum Amount of Stability," thesis for degree of Doctor of Technology, Helsinki.

Rawson, K.J., and Tupper, E.C. (1965), *Basic Ship Theory*, Vol. I.

Sarchin, T.H. and Goldberg, L.L. (1962), "Stability and Buoyancy Criteria for U.S. Naval Surface Ships," *Trans.* SNAME, Vol. 70.

SOLAS (1948, 1960) see IMO (1973).

Trillo, R.L. (1971), "Marine Hovercraft Technology," U.S. Naval Institute *Proceedings*.

Wachnik, Z.G., and Livingston, W.H. (1980), "ACV Lift Systems Flow Simulation Model," AIAA 18th Aerospace Sciences Meeting, Trim Section, Jan., Paper No. 80-0222.

Wise, J.L., and Comiskey, A.L. (1980), "Superstructure Icing in Alaskan Waters," NOAA Special Report, April.

CHAPTER III

George C. Nickum

Subdivision and Damage Stability

Section 1
Introduction

1.1 General. All types of ships and boats are subject to the risk of sinking if they lose their watertight integrity, whether by collision, grounding or internal accident such as an explosion. (Exceptions, of course, are vessels constructed entirely of buoyant materials and having mostly buoyant contents.) Such accidents are frequent enough in practice that some degree of protection against the effects of accidental flooding is an essential feature of the design of any water craft. The most effective protection is provided by internal subdivision by means of watertight transverse and/or longitudinal bulkheads, and by some horizontal subdivision—double bottoms in commercial ships and watertight flats in naval vessels. Such protection is by no means new, for Marco Polo, near the end of the thirteenth century, referred to watertight bulkheads in Chinese junks.

The flooding of a ship's hull can have one or the other of two principal consequences. One is loss of buoyancy and change of trim, which if unchecked will lead to sinking by foundering. The other is loss of transverse stability or build-up of such an upsetting moment that capsizing takes place. The nature and arrangement of the internal subdivision to control these effects will be discussed in this chapter. However, it will be shown that there are many uncertainties in providing adequate subdivision. First of all, the location and extent of damage to be protected against is unknown in advance. Second, the amount, type and location of cargo and liquids in the ship varies both during and between voyages. Finally, the designer cannot be sure that corrective measures that might be followed by the ship's officers in an emergency will be taken—or that hazardous steps might be adopted by mistake. Furthermore, subdivision inevitably adds to the cost of the ship and may interfere with its ability to perform its function economically. In fact, a ship so ideally compartmented as to be virtually unsinkable might be of no economic or military value whatsoever.

Consequently, the subdivision of ships inevitably involves a compromise between safety and cost. This dilemma has been partially resolved for passenger ships by the development of national and international standards of what is considered acceptable, considering the size and type of ship, the number of passengers carried, the nature of the service, etc. Hence, considerable attention will be given in this chapter to the development of such passenger ship standards and to the current status of the various standards now in effect.

It will be seen that for cargo ships subdivision standards have been minimal. However, for these vessels, the above dilemma of cost versus safety can be resolved more scientifically, provided that loss of human life can be ruled out by virtue of reliable provisions for lifesaving—with time to use them provided by some subdivision. This approach is based on probability calculations to minimize the *expected cost*, which is the sum of the extra cost of the subdivision and the so-called *failure cost*, the total value of the ship and its contents multiplied by the probability of its loss during its lifetime. This procedure has been applied to design problems of off-shore structures (Freudenthal, 1969),[1] but has not yet been standardized by regulations.

For passenger ships, where it is unrealistic to assume no risk to human life, the problem is that it is difficult to put a dollar value on a life. However, efforts have been made to evaluate statistically the levels of risk actually found in different modes of transportation and hence to draw conclusions regarding what people generally consider acceptable (W. European Conf., 1977). Furthermore, court decisions have in effect established a value for human life by the awards made in actions brought before them.

It will be shown in this chapter how a trend toward

---
[1] Complete references are listed at end of chapter.

probabilistic solutions is well along in the area of ship subdivision standards. A similar probabilistic approach is being applied in the field of ship structural design, as discussed in Chapter IV. The probability approach can also be extended to cover the cost of environmental damage resulting from accidents involving ships.

1.2 Historical. In the late nineteenth century, classification societies established empirical rules for the installation of bulkheads in merchant ships, primarily fore and after peak bulkheads and bulkheads separating the machinery space from cargo holds. These classification society provisions were not based on any specific degree of floodability (and are not today).

In the late nineteenth and early twentieth century the major maritime nations were studying means for determining the capability of ships to resist flooding and various proposals were put forth by governmental regulatory bodies. This interest was spurred by the increasing number of maritime disasters involving large losses of human life, culminating in the loss of the *Titanic* with 1430 lives in 1912. In 1913 an international conference on Safety of Life at Sea considered studies by the British Board of Trade, regulations established by the Society of German Shipowners and a factorial system proposed by the French. The regulations formulated were a compromise among these systems, but owing in part to the advent of World War I, never came into effect. In 1929 another full International Conference on Safety of Life at Sea was convened (Tawresey, 1929). An agreement was finally reached on a factorial system of subdivision employing a *criterion of service* formulation, aimed at evaluating for any given vessel the relative importance of cargo and passenger carrying functions. This system of subdivision, discussed in Section 7, left much to be desired and, except for vague generalities, stability was not considered.

The United States did not immediately ratify the 1929 convention and until 1936 had essentially no mandatory subdivision requirements. Public opinion was aroused by the loss of the *Mohawk* by collision and the *Morro Castle* by fire. The investigations into these losses led to proposed standards of subdivision and stability which substantially improved on the 1929 convention, and these were put into U.S. regulatory form that year. With modifications, these standards are still incorporated in the present U.S. Regulations for Passenger Vessels. Also in 1936 the Maritime Commission was established to revive the American Merchant Marine and to establish an American fleet capable of supporting our military forces in time of war. Since they were considered as naval auxiliaries, all large ships built under subsidy or with the financial support of the Maritime Commission, were, and still are, required to meet not less than a one-compartment standard of subdivision, i.e., they are designed to stay afloat with any one compartment flooded.

Later International Conferences on Safety of Life at Sea (SOLAS) took place in 1948 and 1960, resulting in only minor modifications, adopting higher standards for vessels carrying large number of passengers on short voyages (cross-channel type), and increasing the proportion of ships required to meet a 2-compartment standard. At the 1960 SOLAS Conference, however, new concepts were proposed for future discussion (Richmond, 1960). One stated that the safety of a ship can be measured by the extent of damage it could survive. Another dealt with the survival of damage on a probability basis. While these new concepts were under discussion the capsizing of the liner *Andrea Doria*, which had been built to the 1948 Standards, gave a shocking demonstration of the inadequacies of practical application of the stability provisions of the 1948 convention. This spurred the maritime nations to agree upon a continued study of this subject under the auspices of the then-called Inter-Governmental Maritime Consultative Organization (IMCO),[2] [See 71A] the United Nations affiliate that had been established in 1958.

During the 1960's there were two significant developments. First, the regular meetings of the Technical Subcommittee of IMCO on Subdivision and Stability brought together the technical representatives of the majority of maritime nations in the world. The exchange of information on casualty data and accident experience of all individual countries stimulated the members to strive for more rational and sounder criteria of subdivision. Second, there was a simultaneous expansion throughout the world in the utilization of computers for the calculation of vessel characteristics and subdivision capabilities. From the days in the 1930's and 1940's when the calculation of floodable length curves and damage stability characteristics took weeks of manual calculations, equal or better results could now be obtained by relatively few man-hours of naval architect's labor and a few minutes of modern computer time.

During these years the dramatic rise in concern over the environment, first evidenced in the 1954 Conference on Marine Pollution (MARPOL), coupled with the above mentioned increase in international activities in the subdivision field, resulted in the inclusion of a permissive standard for tanker subdivision in the 1966 U.S. Load Line Act. This was made a required standard in the 1973 MARPOL, and in the U.S. Ports and Waterways Act of 1972.

Meanwhile, proponents of the probabilistic approach to subdivision were discussing and arguing the matter extensively at IMCO. They were successful in getting support for an IMCO Resolution which in 1973 accepted this new approach to subdivision and adopted new standards based on the probabilistic approach, as a permissive alternative to the 1960 SOLAS Conven-

---

[2] IMCO (Inter-governmental Maritime Consultative Organization) recently has changed its name to IMO (International Maritime Organization).

tion requirements. The 1974 Conference on Safety of Life At Sea, which came into force in 1980, also endorsed this approach as an alternative, but continued to permit the basic factorial provisions of the 1960 Convention. The new alternative regulations are discussed in Robertson, et al (1974), "The New Equivalent International Regulations on Subdivision and Stability of Passenger Ships." See Section 8.

Up to 1970, international agreements on subdivision were limited entirely to passenger ships, which were defined as those carrying more than 12 passengers, and as a permissive rather than a required approach, to tankers. After 1970, however, in rapid succession, the IMCO group put forth international recommendations for the subdivision of bulk chemical carriers and liquefied gas carriers, for the required subdivision of tankers, and for the subdivision of mobile offshore drilling units. Some of these have also been incorporated in U.S. Coast Guard Regulations. Subsequently, a recommendation for subdivision of offshore supply vessels was issued and a recommendation for subdivision of large fishing vessels was contained in the 1977 Convention on the Safety of Fishing Vessels. A recommendation for special purpose craft, including oceanographic ships, training ships, fish processing and research ships, etc. is currently in the process of approval and issuance at IMO, as it is now called, for guidance of designers.

In all of these later conventions and IMO recommendations, the factorial system of subdivision has been dropped completely. The designated ships generally have to meet a one-compartment standard, with some types of ships required to meet a two-compartment standard at certain locations along their lengths. The approach to subdivision has also changed from the previous preoccupation with sinkage and trim to a more realistic treatment, whereby the damaged ship is analyzed for its floodability and stability in both the longitudinal and transverse directions.

With the technical representatives of the maritime nations still meeting regularly at IMO, changes and further development of international regulations and recommendations are inevitable. Naval architects in the future will have to keep in close touch with the regulatory authorities to ensure that their designs meet the regulations in effect at that time.

During the past century, while naval architects were being pushed to develop means of analyzing the effect of flooding on commercial vessels and the means of subdividing these vessels to improve their resistance to sinkage and capsizing after flooding, there was a similar push from naval authorities around the world to improve the ability of naval vessels to survive flooding from the sea due to action by an enemy, as well as by collision. While the causes and extent of the flooding and the required survival conditions are different between commercial and naval vessels, the principles used in analyzing and evaluating the effects of flooding are the same. In the United States, the Navy has been in the forefront of developing means and methods of calculation and analysis to be used in determining the effectiveness of compartmentation in resisting the effects of flooding.

1.3 Need for Standards. Standards are necessary as a basis for regulations if a regulatory body is to administer safety requirements equitably among all commercial vessels under its jurisdiction. National standards are generally established by countries to protect its citizens, its wealth in the form of cargo, and its environment, from the hazards of the sea. It is a truism that the more severe the standard adopted for subdivision and stability, the greater the probability that capital and operating costs will be increased. For example, too close spacing of bulkheads may unnecessarily increase both the first cost and operating costs and may also seriously restrict the vessel's usefulness.

If a nation were to adopt national standards which are in excess of the standards used by competing maritime nations, a competitive advantage would be given to the latter nations. Only by having international standards can these competitive pressures be equalized.

In the early years of international discussions on subdivision and stability, agreements on effective international regulations, particularly for damage stability, were impeded by fear that excessive $\overline{GM}$ and its effect on severity of rolling would seriously curtail the comfort and safety of passengers. This fear has largely been dispelled now as a result of years of satisfactory operating experience with higher $\overline{GM}$, along with the development of both passive and active systems of roll stabilization.

The naval architectural profession welcomes recognized standards which can guide in the many practical compromises between safety and cost which always arise in any new design. They are effective tools for the designers in resisting the demands of non-technical operators who often cannot grasp the reasons for effective compartmentation. They are, furthermore, very good protection against the possibility of liability for damages in the event of a casualty.

Although the U.S. Navy is also concerned with ship survival in case of collision or grounding, the primary consideration in the subdivision of naval combatant and personnel carrying ships is the probability of surviving an enemy hit or hits, with adequate residual stability and buoyancy. Consequently the Navy requires each type of new ship to be able to survive a specific extent of damage (as a percentage of length), although the standards are classified in some cases.

Before dealing with international standards, the fundamental effects of damage and methods of making calculations of subdivision and damage stability will be discussed.

Section 2
Fundamental Effects of Damage

If the shell of a ship is damaged so as to open one or more internal spaces to the sea, leakage will take place between the sea and these spaces until stable equilibrium is established or until the ship sinks or capsizes. It is impractical to design a ship to withstand any possible damage due to collision, grounding or military action. The degree to which a vessel approaches this ideal is the true measure of its safety.

The probability that a ship will survive damage resulting in flooding is dependent upon a number of variable interrelated factors (Comstock & Robertson, 1961). Some consideration of these factors is useful as a basis for design, as well as in making calculations.

2.1 Extent of Damage and Location and Number of Bulkheads. The length and depth of damage, and its location relative to transverse bulkheads, has a strong influence on probability of survival. In general, it might be expected that the more bulkheads the safer the ship. But damage may occur entirely between adjacent bulkheads or may involve one or more bulkheads. Hence, for a given length of damage, any increase in the number of bulkheads may actually increase the likelihood of bulkhead damage, which would reduce rather than increase the chances of survival. International subdivision standards, discussed in Sections 6-8, specify assumptions to be made regarding extent of damage and indicate how each standard deals with the question of possible damage on a bulkhead.

An international SOLAS working group assembled and analyzed statistical data on collisions, particularly of cargo and passenger ships, to determine location and extent of damage. The results indicated that damages may vary from only 1 or 2m to over 30m (100 ft) in longitudinal extent. Many low-energy collisions occur without producing any penetration below the waterline; i.e., without causing flooding. Very short damages are usually very shallow. Damages of intermediate longitudinal extent, say 6m (20 ft) to 15m (50 ft), may vary in penetration from moderate to deep. Such damages usually result from collisions at an angle of incidence near 90 deg. Under this circumstance, a high-energy collision may result in maximum penetration in association with appreciable longitudinal extent. Very long damages are most infrequent and are likely to be quite shallow.

Navy data on the location and extent of damages due to enemy action are classified, but assumptions made in calculations are discussed in Section 7.

2.2 Effects of Flooding. *(a) Change of draft.* The draft will change so that the displacement of the remaining unflooded part of the ship is equal to the displacement of the ship before damage less the weight of any liquids which were in the space opened to the sea.

(b) Change of trim. The ship will trim until the center of buoyancy of the remaining unflooded part of the ship lies in a transverse plane through the ship's center of gravity and perpendicular to the equilibrium waterline.

(c) Heel. If the flooded space is unsymmetrical with respect to the centerline, the ship will heel until the center of buoyancy of the remaining unflooded part of the ship lies in a fore-and-aft plane through the ship's center of gravity and perpendicular to the equilibrium waterline. If the $\overline{GM}$ in the flooded condition is negative, the flooded ship will be unstable in the upright condition, and even though the flooded space is symmetrical, the ship will either heel until a stable heeled condition is reached or capsize (see Section 3.) Trim and heel may result in further flooding through immersion of openings in bulkheads, side shell or decks (downflooding).

(d) Change of Stability. Flooding changes both the transverse and the longitudinal stability. The initial metacentric height is given by,

$$\overline{GM} = \overline{KB} + \overline{BM} - \overline{KG}.$$

When a ship is flooded, both $\overline{KB}$ and $\overline{BM}$ change. Sinkage results in an increase in $\overline{KB}$. If there is sufficient trim, there may also be an appreciable further increase in $\overline{KB}$ as a result. $\overline{BM}$ tends to decrease because of the loss of the moment of inertia of the flooded part of the waterplane. However, sinkage usually results in an increase in the moment of inertia of the undamaged part of the waterplane, thus tending to compensate for the loss. Also, trim by the stern usually increases the transverse moment of inertia of the undamaged waterplane, and vice versa. For ocean-going ships of usual proportions and arrangements, the combined effect of these factors is usually a net decrease in $\overline{GM}$. However, in ships of low beam-draft ratio or in ships having flare above the waterline up to the bulkhead deck, the net effect may be an increase in $\overline{GM}$. In case of damage to a deep tank, fluid runoff from the tank may reduce $\overline{KG}$.

(e) Change of Freeboard. The increase in draft after flooding results in a decrease in the amount of freeboard. Even though the residual $\overline{GM}$ may be positive, if the freeboard is minimal and the waterline is close to the deck edge, submerging the deck edge at small angles of heel greatly reduces the range of positive righting arm, $\overline{GZ}$, and leaves the vessel vulnerable to the forces of wind and sea. Fig. 1 illustrates this point.

Although, as previously noted, increasing the number of bulkheads increases the likelihood of damage to one or more bulkheads, it also reduces the extent of flooding in cases of damage that do not include these bulkheads. For such cases, it increases the freeboard

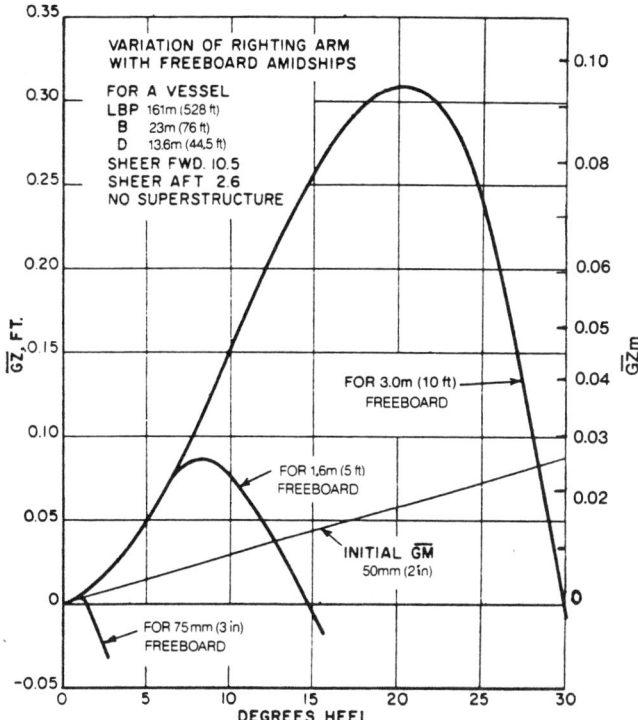

Fig. 1 Variation of righting arm with freeboard amidships

after damage and therefore the likelihood of survival.

For any given number of bulkheads, increasing the freeboard before damage increases the freeboard after damage and improves the chances of survival.

Where portions of the hull and related bulkheads extend watertight above the bulkhead deck, the beneficial effect may be considerable. Even spaces that are not fully tight may contribute to survival if they are tight enough to contribute righting or trimming moment effectively under dynamic conditions of rolling or pitching. For this purpose, tight spaces in the wings of the ship and the ends are most effective. Conversely, reserve buoyancy near the centerline amidships has little effect.

(f) Loss of Ship. Where changes in draft, trim and/or heel necessary to attain stable equilibrium are such as to immerse non-watertight portions of a ship, equilibrium will not be reached because of progressive flooding and the ship will sink either with or without capsizing. Where the loss in $\overline{GM}$ is such that the remaining maximum righting arm is less than any existing heeling arm, capsizing will occur. Even if there were no heeling arm, capsizing could be expected if the $\overline{GM}$ in the damaged condition were negative and if the maximum righting arm were so small as to result in negative dynamical stability. (Section 3.) Practically, even for symmetrical flooding, there is always some heeling arm due to unsymmetrical weights and/or wind.

Where the maximum righting arm in the damaged condition is adequate and where the immersion of non-tight portions of the ship only results in slow extension of flooding, sinking may be quite slow. In such cases, control measures aimed at stopping progressive flooding, either by reducing heel, pumping leakage water or fitting emergency means of checking the flow of water or a combination of such measures may be successful.

In recent deliberations at IMO, looking towards methods of improving the safety of cargo and other vessels where it is impractical to meet any fixed standard of subdivision, the consensus was that providing a master with an instruction manual outlining damage control measures available to minimize flooding would be a valuable contribution to safety.

The U.S. Navy has long recognized this and prepares and issues detailed damage control books and flooding effect diagrams, as do some commercial operators, to aid their crews in damage control. A Flooding Effect Diagram shows by a shading scheme the effect on stability of sea water flooding of various compartments in the ship, including transverse moments for off-center compartments. See Fig. 2.

Some naval ships such as aircraft carriers have layers of side protective tanks of liquid or air to protect the ship's vital ammunition and machinery spaces. In case of damage, flooding of the side protective tanks may produce a large list toward the side of the damage, which requires counterflooding of opposite side tanks to reduce the list. Trim control after damage to aircraft carriers can also be achieved to some extent by flooding tanks at the end opposite to that of the damage.

Rough sea conditions and/or heavy winds substantially increase the reserves necessary to prevent either progressive flooding or capsizing. Fortunately, most collisions occur in harbor approach areas under comparatively favorable sea conditions. An IMO working group has collected data on wind and sea conditions at the time of actual collisions. Damage stability model tests were then carried out in waves. Conclusions were drawn regarding average values of stability necessary for survival as a function of sea state. See Section 8.

2.3 Intact Buoyancy. The general effect of intact buoyancy in a flooded compartment is to decrease both sinkage and trim and therefore to increase the length which may be flooded insofar as sinkage and trim are concerned.

The effect of intact buoyancy on the change in stability with flooding depends upon whether the tops of the intact spaces are above or below the final flooded waterline. If they are above this waterline, the reduction of lost buoyancy, sinkage and rise of $\overline{KB}$, caused by the intact buoyancy, will be accompanied by a reduction in lost waterplane and in lost $\overline{BM}$, and the net effect may be simply that of reducing the effective size of the flooded compartment. If the tops of the W.T. intact spaces are below the final flooded waterline, they still will reduce the rise of $\overline{KB}$, but they will not reduce

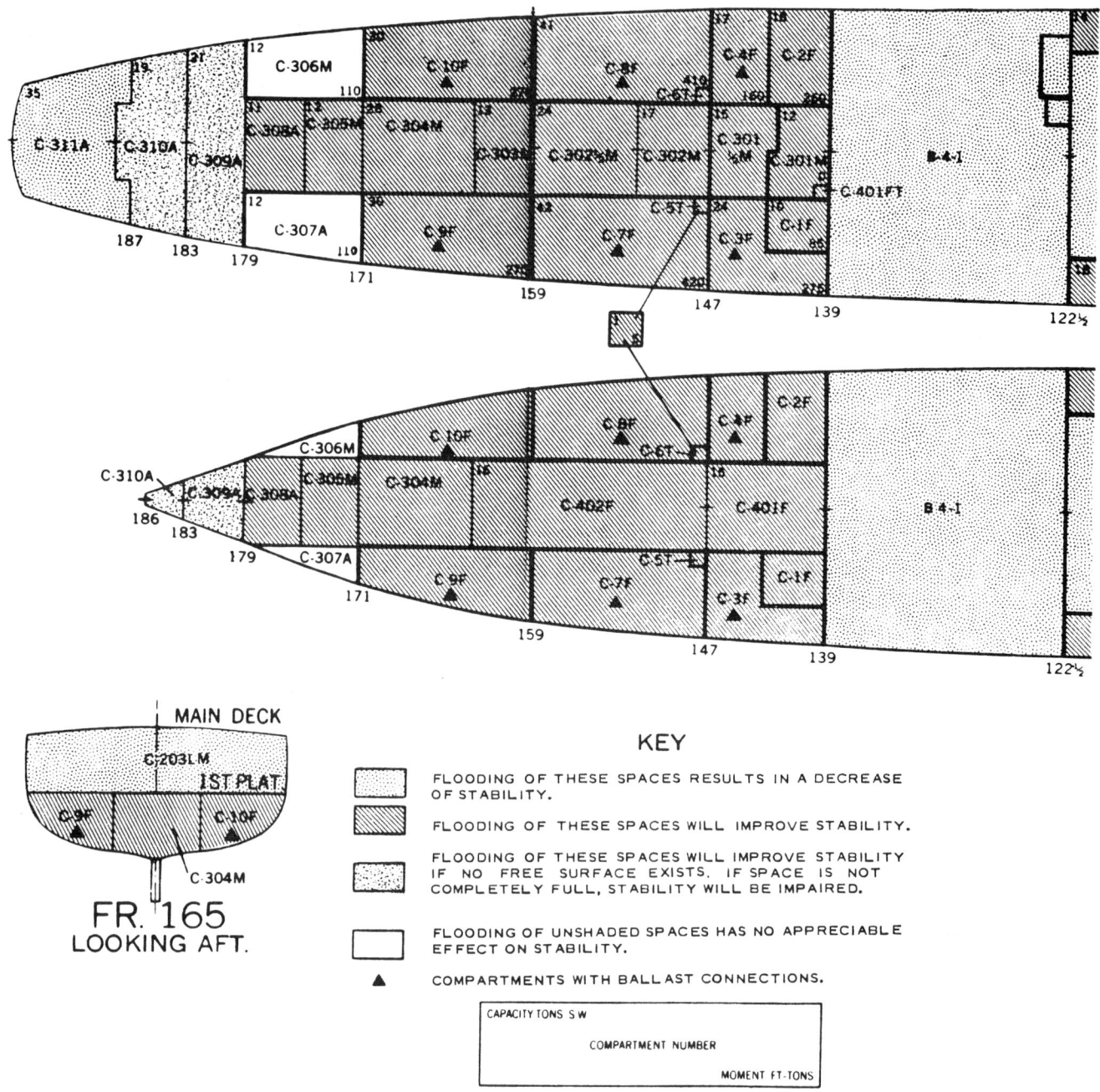

Fig. 2 Portion of Typical Flooding Effect Diagram for Destroyer

the loss of $\overline{BM}$. Hence, submerged intact buoyancy always increases the net loss of $\overline{GM}$. If there is a reserve of stability, but sinkage and trim are critical (as may be the case when end compartments are flooded), intact buoyancy is beneficial. If there is a reserve of freeboard when flooded, but stability-flooded is critical (as is commonly the case when amidships compartments are flooded), submerged intact buoyancy is harmful. If unsymmetrical intact buoyancy extends above the final damaged waterplane its favorable effect on the damaged-condition $\overline{GM}$ may, in some cases, compensate in part for its heeling moment. Thus, the use of high narrow wing tanks extending well above the waterline may be advantageous.

Section 3
Subdivision and Damage Stability Calculations

3.1 General. The *floodable length* at any point in the length of the ship is defined as the maximum portion of the length, having its center at the point in question, that can be symmetrically flooded at the prescribed permeability, without immersing the margin line, which is generally 7.5 cm (3 in.) below the top of the bulkhead deck at the side.

The accurate determination of the floodable length of any portion of a ship requires an analysis of sinkage, trim, and heel to determine that the margin line is not immersed. However, in the following method of calculation, sinkage and trim only are determined. The effects of heel are left to be analyzed by other techniques. When the methods of calculation to determine sinkage and trim were developed in the early part of the century, little thought was given to the effect of heel. The development of techniques to determine heel came later and by common usage the term floodable length has been applied to the curves developed with the understanding that they must be supplemented by damage stability calculations to find the actual allowable length for any compartment.

While the preparation of the curves of floodable length are not essential in regulations using integer compartmentation standards and required residual range and amplitude of righting arm, they are a very valuable tool in the preliminary design stage for any class of ship. They can be very rapidly and thus inexpensively plotted and give the designer a quick visual indication of the probable allowable lengths of compartments. This allows him to select those compartments that may be marginal and perform damage stability and trim checks on them. He can then make any length or weight corrections necessary before calculations are prepared for the rest of the compartments. In ships not required to meet any subdivision standards the curve permits the designer to easily space bulkheads, within the operating constraints imposed by the ship's mission, in such a way that at least certain parts of the ship can be flooded without loss of the ship.

The term *damage stability* is a coined usage having wide acceptance. It is a comprehensive term applying to the calculation of the related changes in draft, trim, heel, and stability as a result of damage to one or more specific compartments of a ship. It also relates to calculation of the intact stability and buoyancy necessary to attain any particular assumed equilibrium damaged condition, as well as to the stability characteristics in that damaged condition.

While the newer standards for damage stability use residual range and amplitude of the righting arm, $\overline{GZ}$, and righting energy represented by the area under the righting arm curves as the basic parameters for compliance with requirements, the use of $\overline{GM}$ is not neglected. The required parameters can be related back to a required $\overline{GM}$ in the intact condition and are invaluable in providing guidance to the operators as to how to load their ships to meet the required safety standards.

3.2 Fundamental Relationships—Floodable Length. A completely rigorous treatment of the sinkage and

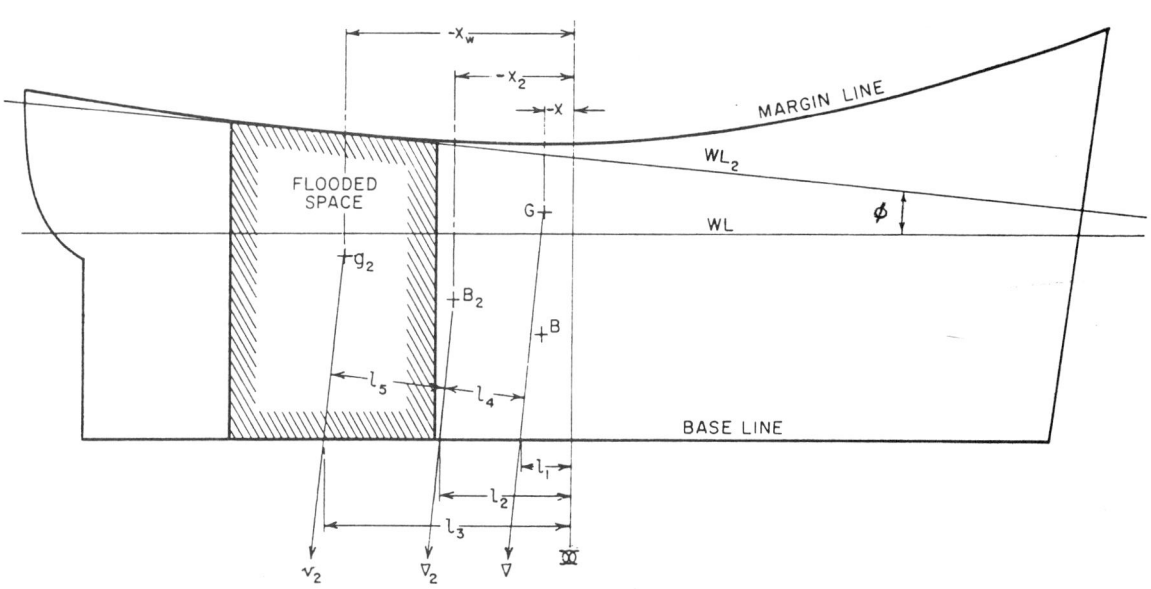

Fig. 3 Diagram for floodable-length calculation

trim from an initial WL to a final waterline WL_2 due to flooding of an unknown part of a vessel's length is illustrated in Fig. 3. In this figure

∇ is displacement volume to WL
B is initial center of buoyancy
G is initial center of gravity
v_2 is net volume of flooding water up to WL_2 (i.e., gross vol. $\times \mu$)
g_2 is center of gravity of v_2
B_2 is center of buoyancy of entire immersed volume up to WL_2
∇_2 is entire displacement volume to WL_2
μ is permeability, the percentage of a space that can be occupied by water.

$\overline{KG}$, $\overline{KB}$, $\overline{Kg}$, and so on are the perpendicular distances of G, B, g, and so on, respectively, from the baseline. For equilibrium,

$$v_2 = \nabla_2 - \nabla; \; v_2 l_5 = \nabla l_4; \; \text{therefore } v_2 = \frac{\nabla l_4}{l_5}$$

$$l_1 = x + \overline{KG} \tan \theta$$

where

$$l_2 = x_2 + \overline{KB_2} \tan \theta$$
$$l_3 = x_w + \overline{KG_2} \tan \theta$$
$$l_4 = [x_2 - x + \tan \theta (\overline{KB_2} - \overline{KG})] \cos \theta$$
$$l_5 = [x_w - x_2 + \tan \theta (\overline{Kg_2} - \overline{KB_2})] \cos \theta$$

Therefore,

$$v_2 = \nabla_2 - \nabla = \frac{\nabla [x_2 - x + \tan \theta (\overline{KB_2} - \overline{KG})]}{[x_w - x_2 + \tan \theta (\overline{Kg_2} - \overline{KB_2})]}$$

Transposing and rearranging, one gets

$$x_w = \frac{(\nabla_2 x_2 - \nabla x)}{(\nabla_2 - \nabla)} - \tan \theta \left[\overline{Kg_2} - \frac{(\nabla_2 \overline{KB_2} - \nabla \overline{KG})}{(\nabla_2 - \nabla)} \right] \quad (1)$$

The $\tan \theta$ [ ] term involving vertical levers is relatively small. Neglecting it, one may write,

$$v_2 = \nabla_2 - \nabla = \frac{\nabla (x_2 - x)}{(x_w - x_2)} \quad (2)$$

and

$$x_w = \frac{(\nabla_2 x_2 - \nabla x)}{(\nabla_2 - \nabla)} \quad (3)$$

Equations (2) and (3) provide a basis for determining the length and location of flooding that would cause the flooded ship to reach a state of equilibrium at WL_2, neglecting heel. Methods of calculation are discussed in Section 4.

3.3 Fundamental Relationships—Damage Stability. It will be explained in Section 4 that stability and heel, as well as sinkage and trim, after flooding can be evaluated by means of either a *lost buoyancy* or a *trimline added weight* approach. Since the latter has wider applicability and is consistent with the floodable length approach just discussed, its fundamental relationships will be given here. As in the case of floodable length calculations, one assumes an equilibrium damaged condition waterline, calculates the flooding water up to this waterline and subtracts it and its moments from the total displacement and moments up to this waterline, in order to obtain a corresponding before-damage condition. The effect of heel can also be taken into account, all as described in Section 4.

(a) Required $\overline{GM}$ for zero-heel condition. Having determined the draft and trim (if any) before damage, one then can determine the $\overline{GM}$ requirements. Dealing first with the zero-heel condition (either actual symmetrical flooding, or unsymmetrical flooding with assumed equalization of heeling moment), it is convenient to think of the flooding water as tankage having the same vertical moment and free surface. The $\overline{GM}$ in the damaged condition, referred to the gross displacement Δ_2 at T_2, then is expressed by the equation

$$\overline{GM}_{2\Delta_2} = \overline{KM_2} - \frac{\mu_s i_T}{\Delta_2/\rho} - \overline{KG_2} \quad (4)$$

where

Δ_2 is mass displacement = W_2/g
μ_s is surface permeability
i_T is inertia of flooded waterplane
$\overline{KM_2}$ is upright metacenter above keel at T_2
$\frac{\mu_s i_T}{\Delta_2/\rho}$ is $\overline{BM_2}$ loss due to free surface of flooding water (if flooded waterplane is unsymmetrical, this is modified to account for TCF of net waterplane)
$\overline{KG_2}$ is vertical center of gravity of vessel including mass of flooding water, w/g
ρ is water density

As already noted, the damaged spaces are in open communication with the sea. Therefore, the true displacement in the damaged condition does not include the water in such spaces. It is customary to regard the displacement in the damaged condition as equal to $\Delta_2 - w/g$, (or $W_2 - w$ if weight units are used). In order to provide a minimum positive damaged condition $\overline{GM}$ of 0.05 m (2 in.), based upon $\Delta_2 - w/g$, one then writes,

$$0.05 \times \frac{(\Delta_2 - w/g)}{\Delta_2} = \overline{KM_2} - \frac{\mu_s i_T}{\Delta_2/\rho} - \overline{KG_2}$$

or

$$\overline{KG}_2 = \overline{KM}_2 - \frac{\mu_s i_T}{\Delta_2/\rho} - 0.05 \times \frac{(\Delta_2 - w/g)}{\Delta_2}$$

The corresponding $\overline{KG}$ before damage (maximum allowable $\overline{KG}$) is then equal to

$$\frac{\overline{KG}_2 \times \Delta_2 - \overline{Kg} \times w/g}{(\Delta_2 - w/g)}$$

where $\overline{Kg}$ is vertical center of gravity of damage water.

Substituting in the preceding equation one obtains

$$\overline{KG} = \frac{\Delta_2 \overline{KM}_2 - \frac{\mu_s i_T}{\Delta_2/\rho} - \overline{Kg}\, w/g}{(\Delta_2 - w/g)} - 0.05 \quad (5)$$

The intact $\overline{GM}$ to meet this condition is therefore $\overline{KM} - \overline{KG}$, where $\overline{KM}$ corresponds to the draft and trim before damage.

(b) Required $\overline{GM}$ to limit heel and righting arms in damaged condition. Referring to Fig. 4, the following definitions and derivation apply:

G is CG of upright ship before damage
G_2 is CG of heeled damaged ship including flooding water

$$t = \frac{(w/g) \times \text{its transverse center of gravity (tcg)}}{\Delta_2}$$

$$\overline{KG}_2 = \frac{\Delta \times \overline{KG} + w/g\, \overline{Kg}}{\Delta_2}$$

where

Δ is displacement before damage $= \Delta_2 - w/g$
$\overline{G_2Z_2}$ is damaged condition righting arm
$\quad = [(\overline{KP} - \overline{KG}_2) \tan\phi - t] \cos\phi$
M_2 is metacenter (upright) of intact ship at Δ_2
$\overline{M_2P} = \frac{\overline{M_2S}}{\sin\phi} = F\,\overline{B_2M_2}$

$$\overline{KP} - \overline{KG}_2 = \overline{KM}_2 + \frac{\overline{M_2S}}{\sin\phi} - \overline{KG}_2$$

$$\overline{G_2Z_2} = \left[\left(\overline{KM}_2 + \frac{\overline{M_2S}}{\sin\phi} - \frac{\Delta \overline{KG} + \overline{Kg}\,w/g}{\Delta_2}\right) \times \tan\phi - \frac{\overline{Kg}\,w/g}{\Delta_2}\right] \cos\phi$$

$$\overline{G_2Z_2} = \sin\phi \left[\overline{KB}_2 + \overline{B_2M}_2(F+1)\right.$$
$$\left. - \frac{(\Delta_2 - w/g)\overline{KG}}{\Delta_2} - \frac{w/g}{\Delta_2}\left(\overline{Kg} + \frac{\text{tcg}}{\tan\phi}\right)\right] \quad (6)$$

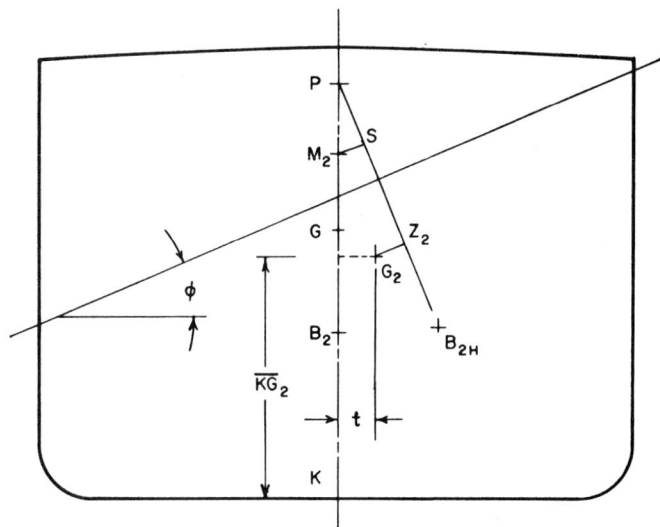

Fig. 4 Diagram for required $\overline{GM}$ to limit heel

Equilibrium exists when $\overline{G_2Z_2} = 0$ and Equation (6) can then be solved for $\overline{KG}$:

$$\overline{KG} = \frac{\Delta_2}{\Delta_2 - w/g}\left[\overline{KB}_2 + \overline{B_2M}_2(F+1)\right] \quad (7)$$
$$- \frac{w/g}{\Delta_2 - w/g}\left(\overline{Kg} + \frac{\text{tcg}}{\tan\phi}\right)$$

Again, the intact $\overline{GM}$ to meet this condition is $\overline{KM} - \overline{KG}$, and the *required $\overline{GM}$* is either this value necessary to limit the heel to the maximum allowable value ϕ, or that derived from equation (5), if greater.

When the position of the pole point P is directly determined by integration of the ship's volume to the flooded waterline, as by computer, $\left(\overline{KM}_2 + \frac{\overline{M_2S}}{\sin\phi}\right)$ is replaced by $\overline{KP}$. Then

$$\overline{G_2Z_2} = \sin\phi\left[\overline{KP} - \frac{(\Delta_2 - w/g)}{\Delta_2}\overline{KG}\right.$$
$$\left. - \frac{w/g}{\Delta_2}\left(\overline{KG} + \frac{\text{tcg}}{\tan\phi}\right)\right] \quad (8)$$

and, for equilibrium:

$$\overline{KG} = \frac{\Delta_2}{\Delta_2 - w/g}\overline{KP}$$
$$- \frac{w/g}{\Delta_2 - w/g}\left(\overline{KG} + \frac{\text{tcg}}{\tan\phi}\right) \quad (9)$$

Required $\overline{GM}$, to limit heel is

$$\overline{KM} - \overline{KG}.$$

See sub-section 4.7 for details of derivation and forms for manual calculation.

Although the U.S. Navy sometimes also calculates required $\overline{GM}$, or $\overline{KG}$, the emphasis in design of combatant ships, in particular, is on determining residual righting arms after damage to various compartments or combinations of compartments. (NAVSEC, 1975).

(c) Calculation of Righting Arms in Damaged Condition In the preceding Equation (8) $\overline{G_2Z_2}$ is the net righting arm after flooding, as related to Δ_2, with $\overline{KG}$ corresponding to the $\overline{GM}$ before damage. However, at large angles of heel a more accurate calculation of damaged-condition righting arms is necessary.

In principle, intact-condition righting arms must be modified by the damage. If trim is neglected, this can be accomplished readily by

$$G_2Z_2 \quad (10)$$
$$= \frac{\text{(Intact vessel rt. mom.)} - (\Delta_2 - w/g)\overline{KG}\sin\phi - \text{(Damage heel mom.)}}{\Delta_2}$$

(where both righting moment and heeling moment correspond to the damaged condition draft, T_2, and are referred to an axis intersecting the centerline at the baseline).

If the damage location is such as to result in appreciable change of trim, the change in draft in way of the damage is more than that at the vessel's LCF, and the amount and moment of flooding water are correspondingly affected.

Since heel, even without damage, may result in some trim, a direct calculation of this effect requires in essence the use of heeled-condition Bonjean curves giving both sectional areas and heeling moments. For each damaged-condition draft and heel, the equilibrium trim corresponding to the condition of draft and trim before damage can then be determined by a cross-plotting, or iterative process.

Nowadays, the calculation of residual righting arms after damage, including the effect of trim, is almost always carried out by computer (Section 5.)

Section 4

Manual Subdivision and Damage Stability Calculations

4.1 Manual Floodable Length Calculations—General. Curves of floodable length (and, the corresponding *permissible length* curves described in Section 7) are usually plotted to a vertical scale equal to the longitudinal scale. With such an arrangement, lines drawn from the intersections of the bulkheads with the baseline at an angle $\tan^{-1}$ of 2 will intersect at the midpoint of the compartment and at a height above the baseline that is equal to the length of the compartment. Such curves are sometimes plotted to a vertical scale equal to half the horizontal scale. In such case, the slope of the inclined lines is 1. In either case the intersection of the inclined lines indicates at a glance whether or not the length of a compartment or a combination of compartments is greater than either the permissible length or the floodable length. Fig. 5 illustrates a typical plot of curves of floodable length. Floodable length calculations are based upon the relationships given in 3.2 of Section 3; specifically, equations (1), (2) and (3). The possible errors in permeability assumptions, etc., may be larger than the errors due to neglecting the $\tan\theta$ [ ] term of Equation (1). Accordingly, for manual calculations they are usually neglected and Equations (2) and (3) are used as a basis.

4.2 Direct Method of Calculation. In this method, points on the floodable length curve are calculated for the actual lines of the ship, using Equations (2) and (3) to determine the volume and location of the flooding water that would immerse the ship to the margin line. In general, the procedure is that proposed by Dipl. Ing. F. Shirokauer (1928).

For the solution, no special data, curves or instruments are necessary. The procedure is not unduly long and the accuracy is better than that of comparable methods.

On a profile drawing showing the margin line and a number of transverse stations, Bonjean curves are plotted from a low draft to the margin line. In Fig. 6 the Bonjean curves are shown plotted for ten stations and various scales have been adopted for length, depth, and areas, but any number of stations and any system of scales may be used.

The subdivision load line is drawn on the profile and the trim line parallel to the subdivision load line is drawn tangent to the margin line at its lowest point. To obtain a satisfactory curve of floodable length with the minimum amount of work, the following procedure, as proposed by Shirokauer (1928), is then used. Let,

D be depth from baseline to margin line (at lowest point)
T be draft from baseline to subdivision load line
$H = 1.6D - 1.5T$

At perpendiculars at the extremities of the subdivision length, the distances $H/3$, $2H/3$, and H are laid off as shown in Fig. 6 and tangents are drawn to the margin line from these points. These tangents are designated as in the figure; for example, $3F$ is the tangent drawn from a point on the after perpendicular at a distance of H below the parallel trim line.

The following calculation is then made for each of

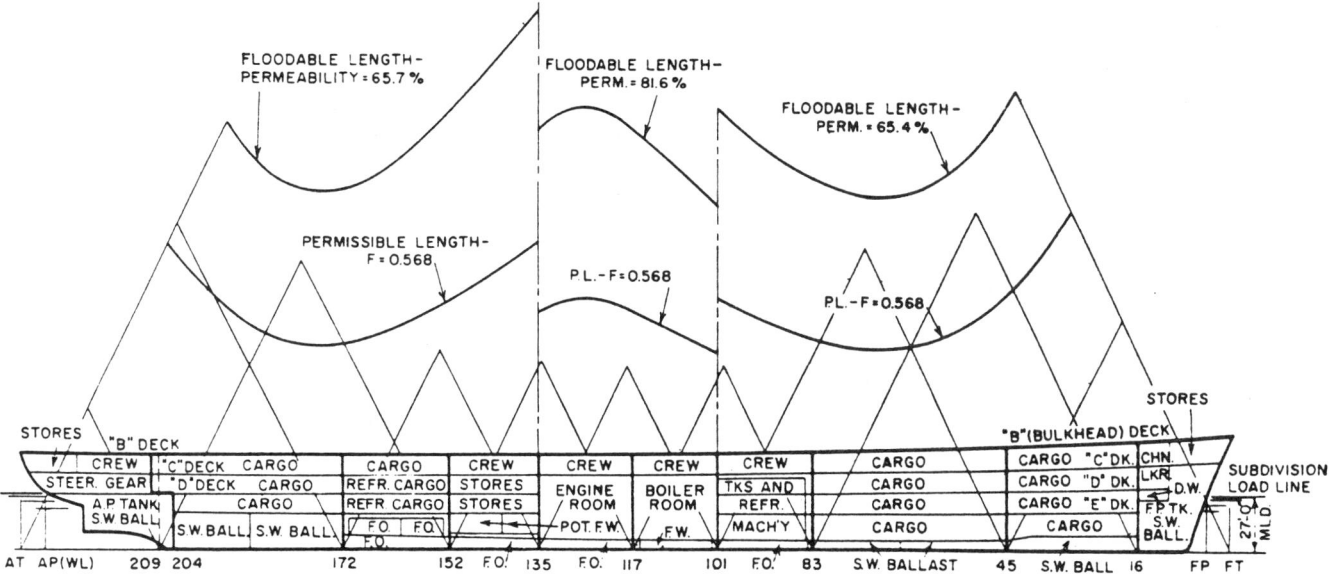

Fig. 5 Floodable-length curves

the three forward trim lines and the subdivision load line. The area of each station up to the trim line is first read from the Bonjean curves. These areas are then integrated by use of Simpson's rule or other rule to obtain the volume of displacement and the distance of LCB from some convenient station such as amidships, Fig. 7. Allowance normally need not be made for appendages. Neglecting them, in fact, tends to compensate for disregarding the effect of the vertical levers in Eq. (1).

The next step is to calculate for each trim line the corresponding volume of damage water v_2 and the longitudinal distance x_w between its center of gravity and amidships. The volume of damage water v_2 is the volume of displacement below the trim line less the volume of displacement of the undamaged ship at the subdivision load line. The distance x_w is calculated from equation (3). Fig. 7 has suitable spaces for the tabulation of the values of v_2 and x_w for each trim line.

It should be noted that, if either B_2 or G lies on a different side of amidships than that shown in Fig. 3, the signs must be changed accordingly.

After v_2 and x_w have been determined for each of the trim lines, the interpolation curves should be drawn as shown in Fig. 8. The parallel lines labelled 3A, 2A, and so on, correspond to the trim lines and are spaced any equal distance apart.

Ordinates on the curves are plotted vertically from the horizontal axis shown. Since the curves of LCG of damage water have considerable curvature in the vicinity of the 3F and 3A trim lines, additional points are necessary to determine their shape. The other curves have very little curvature in this vicinity, and points from them may be used with sufficient accuracy to determine additional points on the LCG curves.

For example, in Fig. 8, another vertical line may be drawn halfway between the 2F and 3F lines. This will correspond to a 2½F trim line. Let v_2' be the value of

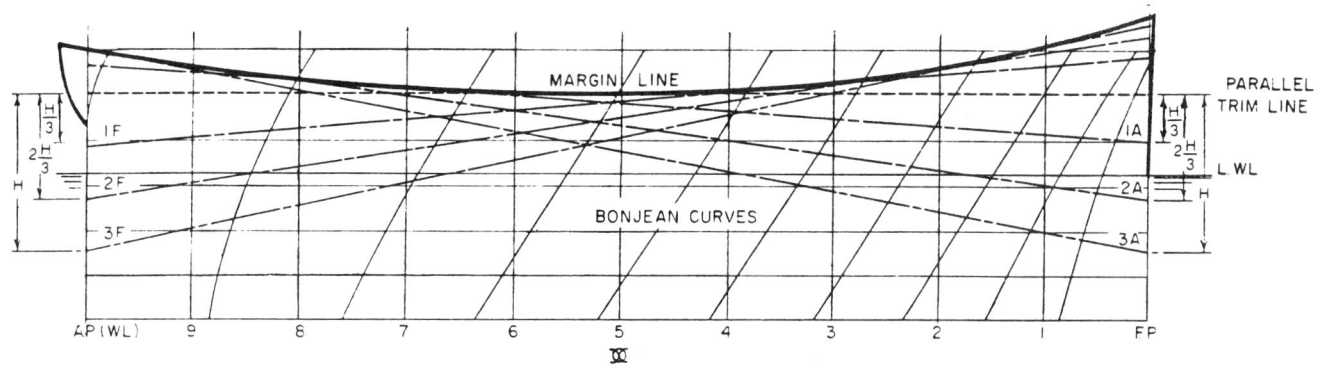

Fig. 6 Bonjean curves and trimlines for floodable-length calculations

TRIMMED DISPLACEMENTS AND CENTERS

S = SPACING OF STATIONS
S =
$\frac{S}{3}$ =

SHEET:
SHIP:
DATE:
BY:

| STATION | SM VOL. | SM MT. | SUBDIVIS. LOAD LINE | \multicolumn{7}{c}{SECTION AREAS TO TRIM LINES} |
				3A	2A	1A	PAR.	1F	2F	3F
0	$\frac{1}{2}$	$2\frac{1}{2}$								
$\frac{1}{2}$	2	9								
1	$1\frac{1}{2}$	6								
2	4	12								
3	2	4								
4	4	4								
5	2	0								
6	4	4								
7	2	4								
8	4	12								
9	$1\frac{1}{2}$	6								
$9\frac{1}{2}$	2	9								
10	$\frac{1}{2}$	$2\frac{1}{2}$								
Σ SM VOL. x A = $\Sigma f(\nabla)$										
Σ SM MT. x A = $\Sigma f(M)$										
① DISPLACEMENT $\Sigma f(\nabla) \times \frac{S}{3}$			∇	∇_2	∇_2	∇_2	∇_2	∇_2	∇_2	∇_2
② DW. VOLUME = $\nabla_2 - \nabla$			✗	v_2	v_2	v_2	v_2	v_2	v_2	v_2
③ L.C.B.= $\frac{\Sigma f(M)}{\Sigma f(\nabla)} \times S$			x	x_2	x_2	x_2	x_2	x_2	x_2	x_2
④ SHIFT OF L.C.B = $x_2 - x$			✗							
⑤ C.G.DW. FROM L.C.B. $\frac{\text{LOAD DISPLT}}{\text{DW. VOLUME}} \times$ ④			✗							
⑥ C.G. DW. FROM ⊥ ③ + ⑤			✗	x_w	x_w	x_w	x_w	x_w	x_w	x_w

∇, THE VOLUME OF DISPLACEMENT AT SUBDIVISION LOAD LINE
∇_2, THE VOLUME OF DISPLACEMENT AT STATED CONDITION OF DAMAGE
x, LCB OF SHIP ON SUBDIVISION LOAD LINE, MEASURED FROM ⊥
x_2, LCB OF SHIP AT STATED CONDITION OF DAMAGE MEASURED FROM ⊥

Fig. 7 Floodable-length calculation form, m or ft

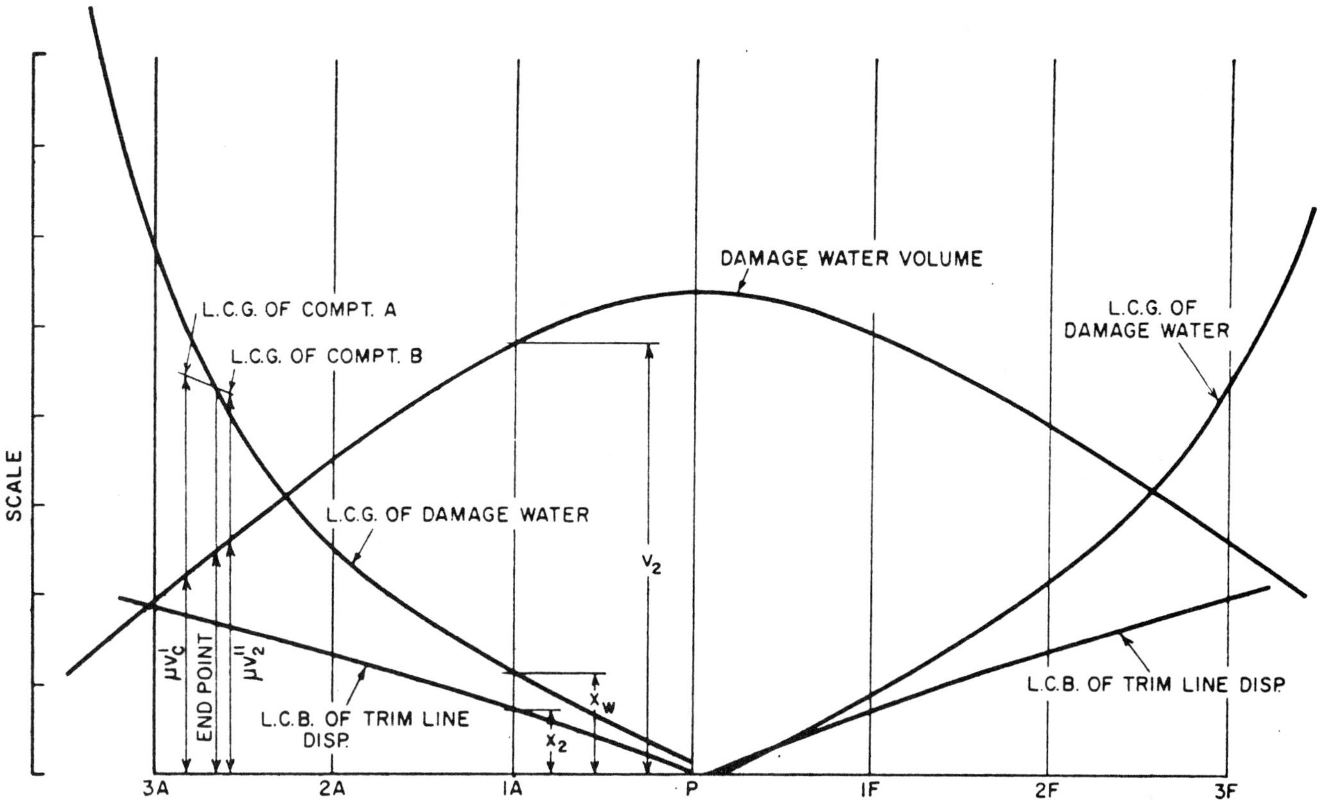

Fig. 8 Interpolation curves

the ordinate of the damage water volume curve and x_2' be the value of the ordinate of the LCB curve. Then, since V, the volume of intact displacement, and x, the LCB of the intact displacement, are known, the value of x_w' will be

$$x_w' = \frac{\nabla}{v_2'}(x_2' - x) + x_2' \qquad (11)$$

The interpolation curves, if they are found to plot fairly, serve as a check on the previous calculations. Their primary purpose, however, is to facilitate calculation of the end points and auxiliary points of the floodable-length curve (following).

Sectional areas read from the Bonjean curves in Fig. 7 are now plotted as a sectional area curve of each trim line as shown in Fig. 9. These sectional area curves are used in conjunction with the interpolation curves, Fig. 8, to determine points on the floodable length curve.

For example, let us take the floodable length calculations for trim line 2A. Let

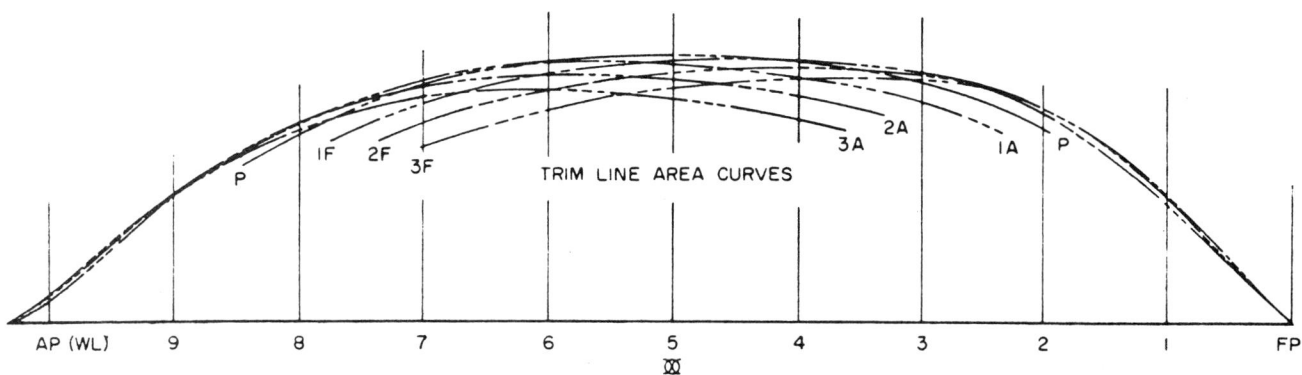

Fig. 9 Section area curves for trim lines

v_2 be net volume of flooding water (from Fig. 8)

x_w be distance of its center of gravity from amidships (from Fig. 8)

θ be assumed permeability in percent

v_c is volume of compartment $= \dfrac{v_2}{\mu} \times 100$

x_m be distance of midlength of compartment from center of gravity of damage water (or compartment)

x_c be distance of midlength of compartment from amidships

l be length of compartment or floodable length at assumed permeability.

The method is one of trial and error; so the first step is to guess as to the distance x_m of the midlength from the center of gravity of damage water and as to the length of compartment which will give a volume $v_c = (v_2/\mu) \times 100$. The length may be estimated by dividing the required volume v_c by an assumed mean section area. This length is laid off from the assumed position of the middle of the compartment, as shown in Fig. 10, and gives $abcd$ as a compartment. By Simpson's rule, Fig. 11, with five ordinates, the volume of this compartment and the distance of its center of gravity from its midlength may be computed. This computed volume should equal the required volume v_c, and the distance of its center of gravity from the midlength should equal the assumed distance x_m. Usually a second try is necessary, such as one which gives the compartment $a'b'c'd'$ of Fig. 10, for example. The correct length of the compartment and the corresponding value of x_c are obtained by interpolation in Fig. 8 and are plotted as a point on the floodable length curve (Fig. 5).

A point on the floodable length curve is calculated for each of the trim lines. On some ships a compartment for the 3A or 3F trim line cannot be found. Such a development indicates that the limiting trim line for the end point of the floodable length curve lies between 2A and 3A or 2F and 3F. An additional point, such as is given by a 2½F or 2½A trim line, is then desirable. The volume and LCG of damage water for such a trim line is taken directly from the interpolation curves, Fig. 8, and the corresponding point on the floodable

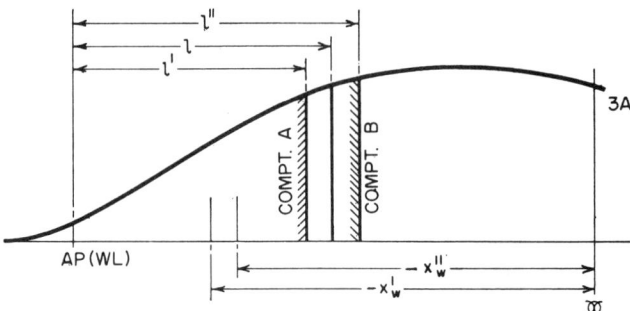

Fig. 12 Diagram for calculating endpoint

length curve is determined in the same way as for any other point. Similarly, data for other additional points required to get a fair floodable length curve may be taken from the interpolation curves.

In using the interpolation curves to determine the end points of the floodable length curve, a somewhat different procedure is used. The exact trim lines which correspond to the end points are not known but obviously they will be close to trim lines 3A and 3F, respectively. Hence, the sectional area curves of these trim lines are used in the end-point calculations. From the trend of the floodable length curve as plotted from the points already determined, reasonably close estimates can be made of the lengths of the ordinates at the ends of the curve. These ordinates are the respective floodable lengths at the extreme ends of the ship. To obtain more accurate values of the ordinates, each end of the ship is treated separately by the following method.

At the after end of the ship, for example, two compartments are laid down on curve 3A of Fig. 9 to form a diagram similar to Fig. 12. One of these compartments A is slightly shorter than the estimated floodable length, while the other compartment B is slightly longer than this estimated length. By Simpson's rule and with sectional areas from the 3A trim line of Fig. 9, calculations are made as indicated in Fig. 13 to get

v_c' = volume of compartment A

v_c'' = volume of compartment B

x_w' = distance from amidships of center of gravity of compartment A

x_w'' = distance from amidships of center of gravity of compartment B

$\mu v_c'$ = damage water volume in compartment A

$\mu v_c''$ = damage water volume in compartment B

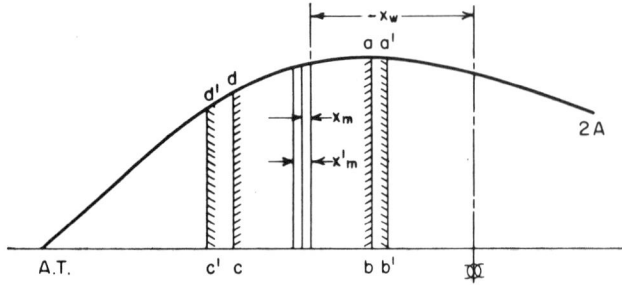

Fig. 10 Volume and C. G. of assumed compartment

In Fig. 8 an ordinate is erected which cuts the curve of damage water volume at height $\mu v_c'$ and upon this ordinate is plotted the LCG of compartment A at the distance x_w above the axis. Similarly an ordinate for compartment B is erected and the LCG of this compartment is plotted on it. Between these two LCG points a straight line is drawn which cuts the curve

SUBDIVISION AND DAMAGE STABILITY

FLOODABLE LENGTH CALCULATION

TRIM LINE No. _____

SHEET _____
SHIP _____
DATE _____
BY: _____

L.C.G. OF DAMAGE WATER FROM ⊗ = x_w = _____
DAMAGE WATER VOLUME = v_2 = _____
COMPARTMENT VOLUME AT ____ % PERMEABILITY = v_c = _____

A

SECTION NUMBER	SECTION AREA	SM VOL.	SM MT.
AFT 0		1	+2
1		4	+4
2		2	0
3		4	−4
FOR'D 4		1	−2
Σf(V)			
Σf(M)			

① TRIAL FLOODABLE LENGTH = $\dfrac{v_c}{\text{SEC. AREA (MEAN)}}$ = ℓ = _____

② TRIAL x_m _____ ; CORRESPONDING x_c _____

③ S (STA. SPACING) = $\dfrac{\ell}{4}$ = _____ ; $\dfrac{S}{3}$ = _____

④ COMPT. VOL. = $\Sigma f(V) \times \dfrac{S}{3}$ = _____

⑤ COMPUTED $x_m = \dfrac{\Sigma f(M)}{\Sigma f(V)} \times S$ = _____

⑥ APPROX. FLOOD. LENGTH = $\ell \times \dfrac{v_c}{④}$ = _____

⑦ APPROX. $x_c = x_w + x_m$ = _____

B

SECTION NUMBER	SECTION AREA	SM VOL.	SM MT.
AFT 0		1	+2
1		4	+4
2		2	0
3		4	−4
FOR'D 4		1	−2
Σf(V)			
Σf(M)			

2nd TRIAL FLOODABLE LENGTH, ⑥ = _____

2nd TRIAL x_c, ⑦ = _____

⑧ S = _____ ; $\dfrac{S}{3}$ = _____

⑨ COMPT. VOL. = $\Sigma f(V) \times \dfrac{S}{3}$ = _____

⑩ $x'_m = \dfrac{\Sigma f(M)}{\Sigma f(V)} \times S$ = _____

⑪ FLOODABLE LENGTH = ⑥ $\times \dfrac{v_c}{⑨}$ = _____

⑫ $x_c = x_w +$ ⑩ = _____

Fig. 11 Calculation form for a point on the floodable-length curve, m or ft.
(Calculations A and B apply to two trial compartments for each of which l and x_m are estimated. Additional trials may be necessary.)

END POINT CALCULATION

L = LENGTH OF VESSEL		SHEET
		SHIP
		DATE
		BY:

ASSUMED COMPARTMENT "A"

SECTION	SM VOL.	SM MT.	SECTION AREA
0	1	0	
1	4	4	
2	2	4	
3	4	12	
4	1	4	
$\Sigma f(V)$			
$\Sigma f(M)$			

ASSUMED FLOODABLE LENGTH = _____ = l'

S (SPACING OF SECTIONS) = $\dfrac{l'}{4}$ = _____ ; $\dfrac{S}{3}$ = _____

$v_c' = \Sigma f(V) \times \dfrac{S}{3} =$ _____

D_W VOLUME = $\mu v_c' =$ _____

C.G. D_W FROM SEC. 0 = C = $\dfrac{\Sigma f(M)}{\Sigma f(V)} \times S =$ _____

C.G. D_W FROM ⊠ = $\dfrac{L}{2} - C =$ _____ = X_w'

ASSUMED COMPARTMENT "B"

SECTION	SM VOL.	SM MT.	SECTION AREA
0	1	0	
1	4	4	
2	2	4	
3	4	12	
4	1	4	
$\Sigma f(V)$			
$\Sigma f(M)$			

ASSUMED FLOODABLE LENGTH = _____ = l''

S = _____ $\dfrac{S}{3}$ = _____

$v_c'' = \Sigma f(V) \times \dfrac{S}{3} =$ _____

D_W VOLUME = $\mu v_c'' =$ _____

C.G. D_W FROM SEC. 0 = C = $\dfrac{\Sigma f(M)}{\Sigma f(V)} \times S =$ _____

C.G. D_W FROM ⊠ = $\dfrac{L}{2} - C \cdot =$ _____ = X_w''

D_W VOLUME FOR FLOODABLE END COMPARTMENT = _____

MOLDED VOLUME OF FLOODABLE END COMPARTMENT = v_c = _____

$l = l' + \dfrac{v_c - v_c'}{v_c'' - v_c'} (l'' - l') = +$ _____ X

FLOODABLE LENGTH = _____

Fig. 13 Form for calculating endpoint (m or ft)

of LCG of damage water at the ordinate corresponding to the desired trim line for the end point of the floodable length curve. At this ordinate the volume of damage water is read from Fig. 8; this volume divided by the permeability gives the volume of the end compartment. The length of the end compartment found by Equation

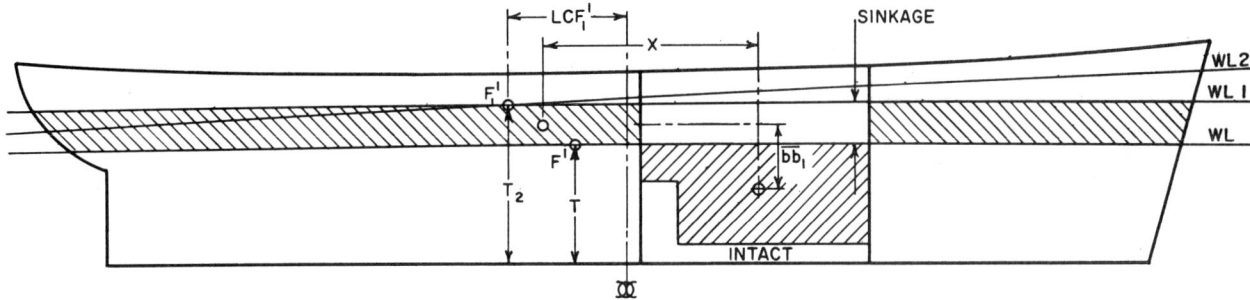

Fig. 14 Sinkage and trim by lost buoyancy method

(12) determines the end point of the floodable length curve. Let

v_c be volume of end compartment
l' be length of compartment A
l'' be length of compartment B
l be length of end compartment

Then

$$l = l' + \frac{(v_c - v_c')}{v_c'' - v_c'}(l'' - l') \qquad (12)$$

The other end point is found in a similar manner. See also Muckle (1963).

When the portions of the hull below the margin line but forward of the FP and abaft the AP are not included in calculating the displacement volumes and the flooding water volumes (see Figs. 7, 11, 12 and 13) the calculated value of the forward-end floodable length is measured aft from the FP and that of the after-end floodable length is measured forward from the AP. However, in constructing the floodable length curves, Fig. 5, the forward terminus of the diagram (FT) should be taken at the forward extremity of the margin line, and the corresponding terminal floodable length is equal to the calculated forward-end floodable length plus the longitudinal distance between the FP and the FT. Similarly, the after terminus of the diagram (AT) should be taken at the after extremity of the margin line and the corresponding terminal floodable length is equal to the calculated after floodable length plus the longitudinal distance between the AP and the AT.

4.3 Manual Damage Stability Calculations—General. The methods of calculation shown in this section were developed early in this century for use on naval ships and commercial passenger vessels, and during the time when manual calculation was the only method available. While the basic principles are the same, modern computer programs, particularly the widely used Ship Hull Characteristics Program (SHCP) (NAVSEA, 1976), use entirely different techniques to determine damage stability. These are described in Section 5. The manual calculations described in this section are applicable only to vessels where the survival criteria (Section 7) specify minimum $\overline{GM}$, maximum heel, and minimum freeboard.

When stability criteria for naval or other ships specify required residual range and amplitude of righting arms, then complete computer calculations of cross curves and/or righting arms after damage are required, as discussed in Section 5.

4.4 Equilibrium Draft, Trim, and Heel After Flooding. The draft, trim and heel in flooded condition after equilibrium has been established can be calculated by starting with an assumed service condition, calculating the lost buoyancy due to a compartment or compartments being opened to the sea, and equating that lost buoyancy and its moments to the buoyancy gain and moments accompanying sinkage, trim, and heel of the remaining intact part of the ship.

This procedure is convenient and simple to use if the form of the vessel and the configuration of the flooded space are such that the resulting sinkage, trim, and heel do not involve extreme or discontinuous changes in the remaining undamaged part of the waterline plane. Consequently, this procedure, commonly known as the *lost-buoyancy* method, is often used for merchant ships.

Alternatively, one may assume an equilibrium damaged condition waterline, calculate the flooding water up to this waterline, and subtract it and its moments from the total displacement and moments up to this waterline in order to obtain a corresponding before-damage condition.

This second procedure is commonly known as the *trim-line added-weight* method. This is a misnomer, since water in spaces open to the sea and free to run in or out does not actually add to a vessel's weight. For calculation purposes it is convenient to regard such flooding water as adding to the displacement; however, it must be remembered that the resulting (virtual) displacement not only differs from the initial displacement but varies with change in trim or heel.

Since the trim-line added-weight method involves a direct integration of volumes up to the damaged condition waterplane, it is just as well adapted to dealing with complex flooding conditions as with simple ones. It does require, however, the determination by iteration of drafts and trims after flooding that correspond to undamaged values covering the range of actual service conditions. As will be shown, this may be accomplished in a systematic manner without great difficulty. It should be noted that the fundamental relationships used in direct floodable length calculations and illustrated by Fig. 3, Section 3, and the accompanying text follow the trim-line added-weight method.

Complete procedures for both methods are described here. In practice some calculators may depart from these procedures for convenience or simplification, or personal preference. However, the principles still apply.

In both methods it is usually assumed that trim and heel are independent functions. This is not exact, but is adequate for ordinary merchant vessels. For vessels with asymmetric waterplanes, such as those having fine bows and transom sterns, the trim-line added-weight method is preferable and can be modified to take account of trim and heel interaction.

4.5 Displacement and $\overline{GM}$ in Flooded Condition. As already noted, liquids in damaged spaces in open communication with the sea obviously do not act as a part of a vessel's weight, i.e., a vessel's displacement in the damaged condition is equal to its initial undamaged displacement less the weight of any liquids which were in breached tanks before damage. This assumption is used when calculations are made by the lost buoyancy method, in which flooding water is considered as lost buoyancy rather than added weight.

It should be noted that when calculations are made by the added weight method, not only is the virtual displacement different from the initial displacement, but the $\overline{GM}$ has a different meaning—since both $\overline{KG}$ and $\overline{KM}$ are different. However, the product of displacement and $\overline{GM}$ should remain unchanged. Hence, if a stability criterion specifies a minimum residual $\overline{GM}$ of 5 cm (2 in.), the lost buoyancy approach is intended, and the corresponding $\overline{GM}$ by added weight is 5 cm x Δ/Δ_v, where Δ is the initial displacement and Δ_v is the virtual displacement after damage.

4.6 Lost-Buoyancy Method. For convenience, the given procedures make maximum use of the quantities ordinarily shown on the curves of form and are sufficiently accurate for ordinary cases. Figs. 15 and 16 illustrate tabular forms for making the following lost-buoyancy-method calculations. For greater accuracy, or where complicated flooding configurations are involved, trim-line added-weight calculations are recommended.

(a) *Determination of sinkage and trim:* The ship in Fig. 14, initially at waterline WL, comes to float at waterline WL_2 because of the sinkage and trim due to flooding of one or more of its compartments.

We may imagine that, when flooding takes place, the ship is restrained temporarily from trimming. The net lost buoyancy below WL, i.e., the net volume v' of damage water below that waterline, must be replaced by an equal gain in buoyancy above WL. Hence, the sinkage or change in mean draft is approximately:

$$\text{Sinkage} = \frac{v'}{A'}$$
$$= \frac{(v - \text{intact buoyancy})\mu - P/\rho}{A - \mu_s a} \quad (12)$$

where

v is molded volume of flooded compartment below WL
v' is net lost buoyancy below WL
μ is permeability of compartment
P is total mass of liquids, if any, in breached tanks before damage[3]
ρ is density of liquid in breached tanks t/m³
μ_s is surface permeability of compartment
A is area of waterplane WL
a is area of portion of waterplane WL within compartment
A' is net area of waterplane WL remaining intact after flooding $= A - \mu_s a$

After the sinkage has been estimated by Equation (12) a second and closer approximation may be made by

$$\text{Sinkage} = \frac{2v'}{A' + A_1'} \quad (13)$$

where

A_1' is remaining area of impaired waterplane $WL_1 = A_1 - \mu_s a_1$
a_1 is area of portion of impaired waterplane WL_1 within compartment

If A' and A_1' are each replaced by the remaining TPcm at WL and WL_1,

$$\text{Sinkage (m)} = \frac{v'\rho}{50 \times \text{sum of remaining TPcm values}} \quad (14)$$

Or in English units (ft and long tons),

$$\text{Sinkage (ft) in SW} = \frac{v'/35}{6 \times \text{sum of remaining TPI values}} \quad (14a)$$

When the imaginary restraint against trimming is removed, the ship comes under the action of a trimming moment which is equal to

$$\text{Trimming moment} = v'\rho x \quad (15)$$

---
[3] When, in accordance with Section 7.6, tanks are taken at zero permeability, $P = 0$.

DAMAGE STABILITY - LOST BUOYANCY METHOD - SINKAGE AND TRIM - BASIC DATA

SHIP _____ DATE _____ PAGE 1
DAMAGE EXTENT _____ BY _____

ITEM No.	SOURCE	DESCRIPTION						
1	————————	T, DRAFT BEFORE DAMAGE						
2	CURVES OF FORM	DISPLACEMENT Δ						
3	CURVES OF FORM	TONNES PER m						
4	CURVES OF FORM	LCF F+ / A−						
5	CURVES OF FORM	$\overline{KM}$						
6	CURVES OF FORM	$\overline{KB}$						
7	COMPARTMENT CALCULATIONS	NET LOST BUOYANCY, $\frac{w}{g}$ ☆						
8	COMPARTMENT CALCULATIONS	VERTICAL CENTER OF ⑦						
9	COMPARTMENT CALCULATIONS	CENTER OF ⑦ FROM MIDSHIPS F+ / A−						
10	COMPARTMENT CALCULATIONS	CENTER OF ⑦ FROM CL P− / S+						
11	COMPARTMENT CALCULATIONS	LOST TONNES PER m						
12	COMPARTMENT CALCULATIONS	CENTER OF ⑪ FROM MIDSHIPS F+ / A−						
13	COMPARTMENT CALCULATIONS	CENTER OF ⑪ FROM CL P− / S+						
14	③ − ⑪	REMAINING TONNES PER m						
15	$\frac{③ \times ④ - ⑪ \times ⑫}{⑭}$	CENTER OF ⑭ FROM MIDSHIPS F+ / A−						
16	$\frac{⑦}{⑭}$	APPROXIMATE SINKAGE						
17	① + ⑯	APPROXIMATE DRAFT AFTER FLOODING						
18	CURVES OF FORM	TONNES PER m AT ⑰						
19	COMPARTMENT CALCULATIONS	LOST TONNES PER m AT ⑰						
20	⑱ − ⑲	REMAINING TONNES PER m						
21	$\frac{2 \times ⑦}{⑭ + ⑳}$	SINKAGE						
22	① + ㉑	DRAFT AFTER FLOODING						
23	CURVES OF FORM	TONNES PER m AT ㉒						
24	CURVES OF FORM	LCF AT ㉒ F+ / A−						
25	CURVES OF FORM	MOM. TO TRIM ONE m AT ㉒						
26	COMPARTMENT CALCULATIONS	LOST TONNES PER m AT ㉒						
27	COMPARTMENT CALCULATIONS	CENTER OF ㉖ FROM MIDSHIPS F+ / A−						
28	COMPARTMENT CALCULATIONS	CENTER OF ㉖ FROM CL P− / S+						
29	$\frac{㉓ \times ㉔ - ㉖ \times ㉗}{㉓ - ㉖}$	CENTER OF REMAINING TPm F+ / A−						
30	$⑨ - \frac{(⑮ + ㉙)}{2}$	TRIMMING LEVER F+ / A−						
31	⑦ × ㉚	TRIMMING MOMENT F+ / A−						
32	$\frac{㉓ \times ㉖ \times (㉔ - ㉗)^2}{L \times (㉓ - ㉖)}$	————————						
33	————————	AVERAGE LENGTH OF LOST WATERPLANE, S						
34	$\frac{㉖ \times ㉝^2}{12 L}$	————————						
35	㉕ − ㉜ − ㉞	NET MOMENT TO TRIM ONE m						
36	㉛ / ㉟	TRIM IN m F+ / A−						

☆ NOTE: ITEM 7, MASS OF NET BUOYANCY = $\Sigma \mu \rho v$ (at T) FOR ALL SPACES OPENED TO SEA LESS TOTAL TONS OF LIQUIDS WHICH WERE IN ANY BREACHED TANKS BEFORE DAMAGE.

Fig. 15 Form for calculation of sinkage and trim, lost buoyancy method

where x is the distance from the center of volume of the net lost buoyancy under WL to the center of volume of the buoyancy gained above WL. The latter point lies about midway between the respective centers of flotation (F' and F_1') of the remaining areas of the impaired waterplanes WL and WL$_1$. The moment to trim 1 cm may then be taken as (from Eq. (6), Chapter II),

$$\mathrm{MTcm}'_1 = \frac{\rho I_{L_1}'}{100\, L} \quad (16)$$

where MTcm'_1, is the moment to change trim 1 cm for the damaged ship and I_{L_1}' is the longitudinal moment of inertia of the intact portion of waterplane WL$_1$ about a transverse axis through its center of area, F_1'. Then

$$I'_{L_1} = I_{L_1} + A_1\,(\overline{\bigcirc\!\!\!\!F_1})^2 - (A_1 - \mu_s a_1)\,(\overline{\bigcirc\!\!\!\!F_1'})^2 - \mu_s\,[a_1(\overline{\bigcirc\!\!\!\!f_1})^2 + i_{L_1}] \quad (17)$$

where

$\quad I_{L_1}$ is longitudinal moment of inertia of undamaged waterplane WL$_1$ about its center of flotation, F_1
$\quad \overline{\bigcirc\!\!\!\!f_1}$ is center of a_1 from amidships
$\quad i_{L_1}$ is longitudinal moment of inertia of area, a_1 about its center

In general,

$\quad I_L = 100/\rho \times$ moment to trim 1 cm, MTcm
$\quad A = 100/\rho \times$ tons per cm, TPcm
$\quad \mu_s a_1 = 100/\rho \times$ lost tons per cm, tpcm,
$\quad \mu_s i_L = \dfrac{\mu_s a_1 S^2}{12} = \dfrac{100}{12\rho} \times \mathrm{tpcm}_1 \times S^2$

where $S =$ mean length of lost area and

$$\overline{\bigcirc\!\!\!\!F_1'} = \frac{\mathrm{TPcm}_1 \times \overline{\bigcirc\!\!\!\!F_1} - \mathrm{tpcm}_1 \times \overline{\bigcirc\!\!\!\!f_1}}{\mathrm{TPcm}_1 - \mathrm{tpcm}_1}$$

Therefore, Equations (16) and (17) can be combined to produce:
Net moment to trim 1 m at d_2

$$= 100\left\{ \mathrm{MTcm}_1 - \frac{\mathrm{TPcm}_1 \times \mathrm{tpcm}_1 \times [\overline{\bigcirc\!\!\!\!F_1} - \overline{\bigcirc\!\!\!\!f_1}]^2}{L \times [\mathrm{TPcm}_1 - \mathrm{tpcm}_1]} - \frac{\mathrm{tpcm}_1 \times S^2}{12L} \right\} \quad (18)$$

In English units (ft and long tons), net moment to trim 1 ft at d_2

$$= 12\left\{ \mathrm{MTI}_1 - \frac{\mathrm{TPI}_1 \times \mathrm{tpi}_1 \times [\overline{\bigcirc\!\!\!\!F_1} - \overline{\bigcirc\!\!\!\!f_1}]^2}{L \times [\mathrm{TPI}_1 - \mathrm{tpi}_1]} - \frac{\mathrm{tpi}_1 \times S^2}{12L} \right\} \quad (18\mathrm{a})$$

The calculations may be carried out conveniently in the following steps:

1. Calculate the net lost buoyancy v' below WL, and the longitudinal center of this volume.
2. Determine the sinkage from Equation (12) and later from Equation (13).
3. Calculate the location of F_1', the center of flotation of the remaining area of the impaired waterplane WL$_1$.
4. From Equation (18) calculate the moment to change trim 1 cm.
5. Calculate the change of trim from Equations (15) and (18).
6. If the change of trim is by the bow (positive) the approximate draft at the bow of the flooded ship is the original draft plus sinkage plus the product of change in trim and $[0.5 - (\overline{\bigcirc\!\!\!\!F_1'}/L]$, and the draft at the stern is the original draft plus sinkage minus the product of change in trim and $[0.5 + (\overline{\bigcirc\!\!\!\!F_1'}/L)]$. These drafts determine the trimmed waterline WL$_2$.

(b) *Change of* $\overline{GM}$. The change of $\overline{GM}$ (gain or loss) due to flooding conveniently may be considered as made up of three parts; i.e., a decrease due to lost waterplane, an increase due to sinkage, and an increase due to trim.

1. The decrease in $\overline{BM}$ due to loss of waterplane moment of inertia is

$$\overline{BM} - \frac{I_{T2} - \mu_s i_{T2}}{\Delta/\rho} \quad (19)$$

where

$\quad \overline{BM}$ is $\overline{BM}$ at initial undamaged condition
$\quad \Delta$ is corresponding displacement
$\quad I_{T2}$ is transverse moment of inertia of undamaged WL$_2$
$\quad \mu_s i_{T2}$ is net lost transverse moment of inertia

The foregoing moments of inertia are taken about the transverse center of gravity of the remaining intact waterplane at WL$_2$. It is somewhat more convenient, however, to take the moments of inertia about the centerline. If

$\quad I_{T2CL}$ is transverse moment of inertia of undamaged WL$_2$ about centerline
$\quad \mu_s i_{T2CL}$ is net lost transverse moment of inertia about CL
$\quad A_2$ is waterplane area at WL$_2$
$\quad \mu_s a_2$ is net lost water plane area, and
$\quad t_2$ is its center from centerline

the transverse center of gravity of the remaining waterplane is

SUBDIVISION AND DAMAGE STABILITY

DAMAGE STABILITY - LOST BUOYANCY METHOD - HEEL AND $\overline{GM}$

SHIP _____ DATE _____ PAGE 2
DAMAGE EXTENT _____ BY _____

ITEM No.	SOURCE	DESCRIPTION					
(1)	————————	T, DRAFT BEFORE DAMAGE					
(37)	$(22) + (36)\left(0.5 - \frac{(29)}{L}\right)$	DRAFT FORWARD AFTER FLOODING (WL_2)					
(38)	$(22) - (36)\left(0.5 + \frac{(29)}{L}\right)$	DRAFT AFT AFTER FLOODING (WL_2)					
(39)	DIRECT CALCULATIONS OR CURVES OF FORM	WATERPLANE AREA AT WL_2					
(40)	DIRECT CALCULATIONS OR CURVES OF FORM	TRANS. MOM. INERTIA OF (39) ABOUT C.L.					
(41)	COMPARTMENT CALCULATIONS	μ_S x LOST WATERPLANE AT WL_2					
(42)	COMPARTMENT CALCULATIONS	CENTER OF (41) P- FROM C.L. S+					
(43)	COMPARTMENT CALCULATIONS	$\mu_{S'T}$ OF (41) ABOUT C.L.					
(44)	(41) x (42)	———————					
(45)	$\frac{(44)^2}{(39) - (41)}$	———————					
(46)	(40) - (43) - (45)	———————					
(47)	$(5) - (6) - \frac{(46)}{(2)\rho}$	$\overline{BM}$ LOSS					
(48)	$\frac{(1) + (22)}{2}$	VERTICAL CENTER OF PARALLEL SINKAGE LAYER					
(49)	(48) - (8)	$\overline{b\,b_1}$					
(50)	$\frac{(7) \times (49)}{(2)}$	VCB RISE DUE TO SINKAGE					
(51)	$(36)^2$	$TRIM^2$					
(52)	$\frac{(35) \times (51)}{2 \times L \times (2)}$	VCB RISE DUE TO TRIM					
(53)	(47) + 0.17 - (50) - (52)	INTACT $\overline{GM}$ FOR 5 cm $\overline{GM}$ DAMAGED					
(54)	TRIM AND HEEL DIAGRAM	MINIMUM TANGENT TO MARGIN LINE					
(55)	———————	CORRESPONDING ANGLE OF HEEL					
(56)	———————	SELECTED ANGLE ϕ					
(57)	———————	TAN (56)					
(58)	$(21) + \left((29) + (27)\right)\frac{(36)}{L}$	HEIGHT WL TO WL_2 IN WAY OF (27)					
(59)	$\frac{-12 \times (26) \times (28) \times (58)}{(7)}$	CENTER OF NET ADDED P- BUOYANCY FROM C.L S+					
(60)	(59) - (10)	TRANSVERSE SHIFT P- OF BUOYANCY S+					
(61)	$-(7) \times (60)$	TRANSVERSE MOMENT P- S+					
(62)	COMPARTMENT CALCULATIONS	MOM. CORRECTION FOR CHANGE P- IN μ ETC. HEELING THROUGH ϕ S+					
(63)	$\frac{(61) + (62)}{(57) \times (2)}$	———————					
(64)	FIGS. 19, 20 OR DIRECT CALCULATIONS	F					
(65)	$\frac{(53) + (63) - (64) \times (46)}{-0.05 \quad (2)}$	INTACT $\overline{GM}$ TO LIMIT HEEL TO (56)					
(66)	(53) OR (65) WHICHEVER IS LARGER	REQUIRED INTACT $\overline{GM}$					
(67)	(66) + 0.05 - (53)	CORRESPONDING $\overline{GM}$ IN DAMAGE CONDITION					

IF (53) IS MORE THAN (65), EQUILIBRIUM ANGLE OF HEEL AND RELATED F ARE THOSE WHICH SATISFY EQUATION: $0.05 + F \times \frac{(46)}{(2)} = \frac{(61) + (62)}{(2) \times \tan \phi}$

Fig. 16 Form for calculation of heel and $\overline{GM}$, lost buoyancy method

and

$$\frac{-\mu_s a_2 t_2}{A_2 - \mu_s a_2}$$

$$I_{T2} = I_{T2CL} + A_2 \times \left(\frac{\mu_s a_2 t_2}{A_2 - \mu_s a_2}\right)^2$$

$$\mu_s i_{T2} = \mu_s i_{T2CL} - \mu_s a_2 t_2^2 + \mu_s a_2 \left(t_2 + \frac{\mu_s a_2 t_2}{A_2 - \mu_s a_2}\right)^2$$

Substituting in Equation (29), one obtains decrease in $\overline{BM}$

$$\overline{BM} - \frac{\left[I_{T2CL} - \mu_s i_{T2CL} - \frac{(\mu_s a_2 t_2)^2}{(A_2 - \mu_s a_2)}\right]}{\Delta/\rho} \quad (20)$$

2. The increase in $\overline{KB}$ due to sinkage is

$$\frac{v' \times \overline{bb}_1}{\Delta/\rho} \quad (21)$$

where $\overline{bb}_1$ = vertical distance from center of lost buoyancy to center of parallel sinkage layer.

3. The increase in $\overline{KB}$ due to trim is

$$\frac{MTm' \times (\text{trim})^2}{2\Delta L} \quad (22)$$

or in English units (ft),

$$\frac{MTF_1' \times (\text{trim})^2}{2WL} \quad (22a)$$

As noted in Section 7.5, the present U.S. and international rules require that the $\overline{GM}$ in the damaged condition (displacement taken constant) be at least 5 cm. The required intact $\overline{GM}$ to meet this condition is therefore

$$\overline{BM} - \frac{\left[I_{T2CL} - \mu_s i_{T2CL} - \frac{(\mu_s a_2 t_2)^2}{(A_2 - \mu_s a_2)} + vbb_1\right]}{\Delta/\rho}$$

$$- \frac{MTm' \times (\text{trim})^2}{2\Delta L} + 0.05 \quad (23)$$

(c) *Unsymmetrical moment and heel.* If the net lost buoyancy is unsymmetrical or if the net added buoyancy between WL and WL$_2$ is unsymmetrical so that, in either case, there is a transverse shift in buoyancy in sinking and trimming from WL to WL$_2$, heel will result. In calculating the transverse center of the net added buoyancy, it is necessary to include the effect of trim, which increases the volume swept by the lost area in sinking and trimming from WL to WL$_2$. Thus

TCG of net added buoyancy is

$$\frac{\frac{-100 \,(\text{tpcm} + \text{tpcm}_2) \times h_s}{2} \times \frac{(t + t_2)}{2}}{\text{net added buoyancy}} \quad (24)$$

where

    tpcm is lost tons per cm at WL
    tpcm$_2$ is lost tons per cm at WL$_2$
    h_s is height WL to WL$_2$ in way of damage
    t is center of tpcm from CL
    t_2 is center of tpcm$_2$ from CL

(The negative sign means that the net added buoyancy is on the opposite side of the centerline from the lost tons per inch.)

Net added buoyancy = net lost buoyancy = v';

$\frac{\text{tpcm} + \text{tpcm}_2}{2}$ = approx tpcm$_1$ = lost tons per cm at WL$_1$;

$\frac{t + t_2}{2}$ = approx t_1 = center of tpcm$_1$ from CL;

h_s may be read from a trim and heel diagram, Fig. 41, or is

parallel sinkage + [LCF$_1'$ - lcf_1] × trim/L

The transverse moment due to this shift in buoyancy is then equal to

$$\rho v' \times (\text{TCG of net added buoyancy} - \text{tcg of } v') \quad (25)$$

(d) *Required $\overline{GM}$ to limit heel.* For final stages of flooding, the allowable angle of heel, ϕ, generally is that to the margin line but in no case more than the fixed heeling limit prescribed by the applicable rules. Tan ϕ corresponding to heel to the margin line can readily be determined from the trim and heel diagram, Fig. 17.

In heeling through the angle ϕ to the final heeled waterplane, changes in the longitudinal or transverse extent of flooding and/or changes in permeability may occur. These normally may be accounted for as illustrated in Fig. 18, treating the resulting change as producing a transverse moment, which is added to that of Equation (25). In the suggested calculation form, Fig. 16, term (62) provides for this correction. (The accompanying effects on the damaged condition draft and VCB are neglected.)

The heeling moment then is equal to the total transverse moment × cos ϕ. At the same time, as shown in Chapter II, the righting moment at moderate angles may be expressed as

$$\Delta \times (\overline{GM}_R + F\,\overline{BM}_R) \sin \phi \quad (26)$$

where both $\overline{GM}_R$ and $\overline{BM}_R$ are the residual values at WL$_2$ and F is a factor dependent upon the vessel's

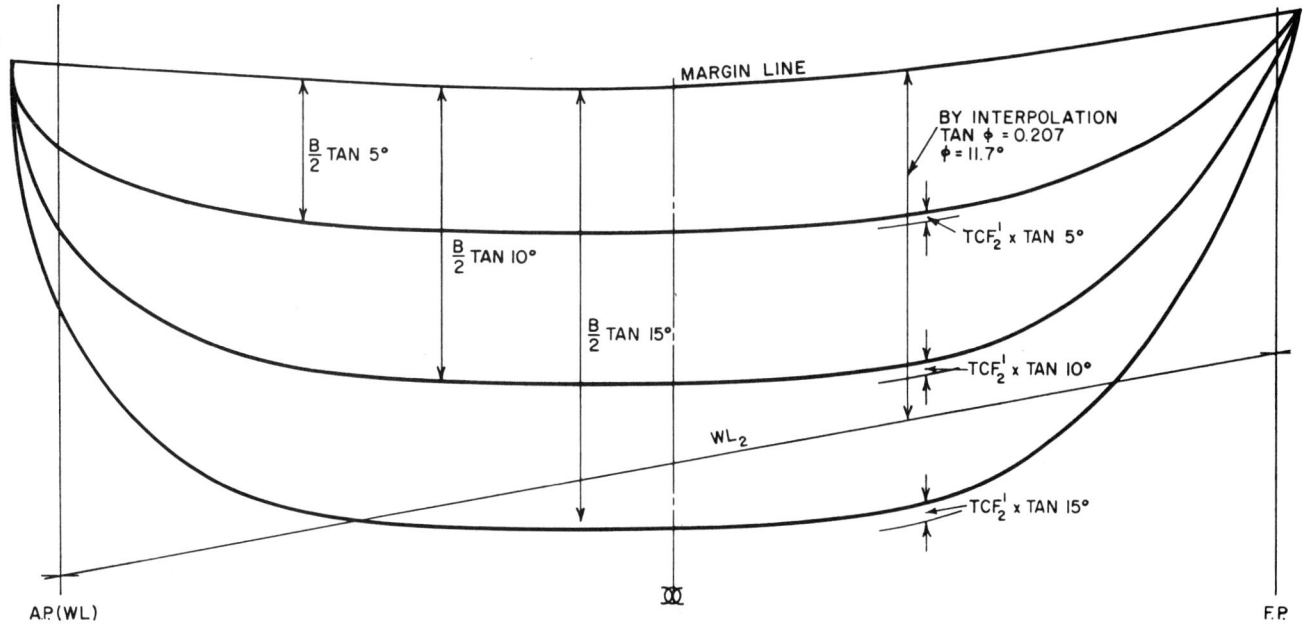

Fig. 17 Trim and heel diagram

proportions and form, such that $F\overline{BM}$ is the vertical distance between the zero-heel metacenter and the point P, at which the vertical through the heeled condition center of buoyancy intercepts the centerline. F is, in effect, the ratio of the average change in transverse waterplane inertia, as the vessel heels to any angle ϕ, to the zero heel transverse waterplane inertia. It is generally assumed that the F-value for a damaged ship at any given draft and heel is the same as for the intact ship at the same draft and heel. This assumption is considered to be sufficiently accurate for ordinary merchant ship forms, at moderate angles of heel, and within ordinary operating ranges of beam-draft ratio. It may not be valid at very light drafts or for unusual hull forms. In applying this assumption, it is also necessary to account for the effect of any local changes in lost waterplane as the vessel heels. In the case of lost-buoyancy calculations this is done as illustrated in Fig. 18 and the related text. If the trim-line added-weight procedure is used, the calculations of flooding water and moment automatically include the effect of any such changes.

Figs. 19 and 20 give approximate F-values for the average ship form, without tumble-home or flare amidships and having a midship coefficient of about 0.96 or greater. For heels not immersing the deck edge or rolling the bilge side tangency out, the solid line portions of these figures give values which are sufficiently accurate for usual intact or damage stability calculations. The dotted line portions are progressively likely to be less accurate as they depart from the solid line portions, and should only be used without further check when the resulting values indicate an appreciable stability margin (Prohaska, 1961).

Equating terms, the intact $\overline{GM}$ necessary to limit heel to any angle ϕ is given by:

$$\overline{GM} \text{ loss} + \frac{\text{transverse moment}}{\Delta \times \tan \phi} - F\overline{BM}_R \quad (27)$$

where $\overline{GM}$ loss = value from equation (23) − 0.05.

The required intact $\overline{GM}$ then is that determined by Equation (23) or (27), whichever is larger.

The required initial $\overline{GM}$s at various drafts are usually plotted along with the available $\overline{GM}$s in the various operating conditions. An example is shown in Fig. 21,

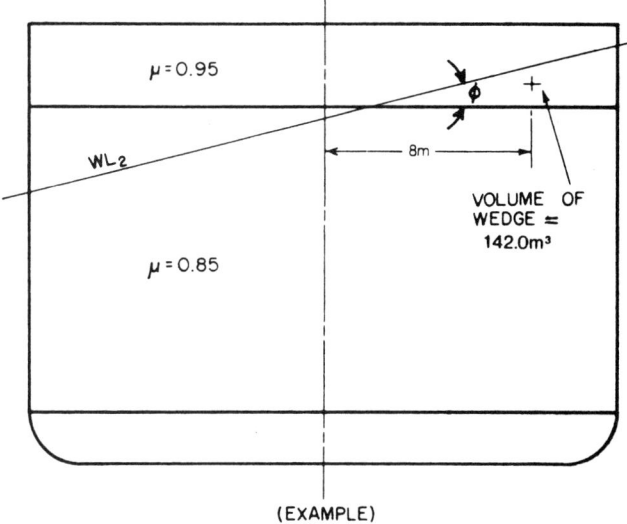

(EXAMPLE)

Fig. 18 Transverse moment due to heel into a level of higher permeability
$(142(0.95-0.85)1.025 \times 8 = 116 \text{ t-m})$

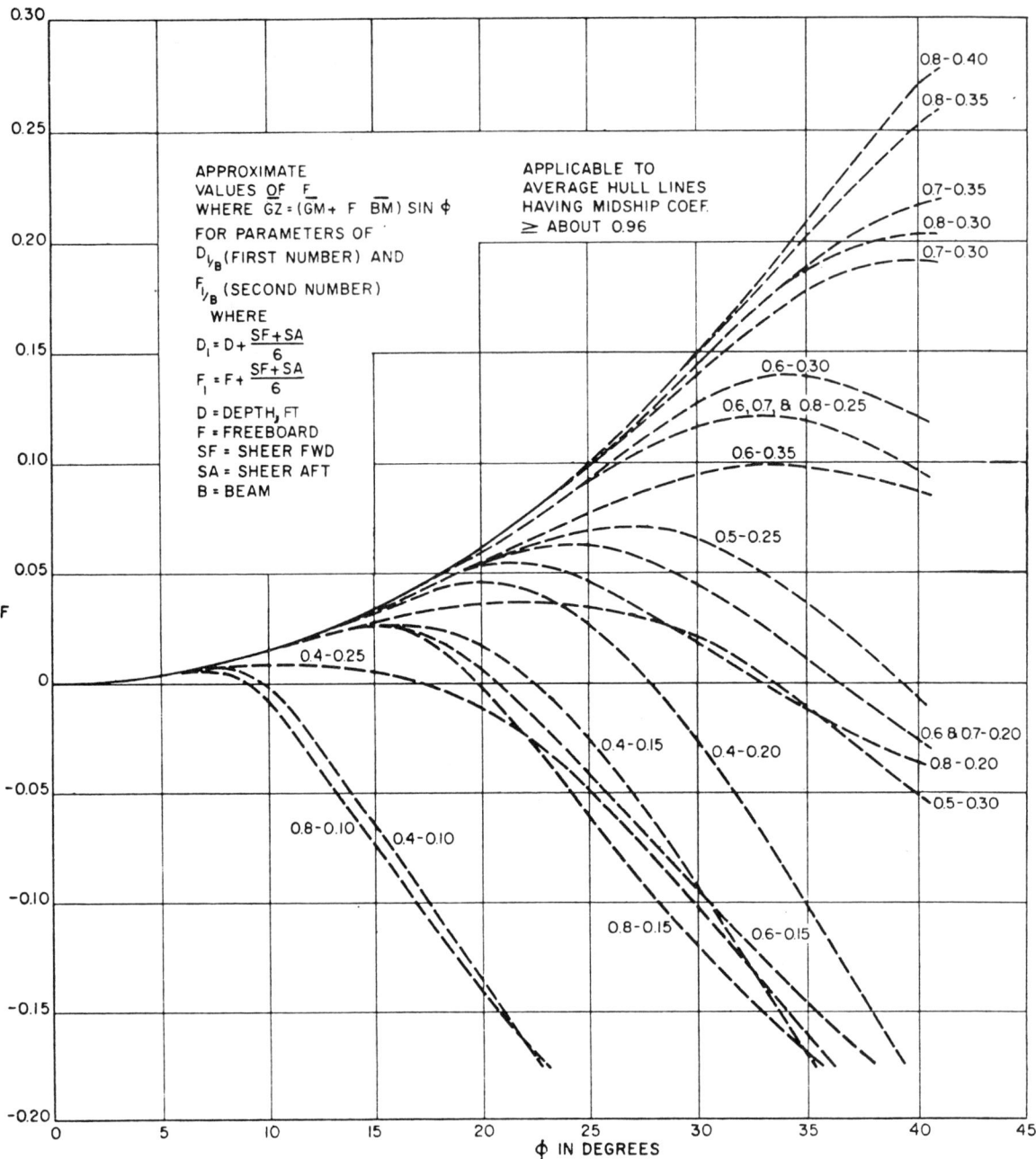

Fig. 19 Approximate values of F, where $\overline{GZ} = (\overline{GM} + F\,\overline{BM})\sin\phi$

which indicates whether the available $\overline{GM}$ in these conditions equals or exceeds the required $\overline{GM}$. Obviously if they all do not, the vessel does not meet the criterion. See also Chapter II.

4.7 Trim-Line Added-Weight Method. The version of this method described in the following text, including Section 4.5, provides sufficiently accurate results for all usual conditions and is more versatile in this respect than the lost buoyancy method. In addition, it is believed generally to involve less computation than trim-line added-weight calculations based upon direct use of Bonjean curves. However, if flooding involves extreme conditions of trim and heel combined, the most accurate results can be attained by obtaining displacements and moments by direct use of Bonjean curves. In such case, it is necessary that these curves be provided for both upright and heeled conditions, and that they include not only the usual area values but also transverse and vertical moment values and, for the upright condition, half-breadths cubed. If this is done, the necessary related calculations differ in detail but agree fully in principle with those described herein.

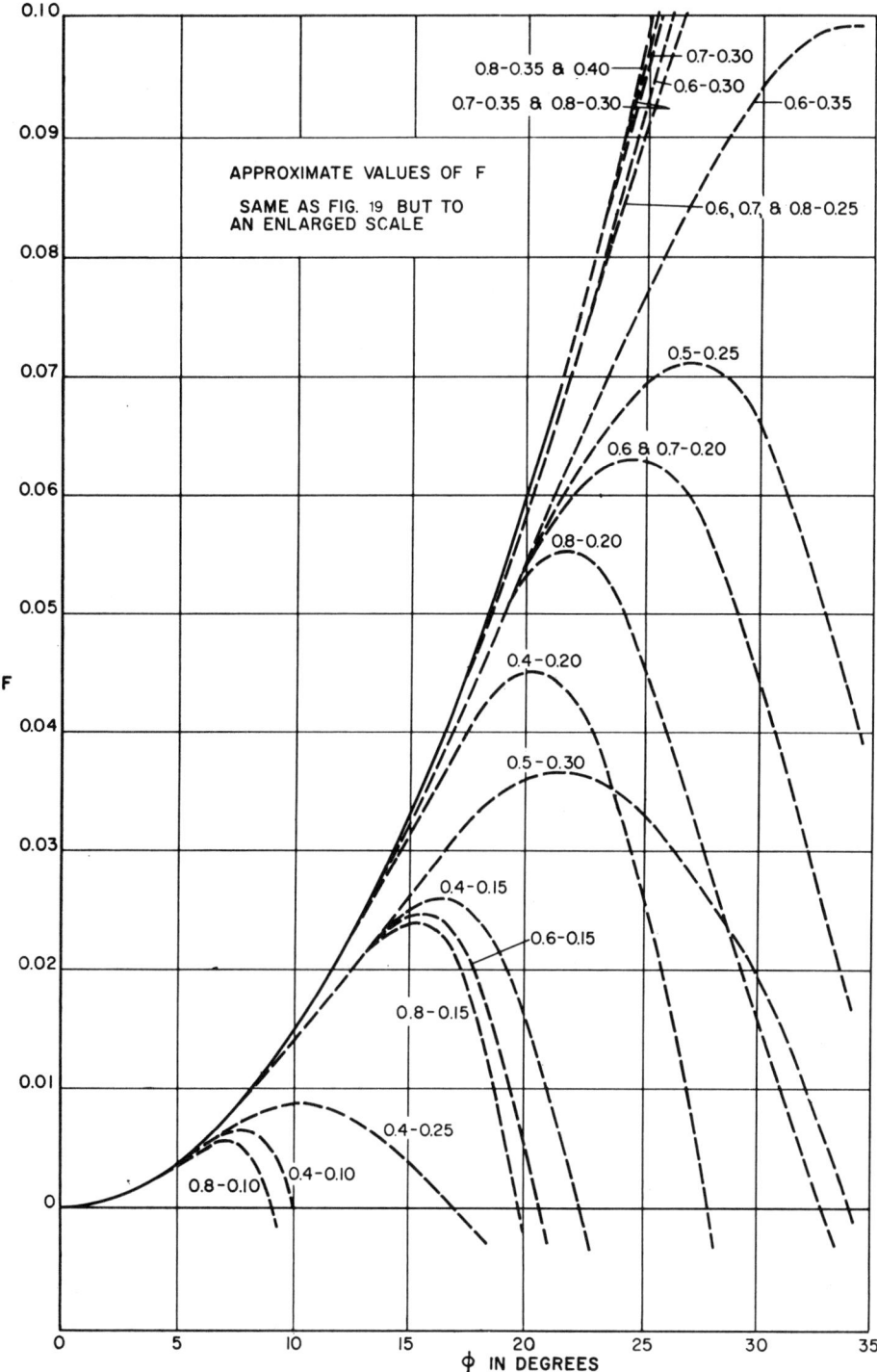

Fig. 20 Approximate values of F, where $\overline{GZ} = (\overline{GM} + F\overline{BM})\sin\phi$

In using the method hereinafter described, all ship hydrostatic values, except where otherwise specially noted, are for an assumed undamaged ship at the applicable draft and trim. If conditions involving appreciable trim exist, convenience and accuracy favor the use of curves of form which give the effective LCF, $\overline{KM}$, $\overline{KB}$, and TPcm values for trims forward and aft as well as level trim. It is usually adequate for this purpose to calculate displacement, $\overline{KM}$, $\overline{KB}$, and TPcm versus mean draft for fixed values of trim, say $L/40$, forward and aft, depending upon interpolation or cross-plotting for intermediate values. The trimmed condition effective LCF is taken as the longitudinal position at which the trimmed waterplane is the same

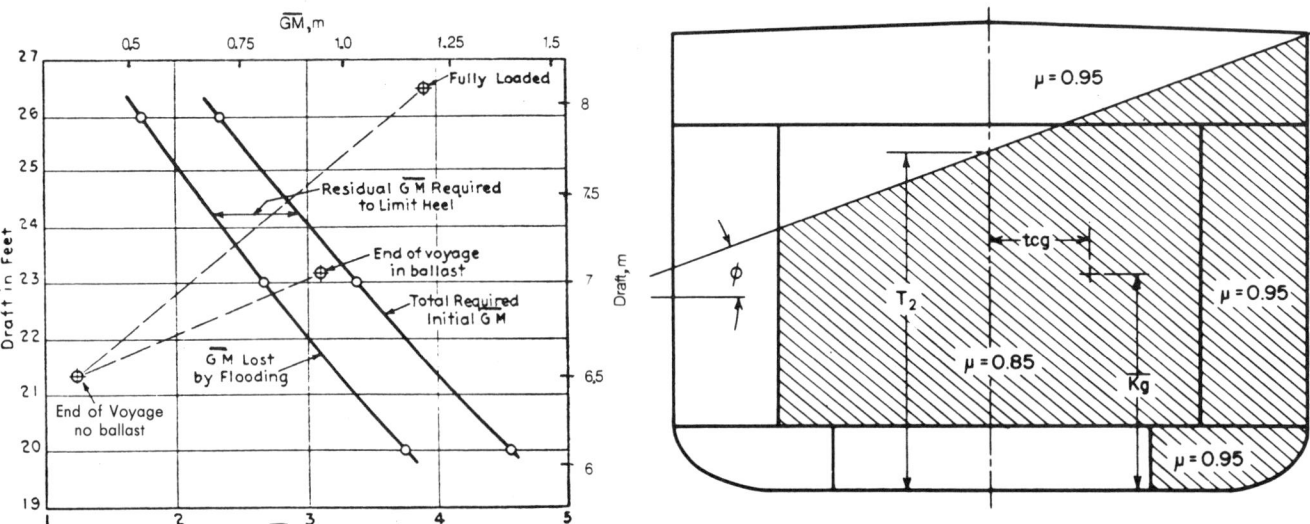

Fig. 21 Diagram Showing Available $\overline{GM}$ at Beginning and End of Voyage and Required Initial $\overline{GM}$

Fig. 22 Section through flooded compartment

distance above the baseline as the level trim draft for the same displacement. That is to say, the effective LCF of any waterplane equals the difference between the amidships draft for that waterplane and the corresponding level trim draft × the cotangent of the trim. Thus, the final plot of displacement versus draft at LCF shows a single line for all trims. At the same time, the calculated mean draft points for the $\overline{KM}$, $\overline{KB}$, and TPcm curves are shifted vertically so that these values are also plotted on the basis of the draft at the LCF. Unless specifically excepted, all drafts used in this method are at the LCF. Moment to trim 1 cm is assumed to be constant over the range of trim concerned and the plotted value may be based either upon the average calculated shift of LCB with trim or upon the level trim waterplane inertia.

A mean transverse section similar to Fig. 22 is prepared for each damaged compartment to be considered. If the compartment is long, has appreciable variation with length, or contains longitudinal discontinuities, convenience and accuracy may require that it be separated into two or more parts. In judging the need for doing this, the use of mean sections will result generally in underestimating the heeling moment due to free surface by less than 1 percent if the breadth of the smaller end is 85 percent of that of the larger, while a breadth ratio of 75 percent may result in an underestimate of about 2 percent.

Taking into account the assumed extent of damage and the allocation of the various spaces represented by the mean transverse section, a summation of the areas and moments of each portion of the section, each multiplied by a suitable constant depending upon the permeability and the flooded length, then gives the mass of flooding water and the moments of flooding water. This integration is performed for the expected range of drafts in way of the compartment and for at least two heeled conditions, usually 7 and 15 deg, for which mass of flooding water w/g and virtual vertical moment of flooding water $\left[w/g\left(\overline{Kg} + \dfrac{tcg}{\tan \phi}\right)\right]$ are obtained. If righting-arm curves in the damaged condition will be required, calculations are also made for additional angles of heel to cover the necessary range. If the amount of unsymmetrical flooding is small and it therefore seems likely that the $\overline{GM}$ necessary to limit heel may be less than that required to provide a 5-cm (2-in) positive $\overline{GM}$ upright in the damaged condition, calculations are also made for the zero-heel condition, in which case the quantities which need to be determined are the mass of flooding water w/g, vertical moment of flooding water $(w/g)\overline{Kg}$, net lost waterplane transverse moment of inertia about centerline $\mu_s i_T$, net lost TP$_{cm}$ $\rho\mu_s a/100$ (or tons per m, $\rho\mu_s a$), and center of same from centerline (tca).

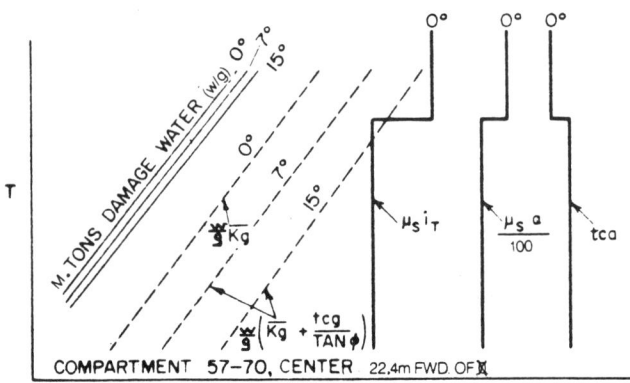

Fig. 23 Plot of damage water quantities with heel

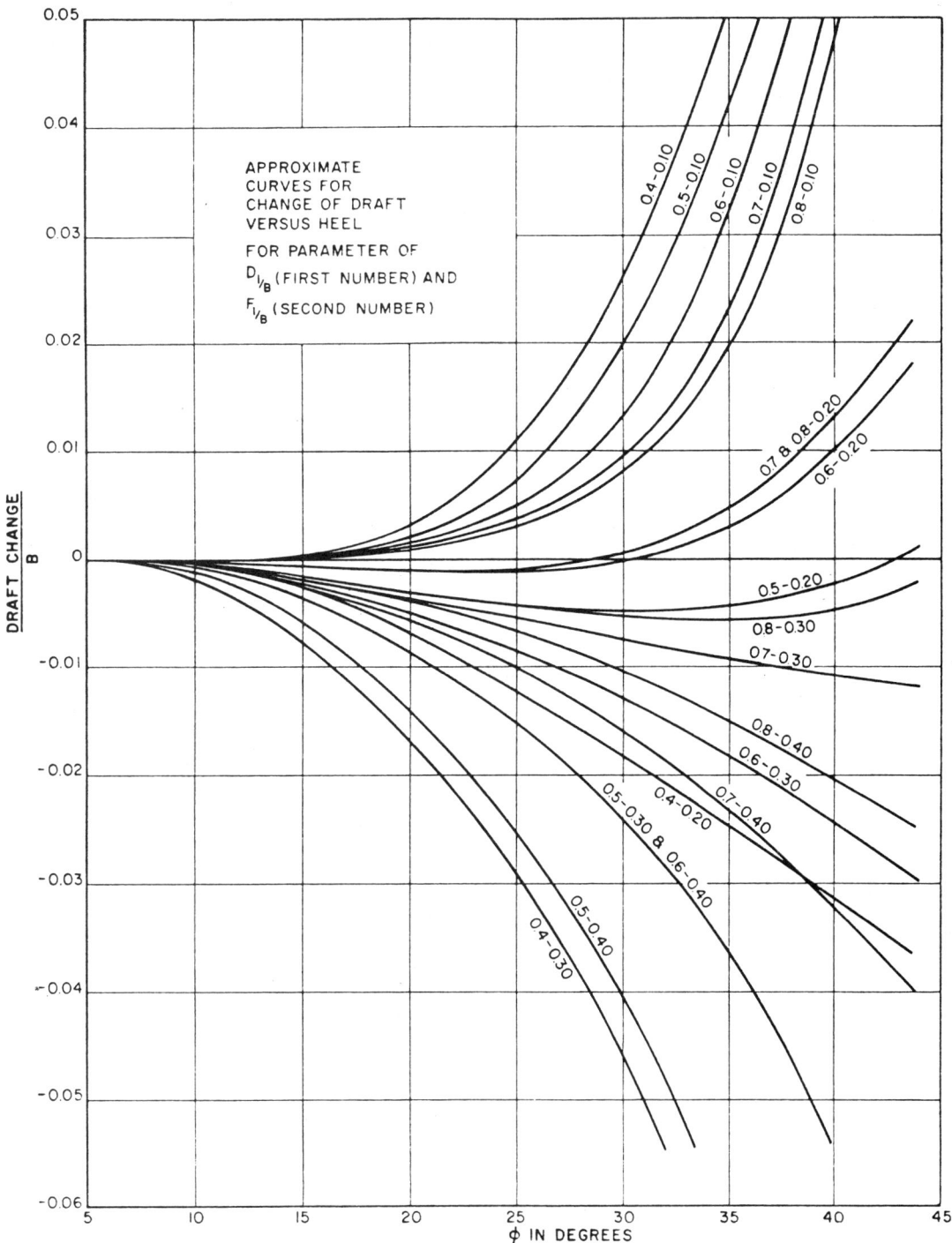

Fig. 24 Change of draft versus heel

In making these calculations, the mass of flooding water, and related quantities are taken to correspond to $\Sigma \mu v \rho$ for all breached spaces up to the assumed final flooding line less the total mass of liquids assumed to be in such spaces before damage. However, where tanks are low in the ship, the most severe condition may be one in which they are not breached and flooding occurs only above them. Where doubt is considered to exist as to the outcome, calculations should be made assuming such tanks both breached and nonbreached. The nonbreached condition corresponds to the zero permeability condition called for by Section 7.5. These calculations are performed and the data are plotted for each section as illustrated in Figs. 22 and 23.

(a) *Draft and trim.* The ship is assumed at a damaged condition draft, T_2. At this draft and an assumed level trim, displacement, LCF, and moment to trim 1 cm are read from the curves of form; and mass of flooding water and lcg of flooding water are read from the curves for the applicable damaged compartment(s). (Because the effects of compartment shape and of trim are partially compensatory, sufficient accuracy is normally attained by assuming the lcg of a compartment or of a part to be at its midlength.)

A first approximate value for the trim due to flooding is

$$\text{Trim, cm} = \frac{(w/g)(\overline{\oslash g} - \overline{\oslash F})}{100 \text{ MTcm}} \quad (28)$$

or in English units,

$$\text{Trim, ft} = \frac{w(\overline{\oslash g} - \overline{\oslash F})}{12 \times \text{MTI}} \quad (28a)$$

Since trim increases the draft in way of the damage and therefore the amount of flooding water, the actual trim due to the damage will, in general, be more than this first approximation. At average damaged condition drafts, it may be about 10 percent more. With this in mind, a trimmed waterline corresponding to the vessel LCF draft, T_2 is assumed.

A trim and heel diagram similar to Fig. 17, but showing both the LCF curve or curves and the longitudinal center or centers of flotation of the damaged compartments, is used. By placing the assumed trimmed waterline on the diagram, the angle of heel to the margin line and the draft at the center of the damaged space or spaces, or of each section thereof, can be read off. Since the compartment characteristics are plotted on a basis of centerline draft, no draft correction for TCF of damaged waterplane is made.

Using the compartment curves, the values corresponding to these damaged space drafts are then totalled for the space or spaces included in the damage. If the unsymmetrical part of the flooding is small, i.e., if a large portion of the heeling moment results from the transference of wedges as the vessel heels, the $\overline{GM}$ necessary to provide a 5-cm (2-in) positive $\overline{GM}$ in the damaged condition may be more than that to limit heel. If this is considered likely, calculations are made both for a zero heel condition and for one or more assumed heeled conditions. (In dealing with the zero heel condition, transverse moment is assumed to be eliminated; i.e., as would be the case if weights elsewhere in the ship were shifted to equalize the heeling moment.)

In either case the amount of flooding water subtracted from the displacement at T_2 then gives the displacement before damage, and the corresponding LCF draft before damage, T, is read from the curves of form.

For angles of heel up to and including 15 deg the change in draft accompanying heel is normally disregarded. However, for higher angles of heel, where the deck edge is appreciably immersed or the bilge emerges, the change may be considerable. Fig. 24, which is based upon the stability model tests reported in Russo & Robertson (1950), provides a means of approximating this change, and thus improving the accuracy of the calculations. The change in draft determined from this figure is applied to T_2, the centerline LCF draft for the assumed trim line. The correct displacement and other hydrostatic properties corresponding to T_2 are then determined by entering the curves of form with the resulting amended draft, not T_2.

For purposes of calculating trim it is usually assumed that the mean emergence draft is equal to the mean between the assumed damage condition draft, T_2, and the mean of the drafts before damage determined for the zero heel and for the limiting heel conditions. It is assumed that addition of a weight equal to the flooding water at the level or normal trim LCF corresponding to this mean emergence draft would result in sinkage from T to T_2 without any change in trim. The trimming lever is therefore taken as the distance between this LCF and the lcg of flooding water, and the trim recomputed by Equation (28). The difference between this computed trim and the assumed trim is the trim before damage. If this trim before damage is within the normal operating range for that draft it may be considered satisfactory, having regard, at the same time, that trim may be more critical when heeling is restricted by damaged condition freeboard. With a little experience in the selection of assumed trims, agreement within 30 cm (about 1 ft) can usually be obtained, and this is entirely adequate.

The calculation of required initial $\overline{GM}$ is then carried out in accordance with the fundamental relationships given in Section 3.3. See Figs. 25 and 26.

4.8 Stability during Intermediate Flooding. During the intermediate stages of flooding the water is continually flowing into the damaged compartment. Hence, the added weight approach is applicable, with allowance for the free surface of the water that has flowed in.

Conditions during intermediate flooding may vary widely. When colliding vessels immediately separate after a collision, the dynamic effect of the roll of the struck vessel as a result of the impact, plus the surge of water into the damaged space, conceivably may be quite unfavorable. On the other hand, if the striking vessel remains engaged with the struck vessel for a time, it may more or less restrain that vessel during the period of intermediate flooding.

In recognition of these extremes, as well as the practical requirements of using known quantities in the calculations, the following assumptions are used in intermediate flooding:

(a) The vessel is assumed to be in static equilibrium

DAMAGE STABILITY – TRIM LINE ADDED WEIGHT METHOD – DRAFT AND TRIM – BASIC DATA

SHIP _____ DATE _____ PAGE 1
DAMAGE EXTENT _____ BY _____

ITEM No.	SOURCE	DESCRIPTION						
①	ASSUMED	T_2 LCF DRAFT IN DAMAGED CONDITION						
②	CURVES OF FORM	GROSS DISPLACEMENT, Δ AT ①						
③	CURVES OF FORM	LCF AT ① AND LEVEL TRIM F+ A−						
④	CURVES OF FORM	MOM. TO TRIM ONE m AT ①						
⑤	DAMAGED COMPARTMENT CURVES	TONNES FLOODING WATER, $\frac{w}{g}$ AT ① AND ZERO HEEL						
⑥	DAMAGED COMPARTMENT CURVES	lcg OF ⑤ F+ A−						
⑦	⑥ − ③	APPROXIMATE TRIM LEVER F+ A−						
⑧	⑤ × ⑦ / ④	APPROXIMATE TRIM F+ A−						
⑨	ABOUT 1.1 × ⑧	ASSUMED TRIM F+ A−						
⑩	CURVES OF FORM	EFFECTIVE LCF AT ① FOR ② AND ⑨ F+ A−						
⑪	CURVES OF FORM	$\overline{KM}$ FOR ② AND ⑨						
⑫	CURVES OF FORM	$\overline{KB}$ FOR ② AND ⑨						
⑬	CURVES OF FORM	TONNES PER m FOR ② AND ⑨						
⑭	① + ⑨ $\left(0.5 - \frac{⑩}{L}\right)$	DAMAGED CONDITION DRAFT FORWARD						
⑮	① − ⑨ $\left(0.5 + \frac{⑩}{L}\right)$	DAMAGED CONDITION DRAFT AFT						
⑯	TRIM AND HEEL DIAGRAM	MINIMUM TANGENT TO MARGIN LINE						
⑰	———————	CORRESPONDING ANGLE OF HEEL, DEG.						
⑱	———————	SELECTED ANGLE ϕ, DEG.						
⑲	TRIM AND HEEL DIAGRAM	T_d DRAFT(S) AT DAMAGED SPACE(S)						
⑳	DAMAGED COMPARTMENT CURVES	TONNES FLOODING WATER, $\frac{w}{g}$ AT ⑲ AND ZERO HEEL						
㉑	DAMAGED COMPARTMENT CURVES	TONNES FLOODING WATER AT ⑲ AND ⑱						
㉒	DAMAGED COMPARTMENT CURVES	lcg OF ⑳ F+ A−						
㉓	DAMAGED COMPARTMENT CURVES	lcg OF ㉑ F+ A−						
㉔	DAMAGED COMPARTMENT CURVES	$\frac{w}{g} \overline{Kg}$ AT ⑲ AND ZERO HEEL						
㉕	DAMAGED COMPARTMENT CURVES	$\mu_s i_T$ AT ⑲ AND ZERO HEEL						
㉖	DAMAGED COMPARTMENT CURVES	$\mu_s a \rho$ AT ⑲ AND ZERO HEEL						
㉗	DAMAGED COMPARTMENT CURVES	tcg AT ⑲ AND ZERO HEEL P− S+						
㉘	DAMAGED COMPARTMENT CURVES	$\frac{w}{g}\left(\overline{Kg} + \frac{tcg}{\tan \phi}\right)$ AT ⑲ AND ⑱						
㉙	② − ⑳	DISP BEFORE DAMAGE FOR ZERO HEEL						
㉚	② − ㉑	DISP BEFORE DAMAGE FOR ⑱						
㉛	CURVES OF FORM	DRAFT BEFORE DAMAGE FOR ZERO HEEL						
㉜	CURVES OF FORM	DRAFT BEFORE DAMAGE FOR ⑱						

NOTE: IN ALL CASES TONNES FLOODING WATER, $\frac{w}{g} = \Sigma \mu \rho v$ FOR ALL BREACHED SPACES, LESS TONNES OF LIQUIDS ASSUMED TO BE IN SUCH BREACHED SPACES BEFORE DAMAGE.

Fig. 25 Form for calculation of draft and trim, added-weight method

172   PRINCIPLES OF NAVAL ARCHITECTURE

DAMAGE STABILITY - TRIM LINE ADDED WEIGHT METHOD - TRIM, HEEL AND $\overline{GM}$

SHIP _____ DATE _____ PAGE 2
DAMAGE EXTENT _____ BY _____

ITEM No	SOURCE	DESCRIPTION						
(31)	PAGE 1	DRAFT BEFORE DAMAGE FOR ZERO HEEL						
(33)	$(2\times(1)+(31)+(32))/4$	MEAN EMERGENCE DRAFT						
(34)	CURVES OF FORM	LCF AT (33) AND LEVEL TRIM F+ A−						
(35)	(22) − (34)	TRIM LEVER FOR ZERO HEEL F+ A−						
(36)	(20) × (35) / (4)	COMPUTED TRIM FOR ZERO HEEL F+ A−						
(37)	(9) − (36)	TRIM BEFORE DAMAGE FOR ZERO HEEL F+ A−						
(38)	(26) × (27)	TRANS. MOM. OF NET LOST AREA						
(39)	$(38)^2$	----------						
(40)	(13) − (26)	NET TONNES PER m						
(41)	(39) / (40)	----------						
(42)	$((25)+(41))/(2)$	$\overline{BM}$ LOSS FOR (2) AND (9)						
(43)	(11) − (42)	NET $\overline{KM}$ FOR (2) AND (9)						
(44)	$\dfrac{(2\times(43)-(24))}{(29)} - 0.05$	MAX. ALLOWABLE $\overline{KG}$ BEFORE DAMAGE						
(45)	CURVES OF FORM	$\overline{KM}$ AT (31) FOR (29) AND (37)						
(46)	(45) − (44)	INTACT $\overline{GM}$ FOR 5 cm $\overline{GM}$ DAMAGED						
(32)	PAGE −1	DRAFT BEFORE DAMAGE FOR (18)						
(47)	(23) − (34)	TRIM LEVER FOR (18) F+ A−						
(48)	(21) × (47) / (4)	COMPUTED TRIM FOR (18) F+ A−						
(49)	(9) − (48)	TRIM BEFORE DAMAGE FOR (18) F+ A−						
(50)	----------	TANGENT (18)						
(51)	----------	SINE (18)						
(52)	FIGS. 19-23 OR DIRECT CALCULATIONS	F FOR (18)						
(53)	(11) − (12)	$\overline{BM}$ FOR (2) AND (9)						
(54)	(11) + ((52) × (53))	$\overline{KP}$						
(55)	(2) × (54) / (30)	----------						
(56)	(28) / (30)	----------						
(57)	(55) − (56)	MAX ALLOWABLE $\overline{KG}$ BEFORE DAMAGE						
(58)	CURVES OF FORM	$\overline{KM}$ AT (32) FOR (30) AND (49)						
(59)	(58) − (57)	INTACT $\overline{GM}$ TO LIMIT HEEL TO (18)						
(60)	(46) OR (59) WHICHEVER IS LARGER	REQUIRED INTACT $\overline{GM}$						

IF (46) IS MORE THAN (59), EQUILIBRIUM ANGLE OF HEEL WHICH CORRESPONDS THERETO, RELATED F, AND $\frac{w}{g}$ ARE THOSE WHICH SATISFY THE EQUATION: $(11) + F \times (53) - (44) = \frac{w}{g}\left(\overline{Kg} + \frac{tcg}{\tan\phi}\right) - \frac{w}{g} \times (44)$
(WHERE $\frac{w}{g}$ IS BETWEEN (20) AND (21))   (2)

Fig. 26   Form for calculation of heel and $\overline{GM}$, added-weight method

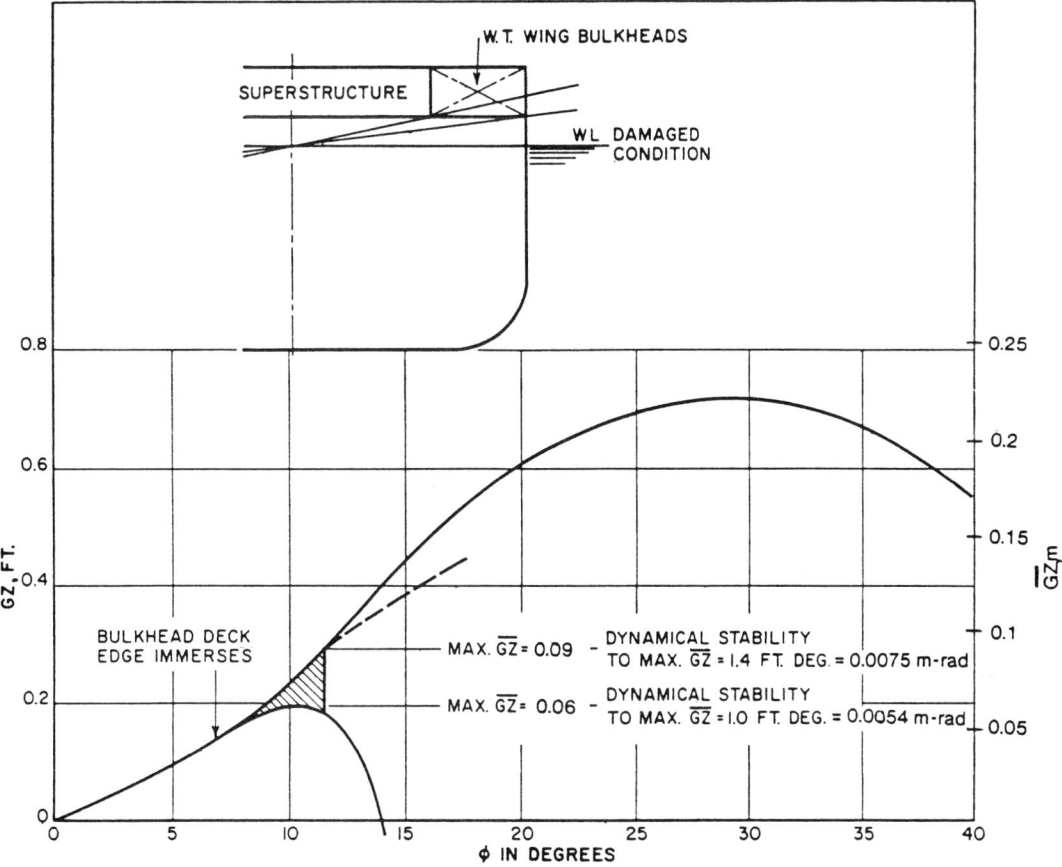

Fig. 27 Effect of watertight wing bulkheads in superstructure

For this example, $B = 25$m; freeboard after damage = 1.5m; $\overline{GM}$ after damage = 0.3m

at every stage of flooding, with the flooding water surface parallel to, but at a lower level than the surface of the sea.

(b) In the case of unsymmetrical flooding involving spaces which are cross-connected by pipes, ducts, or flooding plugs, it is normally assumed that the flooding water in the damaged wing space reaches sea level, and the vessel heels accordingly, before any equalization occurs.

(c) Wing spaces which are freely interconnected by large unobstructed openings are assumed to equalize as they flood.

During intermediate flooding, heel may occur either as a result of negative $\overline{GM}$ or from unsymmetrical flooding. Since the condition is transient, some heel is acceptable provided the accompanying range of stability and maximum righting arm include a sufficient margin against capsizing.

The matter of just how much margin is "sufficient" is rather indeterminate. There may be considerable difference in required margin, depending, for example, on whether equalization takes place in 1.5 min or 1.5 hr (see 7.6). In situations such as this where the safety of people and property are involved, the prudent naval architect will endeavor to arrange compartments so that equalization is either not required or can be at-

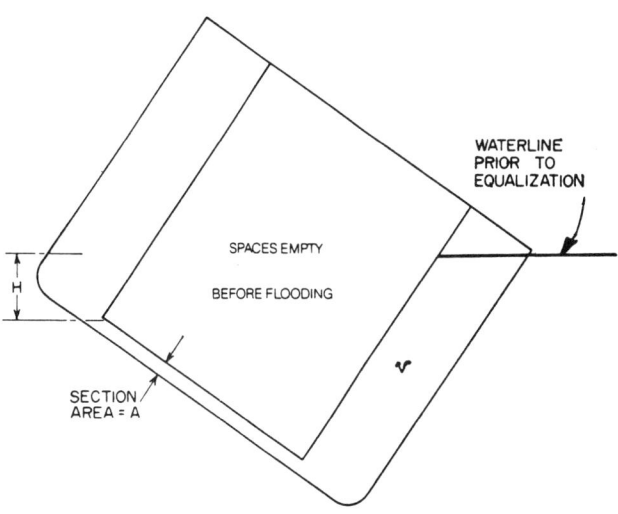

Fig. 28 Equalization of wing tanks

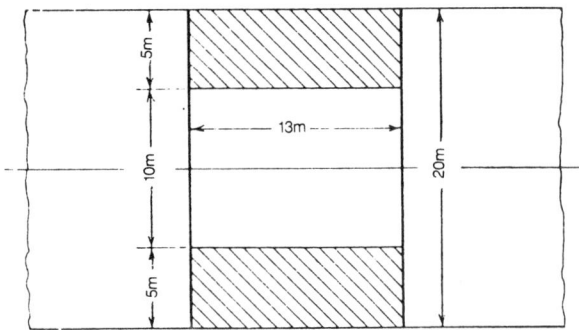

Fig. 29 Assumed wing tanks for free surface calculation

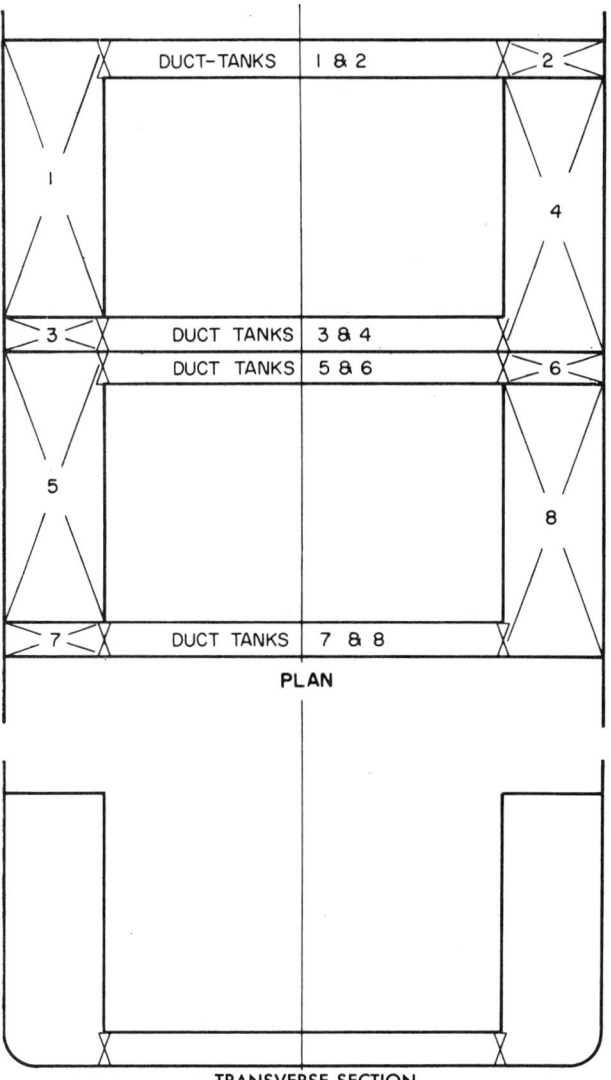

Fig. 32 Alternative cross-connection of wing tanks

tained quickly and will provide, in all stages of equalization, a range and amplitude of righting arm close to that required by the survival criteria. In judging this, the righting arm curve should be assumed to go to zero at the point where downflooding through openings in the bulkhead deck can occur. It is therefore a desirable design feature to have such openings as far inboard as practicable.

It is frequently practicable to have the wing bulkheads in spaces above the bulkhead deck watertight; the beneficial effect of such bulkheads has been referred to in Section 2.2, as illustrated in Fig. 27. At the point where flooding reaches the inboard lower corners of such bulkheads, the righting arm curve theoretically drops to the lower curve, corresponding to the bulkhead deck, and is not reversible. However, where there is a fore and aft passage bulkhead at the inboard edge of the wing bulkhead, outboard extension of flooding may be retarded enough so that the effective dynamic value of $\overline{GZ}$ approaches that indicated by the dotted line. This gain cannot be evaluated specifically, but does have the effect of narrowing the necessary margin between the statical heeling arm and the peak $\overline{GZ}$. It is obviously desirable that the wing bulkheads extend as far inboard, and that their con-

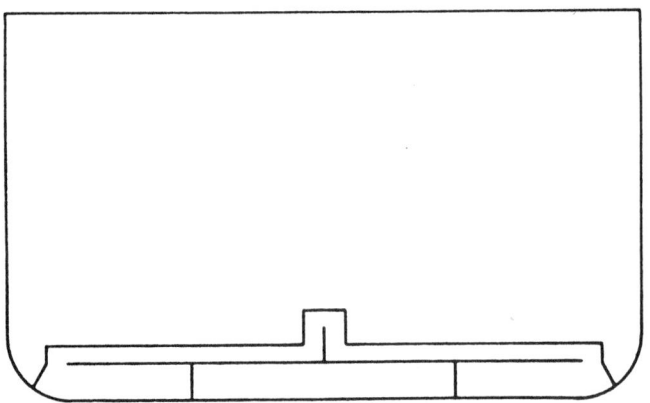

Fig. 30 Cross-connection of double-bottom spaces

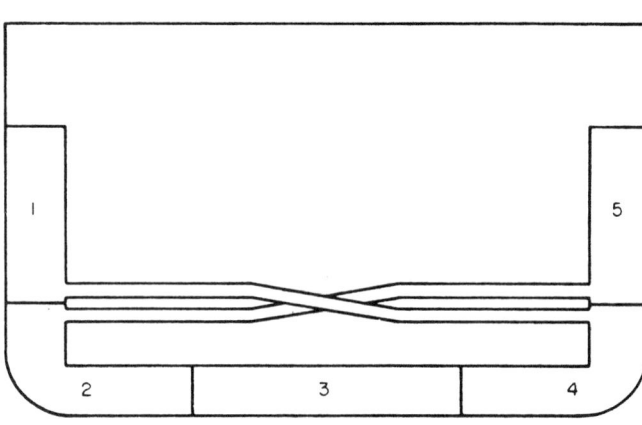

Fig. 31 Cross-connection of wing tanks

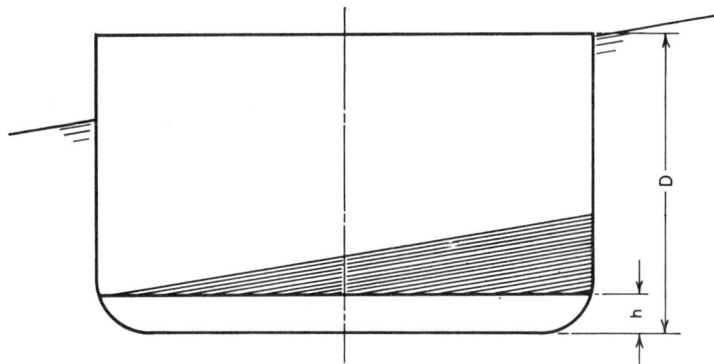

Fig. 33 Intermediate flooding

struction and arrangement be as effective, as is practicable.

When port and starboard spaces are cross-connected to minimize or eliminate unsymmetrical buoyancy, such spaces may be assumed to equalize concurrently with flooding on the damaged side only if they are connected by large unobstructed openings such as open passages or cofferdams, as illustrated in Fig. 28.

When spaces are cross-connected, equalization time, in seconds, may be estimated by the following formula set forth by G. Solda (1961):

$$t = \frac{2v}{fA(2gH)^{1/2}} \qquad (29)$$

where

v is total volume of unequalized flooding, or of liquid to be dumped

f is flow coefficient (0.65 may generally be assumed if connection is short and direct and is fitted with not more than one full area gate valve or equivalent)

A is sectional area of equalizing duct or dump line

H is net head at beginning of equalization or dumping

In English units, this reduces to,

$$t = \frac{v}{157\ AH^{1/2}} \min \qquad (30)$$

Cross connections between tanks present a different problem. Such connections, if of ordinary design and open, may greatly increase the free-surface effect of liquids in the tanks. For example, suppose that the dimensions of the tanks were as shown in Fig. 29. The combined moment of inertia of the surfaces of these two tanks, about longitudinal axes through their respective centers of area, is

$$\frac{2 \times 13 \times (5)^3}{12} = 270.8 \text{ m}^4$$

If the tanks are cross-connected, their surfaces will constitute a single surface the center of area of which is on the centerline of the ship. The moment of inertia about this axis is

$$\frac{13 \times [(20)^3 - (10)^3]}{12} = 7583 \text{ m}^4$$

which is 28 times as much as when the tanks were separate. If the ship should be flooded in compartments other than those containing the cross-connected wing tanks, their large free surface would still exist in the flooded ship. Therefore, a correspondingly increased $\overline{GM}$ before free-surface correction would be required to insure sufficient residual stability after flooding. See Chapter II.

Such operational free-surface effects may be eliminated by providing valves in the cross connections or by using special automatic tank and cross-connection arrangements which permit liquid transfer from a damaged tank to an intact one on the other side, without resulting in appreciable transfer between undamaged tanks.

Where valves are fitted, it is usual to specify that these be maintained closed at all times except when opened and reclosed for drill and test purposes, or when a valve or valves may actually be required to be opened to equalize flooding. This is effective, but is subject to the risk that, in time of emergency, a valve may either not be opened or may be unnecessarily and wrongly opened to the detriment of stability.

No attempt is made here to discuss special automatic tank arrangements and cross connections, since there is a variety of ways in which a designer may reduce the amount of unsymmetrical intact buoyancy and yet maintain satisfactory intact stability. Figs. 30, 31, and 32 illustrate ways of doing this.

With modern computer techniques (Section 5), it is

relatively easy to check the ship's actual heel by the calculation of residual righting arms with varying amounts of water in the compartment at levels of, for example, 1/16, 1/3 and 2/3 of the distances from the top of the double bottom to the intact waterline. See Section 5. From the righting arm curve, the positive or negative $\overline{GM}$ is immediately available.

In the case of symmetrical flooding, the vessel may heel due to negative $\overline{GM}$ in the early stages of flooding. The required intact $\overline{GM}$ to limit heel so as not to immerse any particular reference point, such as the margin line, generally is greatest for the time in flooding at which the double bottom just becomes covered, as illustrated in Fig. 33.

Section 5
Subdivision and Damage Stability Calculations by Computer.

5.1 General. Computer programs are now widely available throughout the world to calculate subdivision and damage stability. In the United States the program most widely used is the Ship Hull Characteristics Program (SHCP). The subroutines available in the basic program are discussed here. Many users, i.e., government agencies, design agents, shipyards, for example, have developed additional subroutines to further reduce manual labor. Thus the use of computers makes possible a more comprehensive and rigorous analysis than would otherwise be possible.

In addition to the Floodable Length Program described below, a Trim Lines Sub-program calculates final damaged ship drafts and trim after flooding of specific compartments or groups of compartments at specified permeabilities. A Limiting Drafts Sub-program determines the maximum drafts forward and aft to which a ship may be loaded prior to flooding and still survive when a specific compartment or group of compartments is flooded.

A fundamental limitation of computers is that the output is only as good as the input and great care and forethought are necessary in subdivision and damage stability calculations to insure that all relevant flooding combinations are included. Not only must longitudinal, transverse and horizontal flooding be carefully defined, but all conditions affecting weight and volume, such as longitudinal, transverse and horizontal changes in permeability, if any, must be allowed for. Dimensioned sketches should be prepared and retained as records of assumed flooding on which input data are based.

In future it may be expected that hand-held and/or microcomputers, which have been programmed for other naval architectural calculations, will be applied to flooding and damage stability.

5.2 Floodable Length. The method used by the Ship Hull Characteristics Program (NAVSEA, 1976) to calculate floodable length by computer is quite different from the method of manual calculation described in Section 4. Instead of using trim lines, SHCP divides the length of the hull into a user-specified number of equal intervals (minimum length LBP/40), and taking each of these points in turn as a center of damaged length, calculates by an iterative process the resulting equilibrium draft and trim, varying the damaged length until the equilibrium waterline just touches the margin line. The damage lengths thus determined represent points on the floodable length curve.

Plotter routines are available which quickly plot curves of floodable length at any number of varied permeabilities.

5.3 Damage Stability. The object of the Damage Stability Sub-program is to calculate righting arm curves for the ship after damage and flooding of various compartments and combinations of compartments. Another sub-program calculates Cross Curves after damage, if desired.

Table 1—SHCP Computer Print Out—One Damage Condition

DAMAGED STATICAL STABILITY CALCULATIONS

CONDITION 2 COMPARTMENTS INCLUDED 40 41 42 43 44 45 46 47 900 901 902 903 904

NET DAMAGED SHIP PROPERTIES

DISP	LCG	POLE HT	HEEL	RA	TCB	VCB	LCB	DRAFT	TRIM
2125.96	−7.404	0.0	0.0	−0.525	−0.525	15.705	−7.720	24.110	−3.863
			2.86	0.676	−0.110	15.716	−7.720	24.106	−3.857
			10.00	3.687	0.951	15.837	−7.717	24.061	−3.793
			15.00	5.845	1.759	16.016	−7.718	24.056	−3.774
			25.00	9.799	3.115	16.506	−7.684	24.193	−3.261
			35.00	13.241	4.178	17.118	−7.532	24.591	−1.431

(a) General. In calculating damage stability using the basic SHCP program, it will be realized that the Ship Data Table and all of the basic hydrostatic data for the hull form have already been entered into the computer and the subroutine simply calls on the computer for information from this data base. The user inputs the offsets of all compartments and/or groups of compartments as well as the appropriate permeabilities, and the assumed pole height. For each damaged condition and each heel angle the program then calculates the volume and LCB of the damaged ship iteratively at varying draft and trim until the net remaining intact portion of the damaged ship has displacement and LCB consistent with the design intact condition.

The machine then calculates the righting arm (RA); the transverse center of buoyancy (TCB), the vertical center of buoyancy (VCB) and the longitudinal center of buoyancy (LCB) for the vessel in each damaged condition at all angles of heel. The user designates the angles (up to 10) so as to ensure that he gets results at points closely spaced above and below changes such as chines, knuckles, and deck edges.

A typical computer printout of the results for one condition is given in Table 1. The numbers shown in the line opposite "Condition" are, of course, keyed to the designations established for the compartment or compartments damaged. Similar data on the intact ship and on individual compartments can also be printed out if desired. Provided input data and instructions are complete and accurate, the results of the computer calculations are more accurate than those of the manual method. This is particularly true for ships of unusual form, or where sinkage, trim and heel involve immersion of decks or where changes in permeability occur. This is because the computer procedure involves the direct integration of all intact and flooded areas and volumes up to the damaged waterline.

(b) Residual $\overline{GM}$, Heel and Freeboard. To determine compliance with survival criteria specifying minimum $\overline{GM}$, maximum heel and minimum freeboard from the data given by the basic SHCP program a manual analysis of the data in the computer printout must be made. This requires:

1. The correction of the righting arm values for the actual $\overline{KG}$ of the ship versus the pole coordinate shown in the printout (see Chapter II).
2. The plotting of the corrected righting arm curve and a determination of the residual heeling angle after damage in the equilibrium condition (i.e. when righting arm = 0).
3. The graphical determination from the righting arm curve of the residual $\overline{GM}$ at the equilibrium angle.
4. The plotting of draft and trim against heeling angles and the determination of final draft and trim at the equilibrium heeling angle.
5. The manual check of the final equilibrium waterline against the lines plan to determine minimum freeboard.

Obviously, the above calculations must be repeated for each damage condition.

Most organizations using SHCP have developed computer subroutines which can perform some or all of the above computations. In addition, subroutines have been developed using iterative techniques to determine the minimum $\overline{KG}$ for each condition required to meet the survival criteria.

Techniques vary among various subroutines. Some, for example, will determine the required $\overline{GM}$ by heeling the vessel one degree past the equilibrium angle and mathematically determining the slope between the righting arms at each angle. With the basic SHCP program, checks of the effects of intermediate flooding must also be done manually. Again, additional subroutines have been developed to perform this computation. While the effects of asymmetric flooding are accounted for in the basic SHCP program, any asymmetrical weight changes must be accounted for manually.

(c) Residual Range and Amplitude of Righting Arm. Where the survival criteria call for minimum range and amplitude of righting arms and minimum values of righting energy, i.e., areas under the righting arm curve, a plot of righting arms from the computer printout provides all the information required to determine compliance or noncompliance with the criterion, except for the location of any point of downflooding, which must be checked manually.

Many subroutines have been developed and are being used to plot the righting arm curve automatically, and to print out the point of downflooding and the amplitude of the maximum righting arm and the righting energy under the curve.

Other subroutines using iterative techniques have been developed to establish minimum $\overline{KG}$'s required to meet the survival criteria at various initial drafts.

Section 6
Definitions for Regulations

6.1 Introduction. The definitions given herein are for reference in connection with the regulations discussed in the following section. They are applicable to vessels complying with the traditional factorial and integer compartment length standards (Section 7) and are in agreement with (but not necessarily worded the same as those of) the U.S. Coast Guard Rules for Subdivision and Damaged Stability, and with the various International Conventions and IMO recommendations. Where differences among definitions applicable to specific classes of vessels occur, the differences are noted.

However, because the new IMO probability-based Equivalent Passenger Vessel Regulations (SOLAS Conference, 1974) contain a number of new and differing definitions, they have been separately set forth in Section 8.

The U.S. regulations for bulk chemical and liquefied gas carriers contain a number of definitions for tanks, spaces, etc., peculiar to those classes of ship that are not included herein. The reader is referred to the Code of Federal Regulations (46 CFR 153 and 154).

Offshore mobile drilling unit require definitions for length, etc., differing widely from standard types of ships and are also not included. The reader is referred to the American Bureau of Shipping *Rules for Building and Classing Offshore Drilling Vessels.*

When the permissive subdivision regulations for tankers were included in the 1966 Load Line Convention, the IMCO delegates adopted the same definition of length that has always been used for load line provisions—essentially length between perpendiculars—rather than length on the subdivision waterline. This made it easier for the regulatory officer preparing a load line certificate, but caused some confusion among naval architects wondering whether to put the end stations for lines drawings and hydrostatic calculations at the ends of the subdivision waterline or at the perpendiculars. However, the procedure of setting the end stations at the waterline is commonly accepted, though some designers still use the length between perpendiculars and account for volumes beyond the perpendiculars by adding additional stations.

6.2 Definitions.

(a) Subdivision Load Line. The subdivision load line is the waterline used in determining the subdivision of the vessel. The deepest subdivision load line is the waterline which corresponds to the greatest draft permitted by the applicable subdivision requirements (including stability considerations). This may or may not be the same as the loadline assigned for freeboard or scantlings.

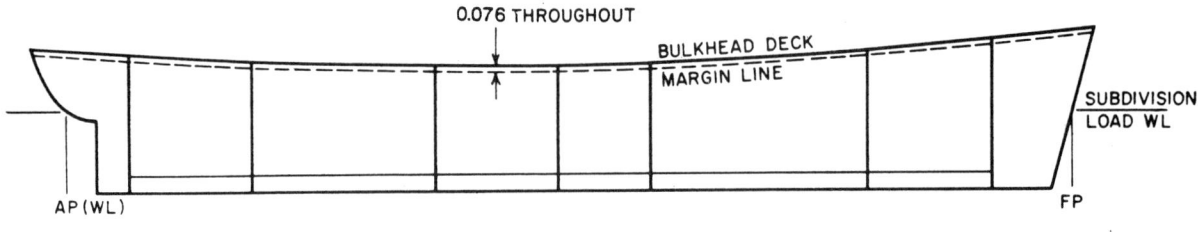

Fig. 34 Margin line with continuous bulkhead deck

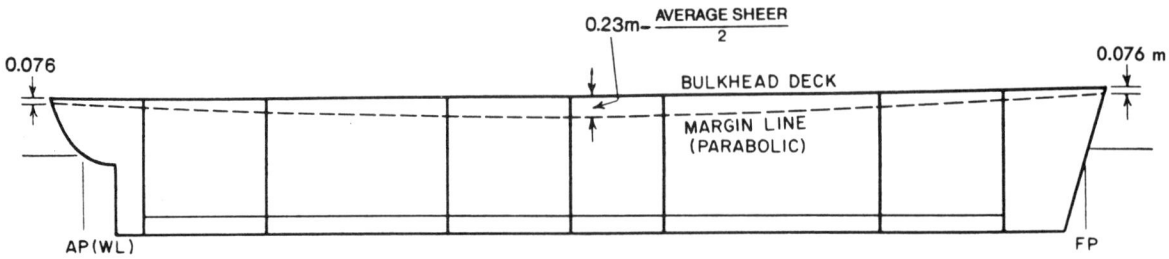

Fig. 35 Margin line with continuous bulkhead deck

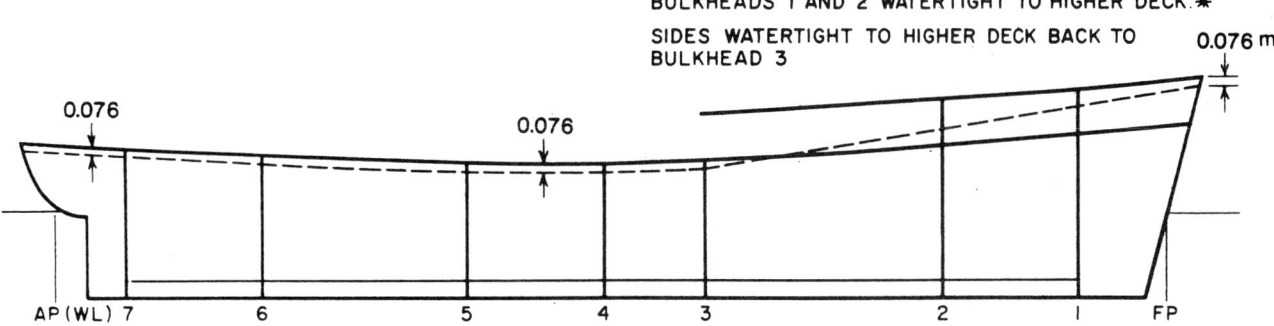

Fig. 36 Margin line where bulkhead deck is stepped

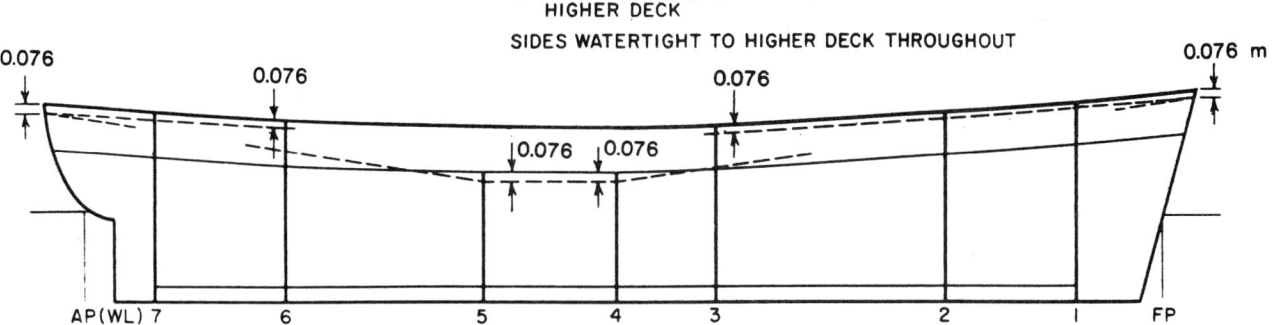

Fig. 37 Case where more than one margin line is required

(b) *Subdivision Length.*

1. *U.S. Passenger Vessels and Oceanographic Vessels*

The subdivision length is the length measured between perpendiculars at the extremities of the subdivision load line.

2. *Tankers, Bulk Chemical and Liquefied Gas Carriers, Large Fishing Vessels and Offshore Supply Vessels*

Length (L) is taken as 96 percent of the total length on a waterline at 85 percent of the least molded depth measured from the top of the keel, or as the length from the fore side of the stem to the axis of the rudder stock on that waterline, if that be greater. In vessels designed with a "rake of keel" (drag), the waterline on which this length is measured is taken as parallel to the designed waterline.

(c) *Breadth of Vessel.* The breadth of the vessel is the extreme molded width at or below the deepest subdivision load line. On wood vessels, breadth is taken to the outside of the planking.

(d) *Bulkhead Deck.* The bulkhead deck is the uppermost deck to which the transverse watertight bulkheads and the shell are carried.

(e) *Margin Line.* The concept of the margin line was adopted in the original 1929 Convention and has been continued for U.S. Passenger Vessels, oceanograhic vessel regulations, vessels under Maritime Administration jurisdiction, and subsequent international conventions on passenger vessels since that date. However, it is not used in the new equivalent regulations for passenger vessels or in the regulations for subdivision of other classes of ships. It is accepted in principle by the U.S. Navy.

The margin line is a line defining the highest permissible location on the side of the vessel of any damaged waterplane in the final condition of sinkage, trim and heel. It is in no case permitted to be less than 7.5 cm (3 in.) below the top of the bulkhead deck at the side. In certain cases the subdivision of one portion of the ship may relate to one margin line while the balance of the ship is calculated to another. The various ways in which margin lines are assigned are illustrated in Figs. 34 through 37.

(f) *Draft.* The draft is the vertical distance from the molded baseline amidships to the waterline in question. In case of floodable length calculations, as required by the traditional factorial standards, this waterline is always the subdivision load line. In the case of all other subdivision standards and all damage stability calculations, the waterline to which the draft is measured is the one at which the vessel is floating in the operating condition.

(g) *Volume.* In all cases volumes are calculated to molded lines. The volumes of flooded spaces are calculated only to the waterline after sinkage.

(h) *Permeability.* The volume permeability of a space is the percentage of the space that can be occupied by water. Surface permeability is the percentage of a waterplane that can be occupied by water.

(i) *Intact Buoyancy.* The term intact buoyancy is used to describe spaces within the limits of damaged compartments that are undamaged and not open to the sea.

(j) *Machinery Space.* The machinery space is taken as extending from the molded baseline to the margin line or deck line and between the extreme main transverse watertight bulkheads bounding the spaces devoted to the main and auxiliary propelling machinery, boilers serving the needs of propulsion and all permanent coal bunkers. In the case of unusual arrangements, special consideration may be necessary.

(k) *Passenger Space.* Passenger spaces are those provided for the accommodation and use of passengers, excluding baggage, stores, provision and mail rooms. For the purpose of permeability calculations, crew spaces and all other spaces which, in the fully-loaded condition, normally contain no substantial quantity of cargo, coal, oil fuel, baggage, stores, provisions, or mail are treated as passenger spaces.

(l) *Floodable Length.* The floodable length at any point in the length of the ship (as previously defined) is the maximum portion of the length, having its center at the point in question, that can be symmetrically flooded at the prescribed permeability, without immersing the margin line.

(m) *Criterion of Service.* The criterion of service is a numeral intended to express the degree to which a vessel is a passenger vessel. In principle, a numeral of 23 corresponds to a vessel engaged primarily in carrying cargo, with accommodations for a small number of passengers, while a numeral of 123 is intended to apply to a vessel engaged solely, or very nearly so, in the carriage of passengers. See Section 7.

(n) *Factor of Subdivision.* The factor of subdivision is a factor prescribed by the applicable regulations and by international convention (Section 7) that depends on ship length and criterion of service. See permissible length.

(o) *Permissible Length.* The permissible length at any point is obtained by multiplying the floodable length at that point by the factor of subdivision.

It should be noted that these paragraphs 12, 13, 14, and 15 contain definitions applicable only to the traditional factorial standards for passenger vessels. See Section 7.

(p) *Use of Positive and Negative Signs.* In addition to their conventional use to indicate addition or subtraction, these signs have the following special usages in this chapter:

	Plus (+)	Minus (−)
Longitudinal levers, trim, etc.	Forward	Aft
Transverse levers, heel, etc.	Starboard	Port
Vertical levers, shifts, etc.	Up or above	Down or below

(q) *Large Fishing Vessel.* A large fishing vessel is defined by IMO as a fish-catching vessel over 100m (300 ft) in length and carrying more than 100 persons on board.

(r) *Administration.* This term is used to indicate the governmental body responsible for administering the particular regulation under discussion.

Section 7
Subdivision and Damage Stability Criteria

7.1 General. All subdivision and damage stability criteria or regulations must provide the answers to the following three key questions:

(a) What is the extent of damage to which the ship is assumed to be subjected?

(b) Where is the damage assumed to be located? Primarily, can it occur between bulkheads, or on one or more specific bulkheads, or on any bulkhead?

(c) What condition is the ship permitted to be in after the assumed damage? What is the permissible sinkage, trim and heel; the residual $\overline{GM}$ and/or $\overline{GZ}$, righting energy and range of stability; and what limits are placed on the location of the waterline in a flooded condition? Or what is the required initial stability?

In addition, all criteria set forth special requirements for such items as assumed permeabilities, minimum bulkhead spacing, local subdivision and equalization after flooding.

Since their inception, criteria for subdivision and damage stability of passenger ships have dealt only with side or end damage. This practice has been followed in the Maritime Administration criteria and in the criteria for large fishing vessels, offshore supply vessels and mobile drilling units. This was undoubtedly due to the belief of the authorities that the requirements for double bottoms in the SOLAS conventions, and their use in general practice, provided adequate bottom protection against flooding. In the case of tankers, bulk chemical carriers and liquefied gas carriers, however, the authorities were concerned about the effect on the environment of escaping cargo and the need for added provisions regarding the extent of bottom damage. There are no requirements in subdivision or damage stability regulations for a ship to comply with any wind heel criterion after damage. However, the U.S. Navy does attempt to allow for the effect of wind on damage stability (section 7.5,(j).

At the time when international standards were being developed, combination cargo and passenger vessels made up the great majority of the world's fleet carrying passengers on international voyages. It was this type of ship that dictated the standards for passenger vessels. Their use has declined steadily and now those remaining are primarily in the pilgrim trade or in out-of-the-way services. The influence of combination cargo and passenger vessels still remains, however, and will continue until they disappear.

After a great deal of preliminary discussion of criteria for subdivision, the 1929 International Conference on Safety of Life at Sea accepted the concept that there is a continuously increasing safety with decreasing bulkhead spacing. Hence, instead of prescribing one, two and three-compartment subdivision, with the bulkhead spacing as long as possible within each grade of subdivision, the Convention gave a permissible length that is obtained by multiplying the floodable length by a factor less than (or, as a limit, equal to) unity, called the factor of subdivision. By this factorial system a factor of subdivision of 1.0 corresponds to a one-compartment standard (i.e., any one compartment can be flooded without the ship sinking—under assumed conditions), 0.5 to a two-compartment standard, etc. An intermediate factor such as 0.75, for example, would correspond to a one-compartment standard under rule assumptions regarding drafts before damage, permeabilities, etc., but it was assumed that under certain more favorable conditions the ship might survive with two compartments flooded (i.e., side damage at a bulkhead.) Hence, the doubtful presumption that a ship with factor of subdivision of 0.75 is safer than one with a factor of 1.0.

As noted in Section 1, an entirely different probabilistic approach to subdivision was introduced at the 1974 SOLAS Conference, and the resulting Alternate Equivalent Passenger Regulations were included in the 1974 Convention. See Section 8. The basic philosophy now is that the true index of safety is the probability of survival after damage occurring anywhere along the length of the ship, between or on a bulkhead.

Under the factorial standards to be summarized in this section, it will be found that a vessel around 180m (600 ft) in length and carrying only six hundred passengers and no cargo is not required to meet a two-compartment standard. Passenger vessels on international voyages in the past twenty years have been constructed either as cruise ships, or as combination automobile and passenger ferries. For both types of ship there is no economic need for large cargo holds. The spaces below the freeboard deck can be subdivided with ample bulkheads to permit them to meet a two-compartment standard without penalizing them unduly. A vessel carrying passengers below the freeboard deck may require a few more bulkheads, a few more sliding watertight doors and a few more stairways, but the costs of these are insignificant in relation to the total cost and relative safety of the ships. Automobile and passenger ferries may easily meet a two-compartment standard with car decks extending below the freeboard deck. See Robertson, et al (1974).

The naval architect should feel a great deal more comfortable if one's passenger ship can sustain damage anywhere in the vessel's length, i.e. meets a two-compartment standard, without the need to worry about the probability of damage occurring at a bulkhead. Such a standard has been easily met on United States inland and coastwise vessels over 100m in length, and it is desirable for it to be met on similar ocean going passenger vessels wherever possible.

On ships travelling on international routes, coming under the jurisdiction of the SOLAS Conventions, any ship carrying more than 12 passengers is classed as a passenger vessel. Ships operating on voyages between any two American ports, which are covered under the U.S. Coast Guard regulations, are classed as passenger vessels if they carry more than 16 persons. Small U.S. passenger vessels, again operating between U.S. ports and which come under SubChapter "T" Regulations, are classed as passenger vessels if they carry more than 6 passengers.

7.2 Summary of U.S. and International Standards. Standards for subdivision and damage stability have been established by International Conventions, by recommendations of IMO, by national regulations and by classification society rules. These standards can be subdivided into two types. The first type uses the traditional *factorial* system. The second type uses the *integer compartmentation* system. In the latter, the standards require the ship to survive damage to 1, 2 or more compartments, in parts or in all of the ship.

In the following summary, under the several standards, the types of ships to which the standards are applicable are listed giving the U.S. (Coast Guard) regulations (if any), the International Convention or IMO recommendation (if any), and the classification society rules (if any), applicable to each type.

(a) Traditional factorial system standards are applicable to all passenger vessels on international voyages, and U.S. passenger vessels over 150 gross tons on domestic ocean and coastwise voyages.

International Standards:	SOLAS Conventions, 1929, 1948, 1960, 1974
U.S. Regulations:	U.S. Code of Federal Regulations: 46 CFR 73.10, 74.10-15

(b) Alternate equivalent passenger vessel standards are also applicable to all passenger vessels on international voyages, and U.S. passenger vessels over 150 gross tons on domestic ocean and coastwise voyages. Note that these are alternate standards which are recommended by IMO and accepted by U.S. and other governments as equivalent to the traditional factorial system standards. See Section 8.

(c) Integer Compartmentation Standards are applicable to various types of ships in accordance with Table 2.

It should be understood that, where U.S. regulations are listed, these are the law of the land and must be complied with on U.S. vessels of the type described. Foreign countries, however, will permit such vessels in their waters only when they comply with the requirements of International Conventions which have been ratified and are in force. All U.S. regulations are equal to, or exceed the requirements of ratified International Conventions. Where IMO resolutions are listed as the International Standards, these are recommendations only and are not legally binding on vessels in international trade. The only case where an International Convention is listed and the United States has no regulations is the Convention on fishing vessels, and this convention has not, at this date, been ratified by a sufficient number of countries to bring it into force.

IMO is working continually toward the goal of uniformity in international standards and consequently various "harmonization proposals" have been made from time to time. Some of these are noted in Section 7.3.

Detailed requirements for the different classes of ships listed in Table 2 are presented in the following sections under these headings:

- 7.3 Extent of damage
- 7.4 Damage location
- 7.5 Damage survival
- 7.6 Special requirements

The American Bureau of Shipping special requirements for mobile offshore drilling vessels, whether they are self-elevating, column stabilizing, or surface drilling units, are too complex to detail here. The Bureau also has a computer program for analyzing the damage stability of the various types of drilling units.

Table 2—Integer Compartment Standards Applicable to Different Types of Ships

	International Standards	U.S. Regulations
U.S. passenger vessels, in other than ocean and coastwise service	None	46CFR 73.15 & 74.10-15
U.S. passenger vessels under 150 GT, on a domestic ocean or coastwise voyage	None	46CFR 73.15 & 74.10-15
Small U.S. passenger vessels (under 100 GT) not on an international voyage	None	46CFR 178
Oceanographic ships	None	46CFR 191
Dry cargo ships (including dry bulk carriers, roll-on-roll-off ships)	None	USCG* U.S. Vessels under MARAD jurisdiction: Des.Letter No. 3A
Tankers	MARPOL'73	46CFR 42 33CFR 157
Bulk chemical carriers	IMO Res. No. 212	46CFR 153
Liquefied gas carriers	IMO Res. No. 328	46CFR 154
Offshore supply vessels	IMO Res. No. A469 (XII)	None
Large fishing vessels	Int. Conf. on Fishing Ves. '77	None
Mobile offshore drilling vessels	IMO Res. No. A414 (XI)	ABS Rules for Bldg. & Classing Mob.Drilling Ves. 46CFR 107-109
Naval vessels	None	Naval Sea Sys. Com., Des. Data Sheet, DDS 079-1

* No regulations, except that U.S. cargo vessels complying with special requirements for hatch covers and other areas through which water could enter the hull, *may* obtain a reduction in freeboard by complying with certain subdivision requirements. See 46CFR 42. These special requirements are readily and therefore commonly adopted in dry bulk carriers.

7.3 Extent of Damage. *(a) General.* While not specifically spelled out in all regulations, it is generally required and commonly accepted that if a lesser extent of damage than those specified in the regulations result in a more severe condition regarding heel or loss of metacentric height, then such lesser extent should be assumed. One example of the need for such a requirement is a vessel with high side tanks with inboard longitudinal bulkheads spaced less than $B/5$ from the shell. Flooding such tanks alone when empty could cause a greater heel than that caused by flooding both the side tank and inboard compartment.

In all cases, the transverse extent of damage is measured from the waterline at the point of minimum beam in way of damage.

In the early passenger vessel regulations, the longitudinal extent of damage was expressed as a fixed amount 3.05m (10 ft), plus a percentage (3.0) of the length. Later regulations adopted a fixed percentage of a power of the length, i.e.: $1/3 L^{2/3}$ or $0.495 L^{2/3}$ ft. or a fixed length, whichever is the least. A comparison of the values obtained by each of these formulas is given in Table 3, for a representative list of ship lengths.

The U.S. regulations on extent of damage for tankers stem from the 1973 Pollution Convention. Bulk chemical ships and liquefied gas carriers stem from IMO Codes. The difference in extent of damage between the two regulations is minor. IMO is now in the process of "harmonizing" these recommendations and it is expected that the U.S. will follow its lead. Notes have been appended to the stated requirements for these vessels in Section 7.3, showing the status of the IMO proposals at the time of writing. Again, the reader is cautioned to check the latest U.S. regulations on any new design.

(b) U.S. Passenger Vessel Regulations. The required assumed extent of side damage in United States passenger vessels on an international voyage, or over 150 gross tons in ocean or coastwise service (assumed to be rectangular both in plan and elevation) is given by Table 4.

The transverse extent of damage (penetration) is measured inboard of the vessel's side and at right angles to the centerline at the level of the deepest subdivision load line. In the United States rules, for vessels on inland waters and for ferry vessels, where the maximum molded beam at the deck and at the load waterline differ appreciably, the transverse extent of damage throughout the ship is taken as the mean between the inboard penetration of the deck, using the maximum beam at the deck, and the inboard penetration at the deepest subdivision load line, using the maximum beam at the load line.

For passenger vessels under 100 gross tons, the Coast Guard has special requirements dependent on length and number of passengers. The reader is referred to the detail regulations in 46 CFR 178 for these requirements.

(c) Oceanographic Vessels. It is assumed that the longitudinal extent of damage does not exceed 3.05m

Table 3—Longitudinal Extent of Damage

Length of Ship		$.03L + 3.05$m		$1/3 L^{2/3}$ (L in m)	
m	ft	m	ft	m	ft
30.49	100	3.96	13.0	3.26	10.7
60.98	200	4.88	16.0	5.15	16.9
91.46	300	5.79	19.0	6.79	22.3
121.95	400	6.71	22.0	8.20	26.0
152.44	500	7.62	25.0	9.51	31.2
182.93	600	8.54	28.0	10.73	35.2
213.41	700	9.45	31.0	11.92	39.1
243.90	800	10.37	34.0	12.99	42.6

Table 4—Assumed Extent of Damage

Vessel Category	Longitudinal Extent	Transverse Extent	Vertical Extent
All vessels	3.05m (10 ft) + $0.03L$ or 10.7m (35 ft) whichever is less—no main bulkhead involved	$B/5$	From baseline upward without limit
Vessels without factor of subdivision, where a two-compartment standard is required	3.05m (10 ft) + $0.03L$ or 10.7m (35 ft) whichever is less, involving one main bulkhead	$B/5$	From baseline upward without limit
Vessels with a factor of subdivision of 0.50 or less	3.05m (10 ft) + $0.03L$) or 10.7m (35 ft) whichever is less, involving one main bulkhead	$B/5$	From baseline upward without limit
Vessels with a factor of subdivision of 0.50 or less	6.1m (20 ft) + $0.04L$ involving not more than one main bulkhead[a]	$B/5$	Top of double bottom to margin line[b]
Vessels with a factor of subdivision of 0.33 or less	6.1m (20 ft) + $0.04L$ but in any case long enough to involve two main bulkheads[a]	$B/5$	From baseline upward without limit

[a] 3.05m (10 ft) plus $0.03L$ in the 1974 International Convention.
[b] From the baseline in the 1974 International Convention.

(10 ft) plus $0.03L$. Transverse or vertical extent of damage is not defined.

(d) *Cargo Ships.* The Maritime Administration assumes, on all major vessels built under government subsidy or mortgage insurance programs, that damage occurs on the side and does not extend to main transverse bulkheads unless they are spaced closer than the required length of damage which is $0.495L^{2/3}$ or 14.5m (47.6 ft), whichever is less. Where U.S. Coast Guard regulations for specific ship types impose more severe requirements, these supersede the Maritime Administration standard. The assumed transverse extent of damage is $B/5$ and the vertical extent is unlimited above and below the baseline.

(e) *Tankers (U.S. Regulations).* For side damage, the extent of damage assumed is:

• Longitudinal extent	$1/3L^{2/3}$ or 14.5 meters (47.5 ft) whichever is less
• Transverse extent (inboard from the vessel's side at the level corresponding to the assigned summer freeboard)	$B/5$ or 11.5 meters (37.7 ft) whichever is less
• Vertical extent	From the baseline upward without limit

For bottom damage, the extent of damage assumed is:

Damage	From $0.3L$ from the forward perpendicular of the ship	Any other part of the ship
• Longitudinal extent	$L/10$ (See Note 1)	$L/10$ or 5 meters, whichever is less. (See Note 2)
• Transverse extent	$B/6$ or 10 meters whichever is less, but not less than 5 meters	5 meters (See Note 3)
• Vertical extent from the baseline	$B/15$ or 6 meters, whichever is less	$B/15$ or 6 meters, whichever is less

Note 1: The IMO harmonization proposal would change this to $1/3L^{2/3}$ or 14.5 meters, whichever is less.
Note 2: The IMO harmonization proposal would change this to $1/3L^{2/3}$ or 5 meters, whichever is less.
Note 3: The IMO harmonization proposal would change this to $B/6$ or 5 meters, whichever is less

(f) *Bulk Chemical Carriers (U.S. Regulations).* It is assumed that damage can occur from either collision or grounding damage, and the damage must consist of the most disabling penetration up to and including penetrations having the following dimensions:

(1) *Collision penetration:*

• Longitudinal extent	$(1/3)L^{2/3}$ or 14.5m (approx. $0.495L^{2/3}$ or 47.6 ft), whichever is less.
• Transverse extent: (inboard from the ship's side at right angles to the centerline at the level of the summer load line assigned)	$B/5$ or 11.5m (approx. 37.7 ft), whichever is less
• Vertical extent	From the base line upwards without limits.

(2) *Grounding penetration:*

Damage	At forward end, but excluding damage aft of point $0.3L$ aft of FPP	At any other longitudinal position
• Longitudinal See Note 1	$L/10$	$L/10$ or 5m (approx. 16.4 ft), whichever is less. See Note 1.
• Transverse See Note 2.	$B/6$ or 10m (approx. 32.8 ft), whichever is less.	5m (approx. 16.4 ft)
• Vertical extent from the base line.	$B/15$ or 6m (approx. 19.7 ft), whichever is less.	$B/15$ or 6m (approx. 19.7 ft), whichever is less

Note 1: The IMO harmonization proposal would change the longitudinal extent of damage in locations forward of the $0.03L$ to $1/3L^{2/3}$ or 14.5m whichever is less. In other longitudinal locations it is $1/3L^{2/3}$ or 5 meters, whichever is less.
Note 2: IMO harmonization proposal would change the transverse extent in "other longitudinal positions" to $\dfrac{B}{6}$ or 5m, whichever is less.

If the damage assumption excludes a transverse bulkhead bounding a machinery space, the machinery space must be assumed to be flooded as a case separate from the damage assumption.

(g) *Liquefied Gas Carriers (U.S. Regulations).*

(1) For side damage, the extent of damage assumed is:

• Longitudinal extent	$1/3L^{2/3}$ or 14.5m, whichever is less.
• Transverse extent (inboard from the ship's side at right angles to the centerline of the level of the summer load line)	$B/5$ or 11.5m, whichever is less
• Vertical extent	From the baseline upward without limit.

(2) Bottom damage:

Damage	At forward end, but not including damage aft of point $0.3L$ aft of FPP	At any other longitudinal location
• Longitudinal extent	$1/3L^{2/3}$ or 14.5m, whichever is less	$L/10$ or 5m whichever is less (See Note 1)
• Transverse	$B/6$ or 10m, whichever is less	$B/6$ or 5m, whichever is less
• Vertical extent from the molded line of the shell at the centerline	$B/16$ or 2m, whichever is less	$B/16$ or 2m, whichever is less

Note 1: the IMO harmonization proposal would change this to $1/3L^{2/3}$ or 5m, whichever is less.

(h) *Offshore Supply Vessels (IMO Recommendations)*

The assumed extent of damage is as follows:

• Longitudinal extent	No provision
• Transverse extent	760mm (30 in.)
• Vertical extent	Full depth of ship

(i) Large Fishing Vessels (IMO Recommendations).
The assumed extent of damage is as follows:

- Longitudinal extent $1/3 L^{2/3}$ m ($0.495 L^{2/3}$ in ft)
- Transverse extent (inboard $B/5$
  from the side at right angles
  to centerline at level of deepest
  operating WL)
- Vertical extent   From the baseline upward without limit

(j) U.S. Naval Vessels.
The extent of damage assumed on U.S. naval vessels depends on the ship type and the ship size. On large combatant vessels with side protection systems such as aircraft carriers, the extent of damage is classified and is based on test data, war damage reports and design experience.

New designs without side protection systems under 91.5m (300 ft) in length are not required to meet a specified longitudinal extent of damage. New designs for vessels over 91.5m (300 ft) are required to withstand flooding from damage equal to 15 percent of the vessel's length if it is a combatant type or personnel carrier. All other types of vessels this size must withstand flooding from damage equal to 12.5 percent of the vessel's length.

The transverse extent of damage for all ships without side protective systems is one-half the beam, but not including a centerline bulkhead. The vertical extent of damage is from the keel upward, except that if *not* flooding the inner bottom results in a worse condition, this is assumed.

Special criteria are assigned to merchant vessels converted to naval auxiliaries, depending on the ship type. If primarily designed for the carriage of cargo, they must meet a two-compartment standard. Other vessels must withstand flooding from an opening in the shell of 12.5 percent of the ship's length.

7.4 Location of Damage.

(a) General. In all criteria the damage is assumed to take place anywhere within the ship's length. However, where all or portions of the ship are only required to meet a one-compartment standard of subdivision, the damage is assumed to take place between watertight transverse bulkheads that are a distance apart equal to or greater than the assumed longitudinal extent of damage. Where a two-compartment standard of subdivision is required by the criterion the damage is assumed to be located anywhere throughout the ship's length, or throughout the specified portion of the ship, including damage at any one of the main transverse watertight bulkheads within these areas. Where three-compartment damage is required, the damaged is presumed to be located so as to include two adjacent main watertight bulkheads.

(b) U.S. Passenger Vessel Regulations (Coast Guard). If the factor of subdivision is above the value of 0.5, the vessel must meet a one-compartment standard and the location of damage is therefore anywhere in the vessel's length, but not including a bulkhead. If the ship must meet a two, three or four-compartment standard, again the damage is assumed to occur anywhere in the ship's length, but to include damage to any one bulkhead or to any two or three adjacent bulkheads, respectively.

For U.S. Passenger Vessels in Service Other than Ocean or Coastwise, or Under 150 Gross Tons in Ocean or Coastwise Service, but not on an International Voyage, the factorial system of subdivision is not used; see 46 CFR 73.15. Instead, for passenger vessels other than automobile ferries, the requirements call for compliance with either a one-compartment or a two-compartment standard of flooding, including stability, depending upon the number of passengers. All passenger vessels must meet a one-compartment standard throughout their length. Vessels carrying more than 400 passengers must not submerge the margin line with the forepeak and one adjacent compartment flooded; vessels carrying more than 600 passengers are required to meet a two-compartment standard forward within at least a full 40 percent of their length; vessels carrying more than 800 passengers forward within at least a full 60 percent of their length; and vessels carrying more than 1000 passengers are required to meet a two-compartment standard throughout their length.

All ferry vessels under 46m (150 ft) in length must meet a one-compartment standard; all vessels over 46m (150 ft) in length must meet a one-compartment standard and additionally be able to withstand flooding of the peak compartment and one adjacent compartment; and all ferry vessels over 61m (200 ft) in length must meet a two-compartment standard.

(c) Oceanographic Ships and Dry Cargo Ships. Damage can occur anywhere in the vessel's length, but is assumed to occur between main transverse bulkheads.

(d) Tankers. The location of damage in tankers over 225 meters (782 ft) in length is anywhere in the vessel's length. Thus, the vessel has to meet at least a two-compartment subdivision standard.

For tankers between 150 meters (492 ft), but not exceeding 225 meters (738 ft) in length, the damage location is anywhere in the vessel's length outside of the machinery space. It is assumed not to occur on either the forward or after machinery space bulkhead. Therefore, the machinery space compartment has to meet only a one-compartment standard of subdivision, but all of the other spaces have to meet a two-compartment standard. For tankers not exceeding 150 meters (492 ft) in length, damage is assumed to occur anywhere in the ship's length between adjacent transverse bulkheads with the exception of the machinery space. Thus the vessel has to meet the one-compartment standard in compartments outside of the machinery space, but does not have to survive damage occurring between the machinery space bulkheads.

Since the Maritime Administration requires a one-compartment standard on all vessels under its jurisdiction, its requirements exceed the U.S. Coast Guard on tankers under 150 meters (492 ft) in length in that the machinery space must be capable of being flooded.

(e) *Bulk Chemical Carriers and Liquefied Natural Gas Carriers.* Vessels in these categories are broken down into sub-classes, based on the severity of the hazard to the crew and to the environment if a compartment is flooded. The damage assumptions for each class vary widely from minimum location between transverse bulkheads to locations at single or multiple combinations of bulkheads. The reader is advised to check the detailed regulations applying to the particular class of ship for the exact requirements for that particular class.

(f) *Large Fishing Vessels and Offshore Supply Vessels.* Damage is assumed to occur anywhere in the vessel's length, but only between transverse bulkheads.

(g) *Naval Vessels.* On vessels with side protective systems the damage is assumed to occur at any place in the ship's length. Other vessels under 30.5m (100 ft) in length must withstand damage anywhere between watertight bulkheads or meet a one-compartment standard. Other vessels between 30.5m (100 ft) and 91.5m (300 ft) in length must withstand damage anywhere in the ship's hull, including one watertight bulkhead or meet a two-compartment standard.

Vessels over 91.5m (300 ft) must withstand damage of the extent given in 7.3 (10), anywhere in the vessel's length.

7.5 Damage Survival.

(a) *Operating Drafts.* The calculations for floodable length under the traditional passenger vessel regulations are performed at the subdivision draft only. This is logical because the effect of sinkage and trim, and the probability of submerging the margin line due to sinkage and trim, is much greater in the fully loaded condition. In calculating damaged stability, however, the benefits of increased freeboard at light draft can be offset by changes in $\overline{KG}$ in the various operating conditions. Therefore, to insure that the worst operating condition is covered, all other criteria call for the vessel to meet the damage survival requirement at the full range of operating drafts and trims. The detailed regulations for each case should be checked for the conditions specified.

(b) *U.S. Passenger Ship Regulations.* After damage, the vessel must meet the following survival conditions:

1. *Margin Line.* In the final flooded condition, the margin line must not be submerged at any place in the ship's length.

2. *Metacentric height.* In the final flooded condition there should be a positive residual metacentric height of at least 0.05 m (2 in.).

3. *Heel must not exceed specified limits.* The present regulations concerning limits of permissible heel deliberately incorporate some flexibility. Consideration of the factors affecting the relationship of heel to risk provides some guidance as to their application and explains why they leave, in some cases, latitude for administrative interpretation. The considerations involved in dealing with heel may be grouped as follows:

• The effect of heel on the safe movement and control of persons on board the vessel, on the risk of shifting of weights, and on the ability to launch lifeboats;
• The risk relative to flooding through side or deck openings;
• The relationship of heel to the range of stability and the angle and value of the maximum righting arm;
• The degree of accuracy with which the heel can be estimated.

It can be seen that consideration of heel is a complex matter, involving at this stage of knowledge, some exercise of judgment within the limits prescribed by the Rules, which are as follows:

For unsymmetrical flooding with assumed side damage not more than 3m (10 ft) plus $0.03L$, the remaining heel due to unsymmetrical moment, after equalization, shall not exceed 7 deg. However, where equalization is fully automatic and by open cross connections of large area, or where no equalization is involved, and in any case, the range of stability in the damaged condition is considered adequate, a greater heel up to but not in excess of 15 deg may be allowed. (The first part of this regulation refers basically to the case where a part of the unsymmetrical flooding is equalized through manually operated cross connections. In such case, the risk of delay or improper operation of these connections is the basis for the 7-deg limit. The relaxation permitted by the second part of the regulation is on the basis that there is no risk of malfunction of cross-connections and is on the further basis that damaged-condition range of stability and maximum righting arm are sufficient for the greater heel.)

Where the assumed side damage is more than 3m (10 ft) plus $0.03L$, the final heel due to unsymmetrical moment, after equalization, may be 15 deg, unless an insufficiency of righting arms would be cause for limiting the heel to a lesser value.

4. *Residual Righting Arm.* While a range of residual righting arm is not specified in these regulations, the regulations state that the range should be examined and the cognizant administration satisfied that they are adequate. This was traditionally done by approximate methods. At the present time an accurately calculated residual righting arm is generally required.

(c) *Oceanographic Ships.* Damage to any one compartment must not submerge the margin line.

(d) *Cargo Ships.* For ships under its jurisdiction the Maritime Administration requires, after damage, the following survival conditions:

1. The equilibrium heel angle θ_1 must be less than 15 deg.

2. Downflooding points must not be submerged at θ_1, unless fitted with watertight closing appliances, which then must remain shut at sea and be so logged.

3. The margin line must not be submerged at θ_1 unless it can be clearly shown that downflooding will not occur.

4. There must be a range of positive stability of at least 20 deg beyond the equilibrium heel angle and no downflooding openings may be within this 20 deg range unless they are fitted with watertight closing appliances.

5. The maximum residual righting arm, within the 20 deg range, must be last least 0.1m (4 in.).

6. For cases of symmetrical damage, the vessel must have 0.05 meters (2 in.) of positive $\overline{GM}$ in the upright condition after damage.

(e) *Tankers (MARPOL, 1973).* Oil tankers shall be regarded as complying with the criteria if the following survival requirements are met:

1. *The final waterline*, taking into account sinkage, heel and trim, shall be below the lower edge of any opening through which progressive flooding may take place. Such openings shall include air pipes and those which are closed by means of weathertight doors or hatch covers and may exclude those openings closed by means of watertight manhole covers and flush scuttles, small watertight cargo tank hatch covers which maintain the high integrity of the deck, remotely operated watertight sliding doors, and side scuttles of the non-opening type.

2. *The angle of heel* due to unsymmetrical flooding shall not exceed 25 deg in the final stage of flooding, provided that this angle may be increased up to 30 deg if no deck edge immersion occurs.

3. *The righting lever curve* for acceptable stability in the final stage of flooding must have a range of at least 20 deg beyond the position of equilibrium in association with a maximum residual righting lever of at least 0.1 m (4 in.). For the calculations required in this section, weathertight openings or openings fitted with automatic closures (e.g. a pressure, vacuum relief valve or a vent fitted with a ball-check valve), need not be considered as points of downflooding within the range of residual stability, but other openings must be included in the calculations.

(f) *Bulk Chemical Carriers.* A bulk chemical carrier is presumed to survive if it meets the following conditions:

1. *Heel Angle.* Except as indicated below, in the final condition of flooding the angle of heel must not exceed 15 deg (17 deg if no part of the freeboard deck is immersed).

The cognizant Administration should consider on a case-by-case basis vessels 150 m or less in length having heel angles greater than 17 deg, but less than 25 deg.

2. *Final Waterline.* The final waterline, taking into account sinkage, heel, and trim, must be below the lower edge of openings such as air pipes and openings closed by weathertight doors or hatch covers. The following types of openings may be submerged when the tankship is at the final waterline:

- Openings covered by watertight manhole covers or watertight flush scuttles.
- Small watertight cargo tank hatch covers.
- Remotely operated watertight sliding doors.
- Side scuttles of the non-opening type.

3. *Range of Stability.* Through an angle of 20 deg beyond its position of equilibrium after flooding, a tankship must meet the following conditions:

- The righting lever curve must be positive.
- The maximum of the righting lever curve must be at least 10 cm (approx. 4 in.).
- Each submerged opening must be weathertight.

4. *Metacentric Height.* After flooding, the tankship's metacentric height must be at least 5cm (approx. 2 in.) when the ship is in the upright position.

(g) *Liquefied Natural Gas Carriers.* A vessel is presumed to survive assumed damage if it meets the following conditions in the final stage of flooding:

1. *Heel Angle.* The maximum angle of heel must not exceed 30 deg.

2. *Final Waterline.* The waterline, taking into account sinkage, heel and trim, must be below the lower edge of openings such as air pipes and openings closed by weathertight doors or hatch covers, except openings closed by means of watertight manhole covers and watertight flush scuttles, small watertight cargo tank hatch covers that maintain the high integrity of the deck, remotely operated watertight sliding doors, and side scuttles of the non-opening type.

3. *Range of Stability.*

- The righting lever curve must be positive and have a minimum range of 20 deg beyond the angle of equilibrium.
- The maximum righting lever within the above range must be at least 100 mm (4 in.).
- Each opening within the 20 deg range beyond the angle of equilibrium must be at least weathertight.

4. *Metacentric Height.* After flooding the vessel's metacentric height must be at least 50 mm (2 in.) when the vessel is in the upright position.

(h) *Offshore Supply Vessels* must meet the following survival requirements:

1. *The final waterline*, taking into account sinkage, heel and trim, should be below the lower edge of any opening through which progressive flooding may take place. Such openings include air pipes and those which are capable of being closed by means of weathertight doors or hatch covers and may exclude those openings closed by means of watertight manhole covers and flush scuttles, small watertight cargo tank hatch covers which maintain the high integrity of the deck, remotely operated watertight sliding doors, and side scuttles of the non-opening type.

2. *The angle of heel* due to unsymmetrical flooding

should not exceed 15 deg in the final stage of flooding. This angle may be increased up to 17 deg if no deck immersion occurs.

3. *The stability* in the final stage of flooding should be investigated and may be regarded as sufficient if the righting lever curve has at least a range of 20 deg beyond the position of equilibrium in association with a maximum residual righting lever of at least 0.1 m (4 in.) within this range. Unprotected openings should not become immersed at an angle of heel within the prescribed minimum range of residual stability unless the space in question has been included as a floodable space in calculations for damage stability. Within this range, immersion of all the openings listed in (a) need not be considered as downflooding points providing air pipes are fitted with ball-check valves.

(*i*) *Large Fishing Vessels.* The vessel is considered to survive the conditions of damage provided the vessel remains afloat in a condition of stable equilibrium and satisfies the following stability criteria:

1. *The stability* in the final condition of flooding may be regarded as sufficient if the righting lever curve has a minimum range of 20 deg beyond the position of equilibrium in association with a residual righting lever of at least 0.1 m (4 in.). The area under the righting lever curve within this range should be not less than 0.0175 meter-radians (3.25 deg-ft). Consideration should be given to the potential hazard presented by protected or unprotected openings which may become temporarily immersed within the range of residual stability. The unflooded volume of the poop superstructure around the machinery space casing, provided the machinery casing is watertight at this level, may be taken into consideration in which the case the damage waterline should not be above the after end of the top of the poop superstructure deck at the centerline.

2. *The angle of heel* in the final condition of flooding should not exceed 20 deg.

3. *The initial metacentric height* of the damaged vessel in the final condition of flooding for the upright position should be positive and not less than 5 cm (2 in.).

(*j*) *U.S. Naval Vessels.* On vessels with side protective systems the emphasis of damage survival is to maintain the vessel after damage at a static heel (at GZ = 0) not to exceed 15 deg, which is the limiting angle at which all machinery and equipment is designed to operate, with a 20 deg list the maximum, since it is assumed that a vessel with such a list can be safely towed back to port. Naval design and operational procedures also require that arrangements be provided for rapidly correcting list to less than 5° by counterflooding from the sea, assuming that pumping equipment is operational.

On all vessels the degree of list or trim after damage must be such that the margin line (7.5 cm, or 3-in. below the bulkhead deck) is not submerged. This, as in all criteria, assumes a level, calm sea. The Navy, in addition, has certain minimum criteria for survival under specified wind and wave forces, which may involve submerging the margin line.

The stability after damage is considered adequate if the areas A_1 and A_2 on Fig. 38 have the relationship $A_1/A_2 \geq 1.4$. The point ϕ is the angle of down flooding or 45 deg whichever is less. Point C is the initial static angle of heel after damage.

The angle ϕ_r is the expected roll angle due to wind and waves. The righting arm curve is reduced by an amount equal to 0.05 cosine ϕ to account for unknown unsymmetrical flooding or transverse shift of loose material. Values of the constant for calculating the wind heel curve and the amplitude of the rolling angle ϕ_r are given in DDS-079-1. (See Chapter II).

(*k*) *Residual $\overline{GM}$ and Residual Righting Arms.* It will be noted that the passenger vessel criteria, including those in the alternate equivalent regulations, call for a minimum residual $\overline{GM}$ for survival after damage. All other international criteria—such as the International Regulations recently adopted for tankers, bulk chemical carriers, liquefied natural gas car-

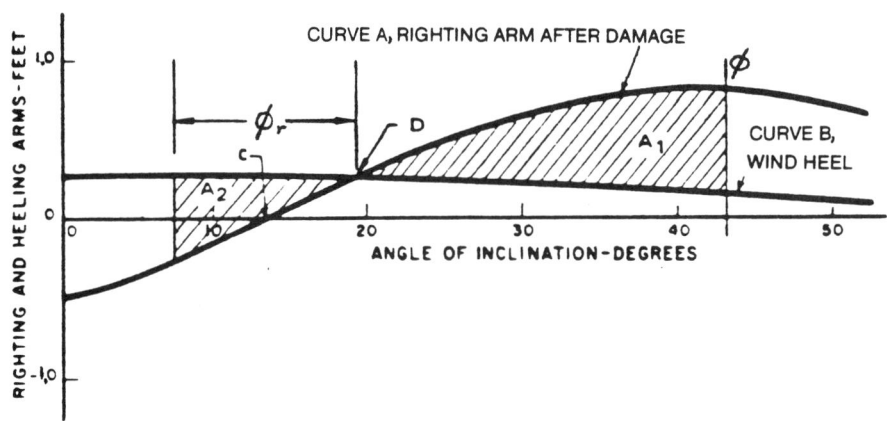

Fig. 38 U.S. Navy standard for stability after damage

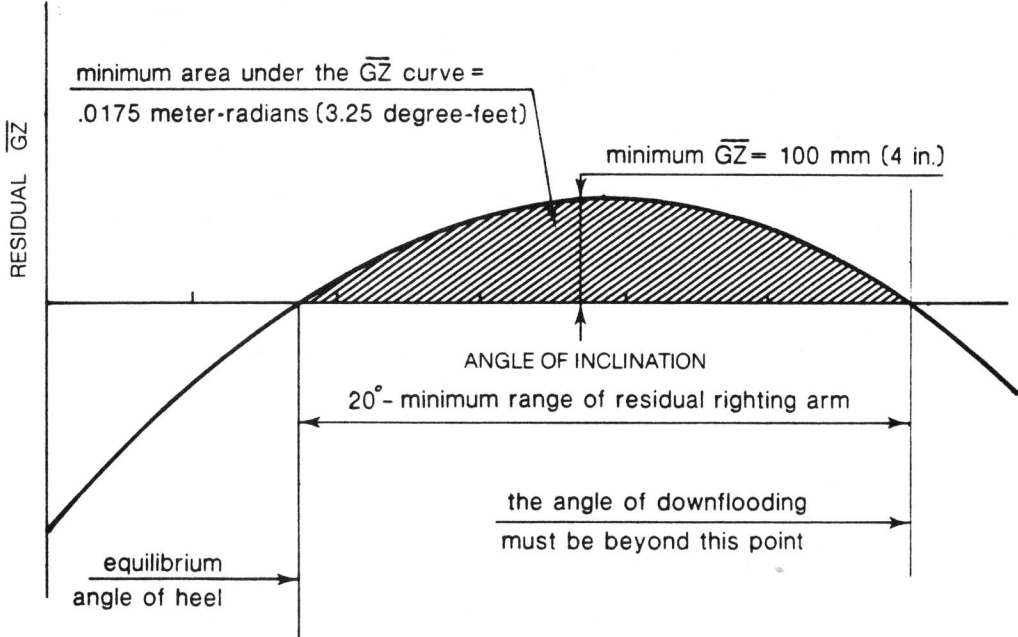

Fig. 39 Typical standards for required residual righting arms

riers, offshore supply vessels and large fishing vessels—specify minimum amplitude and range of righting arm for survival. The U.S. Coast Guard regulations additionally include minimum $\overline{GM}$ requirements for chemical and liquefied gas carriers.

Typical standards for the residual righting arm curve are shown on Fig. 39. Naval architects have long considered righting arm amplitude and range to be the best indices of stability. Nevertheless, for purposes of simplicity, the requirements for passenger ships laid down in the Regulations of the 1960 and 1974 International Conventions for Safety of Life at Sea, only called for minimum residual $\overline{GM}$'s and freeboards (both very small).

7.6 Special Requirements.

(a) Passenger Vessels—Traditional Factorial Standards. As previously stated, the factorial system of subdivision was intended to take into account the relative importance of the cargo and passenger carrying functions of a vessel. The intent was that vessels having a large volume of space below the margin line allocated to passengers and a greater total number of passengers on board, would require a greater degree of safety than a similar vessel with the entire volume of space under the margin line devoted to cargo and with fewer total passengers on board. They therefore devised two formulas to take these variations into account. The first is the *Criterion of Service,* and the second the *Factor of Subdivision.*

The formula for Criterion of Service uses length, number of passengers, total volume of the ship below the margin line, volume of the machinery space, and volume of the accommodation spaces below the margin line and combines them so that the lower the value of the criterion of service, the farther apart the watertight bulkheads may be spaced.

The formula for Factor of Subdivision uses factors of length and the Criterion of Service, and the result is a percentage running from 30 to 100 percent (0.3 to 1.0). This Factor of Subdivision establishes the permissible length between watertight bulkheads. A factor of 1.0 means the bulkheads may be spaced apart a distance equal to 100 percent of the floodable length. A factor of 0.3 means the spacing can only be 30 percent of the floodable length. For methods of determining the criterion of service and factor of subdivision and the special provision required in their application, the reader is referred to the U.S. Regulations which vary slightly and are more severe than the requirements of the SOLAS Conventions (46 CFR 73.10, 74.10-15).

(b) Permeabilities. Before the effects of flooding can be calculated, definite values for the permeabilities of the spaces involved must be assumed. The actual permeabilities in service are not accurately assessed and, in the case of cargo spaces, vary from space to space and from voyage to voyage. Consequently, all calculations of the affect of flooding based on the assumed permeabilities are unavoidably only approximations.

All criteria, with the exception of the traditional regulations for subdivision (not damaged stability) of passenger vessels on international voyages, and the regulation for permeability of cargo spaces for offshore supply vessels, assume a permeability for individual spaces as given in Table 5.

Table 5—Assumed Permeabilities

	Permeability Percent
Appropriated to cargo hold or stores	60
Appropriated to accommodation and voids	95
Appropriated to machinery	85
Appropriated for consumable liquids (using the value resulting in the most severe requirement)	0 to 95
Appropriated for liquid cargo tanks	0 or 95

The two table values of permeability for consumable liquid and liquid cargo tanks were put in for a specific reason. In vessels with high side tanks extending well above the waterline, for example, the stability effect of run off from initially full tanks (zero permeability), may be worse than that due to the ingress of damage water if the tanks are initially empty, i.e., (95 permeability). It is particularly important to investigate the stability effect of tanks having either zero or 95 permability during the period between initial damage and the final equalization. See Section 4.8. For example, transverse pairs of side tanks located outboard, port and starboard, which are interconnected by equalizing ducts and in which the tops of the tanks are below the waterline, can cause a large $\overline{GM}$ loss due to free surface during the time the tanks are being filled from the sea. This could conceivably cause the vessel to capsize before the tanks are filled and equalized.

In all cases where spaces between main transverse watertight bulkheads are occupied by spaces having different permeabilities, an average permeability is used based on the percentage of the total space occupied by each category or permeability.

On offshore supply vessels the permeability of dry cargo spaces is assumed to be 95.

Volume and surface permeabilities are normally required to be the same. However, most criteria make a general statement that higher surface permeabilities should be assumed in respect of spaces which, in the vicinity of the damaged waterplane, contain no substantial quantity of accommodation or machinery, and spaces which are not generally occupied by any amount of cargo or stores. The maximum assumed surface permeability need not exceed 95.

All regulations, with the exception of the alternate equivalent passenger vessel regulations (Section 8) are silent on the matter of varying permeabilities at varying drafts. The Maritime Administration does, however, state that the permeabilities given in this section are average permeabilities and that the permeabilities should be routinely chosen so as to agree with the vessel's operating condition. For example, if the vessel has a cargo hold essentially empty in a light condition, they recommend a permeability of 95 rather than 60.

The regulations for bulk chemical and liquefied gas carriers both state that it shall be assumed that all liquids that are in a tank prior to damage have been completely replaced by salt water after damage. Other regulations are silent on this point, though the substitution seems to be implied. The tanker regulation makes the statement that the permeability of partially filled compartments shall be consistent with the amount of liquid carried.

While all regulations for damaged stability call for an investigation at a range of operating drafts, only the new equivalent passenger vessel regulations makes a provision for varying the permeability of cargo spaces depending upon the draft. See Section 8.

For some time, it has been suspected that containerships, roll-on roll-off ships and barge-carrying ships have higher permeabilities in their cargo holds than traditional bulk cargo ships. In two recent studies sponsored by the Coast Guard and performed by the Maritime Administration, containerships had average permeabilities of 75, with a value forward of 80, a value amidship of 70, and a value aft of 75. The assumption was made that the containers were nontight and their interiors were flooded. The differences in values can be accounted for by the flare of the hulls forward and aft, with a greater percentage of the width of the holds not being occupied by containers.

On roll-on, roll-off ships with containers stacked on deck the permeabilities were 80 and with the containers rolled on and left on a chassis the permeability was 90. On the barge-carrying ships the cargo holds with barges assumed to be watertight, had a permeability of only 30, but with the barges not assumed to be watertight, the permeability of the space was 76.

No changes in U.S. regulations or international agreements have as yet resulted from the above studies. However, the Maritime Administration adopted a standard of 70 for container holds and for Ro-Ro ships a value of 80 for containers stowed on deck and 90 for containers on wheels.

The permeability of all spaces and particularly that of cargo spaces tends to be higher in the upper parts of the spaces. In accommodation and machinery spaces, this variation may not be enough to affect seriously the validity of calculations based on a uniform value. However, in cargo spaces, calculations based on a uniform value may grossly underestimate the $\overline{GM}$ loss.

Because of the variety in general cargo loadings, it is impracticable to attempt to account accurately for these variations. However, vertical variations in the permeability of cargo spaces may be compensated for partially by assuming, when calculating the available $\overline{GM}$, that the cargo is at the homogeneous center even when actually it may be only in the lower part of the space. In this way, underestimating the available intact $\overline{GM}$ tends to compensate for the underestimate in the $\overline{GM}$ loss due to disregarding the actual cargo distribution.

In determining the floodable length of passenger vessels on international voyages, a uniform average permeability is used throughout the whole length of

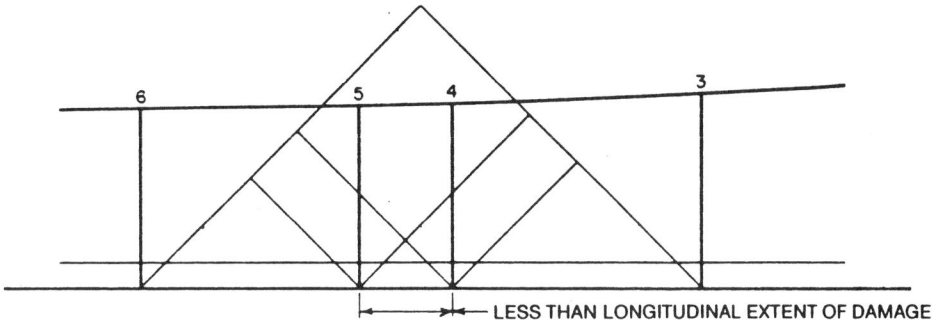

Fig. 40 Minimum spacing of bulkheads

each of the following portions of the vessel below the margin line:

1. The machinery space.
2. The portion forward of the machinery space.
3. The portion aft of the machinery space.

Formulas are given in the U.S. and SOLAS Regulations for determining the permeability in each of these areas. The assumption that permeabilities are uniform in each is illogical and unnecessary. More accurate results can be achieved readily by calculating the individual average permeability of each compartment.

(c) Minimum Spacing of Bulkheads. All criteria call for a minimum spacing of main transverse watertight bulkheads if two adjacent bulkheads are to be considered intact with the exception of the Regulations for Offshore Supply Vessels. This minimum distance is equal to the longitudinal extent of damage specified for the particular class of vessel. Since, for reasons of practicality, offshore supply vessels are only required to survive a collision which has a penetration of 760 mm (30 in.) it was evidently not felt necessary to put a minimum spacing on these vessels.

The method of handling bulkheads which are spaced closer together than the longitudinal extent of damage is illustrated in Fig. 40.

In U.S. passenger ship regulations the combined length of the forepeak compartment and the compartment just abaft the forepeak compartment in vessels over 100 m (330 ft) in length is required to be not greater than the permissible length. This causes difficulties in vessels that must meet a two or three-compartment standard, in that the after-most bulkhead is located so close to the forepeak bulkhead that the distance between them cannot comply with the requirement for the minimum spacing of bulkheads to be greater than the longitudinal extent of damage. This problem is corrected in the new Alternate Equivalent Passenger Vessel Regulations. (See Section 8.5).

(d) Recessed Bulkheads. In general, all criteria do not permit recessed bulkheads unless the entire recess is located inboard of the side of the vessel at a distance not less than the assumed transverse extent of damage. Fig. 41 shows permissible recessed bulkheads.

In offshore supply vessels where bulkheads in side tanks need to be extended only 760mm (30 in.) in from the shell, bulkheads recessed inboard of that point can be considered as main transverse watertight bulkheads. A bulkhead inboard of the side tank may be stepped, but the double bottom tank extending between the inboard transverse bulkhead and the outboard bulkhead must have its volume added to the volume of the outboard compartment.

It frequently is not recognized that a shaft alley may constitute a recess. If all parts of a shaft alley are inboard of the aforesaid one-fifth beam line, it may be regarded as providing intact buoyancy. However, if within a portion of its length a shaft alley is nearer to a vessel's side it should be regarded as liable to be damaged within that part of its length. In such case, it will flood for its entire length and therefore should be regarded as constituting a recess into the full length of the adjoining compartment or compartments which it penetrates.

(e) Stepped Bulkheads. From the viewpoint of safety it is desirable for bulkheads to be fabricated in a single plane. Steps increase the likelihood that a bulkhead may be damaged. Also the watertight portions of decks required by steps are liable to have their integrity violated by nontight penetrations made during the life of the vessel. This is because such portions of decks are liable not to be recognized as part of a vessel's watertight subdivision when such alterations are made. For this reason, it is important that the

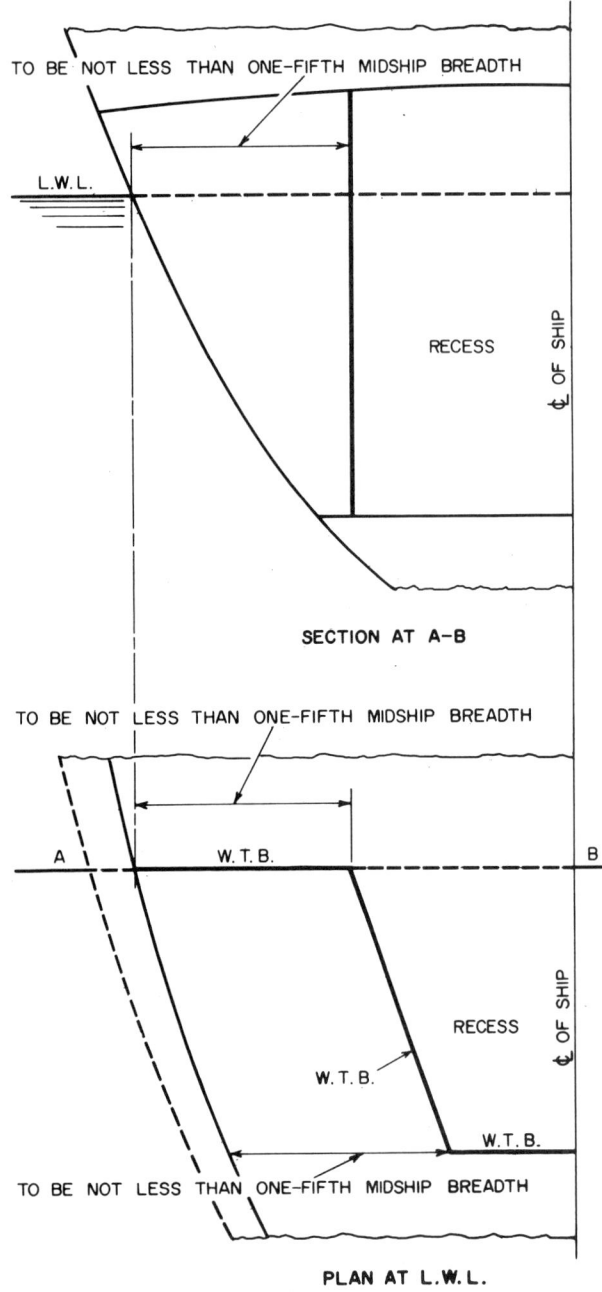

Fig. 41 Recessed bulkheads

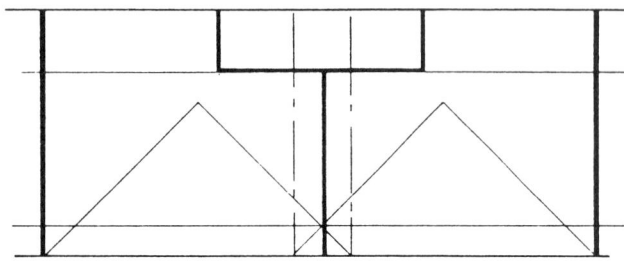

Fig. 42 Stepped bulkheads

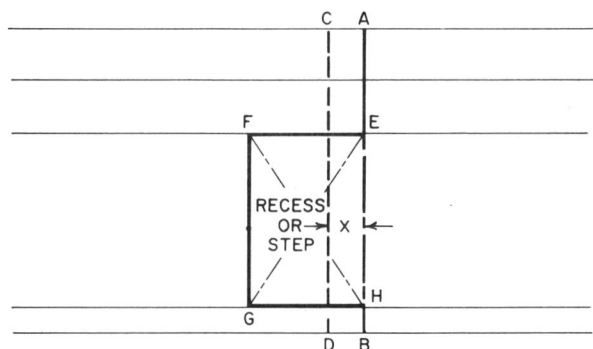

Fig. 43 Equivalent plane bulkhead

Bulkhead AB is recessed or stepped as shown. The location of the equivalent plane bulkhead is given by $X = $ Volume $EFGH$/area A, where A is the sectional area to the margin line approximately midway between AB and CD

extent of any watertight steps be clearly indicated on a plan carried on the vessel.

U.S. Passenger Vessel Rules permit a main transverse bulkhead to be stepped under certain conditions. See 46 CFR 73.10, 74.10-15.

1. *The combined length* of the two compartments, separated by the bulkhead in question, does not exceed either 90 percent of the floodable length or twice the permissible length, except that in ships having a factor of subdivision greater than 0.9, the combined length of the two compartments in question shall not exceed the permissible length; or

2. *Additional subdivision* is provided in way of the step to maintain the same measure of safety as that secured by a plane bulkhead, Fig. 42; or

3. *Length of the compartment* over which the step extends does not exceed the permissible length corresponding to a margin line taken 7.5 cm (3 in.) below the step.

Where a bulkhead is stepped or recessed, an equivalent plane bulkhead (or bulkheads) is used. The usual case is illustrated in Fig. 43. Where additional subdivision is provided in way of the step, such as by stepping the bulkhead both ways, the appropriate equivalent plane bulkheads are as illustrated in Fig. 42.

Where the distance between equivalent plane bulkheads or that between transverse planes passing through the nearest portions of stepped bulkheads is less than the assumed longitudinal extent of damage, the bulkheads concerned are dealt with the same as are plane bulkheads at less than this spacing, as illustrated in Fig. 40.

When a main transverse watertight compartment contains local subdivision, and it can be shown that, after any assumed side damage of the required transverse and longitudinal extent, the whole volume of the main compartment will not be flooded, a proportionate allowance may be made in the permissible length otherwise required for the compartment by the passenger vessel factorial standards. The manner of allowing for these unflooded spaces is shown in Fig. 44. In the integer compartmentation standards there are obviously no reductions allowed in the minimum spacing of bulkheads and the intact buoyancy is taken care of by corrections to the volumes and centers of the lost buoyancy or mass and centers of the added weight.

(f) Independent Tanks, Bulk Chemical and Liquefied Gas Carriers. Detail requirements for the location of independent tanks in these classes of vessels with respect to distance from the ship's side or from the bottom are given in the U.S. Coast Guard Regulations and should be carefully checked by designers.

(g) Internal Non-watertight Compartmentation. Where internal divisions, particularly longitudinal bulkheads, are non-watertight, but are so constructed as to permit water only to leak through slowly when a space adjacent to the bulkhead is flooded, then the heeling moment of the water must be accounted for. A large mass of water may be retained on one side of the bulkhead for a considerable length of time before equalization on both sides of the bulkhead takes place.

Examples of such bulkheads are steel structural bulkheads built to A-0 fire classification standards with metal joiner doors, and steel insulated bulkheads in refrigerated spaces. So called "flooding plugs" designed to blowout at a nominal low head are often installed in such bulkheads, but are not looked upon with favor because there may be sticking due to rust or excess paint and can often be blocked from opening by cargo carelessly stowed against them.

The vessel must have sufficient residual stability after damage that the heeling moments resulting from such bulkheads will not capsize the ship or reduce its survival capability below the criteria requirements during the time that equalization takes place.

(h) Superstructures. The many standards for damaged stability have never been completely consistent regarding the effect of superstructures on residual stability of damaged vessels. It is agreed and generally accepted that a superstructure directly over the location of collision damage can also be opened to the sea by the damage and its buoyancy in that area must be neglected in calculating righting moments. However, the use of terminology and practice from the load line conventions confuses the issues, i.e., the requirement that houses have a certain required deck height to be considered effective. *All* enclosures above the bulkhead deck, whether they are trunks or enclosures at the sides or center, only are effective if their buoyancy is maintained intact by watertight structures and watertight closures (weathertight if they are not submerged at the equilibrium waterline and come into effect only in the residual range of stability). Common sense in including only the buoyancy of those structures that are demonstrably effective in providing buoyancy moments in a heeled condition should be, and generally is, the basic criterion used by the competent naval architect.

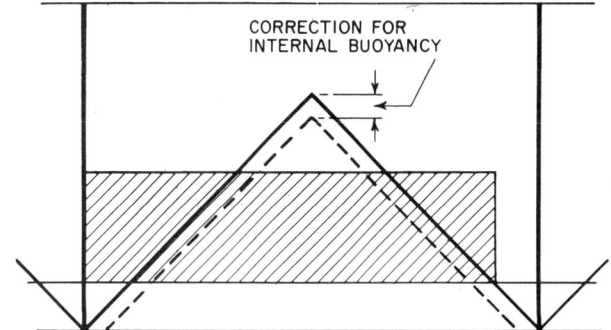
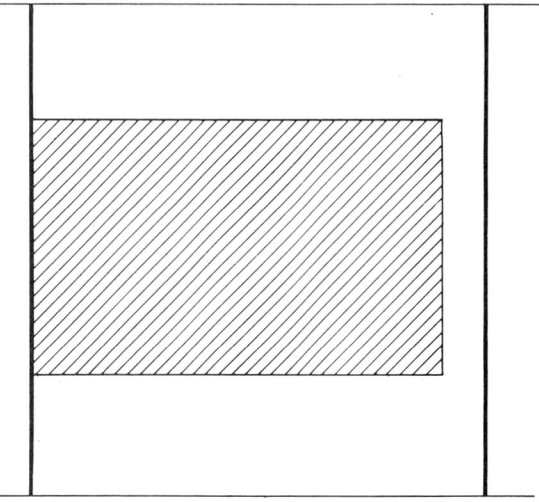

Fig. 44 Correction for internal buoyancy

(i) Unsymmetrical Flooding is required to be kept to a minimum consistent with efficient arrangements. On passenger vessels on international voyages, where it is necessary to correct large angles of heel, the means adopted are required, where practicable, to be self-acting, but in any case, where controls to cross-flooding fittings are provided, they must be operable from above the bulkhead deck. The construction and arrangement of such fittings and of their controls, together with the estimated maximum heel before equalization, are subject to special approval. The time for equalization to acceptable heel limits, as given by the Solda equation (Section 4.8), is not permitted to be more than 15 min. On all other vessels, the equalizing arrangements, where required, must not depend upon either manual or automatic operation of valves or similar appliances. (Note that in Section 8 the equalization time for the alternate equivalent subdivision regulations is reduced to 10 min.).

Even though means for equalizing unsymmetrical intact buoyancy or for flooding of unfavorable symmetrical intact buoyancy are provided, it is important to consider the intermediate condition of the ship after damage, but prior to completion of final flooding. With the exception of those cases where the use of large unobstructed openings such as open passages or cofferdams is possible, it must be assumed that the ship will experience a transient condition during which primary flooding becomes essentially complete before equalization has occurred.

Survival of this transient condition requires: (a) that the static heel be not so great as to eliminate the hydrostatic head necessary for compensatory flooding; (b) that the static heel be not so great as to result in the progressive extension of the flooding (submergence of the margin line during an intermediate stage of flooding is permissible provided there is sufficient watertight integrity above the margin line to insure that progressive extension of flooding will not occur); and (c) that the range of stability in the transient condition be sufficient for the vessel to survive the compensatory period.

Where compensatory flooding or equalization is necessary, calculations should be made to ascertain that requirements (a) and (b) are met. Unless transient condition $\overline{GM}$ and freeboard are large enough obviously to provide a sufficient range of stability, (c) should also be checked by calculation of righting arms in the transient damaged condition.

(j) *Dump Valves.* Where flume stabilizer tanks or swimming pools are fitted with dump or flooding valves and it is desired to credit these valves in determining the final stage of flooding, the maximum time for all required equalization action plus such dumping or flooding must not be more than 15 min. (10 min. in equivalent passenger vessel regulations).

Section 8
Alternate Equivalent Passenger Vessel Regulations

8.1 General. The loss of the *Andrea Doria* in 1956 not only caused a complete discrediting of the theory that tanks designed for the carriage of bunker oil could or would be filled with ballast water to meet damaged stability requirements, but it caused a new critical look to be taken at all of the assumptions implicit in the standards for subdivision and damage stability contained in the existing international passenger vessel regulations.

Among the weaknesses found in the then existing standards, were:

(a) The formulas for Criterion of Service involving the relationship between the volume of different spaces within the ship's hull were out of date. Improvements in design had permitted higher power within less volume in the machinery spaces, and modern requirements for more spacious accommodations resulted in more passenger space above the bulkhead deck. The safety standard applied to ships primarily carrying passengers had therefore depreciated.

(b) The existing regulations did not take into account the fact that for any given bulkhead arrangement quite different extents of flooding could occur as a result of varying assumed damage lengths.

(c) The method employed in the existing regulations did not take into full account the effect of ship proportions, of varying drafts and permeabilities, or of stability when flooded, on the degree of safety. Two vessels might be judged equally safe although they might have very different actual capabilities to survive.

The realization of these and other weaknesses during the 1960 SOLAS Conference gave impetus to the consideration by IMCO of possible new regulations based on probability principles.

As explained by Robertson, et al (1974), casualty data regarding hull damages were collected and analyzed. Voyage data relative to loading conditions, draft, operating $\overline{GM}$, etc., were also sought and analyzed. Model tests simulating damaged ships in a seaway were conducted. In 1967 an ad hoc group was assigned to prepare new regulations based upon these studies. Several different forms of regulations were considered. In 1971, after extensive study of possible regulations, principally involving separate floodable length and damage stability calculations, a revised format based directly on use only of damage stability calculations was submitted. This formed the basis for the new regulations as now adopted. See Appendix 1 of Robertson, et al (1974).

While these new regulations are administratively equivalent and alternative to those of Chapter II of the 1960 International Convention for the Safety of Life at Sea, and provide on the average about the same degree of safety, they are considered vastly superior thereto from the viewpoint of logic and consistency. Calling the new provisions "equivalent" permits a period during which administrations may assess and gain familiarity with them prior to adoption by the formal ratification and amendment procedures of the Safety of Life at Sea Convention.

8.2 Basis of the new regulations. Fundamentally, three probabilities relate to subdivision and damage stability requirements:

(a) Probability that a ship may be damaged.

(b) If the ship is damaged, the probability as to location and extent of flooding.

(c) Probability that the ship may survive such flooding.

The probability that a ship may be damaged is relevant to the required degree of ability to survive damage and to the determination of insurance premiums. It is conditional upon navigational conditions, traffic density, visibility, effectiveness and reliability of navigational aids, ship speeds and maneuvering capabilities, and the judgment, competence and dependability of personnel involved.

From the purely theoretical viewpoint, evaluation of the effect of each of these factors would permit determination of probability of loss for each ship, or at least for each class of ship. Practically, the available statistical and other necessary information is not sufficient for this purpose. However, some useful deductions are possible. Examination of casualty data confirms that damages due to collisions (and strandings) are more prevalent in harbor approach areas and areas of especially high traffic density such as the English Channel. Ships whose operation is principally or exclusively in such areas are more likely to be damaged. Passenger ships in this category tend to carry a higher density of persons than ships on longer voyages outside such waters. In the new regulations, the required degree of safety is dependent principally upon the number of persons, and is increased when the density is such that all persons cannot be accommodated in lifeboats.

So long as a ship remains undamaged, there is no need whatsoever for any subdivision or damage stability. The need for evaluation of subdivision and damage stability stems from the knowledge that risk of damage does exist, and this leads to consideration of probabilities 2 and 3.

Probability 2 is dependent upon the location and extent of hull damage and upon the arrangement of watertight divisions within the ship. This general relationship was recognized as early as 1919, and a relatively rigorous procedure for dealing with it was first presented by Wendel (1960).

Probability 3 is dependent upon buoyancy and stability in the flooded condition. This in turn is subject to the following variable factors:

(a) The location and extent of flooding.
(b) The permeability of flooded spaces.
(c) The draft and stability before flooding.
(d) Applied forces and moments.

Data for a total of approximately 860 ship damage cases attributable to strandings and collisions were compiled. After preliminary review, it was decided that probability as to location and extent of damage might best be evaluated by limiting study to collision cases involving only passenger and cargo ships, excluding tankers. Data on vertical location and extent were also available, but it did not seem feasible to deal with these variables probabilistically.

The probability of survival in any flooded condition is also dependent upon the probability in respect to applied forces and moments in that condition. These may be due to wind and sea, or to location or movement of tankage, persons, or other weights.

To evaluate the relationship between sea state, stability and buoyancy when flooded, and survival probability, damage stability model tests in waves were conducted. Test models included a long-voyage passenger-cargo type ship (Middleton and Numata, 1970) and a short-voyage passenger-vehicle ferry (Bird and Brown, 1973). Supplemental damage stability tests of simplified models dealing particularly with the effects of variations in beam-depth ratio and of internal arrangements were also conducted (Stahlschmidt, 1972.) The mechanism of capsize observed in all tests was similar in similar situations. Despite the wide difference in ship type and proportions, the references showed some degree of agreement as to the stability necessary to survive a given sea state. For both of these series of tests, there was sometimes a substantial difference between the stability necessary to avoid capsize when the damage was to windward and when it was to leeward. However, with that exception, in any given condition, the difference between the stability at which repeated capsizes occurred and that at which they did not occur was small. Therefore, for simplicity, the values of stability necessary for survival at any sea state were averaged and treated as deterministic rather than as random variables, as might theoretically be expected. The model tests were very interesting and the references cited provide much food for thought.

To evaluate actual service variations in draft and permeability and in stability before flooding, relevant ship voyage data were reviewed. The ships for which the most complete data were available were of passenger-cargo and cruise type. Some data on the operation of both short-voyage passenger-vehicle type ships and long-voyage cargo-passenger ships were also included.

Draft distributions were found to differ considerably. Despite these variations, the new regulations assume a standard form of draft distribution to apply to all passenger ships. While this standard draft distribution was based upon the ship draft data, its median value is somewhat higher than the average median of the reported values. Use of such a single standardized form of assumed draft distribution is considered justified for the following reasons:

(a) For uniformity in administration and because of lack of a clear indication as to how a ship's service draft distribution might better be anticipated in the design state.

(b) Because example calculations have indicated that the Subdivision Index A, according to Regulation 6 of the new regulations, is not too sensitive to differences in the form of draft distribution.

The ship voyage data also included information on the kind and quantity of cargo carried, on the cargo space volume, and on the percentage of that volume occupied by cargo. Utilizing that information and other data, the average cargo space permeability for each

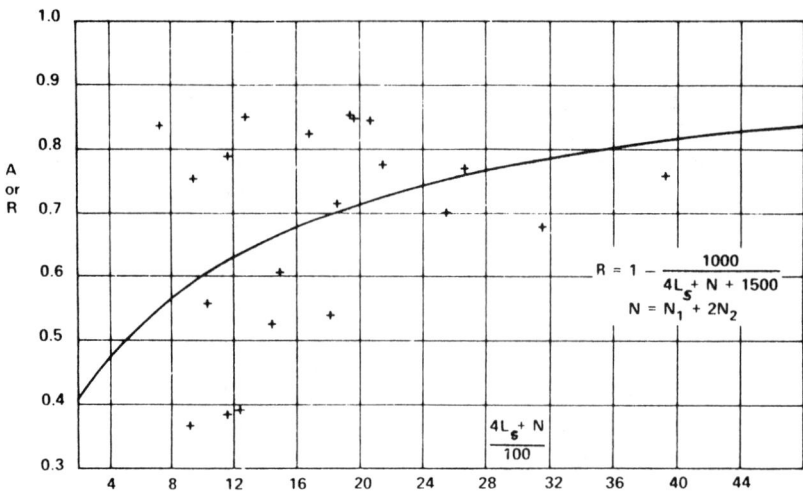

Fig. 45 Comparison of R with the calculated A values of existing ships which comply with Chapter II of the 1960 Safety Convention. L_s and N are as defined in Regulations 1(b) and 2(c) respectively

loading condition was estimated and related to the draft or drafts corresponding to that condition. As might be expected, the cargo space permeability at any draft is a variable. However, for purposes of simplifying the calculations, the cargo space permeability formula given by new Regulation 4(b) treats it as deterministic.

It may be demonstrated by means of probability theory that the probability of ship survival can be calculated as a sum of probabilities of its survival after flooding each single compartment, each group of two, three, etc., adjacent compartments multiplied, respectively, by the probabilities of damage leading to such flooding.

The new regulations prescribe that a ship's Attained Subdivision Index A must be equal to or greater than its Required Subdivision Index R. A is equal to the summation over the ship's length of the expression Σaps, where

a accounts for the probability of damage as related to the position of the damage in the ship's length.

p evaluates the effect of the variation in longitudinal extent of damage on the probability that only the compartment or group of compartments under consideration may be flooded.

s evaluates the effect of freeboard, stability, and heel in the final flooded condition for the compartment or group of compartments under consideration.

If it were possible to accomplish this summation, including evaluation of each of the terms, rigorously and completely, A would be the probability that the ship would survive any and all damages resulting in breaching of the hull which might occur during her lifetime. However, for reasons already mentioned in the foregoing general discussion of the concept of the regulations and of the background studies, approximations and simplifications have been necessary.

Therefore, A should be regarded not as the actual survival probability, but rather as a number which is dependent upon the assessable principal factors and which is approximately proportional to that probability.

The Required Subdivision Index R is dependent upon the length of the ship and the number of persons, and is thus related to the property and lives at stake in case of a casualty. This relationship is complex and not explicit. Furthermore, it is dependent upon the effect of subdivision and stability provisions on economic and other operational conditions. For the time being, a suitable level for R could only be determined from consideration of the A values of existing ships calculated according to the new regulations. Figure 45 shows the A values of some existing ships that comply with the 1960 regulations and R as contained in the new regulations. The wide scatter in the A values of existing ships is at first startling and was a source of considerable discussion. However, it is not so surprising when one considers the illogical aspects of the 1960 SOLAS regulations.

The relation between R and the A values of existing ships reflects the decision reached early in the IMO Subcommittee deliberations that the average level of safety attained under the new regulations should only be equal to that attained under the 1960 SOLAS regulations. This decision was not consistent with the views of the U. S. and some other delegations to the 1960 Safety of Life at Sea Conference, who thought a higher standard was practical and necessary. However, it contributed appreciably to establishment of the degree of rapport within the the Subcommittee essential to completion of the very extensive and drawn-out work culminating in these new regulations. Many believe that a higher standard of safety is practicable and are hopeful that experience in design of ships to the new regulations will lead to eventual agreement on this point.

In addition, such experience should lead to procedural improvements in the provisions of the regulations and related calculations. It is hoped that adequate related systematic study of casualties, of ship operating data, of waves and other sources of upsetting moments, and of relevant ship response will be continued. While these new regulations, *per se*, apply only to passenger ships, the principles upon which they are based are such that similar procedures may well be developed, not only for evaluation of the subdivision of other types of ships, but also for logical evaluation and control of other damage-related ship risks.

These alternate regulations have not as yet been incorporated in the published U.S. Coast Guard Regulations, although they are accepted and endorsed by the U.S. Coast Guard. The complete regulations are given in IMO Resolution A265 (VIII), and explanatory notes to the regulations are given in IMO MSC/CIRC. 153. Both of these are reproduced in the Coast Guard Commandant's International Series (1974). Further explanation of the reasoning behind the regulations is given in Robertson, et al (1974).

There is general acceptance among the leading authorities of the maritime nations that the probability principles on which the alternate regulations are based are sound and their application to other classes of vessels offers the possibility of obtaining increased safety without substantially increasing construction and operating costs. Since the promulgation of these alternative regulations occurred at the same time as the passenger airlines were sweeping the bulk of ocean going passenger vessels from the seas, few naval architects have had any reason to use and understand these regulations and to explore the possibilities of improving the safety of other classes of vessels by their use. It is hoped that in the coming years, this situation will gradually change and experience will lead to the adoption of these principles for a larger portion of the world's seagoing fleet. See Tagg (1982).

8.3 Definitions. Regulation 1 of the IMO new equivalent subdivision regulations gives the following definitions. For the purpose of these Regulations, unless expressly provided otherwise:

(a) (i) A *'subdivision load line' is a waterline used in determining the subdivision of the ship; and*
(ii) *the 'deepest subdivision load line' is the waterline which corresponds to the greatest draught permitted by the subdivision requirements which are applicable.*
(b) *the 'subdivision length of the ship' (L_s) is the extreme moulded length of that part of the ship below the immersion limit line.*

In effect, the subdivision length L_s is the molded overall length of the buoyant part of the ship throughout which damage may affect buoyancy and stability. Thus, in a flush deck vessel, the length is measured between the intersections of the deck with the stem and stern. In vessels with raised forecastles and/or poops, the length is measured between the intersection of the forecastle deck with the bow and the intersection of the poop deck with the stern. In contrast to the length, as defined in the U.S. traditional passenger ship regulations, it is virtually always greater and is unaffected by changes in subdivision draft. While the "subdivision length of the ship" L_s is always used in the formulas for required and attained subdivision indices, it is common practice to use hydrostatic data from lines plans drawn with end ordinates at the load waterline or the ship's perpendiculars. This practice is accepted because it is conservative. Neglecting the buoyancy between the end ordinates and the end points for subdivision length, results in giving a trim for compartments inboard of the ends which is greater than the actual trim. In cases where the attained index is marginally below the required index using the end ordinates on the lines plan, modification can be made treating the added buoyancy at the ends as appendages.

(c) *'midlength' is the midpoint of the subdivision length of the ship (L_s).*
(d) (i) *the 'breadth' (B_1) is the extreme moulded breadth of the ship at midlength at or below the deepest subdivision load line;*
(ii) *the 'breadth' (B_2) is the extreme moulded breadth of the ship at midlength at the relevant bulkhead deck.*

Together with freeboard (and the ship's form) B_2 affects the heel line at which the "relevant bulkhead deck" will immerse. It is possible for more than one deck to be the "relevant bulkhead deck". Therefore in the case of some ship's having tumblehome or flare to the topsides, B_2 may have more than one value.

(e) *The 'relevant bulkhead deck' is the uppermost deck which, together with the watertight bulkheads bounding the extent of flooding under consideration and the shell of the ship, defines the limit of watertight integrity in the flooded condition.*

Where watertight bulkheads and the associated ship's shell are watertight to different levels in different parts of the ship, the "relevant bulkhead deck" in respect to some flooding situations will be different from that of other flooding situations.

(f) *The 'immersion limit line' at any point in L_s is defined by the highest relevant bulkhead deck at side at that point.*
(g) *The 'draught' (d_i) is the vertical distance from the moulded base line at midlength to the waterline in question.*
(i) *The 'subdivision draught' (d_s) is the sub-*

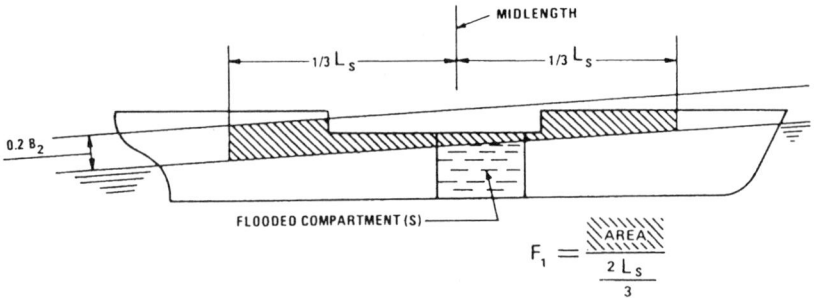

Fig. 46 Illustration of "effective mean damage freeboard" (F_1)—Regulation 1 (h)

division load line in question.

(ii) The 'lightest service draught' (d_o) is the service draught corresponding to the lightest anticipated loading and associated tankage, including, however, such ballast as may be necessary for stability and/or immersion.

(iii) Intermediate draughts between d_s and d_o are:

$$d_1 = d_s - 2/3(d_s - d_o)$$

$$d_2 = d_s - 1/3(d_s - d_o)$$

$$d_3 = d_s - 1/6(d_s - d_o)$$

(h) The 'effective mean damage freeboard' (F_1) is equal to the projected area of that part of the ship taken in the upright position between the relevant bulkhead deck and the damage waterline and between $1/3L_s$ forward and abaft the midlength divided by $2/3L_s$. In making this calculation no part of the area which is more than $0.2B_2$ above the damage waterline shall be included. However, if there are stairways or other openings in the bulkhead deck through which serious downflooding could occur F_1 shall be taken as not more than $1/3(B_2 \tan \theta_F)$, where θ_F is the angle at which such openings would be immersed.

The effective mean damage freeboard and the limitations on F_1 are illustrated in Figures 46 and 47.

(i) The 'permeability' (μ) of a space is the proportion of the immersed volume of that space which can be occupied by water."

8.4 Required Subdivision Index. IMO regulation number 2 reads as follows:

(a) To provide for buoyancy and stability after collision or other damage, ships shall have sufficient intact stability and be as efficiently subdivided as is possible having regard to the nature of the service for which they are intended.

(b) The subdivision of a ship is considered sufficient if:

(i) the stability of the ship in damaged condition meets the requirements of Regulation 5 [Section 8.7]; and

(ii) the attained Subdivision Index A according to Regulations 6 and 7 [Sections 8.8, 8.9] is not less than the required Subdivision Index R calculated in accordance with paragraph (c) of this Regulation.

(c) The degree of subdivision is determined by the required Subdivision Index R, as follows:

$$R = 1 - \frac{1000}{4L_s + N + 1500} \quad (m)$$

$$R = 1 - \frac{1000}{1.22L_s + N + 1500} \quad (ft)$$

(1)

Where:

$$N = N_1 + 2N_2$$

$N_1 =$ *number of persons for whom life boats are provided.*

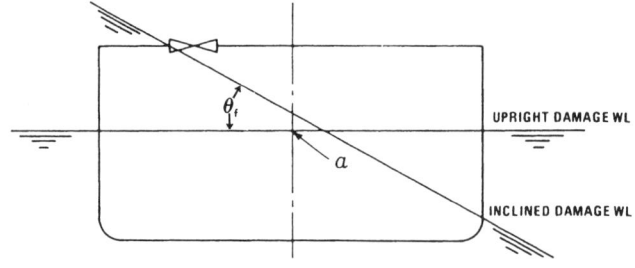

Fig. 47 Illustration of limit on F_1 imposed by openings in the bulkhead deck

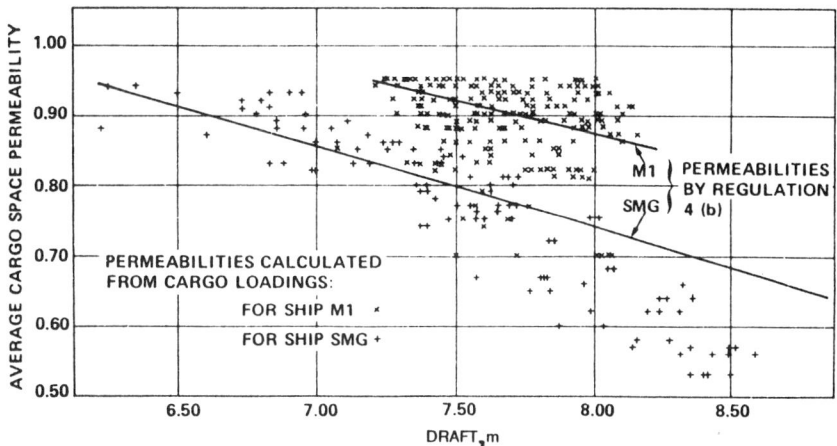

Fig. 48 Permeability vs. draft for two passenger-cargo ships:
M1—180 m, 557 pass.; SMG—156 m, 121 pass

N_2 = *number of persons (including officers and crew) that the ship is permitted to carry in excess of N_1.*

(d) Where the conditions of service are such that compliance with paragraph (b) of this Regulation on the basis of $N = N_1 + 2N_2$ is impracticable and where the Administration considers that a suitably reduced degree of hazard exists, a lesser value of N may be taken but in no case less than $N = N_1 + N_2$.

As an example of how the Required Subdivision Index R increases as the number of passengers increases and also as the length increases, the following data are given:

REQUIRED SUBDIVISION INDEX R

Length of Ship	500 Pass.	1000 Pass.	1500 Pass.
122 m (400 ft)	0.598	0.665	0.713
183 m (600 ft)	0.634	0.691	0.732
244 m (800 ft)	0.664	0.712	0.748

It will be seen later that the higher the subdivision index, the greater the subdivision requirements. Obviously, the increase due to the number of passengers is intended to take into account the need for greater safety with greater numbers; the increase due to length, takes into account the ability of a longer ship to be subdivided in such a way as to decrease the probability of damage occurring that will sink the ship.

8.5 Special Requirements. The IMO Regulations have a number of regulations concerning double bottoms and bulkhead arrangements and watertightness, generally consistent with the existing U.S. Regulations. (See IMO Regulations 3 and 9 through 19). The major variation occurs in Regulation 3, concerning the bulkhead just abaft the forepeak bulkhead, which only requires that the forepeak and the next adjacent compartment have an "s" value of not less than one. Also, an additional requirement is that the distance between the forepeak bulkhead and the next bulkhead shall not be less than the required longitudinal extent of damage.

8.6 Permeability. The permeabilities of various spaces set forth in IMO Regulation 4 are as follows:

(a) Spaces	*Permeability (μ)*
Appropriated as accommodation for passenger and crew, or other spaces not specifically herein designated	*0.95*
Appropriated for machinery	*0.85*
Normally occupied by stores	*0.60*
Intended for consumable liquids	*0.00 or 0.95**

* whichever results in the more severe requirement.

(b) The permeability μ of any space appropriated for cargo shall be assumed to vary with the draught before damage in such a way that for any initial draught d_i the permeability μ_i of any cargo space shall be taken as:

$$\mu_i = 1.00 - \frac{1.2(d_i - d_o)}{d_s} - \frac{0.05(d_s - d_i)}{(d_s - d_o)}$$

but not more than 0.95 nor less than 0.60.

It should be noted that here, for the first time, a subdivision and damage stability standard varies the permeability used for a given compartment depending on the draft, assigning a higher permeability at the lower drafts. This is consistent with the fact that cargo holds generally will not be filled with cargo at the light drafts. Fig. 48 shows a plot of permeability versus draft calculated from paragraph (*b*) above for two of the ships analyzed in preparing the regulations. The detail permeability calculations for these ships on various voyages are also indicated.

It will also be noted that the use of average permeabilities forward and aft of the machinery space is eliminated and only the individual permeabilities for each space are used.

8.7 Required Subdivision and Damage Stability. The requirements for subdivision and damaged stability are given in this section and in Sections 8.8 and 8.9, corresponding to IMO Regulations 5, 6 and 7. Regulation 5 is set forth below:

(a) Sufficient intact stability shall be provided in all service conditions so as to enable the ship to comply with the provisions of this Regulation. Before certification of the ship, the Administration shall be satisfied that the required intact stability can practicably be obtained in service.

(b) (i) All ships shall be so designed as to comply with the provisions of this Regulation in the event of flooding due to one side damage with a penetration of $0.2B_1$ from the ship's side at right angles to the centerline at the level of the subdivision load line and a longitudinal extent of $3 m (9.8 ft) + 0.03L_s$, or $11 m (36 ft)$ whichever is the less, occurring anywhere in the ship's length, but not including a transverse bulkhead. However, where a bulkhead is stepped it shall be assumed as subject to damage.

(ii) Ships for which N is more than 600 shall additionally be able to comply with this Regulation in the event of flooding, due to side damage including transverse bulkheads occurring anywhere within a length equal to (N/600 − 1.00)L_s, measured from the forward terminal of L_s, where N is as defined in Regulation 2(c) and (d). The value of (N/600 − 1.00) shall not be more than one.

(iii) In any calculation required under this paragraph, the damage shall be assumed to extend from the base line upwards without limit. However, if flooding due to a lesser extent of damage either vertically, transversely or longitudinally results in a higher necessary intact metacentric height, such a lesser extent of damage shall be assumed. In all cases, however, only one breach in the hull and only one free surface need be assumed. For the purpose of assessing heel prior to equalization, the bulkheads and deck bounding refrigerated spaces and other decks or inner divisions which in the opinion of the Administration are likely to remain sufficiently watertight after damage, shall be regarded as limiting flooding. Otherwise, flooding shall be assumed as limited only by undamaged watertight structural divisions.

(c) (i) In the final stage of flooding:
  (1) there shall be a positive metacentric height, $\overline{GM}$, calculated by the constant displacement method and for the ship in upright condition, of at least

$$\overline{GM} = 0.003 \frac{B_2{}^2(N_1 + N_2)}{\Delta F_1} \quad or$$

$$\overline{GM} = 0.049 \frac{B_2}{F_1} \quad (ft)$$

$$\overline{GM} = 0.015 \frac{B_2}{F_1} \quad (m) \quad or$$
$$\overline{GM} = 0.05 \ m \ (2 in.) \ whichever \ is \ the \ greater$$

Where Δ = displacement of the ship in the undamaged condition (in long tons or metric tons respectively);

(2) the angle of heel in the case of one compartment flooding shall not exceed 7 deg. For the simultaneous flooding of two or more adjacent compartments, a heel of 12 deg may be permitted unless the Administration considers a lesser heel necessary to ensure an adequate amount and range of residual stability;

(3) except in way of the flooded compartment or compartments no part of the relevant bulkhead deck at side shall be immersed.

(ii) Unsymmetrical flooding shall be kept to a minimum consistent with efficient arrangements. If any equalizing arrangements are necessary to ensure that the angle of heel in the final stage of flooding does not exceed the limits specified in sub-paragraphs (i)(2) and (3) of this paragraph, these arrangements shall, where practicable, be self-acting. However, if controls are necessary, they shall be operable from above the highest relevant bulkhead deck. All such arrangements shall be acceptable to the Administration.

(iii) The Administration shall be satisfied that stability prior to equalization is sufficient. However, in no case shall the maximum heel before equalization exceed 20 deg. nor shall it result in progressive flooding. Additionally, the time for equalization of cross-connected

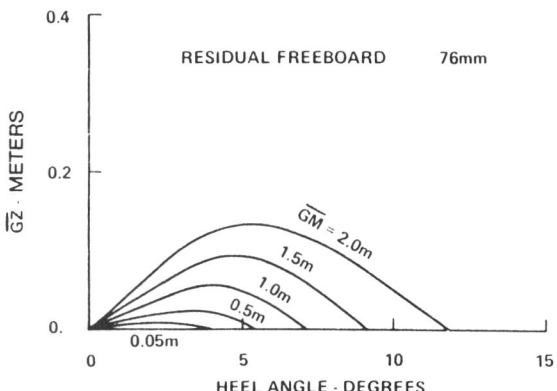

Fig. 49 $\overline{GZ}$ curves for a 100-m vehicle ferry

termediate draught. However, in all cases, where there are vertical discontinuities in permeabilities or in free surfaces which may result in discontinuities in the necessary intact metacentric height, damage stability calculations shall be made for the corresponding draughts in order to define such discontinuities."

It will be seen from the above, particularly from the formulas given in subparagraph (c)(i)(1) that while 0.05 m (2 in.) of residual $\overline{GM}$ is permitted in the damaged condition, such a minimum $\overline{GM}$ must be accomplished by a high freeboard. For example, for a ship with a beam of 18.3m (60 ft), a $\overline{GM}$ of 0.05m (2 in.) requires a freeboard of 5.5m (18 ft). Increasing the $\overline{GM}$ to 0.30m (12 in.), permits the freeboard to be reduced to 0.9m (2.95 ft).

While the traditional regulations permit a residual $\overline{GM}$ of 0.05m (2 in.) with a freeboard after flooding of 0.075m (3 in.), such a condition has virtually no survival probability. This is illustrated in Fig. 49.

Fig. 50 shows clearly the increase in initial $\overline{GM}$ required for a given ship by the provision in these regulations for increased residual $\overline{GM}$ after damage. The left hand curve in the figure shows the $\overline{GM}$ required for compartments 8 and 9 to comply with a two-compartment standard of subdivision under the existing Coast Guard rules. The curves on the right for compartments 10 and 11, and compartment 8, are the required $\overline{GM}$ under the equivalent regulations.

The alternate equivalent international passenger

spaces to at least the limits specified in subparagraphs (i)(2) and (3) of this paragraph shall not exceed 10 min.

(iv) The Administration shall be satisfied that the residual stability is sufficient during intermediate flooding and that progressive flooding will not take place. Calculations relative thereto shall be in accordance with the provisions of subparagraph (b)(iii) of this Regulation, respecting the assumed extent of damage and resulting extent of flooding. Heel during intermediate flooding due either to negative metacentric height alone or in combination with unsymmetrical flooding shall not exceed 20 deg.

(b) Damage stability calculations performed in compliance with this Regulation shall be such as to take account of the form and the design characteristics of the ship and the arrangements, configuration and probable contents of the compartments considered to be flooded. In making calculations for heel prior to equalization and for equalization time, the flooding of that portion of the ship opened to the sea shall be assumed to be completed prior to commencement of equalization. For each initial draught condition, the ship shall be at the most unfavorable intact service trim anticipated at that draught having regard to the influence of the trim on the freeboard in the flooded condition.

(e) The intact metacentric height, and corresponding vertical center of gravity, necessary to provide compliance with the requirements specified in paragraphs (b) and (c) of this Regulation shall be determined for the operating range of draughts between d_s and d_o. If $(d_s - d_o)$ does not exceed 0.1 d_s, damage stability calculations may be made only for d_s and d_o, and the intermediate values may be obtained by linear interpolation. If $(d_s - d_o)$ exceeds 0.1 d_s, damage stability calculations shall also be made for at least one additional in-

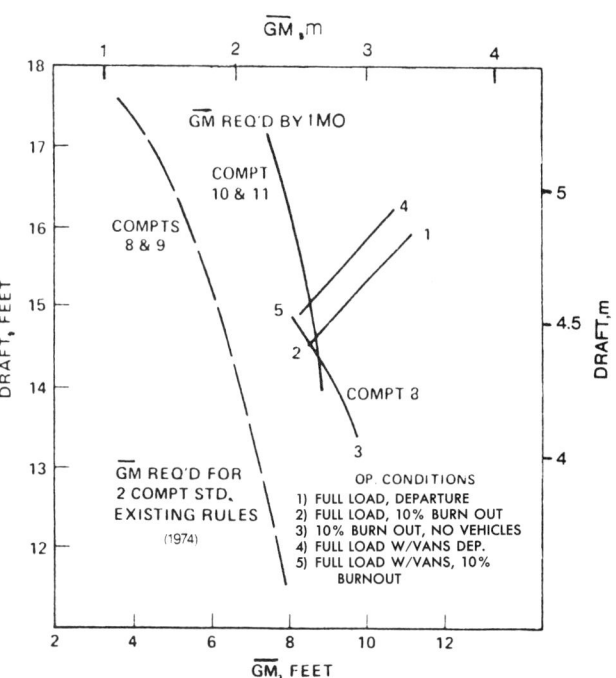

Fig. 50 Initial $\overline{GM}$ requirements for 128m (420-ft.) passenger/vehicle ferry and available $\overline{GM}$s

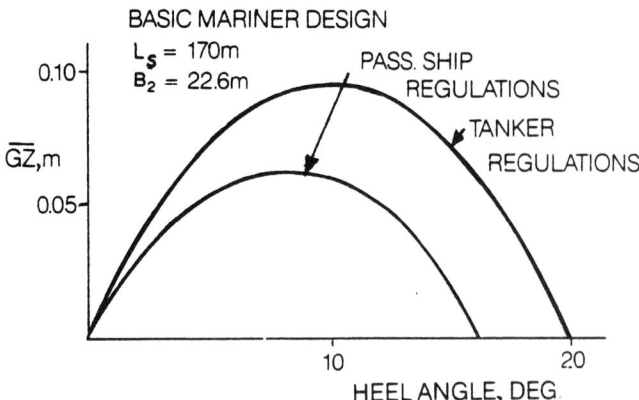

Fig. 51 Comparison of residual $\overline{GZ}$ curves required to meet survival criteria for tankers and for passenger ships under the new equivalent regulations

ship regulations specify only required residual freeboard and $\overline{GM}$ versus beam. This was done advisedly and because model tests both here and abroad indicated that capsizing of a damaged ship lying in a seaway is a quasi-static phenomenon, at least for passenger type ships. For such ships it appeared that freeboard, $\overline{GM}$ and beam, taken together, provided at least as good an index of survival as any simple righting arm and amplitude standard (Robertson, et al, 1974).

Fig. 51 shows that, as applied to the basic C4-5-1a *Mariner* design hull form, the new equivalent passenger ship regulations result in lesser righting arms than called for by the tanker regulations. It thus appears that the latter are more conservative. However, a much more extensive investigation and comparison would probably be necessary to be conclusive.

Subparagraphs (b)(i) and (b)(ii) above require all ships to meet a one-compartment standard, but when the number of passengers exceeds 1200, a two-compartment standard is required throughout. Between 600 and 1200 passengers, a portion of the ship, starting at the forward end is also required to meet a two-compartment standard. Fig. 52 illustrates this requirement.

8.8 Attained Subdivision Index. IMO Regulation 6 calls for the following:

(a) (i)

In addition to complying with Regulation 5 [Section 8.7] the attained Subdivision Index A shall be determined for the ship by formula (II):

$$A = \Sigma\, aps \quad\quad\quad\quad\quad\quad\quad (II)$$

Where:

a accounts for the probability of damage as related to the position of the compartment in the ship's length,

p evaluates the effect of the variation in longitudinal extent of damage on the probability that only the compartment or group of compartments under consideration may be flooded, and

s evaluates the effect of freeboard, stability and heel in the final flooded condition for the compartment or group of compartments under consideration.

(ii) The summation indicated by formula (II) is taken over the ship's length for each compartment taken singly. To the extent that the related buoyancy and stability in the final condition of flooding are such that s is more than zero, the summation is also taken for all possible pairs of adjacent compartments, and may be taken for all possible groups of a higher number of adjacent compartment if it is found that such inclusion contributes to the value of the attained Subdivision Index A.

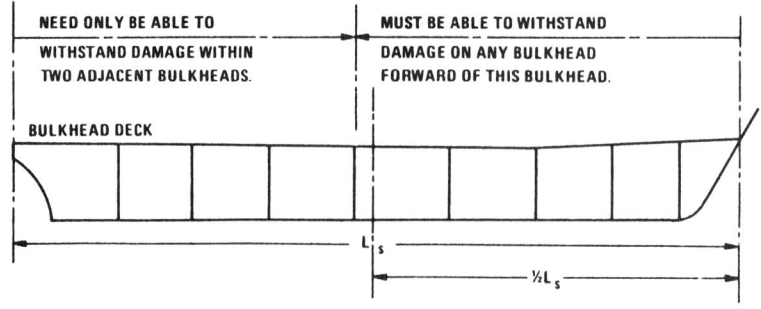

TWO COMPARTMENT STANDARD IS REQUIRED FOR A PROPORTION OF THE SHIP LENGTH ABAFT THE FORWARD TERMINAL, GIVEN BY

$$\left(\frac{N}{600}-1\right)L_s = \frac{L_s}{2} \quad \text{IN THIS CASE.}$$

Fig. 52 Illustration of the application of Subparagraph 5(b)(ii) to a ship with $N = 900$

(iii) Wherever wing compartments are fitted and where the assumed damage used in the damage stability calculations according to Regulation 5 forming the basis for the s calculation does not result in flooding of the associated inboard spaces, p shall be multiplied by r as determined in Regulation 7(b).

IMO Regulation 6 continues by giving details of formulas for determining the above factors for each compartment and group of compartments:

a given by Formula III
p given by Formulas IV-VII
s given by Formulas VIII-IX

A detailed explanation of the meaning of the above factors a, p and s is given by Robertson, et al. (1974). Formula III for a given in the Regulation 6 corresponds to an assumed probability density function of longitudinal location of damage that has a constant value of 1.2 throughout the forward half, length and then decreases linearly to 0.4 at the extreme stern. This assumed density function conforms approximately to a histogram of damage location established from casualty records. Formula IV for p corresponds approximately to a density function of damage length which varies linearly from a maximum value at zero damage length to zero at a damage length of $0.24L_s$, but not more than 48 m (157 ft).

Finally, formula VIII for s is based on an empirical expression of the form,

$$s_i = k \left[\frac{F_e \overline{GM_e}}{B_2} \right]^{0.5}$$

where F_e is effective freeboard flooded and GM_e is effective metacentric height flooded. In the notation of the formula,

$$\frac{F_e}{\overline{GM_e}} = \frac{F_1}{\overline{GM_R}} - \frac{(B_2/2)}{\overline{MM_s}} \tan \theta$$

Although no simple formula applied to different types of ship could be expected to estimate closely the actual probability of survival, the accepted formula was found to give reasonable trends in comparison with model test results.

See IMO Regulation 6 for details of the formulas and their use in calculating the factors discussed above.

IMO Regulation 7 gives detailed requirements applicable to combined longitudinal and transverse subdivision.

After the basic IMO regulations were issued it was felt that certain additional explanatory notes were necessary. Accordingly, IMO Resolution A 265 (VIII), contained in C.G. Commandant's International Technical Series (1974), has attached to it a document, MSC/CIRC 153, that gives explanatory notes to the regulations. Part III of these notes titled "Guidance for Assembling Input Data and Structuring Output Data" contains valuable information which should be studied by anyone performing these calculations. Appendix II to the notes, titled "Combined Transverse and Longitudinal Subdivision," gives detailed information on nomenclature and methods of calculating values of p and a for combined longitudinal and transverse subdivision. Methods of handling recesses are also discussed in Appendix II.

In case of an actual application of the Alternate Equivalent Passenger Vessel Regulations the designer is advised to refer to the current IMO publications for up-to-date guidance information on the practical application of the regulations, as well as for possible changes and extensions to other ship types.

REFERENCES

American Bureau of Shipping (1986), *Rules for Building and Classing Mobile Offshore Drilling Vessels.*

Bird, H. and Brown, R.P. (1973), "Damage Stability Model Experiments," *Trans.* RINA.

Coast Guard Commandant's International Technical Series (1974), Vol. IV, "Regulations on Subdivision and Stability of Passenger Ships as Equivalent to Part B of Chapt. II of the INT. CONV. FOR THE SAFETY OF LIFE AT SEA, 1960, "USCG CITS-74-1-1. (Available from NTIS). See also CFR, 1983, Vol. 48, "Subdivision and Stability Regulations," D.O.T., Coast Guard.

Code of Federal Regulations (CFR), Volume 46, various sections and dates, as indicated.

Comstock, J.P. and Robertson, J.B., Jr. (1961), "Survival of Collision Damage Versus The 1960 Convention on Safety of Life at Sea," *Trans.* SNAME.

Freudenthal, A.M., and Gaither, W.S. (1969), "Probabilistic Approach to Economic Design of Maritime Structures," XXIInd International Navigation Congress, Section 11, Subject 5, Paris.

Middleton, E.H., and Numata, E. (1970), "Tests of a Damaged Stability Model in Waves," *Trans.* SNAME Spring Meeting.

Muckle, W. (1963), "The Determination of the End Drafts of a Ship After Damage," *The Shipbuilder and Marine Engine Builder*, April.

NAVSEA (1975), Design Data Sheet, "Stability and Buoyancy of U.S. Naval Surface Ships," DDS 079-1, Dept. of the Navy, Washington, D.C.

NAVSEA (1976), "Ship Hull Characteristics Program-SHCP," *Users Manual*, CASDAC No. 231072, Dept. of the Navy, Washington, D.C.

Prohaska, C.W. (1961), "Results of Some Systematic Stability Calculations," *Trans.* IESS.

Richmond, A.C. (1960), "The International Conference on Safety of Life at Sea," *Trans.* SNAME.

Robertson, J.B., Nickum, G.C., Price, R.I., Middleton, E.H. (1974), "The New Equivalent International Regulations on Subdivision and Stability of Passenger Ships," *Trans.* SNAME.

Russo, V.L., and Robertson, J.B. 1950, "Standards for Stability of Ships in Damaged Condition," *Trans.* SNAME, vol. 58.

Shirokauer, F. (1928), "A Simplified Method for Exact Arithmetical Determination of Bulkhead Curves," *Schiffbau*, July (translated in *The Shipbuilding and Shipping Record*, September 6).

Solda, G. (1961), "Equalization of Unsymmetrical Flooding," *Trans.* RINA.

Stahlschmidt, E. (1972), "Modellversuche zur Untersuchung der Kentersicherheit lecker Fahrgastschiffe in regelmässigen and unregelmässigen Wellen," *Schiff und Hafen*, vol. 11.

Tagg, Robert D. (1982), "Damage Survivability of Cargo Ships," SNAME, Northern Calif. Section, March 11.

Tawresey, J.G. (1929), "The International Conference of 1929, and the New Convention for Safety of Life at Sea," *Trans.* SNAME.

Wendel, K. (1960), "Die Wahrscheinlichkeit des Ueberstehens von Verletzungen," *Schiffstechnik*.

West European Conference on Marine Technology (1977), *Safety at Sea*, London.

CHAPTER IV

J. R. Paulling | **Strength of Ships**

Section 1
Introduction

1.1 Nature of Ship Structures. The size and principal characteristics of a new ship are determined primarily by its mission or intended service. In addition to basic functional considerations there are requirements such as stability, low resistance and high propulsive efficiency, and navigational limitations on draft or beam, all of which influence the choice of dimensions and form. The ship's structure must be designed, within these and other basic constraints, to sustain all of the loads expected to arise in its seagoing environment. As a result, a ship's structure possesses certain distinctive features not found in other man-made structures.

Among the most important distinguishing characteristics of ship structures are the size, complexity, and multiplicity of function of structural components, the random or probabilistic nature of the loads imposed and the uncertainties inherent in our ability to predict the response of the structure to those loads. In contrast to land-based structures, the ship does not rest on a fixed foundation but derives its entire support from buoyant pressures exerted by a dynamic and ever-changing fluid environment.

The methods of analysis employed by the naval architect in designing and evaluating the structure of a ship must be selected with the above characteristics in mind. During the past twenty years ship structural design and analysis have undergone far-reaching changes towards more rationally founded practices, and the development of readily available computer-based analytical tools has relieved the naval architect of much of the routine computational effort formerly involved in the analysis of a ship's structural performance. Nevertheless, many aspects of ship structures are not completely amenable to pure analytical treatment, and consequently the design of the structure continues to involve a judicious and imaginative blend of theory and experience.

This chapter will deal in detail with the loads acting on a ship's hull, techniques for analyzing the response of its structure to these loads and both current and evolving new methods of establishing criteria of acceptable structural design. A detailed description of ship structures and a discussion of the practical aspects of the structural design of ships, as they are influenced by the blending of experience and analysis embodied in classification society rules, is given in Chapters VI and VII of Taggart (1980).[1] This work should be treated as a companion piece to this chapter.

In order to aid in understanding the nature of the behavior of ship structures, further details of some of their most important distinguishing characteristics will now be given. In some cases, it is helpful to compare the ship and its structure with other man-made structures and systems.

1.2 Size and Complexity of Ships. Ships are the largest mobile structures built by man, and both their size and the requirement for mobility exert strong influences on the structural arrangement and design. Today there are in operation large oil tankers having fully loaded displacements exceeding 600,000 long tons and dimensions of 400m (1312 ft) in length, 63m (207 ft) in breadth, 35.9m (118 ft) in depth, with a loaded draft of 28.5m (93 ft). They are among the most complex of structures and this is due in part to their mobility. Good resistance and propulsive characteristics dictate that the external surface of the hull, or shell, must be a complex three-dimensional curved surface and, since the shell plating is one of the major strength members, the structural configuration may not always be chosen solely on the basis of optimum structural performance. Furthermore, the structural behavior of the many geometrically complex members that constitute a ship's hull is difficult to analyze, and the construction of the vessel may be complicated because there are few members of simple or repetitive shape.

---

[1] Complete references are listed at end of chapter.

1.3 Multipurpose Function of Ship Structural Components. In contrast to many land structures, the structural components of a ship are frequently designed to perform a multiplicity of functions in addition to that of providing the structural integrity of the ship. The shell plating, for instance, serves not only as the principal strength member, but also as a watertight envelope of the ship, having a shape that provides adequate stability against capsizing, low resistance to forward motion, acceptable controllability and good propulsive characteristics.

Internally, many strength members serve dual functions. For example, bulkheads that contribute substantially to the strength of the hull may also serve as liquid-tight boundaries of internal compartments. Their locations are dictated by the required tank volume or subdivision requirements (See Chapter III). The configuration of structural decks is usually governed by the arrangement of internal spaces, but they may be called upon to resist local distributed and concentrated loads, as well as contributing to longitudinal and transverse strength.

Whereas in many instances structural efficiency alone might call for beams, columns, or trusses, alternative functions will normally require plate or sheet-type members, arranged in combination with a system of stiffeners, in order to provide resistance to multiple load components, some in the plane of the plate and others normal to it. An important characteristic of ship structure is its composition of numerous stiffened plate panels, some plane and some curved, which make up the side and bottom shell, the decks and the bulkheads. Much of the effort expended in ship structural analysis is, therefore, concerned with predicting the performance of individual stiffened panels and the interactions between adjoining panels.

1.4 Variability of Ship Structural Loads. The loads that the ship structure must be designed to withstand have many sources. There are static components, which consist principally of the weight and buoyancy of the ship in calm water. There are dynamic components caused by wave-induced motions of the water around the ship and the resulting motions of the ship itself. Other dynamic loads, usually of higher frequency than the simple wave-induced loads, are caused by slamming or springing in waves and by the propellers or propelling machinery. These sometimes cause vibration of parts or the whole of the ship. Finally, there may be loads that originate in a specialized aspect of the ship's function, such as ice breaking, or in the cargo it carries, as in the case of thermally-induced loads associated with heated or refrigerated cargos.

An important characteristic of all of these load components is their variability with time. Even the static weight and buoyancy vary from voyage to voyage, and within a voyage, depending upon the amount and distribution of cargo and consumables carried. In order to design the structure of the ship for a useful life of twenty years or more, this time dependence of the loading must be taken into consideration.

The loads imposed by the sea, like the sea itself, are random in nature, and can therefore be expressed only in probabilistic terms. As a result, it is generally impossible to determine with absolute certainty a single value for the maximum loading that the ship structure will be called upon to withstand. Instead, it is necessary to use a probabilistic representation, in which a series of loads of ascending severity is described, each having a probability corresponding to the expected frequency of its occurrence during the ship's lifetime. When conventional design methods are used, a design load may then be chosen as the one having an acceptably low probability of occurrence (Section 2.3). In more rigorous reliability methods (Section 5) the load data in probability format can be used directly.

1.5 Probabilistic Nature of Structural Behavior. As a consequence of the complexity of the structure and the limitations of our analysis capabilities, it is seldom possible to achieve absolute accuracy in predicting the response of the structure even if the loading were known exactly. As in the case of the uncertainties present in the predictions of structural loading, it is necessary for the designer to take into consideration the probable extent and consequences of uncertainties in the structural response prediction when making a judgment concerning the overall acceptability of the structure. One of the most important tasks facing the engineer is to choose the proper balance between the acceptable level of uncertainty in his structural response predictions and the time and effort that must be expended to achieve a higher level of accuracy. The existence of this uncertainty, then, is acknowledged and must be allowed for in design.

In ship structural performance prediction there are at least three sources of uncertainty. First, the designer's stress analysis is usually carried out on an idealization of the real structure. For example, beam theory may be used to predict the stress distribution in part or the whole of the hull girder, even though it is known that the ship geometry may not follow exactly the assumptions of beam theory.

Second, the actual properties of the materials of construction may not be exactly the same as those assumed by the designer. Steel plates and shapes, as delivered from the mill, do not agree precisely with the nominal dimensions assumed in design. Similarly the chemical and physical properties of the materials can vary within certain tolerance limits. The rules of classification societies specify both physical and chemical standards for various classes of shipbuilding materials, either in the form of minimum standards or a range of acceptable values. The materials actually built into the ship should have properties that lie within these specified limits, but the exact values depend on quality control in the manufacturing process and are not known in advance to the designer. Furthermore, there will inevitably be some degradation of material physical properties caused by corrosion, for example, over the lifetime of the ship.

Third, the integrity of ship construction contains a significant element of skill and workmanship. The designer, in performing a stress analysis, may assume perfect alignment and fitup of load-carrying members and perfectly executed welds. This ideal may be approached by the use of a construction system involving highly skilled workmen, and high standards of inspection and quality control, but an absolutely flawless welded joint, or a plate formed precisely to the intended shape and fabricated with no weld-induced distortion or joint misalignment, is a goal to be striven for, but never attained in practice.

It will be obvious that the uncertainties involved in the determination of both the loads and the structural responses to these loads make it difficult to establish criteria for acceptable ship structures. In the past allowable stress levels or safety factors used by the designer provided a means, based upon past experience with similar structures, of allowing for these uncertainties. In recent years, reliability principles have been applied, using probability theory and statistics, to obtain a more rational basis for design criteria. In the reliability approach to design, structural response data, as well as load data, can be expressed and used in probability format. These principles are discussed in Section 5.

1.6 Modes of Ship Structural Failure. Avoidance of structural failure is an overriding goal of all structural designers, and in order to achieve this goal it is necessary for the naval architect to be aware of the possible modes of failure and the methods of predicting their occurrence. The types of failure that may occur in ship structures are generally those that are characteristic of structures made up of stiffened plate panels assembled through the use of welding to form monolithic structures with great redundancy, i.e., having many alternative paths for lines of stress.

It should be noted that structural failure may occur in different degrees of severity. At the low end of the failure scale, there may be small cracks or deformations in minor structural members that do not jeopardize the basic ability of the structure to perform its function. Such minor failures may have only esthetic consequences. At the other end of the scale is total catastrophic collapse of the structure resulting in the loss of the ship. Between these extremes are several different modes of failure that may reduce the load-carrying ability of individual members or parts of the structure but, as a result of the highly redundant nature of the ship structure, do not lead to total collapse. Such failures are normally detected and repaired before their number and extent grow to the point of endangering the ship.

Four principal mechanisms are recognized as causing most of the cases of ship structural failure, aside from collision or grounding. These modes of failure are as follows:

- Excessive tensile or compressive yield.
- Buckling due to compressive or shear instability.
- Fatigue cracking.
- Brittle fracture.

The first three modes of failure are discussed in more detail in Section 4. The last of them, brittle fracture, was found to play a major role in the failure of many of the emergency cargo ships built during World War II. The causes of these failures ultimately were traced to a combination of factors associated with the relatively new techniques of welded construction employed in building the ships. The solution to the problem was obtained through the development of design details that avoided the occurrence of notches and other stress concentrations, together with the selection of steels having a high degree of resistance to the initiation and propagation of cracks, particularly at low temperatures. Features termed *crack arrestors* were incorporated in order to provide *fail-safe* design by limiting the extent of propagation of any cracks that might actually have occurred.

Since the control of brittle fracture is accomplished principally through detail design and material selection, it is considered only briefly in this chapter. Information on these topics may be found in Taggart (1980), Chapters VII and VIII.

1.7 Design Philosophy and Procedure. The development of completely rational structural design procedures is being pursued in several disciplines, including civil, aeronautical and mechanical engineering, as well as in naval architecture. Using such procedures it should be possible first to formulate a set of requirements or criteria to be met by the structure, then, through the application of fundamental reasoning and mathematical analysis, augmented by the introduction of certain empirical information, to arrive at a structural configuration and a set of scantlings that simultaneously meet all of the criteria. Although this ideal has not yet been attained, steady progress is being made in that direction.

The original set of requirements imposed upon the ship will include the functional requirements of the owner and, in addition, institutional requirements such as those established by governmental and other regulatory bodies concerned with safety of life, navigation, pollution prevention, tonnage admeasurement and labor standards. The methods of selecting the overall dimensions and arrangement of the ship to meet these requirements have been dealt with in Taggart (1980). Thus, in designing the principal members of the ship structure, it may be assumed that the overall dimensions of the ship and the subdivision of its internal volume by bulkheads, decks and tank boundaries has already been determined to meet these various requirements. The problem of structural design then consists of the selection of material types, frame spacing, frame and stiffener sizes, and plate thicknesses that, when combined in this geometric configuration, will enable the ship to perform its function efficiently for its expected operational lifetime.

At this point, in selecting the criteria to be satisfied by the structural components of the ship, the designer must rely on either empirical criteria, including factors of safety and/or allowable stresses, or on the use of reliability principles discussed in Section 5. The term *synthesis*, which is defined as the putting together of parts or elements so as to form a whole, is often applied to the process of ship structural design.

An additional element, however, is needed in order to complete the design synthesis: finding the optimal combination of the various elements. By reason of the complexity of ship structures, as well as the probabilistic nature of available information needed for certain vital inputs to the design process, it is usually impossible to achieve an optimum solution in a single set of calculations. Instead, some sort of iterative procedure must be employed. The traditional method of ship structural design, involving the extrapolation of previous experience, can even be thought of as an iterative process in which the construction and operational experience of previous ships form essential steps. In each new design the naval architect takes into consideration this past experience and modifies the new design intuitively in such a way as to achieve an improved configuration. The successful designer has been one whose insight, understanding and memory, along with skill in manual methods of structural analysis, resulted in consistent improvement over previous designs in successive ships.

Much of structural design, even when the most advanced methods are used, consists of a stepwise process in which the designer develops a structural configuration on the basis of experience, intuition and imagination, then performs an analysis of that structure in order to evaluate its performance. If necessary, the scantlings are revised until the design criteria are met. The resulting configuration is then modified in some way that is expected to lead to an improvement in performance or cost, and the analysis is then repeated to insure that the improved configuration again meets the design criteria. Thus a key element in structural design is the process of analyzing the response of an assumed structure. The process of finding by synthesis a structural configuration having the desired performance is the inverse of analysis, and is not nearly so straightforward, especially in the case of complex structures. As a consequence, it is only after completing several satisfactory design syntheses that the process of optimization can take place.

In summary, four key steps may be identified to characterize the structural design process, whether it be intuitive or mathematically rigorous:

(a) Development of the initial configuration and scantlings.

(b) Analysis of the performance of the assumed design.

(c) Comparison with performance criteria.

(d) Redesign the structure by changing both the configuration and scantlings in such a way as to effect an improvement.

(e) Repeat the above as necessary to approach an optimum.

Formally, the final optimization step consists of a search for the best attainable (usually minimum) value of some quantity such as structural weight, construction cost, overall required freight rate for the ship in its intended service (see Taggart, 1980) or the so-called *total expected cost* of the structure. The last of these quantities, as proposed by Freudenthal (1969), consists of the sum of the initial cost of the ship (or other marine structure), the anticipated total cost of complete structural failure multiplied by its probability, and a summation of lifetime costs of repair of minor structural damages (See also Lewis, et al, 1973).

The search is performed in the presence of constraints which, in their most elementary form, consist of the requirement that each member of the structure not fail under the expected loadings (steps b and c above). Such an optimization procedure forms the basis for a sound, economical design, whether it be carried out automatically, using one of the formal mathematical optimization schemes, or manually (with or without machine computational assistance for some parts of the process).

Section 2
Ship Structural Loads

2.1 Classification of Loads. It is convenient to divide the loads acting on the ship structure into four categories as follows, where the categories are based partly upon the nature of the load and partly upon the nature of the ship's response:

(a) Static loads are loads that change only when the total weight of the ship changes, as a result of loading or discharge of cargo, consumption of fuel, or modification to the ship itself.

1. Weight of the ship and its contents.
2. Static buoyancy of the ship at rest or moving.
3. Thermal loads resulting from nonlinear temperature gradients within the hull.
4. Concentrated loads caused by drydocking and

grounding.

(b) Low-frequency dynamic loads are loads that vary in time with periods ranging from a few seconds to several minutes, and therefore occur at frequencies that are sufficiently low compared to the frequencies of vibratory response of the hull and its parts that there is no appreciable resonant amplification of the stresses induced in the structure. The loads are called dynamic because they originate mainly in the action of the waves through which the ship moves and are, therefore, always changing with time. They may be broken down into the following components:

1. Wave-induced hull pressure variations.
2. Hull pressure variations caused by oscillatory ship motions.
3. Inertial reactions resulting from the acceleration of the mass of the ship and its contents.

(c) High-frequency dynamic loads are time-varying loads of sufficiently high frequency that they may induce vibratory response of the ship structure. Some of the exciting loads may be quite small in magnitude but, as a result of resonant amplification, they can give rise to large stresses and deflections. Examples of such dynamic loads are the following:

1. Hydrodynamic loads induced by propulsive devices on hull or appendages.
2. Loads imparted to the hull by reciprocating or unbalanced rotating machinery.
3. Hydroelastic loads resulting from interaction of appendages with the flow past the ship.
4. Wave-induced loads due primarily to short waves whose frequency of encounter overlaps the lower natural frequencies of hull vibration and which, therefore, may excite appreciable resonant response, termed *springing*.

(d) Impact loads are loads resulting from slamming or wave impact on the forefoot, bow flare and other parts of the hull structure, including the effects of green water on deck. In a naval ship, weapon effects constitute a very important category of impact loads. Impact loads may induce transient hull vibration that is termed *whipping*.

The most important classes of loads are the static loads resulting from the ship's weight and buoyancy, categories (a) 1 and 2, and the low-frequency dynamic loads, categories (b) 1, 2 and 3. In the following sections, attention will be devoted mainly to the methods currently in use for the determination of these loads, with brief discussion of impact loads, (d), and springing loads, (c) 4, which are usually found to be important only in very long flexible ships such as the U.S. and Canadian Great Lakes iron ore carriers. A discussion of thermal loads, (a) 3, may be found in Taggart (1980), Chapter VI. The determination of vibratory forces and responses, (c) 1, 2 and 3, is discussed generally in Vol. II, Chapter VI of this edition of PNA.

In addition to the above categories, there may be specialized operational loads, which part or all of the structure may be called upon to withstand, and which may be the dominant loads for some ships. These loads, which may be either static or dynamic, are not considered here. Some examples are:

- Ice loads in the case of a vessel intended for icebreaking or arctic navigation.
- Loads caused by impact with other vessels, piers, or other obstacles, as in the case of tugs and barges.
- Impact of cargo handling equipment, such as grabs or clamshells used in unloading certain bulk commodities.
- Structural thermal loads imposed by special cargo carried at nonambient temperature or pressure.
- Sloshing and impact loads on internal structure caused by movement of liquids in tanks.
- Landing of aircraft or helicopters.
- Accidental loads caused by collision or grounding.

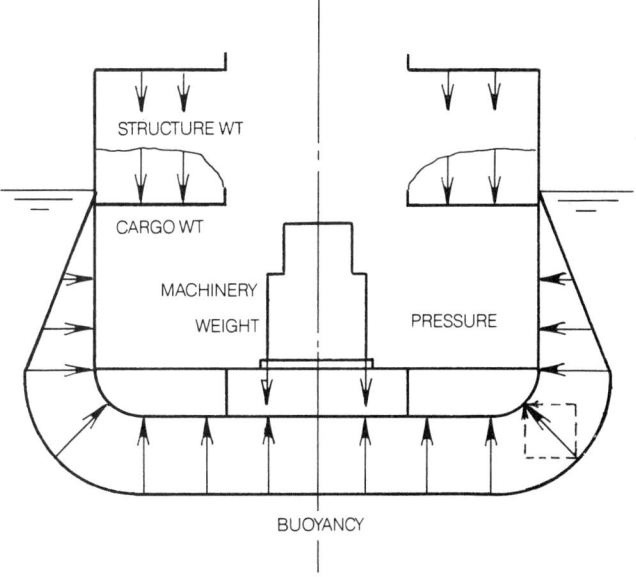

Fig. 1 Static load components on hull

As may be seen from the brief descriptions given above, some of these loads may be of importance in all ships and other loads may be encountered only in specialized ships or circumstances.

2.2 Static Loading on a Ship Afloat in Still Water. The static loads acting on a ship afloat in still water consist of two parts: buoyancy forces and gravity forces, or weights. The buoyancy is the resultant of the hydrostatic pressure distribution over the immersed external area of the ship. This pressure is a surface force whose direction is everywhere normal to the hull. The buoyant force is, however, the resultant perpendicular to the water surface and directed upwards. The weights are body forces distributed throughout the ship and its contents, and the direction of the weight forces is always vertically downward.

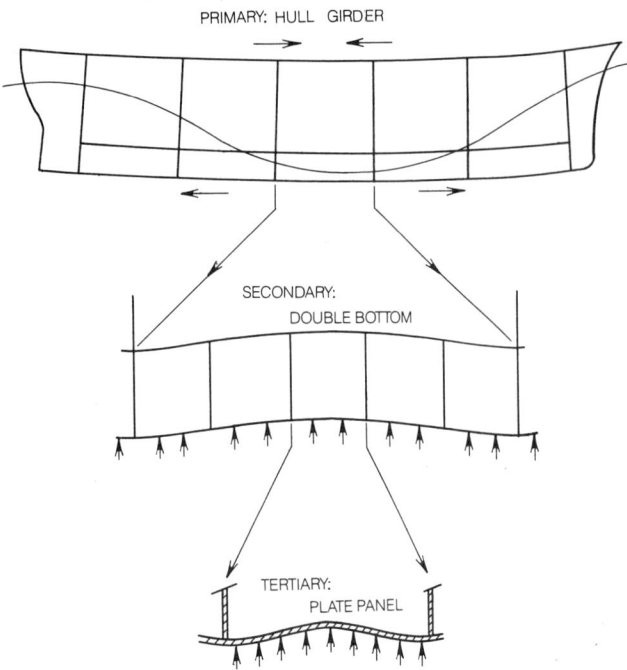

Fig. 2 Primary, secondary and tertiary structure

ered when designing the bulkheads and other tank boundary members.

The geometrical arrangement and resulting stress or deflection response patterns of typical ship structures are such that it is usually convenient to divide the structure and the associated response into three components, which are labelled *primary*, *secondary* and *tertiary*. These are illustrated in Fig. 2 and described as follows:

Primary response is the response of the entire hull, when bending and twisting as a beam, under the external longitudinal distribution of vertical, lateral and twisting loads.

Secondary response comprises the stress and deflection of a single panel of stiffened plating, e.g., the panel of bottom structure contained between two adjacent transverse bulkheads. The loading of the panel is normal to its plane and the boundaries of the secondary panel are usually formed by other secondary panels (side shell and bulkheads).

Tertiary response describes the out-of-plane deflection and associated stress of an individual panel of plating. The loading is normal to the panel, and its boundaries are formed by the stiffeners of the secondary panel of which it is a part.

From the above description it is seen that it is sometimes necessary to know the localized distribution of the loads, and in other cases, depending upon the structural response being sought, to know the distribution of the resultants of the local loads—for example, the load per unit length for the entire hull. The primary response analysis is carried out by hypothesizing that the entire hull of a ship behaves like a beam whose loading is given by the longitudinal distribution of weights and buoyancy over the hull. As in any beam strength computation, it is necessary first to integrate the loads to obtain the longitudinal distribution of the total shear force and then to integrate again to obtain the bending moment. The still-water loads contribute an important part of the total shear and bending moment in most ships, to which wave-induced effects must be added later.

These component force systems are illustrated schematically in Fig. 1.

If we integrate the local buoyant pressures over a unit ship length around a cross section at a given longitudinal position, the resultant is a vertical buoyant force per unit length whose magnitude is given by $\rho g A$, where ρg is the *weight density* of water (ρ is the *mass density*, or mass per unit volume) and A is the immersed sectional area. Similarly, we may add together all of the weights contained in a unit length of the ship at this same section, resulting in a total weight per unit length. The net structural load per unit length is the algebraic sum of the unit buoyancy and unit weight. For convenience, when S.I. units are used in this static case, all loads—both buoyancy and weights—can be expressed in terms of mass. See Chapter I.

The individual loads may have both local and overall structural effects. A very heavy machinery item induces large local loads at its points of attachment to the ship, and its foundations must be designed to distribute these loads evenly into the hull structure. At the same time, the weight of this item contributes to the distribution of shear forces and bending moments acting at all locations along the length of the hull. If a part of the contents of the ship is made up of liquids, e.g., fuel or liquid cargo, there will be hydrostatic pressure forces exerted by such liquids that are normal to the boundary surfaces of the tanks within which they are contained. These internal pressure loads may have important local structural effects and must be consid-

Fig. 3 illustrates a typical longitudinal distribution of weight and buoyancy for a ship afloat in calm water. In the lower part of this figure is plotted a curve (1) of buoyancy force per unit length, which as noted previously is equal to the *weight density*, ρg, of water times the sectional area. The upper curve (2) in this figure shows the longitudinal distribution of the weight force plotted according to a commonly employed convention. In this procedure, the length of the ship is divided into a number of equal station spaces, for example the twenty or so station subdivisions that were used in preparing the lines drawing. All of the weights of hull, equipment and contents lying in the interval between station i and station i + 1 are added together and treated as a single uniformly distributed load over this station interval. This is essentially an accounting process in which every item in the ship—

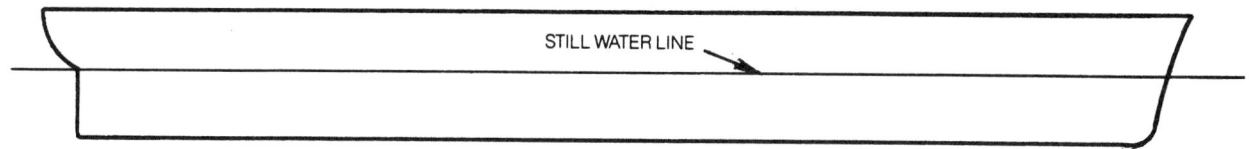

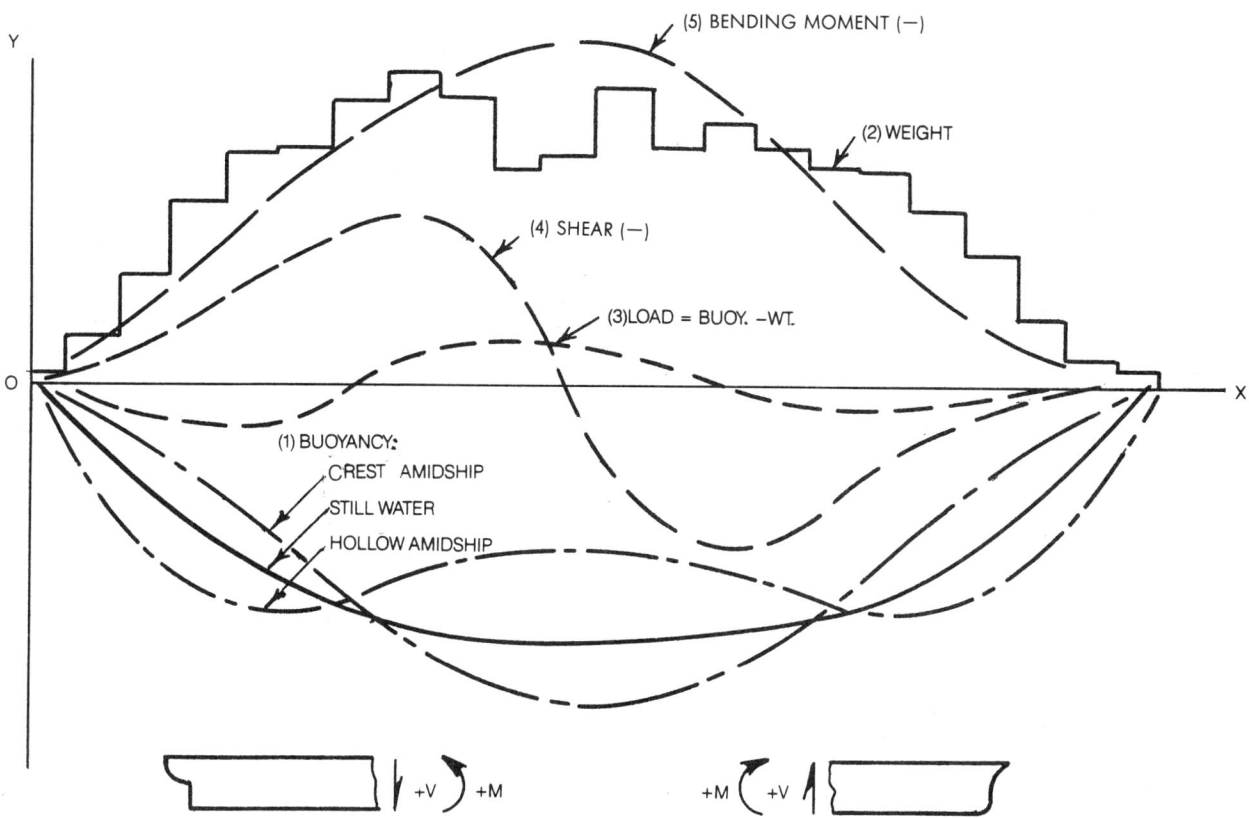

Fig. 3 Static loads, shear and bending moment

for example, hull structure (plating, frames, weld material), outfit (piping, deck covering, cargo gear), propulsion machinery, cargo and so on—is recorded and assigned to a station interval. The procedure must be performed with meticulous care and in great detail in order to assure accuracy. As is the case with most repetitive computations, it lends itself easily to machine solution.

The assumption of a uniform distribution of the sectional weights over the station intervals, which is implied in this step, is only an approximation to the actual weight distribution. Some weight items will occur as nearly concentrated weights in this longitudinal distribution. For example, the weight of a transverse bulkhead will, in reality, be distributed longitudinally over a very short portion of the ship length equal to the thickness of the bulkhead plating. The weights of certain items such as large machinery components (turbines, diesel engines) may be transmitted to the ship structure as point loads at the locations of the foundation bolt-down points. Similarly, cargo containers are usually supported on fittings located under their corners, and their total weight is transmitted to the hull structure as point loads at these locations. The true weight distribution will, therefore, be a much more irregular graph than is shown in Fig. 3, and will consist of some distributed items and some point weights. It may be shown, however, that the integrations that are performed in order to obtain the shear and bending moment distributions from the loads tend to smooth out the effects of these local irregularities. As a consequence, any reasonably accurate loading distribution that maintains the correct magnitude of the force over a local interval that is small compared to the total ship length will generally lead to the correct shear and bending moment distributions within acceptable error limits. However, localized structural effects caused by large point loads of especially heavy items may be analyzed separately and their effects superimposed on the effects of the remaining loads.

Having determined the buoyancy and weight distributions, the net load curve (3) is the difference between the two. This is plotted as the third curve in Fig. 3, with buoyancy positive upwards. The conditions of static equilibrium require that the total weight and buoyancy be equal, and that the center of buoyancy be on the same vertical line as the center of gravity. In terms of the load curve, this requires that the integral of the total load over the ship length and the integral of the longitudinal moment of the load curve each be equal to zero.

As in any beam calculation, the shear force at location x, equal to $V(x)$, is obtained as the integral of the load curve, and plotted as the fourth curve of Fig. 3

$$V(x_1) = \int_0^{x_1} [b(x) - w(x)] \, dx \qquad (1)$$

where

$b(x)$ = buoyancy per unit length,
$w(x)$ = weight per unit length.

The bending moment at location x_1, $M(x_1)$, is the integral of the shear curve, and is plotted as the fifth curve in Fig. 3,

$$M(x_1) = \int_0^{x_1} V(x) \, dx \qquad (2)$$

In the lower parts of Fig. 3 the significance of the shear and bending moment, together with their sign conventions, is shown. If we consider a given longitudinal location, x_1, the shear force is the upward force that the left hand portion of the ship exerts on the portion to the right of this location. Similarly, the bending moment is the resultant moment exerted by the left hand portion on the portion of the ship to the right of location x_1. The conditions of static equilibrium are seen to require that the shear force be equal to zero at both ends of the ship and the bending moment, similarly, must be zero at both ends of the ship.

In the practical execution of the still water loading computation, a general ship hydrostatics digital computer program is today almost invariably employed. Programs, such as the U.S. Navy's SHCP (see Chapter I), and others available through computer service bureaus, contain modules for performing computations of such quantities as hydrostatic properties, static stability, flooding and static hull loading. A common data base is normally employed containing the offsets or other description of the hull geometry, which is required for computing the buoyancy distribution. Supplementary data, including the weight distribution, are then entered in connection with the specific computation of the load, shear and bending moment. The principal task confronting the naval architect lies in preparing and checking the input data, and in evaluating the results of the computation. The importance of complete and accurate input cannot be overemphasized, and it may be readily perceived that the compilation of the complete weight data required for the computation of the shear and bending moment at the final design stage is not a trivial task. This is often incorporated into a computer-based weight control and accounting system.

In concluding this section, it should be observed that the static loading usually must be computed for several different distributions of cargo and other variable weights in order to obtain the extreme values of shear and bending moment, to be combined with other loads upon which to base the design of structural members. Furthermore, it must be borne in mind that the static loading will change during the course of a single voyage as fuel is consumed, ballast is shifted and cargo is loaded and discharged at the ports visited. A time history of the changes in static midship stress during the course of an outbound voyage of a large tanker may be seen in Fig. 4 (Little and Lewis, 1971). Although it also shows stress variations due to waves and thermal effects, the large shifts in the heavy lines are the results of changes in salt water ballast amount and distribution. The recorded variations in still water stress, excluding temperature and wave effects, range from about 27.6 MPa (4 kpsi) tension to 48.3 MPa (7 kpsi) compression.

2.3 Wave-Induced Loads. The principal wave-induced loads are those previously referred to as *low-frequency dynamic loads* or loads involving ship and wave motions that result in negligible dynamic stress amplification. Once these quasi-static loads are determined, the structural response in terms of stress or deflection may be computed by methods of static structural analysis.

At least four procedures of varying degrees of sophistication may be used in estimating the wave-induced loads and their resultant bending moments and shear forces.

(a) Approximate methods.

In the preliminary design process it is often desirable to make an early estimate of the hull structural loading by some approximate method, perhaps before detailed information concerning the weight distribution or hull lines have been developed. Approximate methods are available that include semi-empirical formulations and quasi-static computations.

Earlier texts on naval architecture contain descriptions of a procedure whereby the ship is imagined to be in a state of static equilibrium on either the crest or trough of a wave whose length is equal to the ship length, L, and height is L/20. Using the longitudinal distribution of buoyancy up to such a wave profile and an assumed weight distribution, curves of the longi-

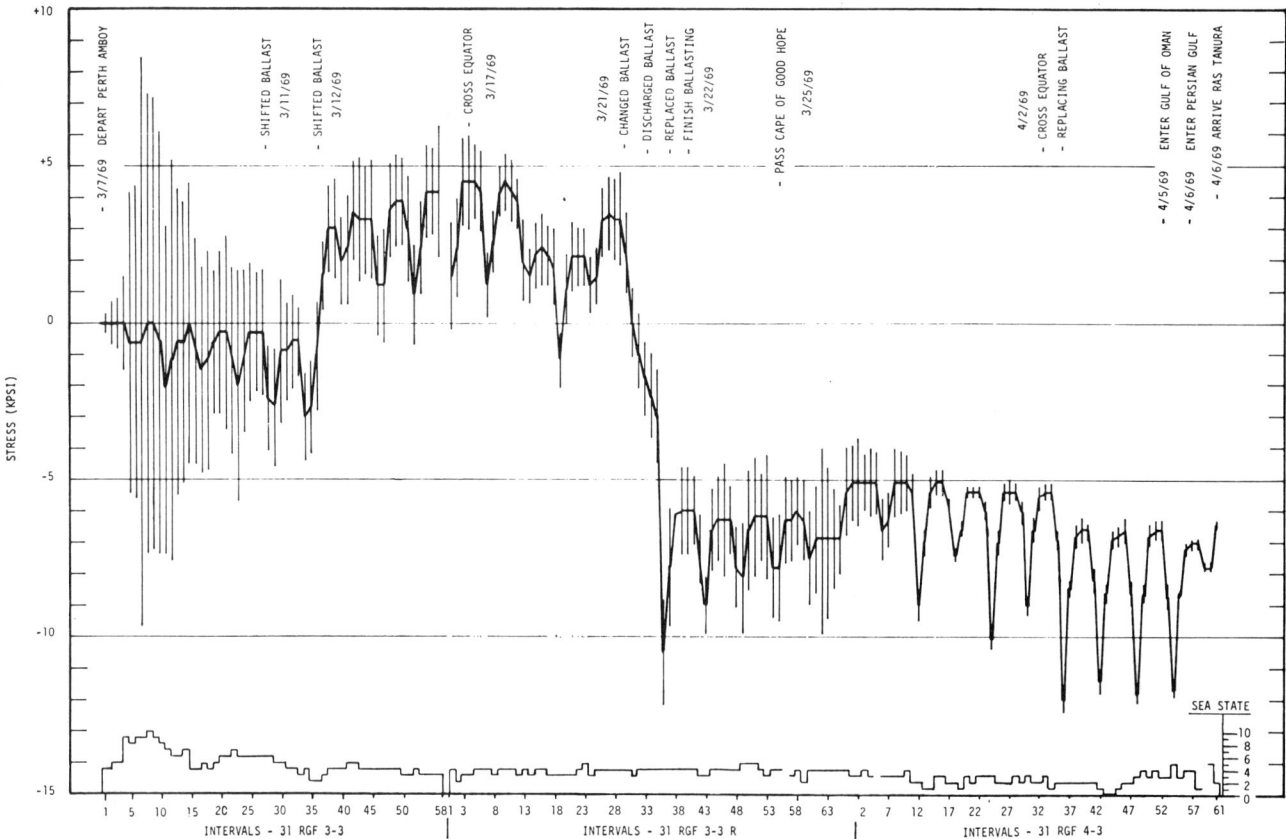

Fig. 4 Typical voyage variation in stresses, R. G. Follis, in ballast

tudinal distribution of shear force and bending moment may be computed, just as in the still water case. Experiments and more exact computational methods have shown that this highly simplified procedure overestimates the actual wave-induced bending moment for any given wave height by a substantial margin as a result of the neglect of dynamic and hydrodynamic effects associated with wave pressures and ship motions. The procedure is of value chiefly when used in comparison with previous design data. Some of the currently available hydrostatics computer programs, such as the previously mentioned U. S. Navy's SHCP, include the static wave bending moment computation as an option.

Most of the classification societies now employ a simplified formulation to be used in estimating the wave-induced bending moment and shear force to be added to the still water moment and shear. However, the *effective wave heights* used in these formulations have been derived from detailed analyses that include the dynamic and hydrodynamic effects, and, therefore, are not subject to the limitations of the static wave computation just described. The *Rules* of the American Bureau of Shipping (1987a) define an extreme wave bending moment which, in conjunction with the still water moment and the rule permissible stress, is intended to be used in determining the required section modulus,

$$M_w = C_2 L^2 B H K_b \qquad (3)$$

where

C_2 and K_b are tabulated coefficients depending on the block coefficient,

L, B are length and breadth of the ship, respectively.

H is a wave height or bending moment coefficient dependent upon L.

Values given in the *Rules* for H show a gradual upward trend with ship length, leveling off to a constant value at 305m (1000 ft).

All the constants in Equation (3) are based upon extensive data obtained for a large number of ships by a combination of computation, model tests and full-scale measurement. The use of a nominal wave height takes into consideration the response of the ship, dependent upon ship size, in combination with the severity of expected waves over the ship's lifetime. It has been shown (Liu, et al, 1981) that the values predicted by the formula are in close agreement with analytic predictions using North Atlantic waves and in agreement with other long-term estimates of the maximum wave bending moment for ships of average proportions and form, having no unusual features of geometry or longitudinal weight distribution. Since it may be evaluated without detailed knowledge of weight distribution or hull geometry, it is useful for preliminary design estimates. When used with suitable allowable stresses (as given in the ABS *Rules*) it provides a satisfactory empirically-based longitudinal strength standard for conventional ships. For further discussion see Taggart (1980).

Details of U.S. Navy standards for longitudinal strength are classified, but a general statement is given by Sikora, Dinsenbacher and Beach (1983), "The primary hull girders of mild steel naval vessels are designed to a stress level of 8.5 t/in^2 (130 MPa) single amplitude by placing the ship on a trochoidal wave of height 1.1 $\sqrt{LBP}$ and length $= LBP$," and then by carrying out a conventional quasistatic calculation. However, in the design of unusual naval craft advanced reliability techniques have been applied, as discussed subsequently.

(b) Strain and/or Pressure Measurements on Actual Ships.

Full-scale measurements obviously cannot be used to obtain specific data for a new ship design. But, although the results apply only to the specific ships studied, they are of great value in testing probability-based prediction methods that are described in Section 2.7. Full-scale measurements suffer from a serious drawback, in addition to expense, and that is the difficulty in accurately measuring the sea environment for correlation with the measured loads. While numerous attempts have been made to develop inexpensive expendable wave buoys, or shipborne wave instruments, a completely satisfactory instrument has not yet been achieved. The principal value of full-scale load response (stress or strain) measurements, therefore, lies in the development of long-term statistical trends of seaway-induced hull loads from measurements carried out over a multi-year period. Since these trends can be related to more general long-term climatological wave data, the problem of wave sensing in the ship's immediate vicinity is of less importance.

Long-term continuous full-scale measurements on ships of various types and sizes have been conducted by several ship classification societies and research organizations around the world, and descriptions of such work may be found in Little and Lewis (1971), Boentgen (1976), Nordenström (1973) and Stambaugh and Wood (1981). These long-term, full-scale measurements are being used to verify theoretical predictions.

(c) Laboratory Measurement of Loads on Models.

In this procedure, a model geometrically and dynamically similar to the ship is equipped with instruments that measure vertical or horizontal shear and bending moment, or torsional moment, amidships and at other sections. This may be accomplished by recording the forces or deflections between several segments produced by transverse cuts through the model. Impact loads can also be determined by recording pressures at several points distributed over the model surface. The experiments are conducted in a towing tank that is equipped to produce either regular or random waves. The most versatile tanks are wide in relation to their length and the model may, therefore, be tested in oblique as well as head or following seas.

Model test results on a model of a T-2 tanker (Numata, 1960) show that the amidship lateral longitudinal bending moment may approach or exceed the magnitude of the vertical longitudinal bending moment when running at oblique heading in regular waves.

The model tests were run at 10 knots vessel speed on courses oblique to waves having an effective length equal to the model length, the wave length being equal to the model length times the cosine of wave-to-course angle. A wave height of 1/48 of the model length was used for all wave lengths to avoid excessive model wetness.

It was found that the lateral bending moments were quite sensitive to change in wave direction and effective wave length. The bending moment increased approximately linearly as the heading varied from 180 to 120 deg. The maximum moments for zero and forward speeds are at a wave direction of approximately 135 deg. The phase lag between the lateral and vertical bending moments was in the region of one quarter cycle.

It should be borne in mind that the effective wave steepness increases in changing from a head sea to an oblique sea. In a head sea with model-length waves having a ratio of 1/48, the steepness changes to 1/24 for effective wave lengths at a 120-deg wave-to-course angle.

Fig. 5 from Numata (1960) shows the relationship between the vertical and lateral longitudinal bending moments for a 120-deg wave-to-course angle, 10-knot vessel speed and a wave-height to model-length ratio of 1/48, corresponding to a full-size wave height of 10.5 ft.

Model tests were also run in irregular waves with the average height of the 10 percent highest waves 23 ft, full size. For the same speed and wave-to-course angle as for regular wave tests, and an apparent wave length of 0.65 times the ship length, Table 1 was prepared. For comparison with the vertical bending moments in regular waves, the values in Fig. 5 are increased in the ratio of the wave heights, or 23/10.5.

Table 1—Bending moments on T-2 tanker from model tests

	Irregular waves average of 10% highest, Ship t-m	Regular waves Ship, t-m
Vertical moment:		
Sag:	23,600	28,500
Hog	26,700	30,400
Lateral moment:		
Weather	24,500	
Leeward	23,600	

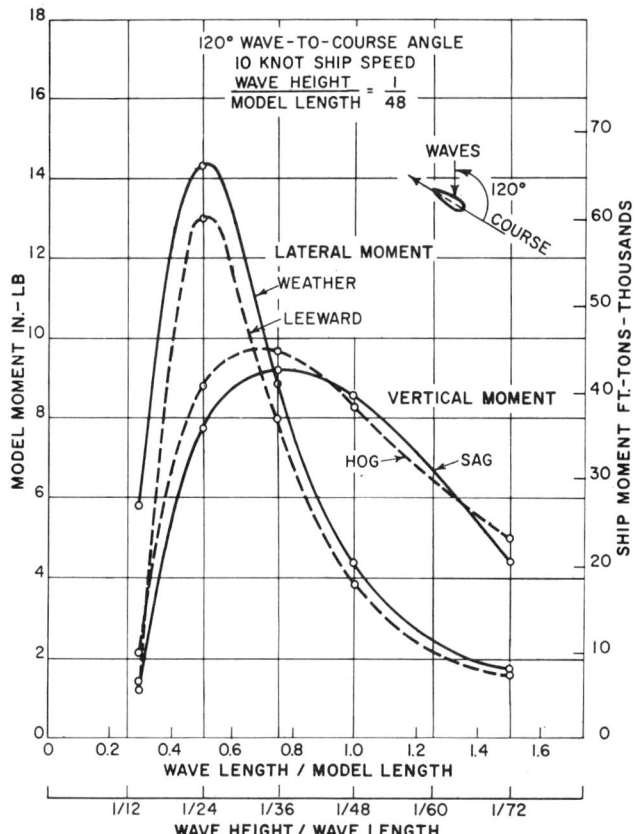

Fig. 5 Lateral and vertical wave bending moments versus wavelength on a T-2 tanker model (English units) (Numata, 1960)

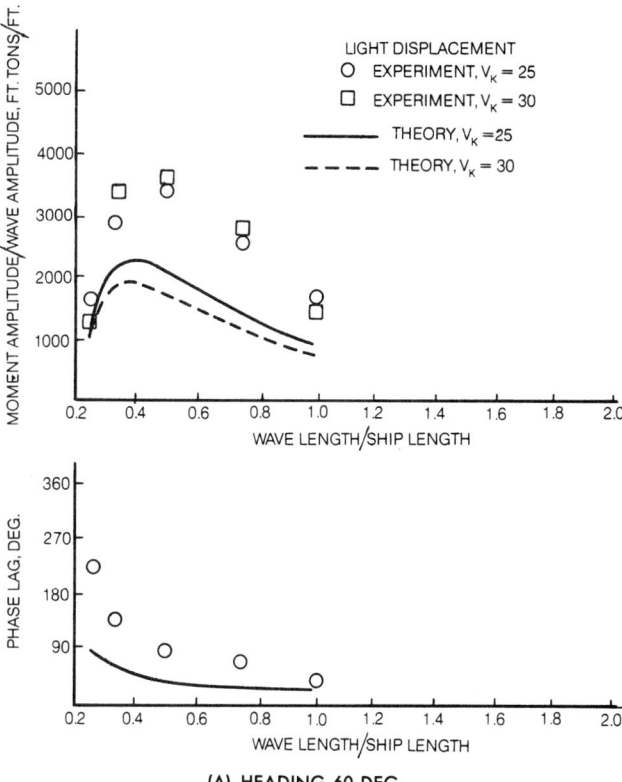

(A) HEADING 60 DEG

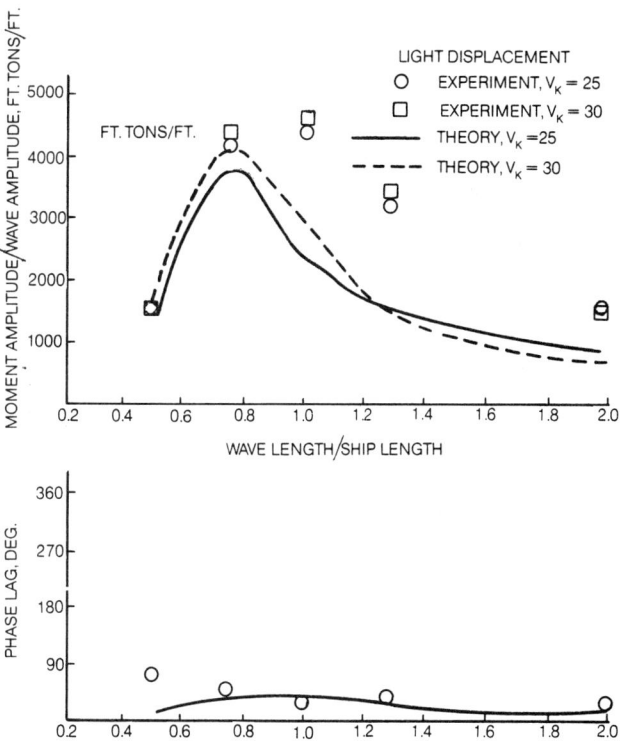

(B) HEADING 180 DEG. (HEAD SEAS)

Fig. 6 Comparison of theoretical and experimental midship vertical wave bending moments and phase lags for SL-7 cargo ships (English units) (Kaplan, et al., 1974)

The lateral and vertical longitudinal bending moments in this model test are of approximately the same magnitude and do not represent the worst conditions that might be experienced at sea.

The *Ocean Vulcan* (Admiralty Ship Com., 1953) observed lateral longitudinal bending moments of similar magnitude to the vertical moments in nearly all wave conditions. The maximum moments occurred at a wave-to-course angle of 110 to 140 deg. The maximum range of moments was 24,800 t-m (80,000 ft-tons), corresponding to a stress range of 39 MPa (2.5 tons per sq. in.), and these moments were frequently in phase with the vertical bending moments.

Early experiments of this type were intended to shed light on the fundamental nature of the dynamic wave-induced loads. More recent experiments have had the principal objective of providing data with which to test or calibrate theoretical calculation procedures of the type referred to in category (d) below. Such experiments are described in Lewis (1954), Gerritsma and Beukelman (1964) and Kim (1975), and comparisons of experimentally and theoretically derived structural loads for a high-speed ship, taken from a report by Kaplan et al. (1974), as shown in Fig. 6.

While, in principle, experiments of this type could be carried out to evaluate the structural loads on a new ship design, this is seldom done because of the time and expense involved. Furthermore, a number of computer programs are now available, based upon procedures described in the next section. These offer the possibility of studying a much broader range of sea and load conditions than would be possible in a model test program and of doing so at considerably less cost. Hence, the principal use for model testing is to provide background data and verification for such computer techniques.

(d) Direct Computation of the Wave-Induced Fluid Loads.

In this procedure, appropriate hydrodynamic theories used in calculating ship motions in waves are applied to computing the pressure forces caused by the waves and by the ship motion in response to those waves. In determining the structural loads, the forces resulting from fluid viscosity may usually be neglected in comparison with the pressure forces, except for the case of rolling. The total structural loading at any instant is then the sum of the wave pressure forces, the ship motion-induced pressures and the reaction loads due to the acceleration of the ship masses.

Note that a preliminary step in the computation of the motion-dependent part of the loads is the solution for the rigid-body motion response of the ship to the wave exciting forces. Both the analysis of the hydrodynamic forces and the solution for the motion response are discussed in detail in Chapter VII. The application of these techniques to the computation of the wave-induced shear force and bending moment in regular waves is outlined in the next subsection. It will also be shown that results for regular waves can be applied to the prediction of forces and moments in realistic irregular seas.

In order to achieve an accurate, purely analytical prediction of the long-term structural behavior of a ship, research is continuing in the disciplines of fundamental hydrodynamics, stress analysis, material fatigue, and in the accumulation of long-term wave statistics for the world's oceans, as discussed in the following sections.

2.4 Deterministic Evaluation of Wave-Induced Loading. The calculation of the bending moment, shear force and torsional loading on a ship hull in waves requires a knowledge of the time-varying distribution of fluid forces over the wetted surface of the hull together with the distribution of the inertial reaction loads. The fluid loads depend on the wave-induced motions of the water and the corresponding motions of the ship. The inertial loads are equal to the product of the local mass of the ship and the local absolute acceleration. The shear force and bending moment are then obtained at any instant by evaluating the first and second integrals of the longitudinally distributed net vertical or horizontal force per unit length. The expressions for these integrals are similar to those used in the calm water case, Equations (1) and (2), with the buoyancy term replaced by the time-varying fluid force per unit length and the weight term replaced by the inertial reaction force per unit length. The results are additive to the calm water values.

As previously noted, the computation of the inertial loads and a part of the fluid loads requires that we first determine the wave-induced motions of the ship. The solution for these ship motions and the system of fluid loading is most frequently accomplished today through the use of a procedure based on so-called *strip theory*. The details of strip theory, including the underlying assumptions and the limitations of the results, are developed in detail in Chapter VII. This procedure has been implemented in several computer programs, examples of which are described by Raff (1970), Salvesen, Tuck and Faltinsen (1970) and Meyers, Sheridan and Salvesen (1975). Strip theory programs are now in common use by design firms, classification societies and government agencies for both routine design investigations of ship wave loading and for the investigation of unusual loading situations that fall outside the range of the simplified formulas and procedures.

The results predicted by strip theory appear to be in good agreement with experiments for the vertical motions of pitch and heave, but a somewhat lower degree of correlation is usually observed for the lateral motions of sway, roll and yaw. Corresponding to these motion predictions, the vertical loads, shears and bending moments are predicted somewhat more accurately than the horizontal and torsional loads, shears and moments. It may be expected that ongoing research in ship hydrodynamics will result in continuing improvements in the capabilities of such programs.

We shall summarize here the principal features of the strip theory as it is applied to the prediction of the structural loading of a ship, and the reader may refer to Chapter VII for details of the theory as applied to the more general aspects of ship motion computations. For simplicity, we shall consider only the vertical load components that act on a ship proceeding into regular head seas, as illustrated in Fig. 7. As a result of symmetry about the longitudinal vertical plane of ship and waves, the motions and loads will have components only in this plane. At any instant of time, the motion of the ship will consist of the time-varying motions of pitch, heave and surge superimposed on a mean forward velocity, U_o.

One of the important assumptions of linear strip theory is that both the wave and ship motion amplitudes are, in some sense, small. As a result, it is possible to consider the total instantaneous vertical force on a thin transverse strip or element of length, dx, to be composed of the sum of several terms that are computed independently of each other. Two of these elementary forces are the still water buoyancy and weight of the element of the ship length. These are the same forces that appear in Equations (1) and (2) for the still water loads, shear and bending moment, and we need not consider them here. The remaining time-varying forces result from inertial reactions and

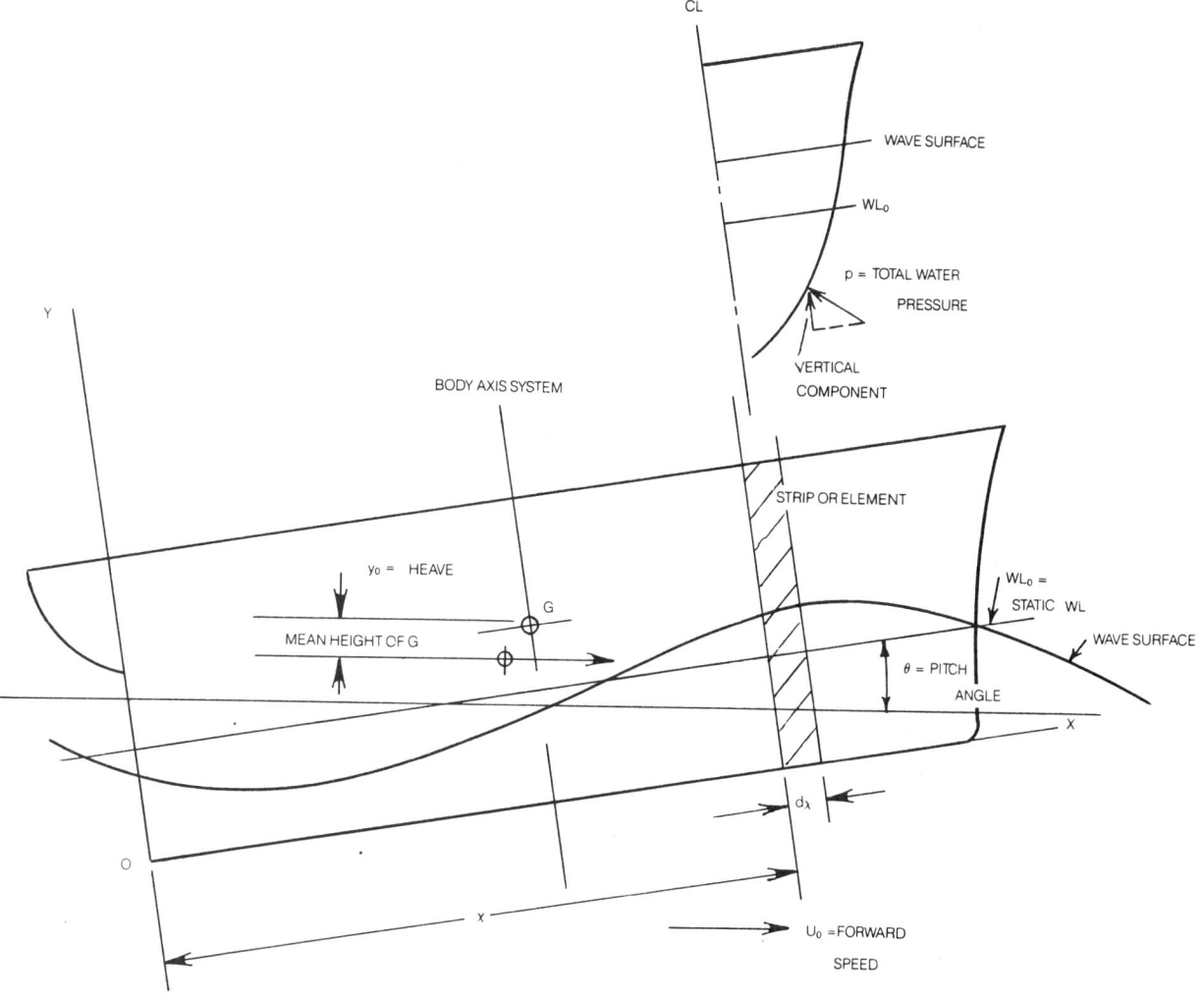

Fig. 7 Nomenclature for stripwise force computations

from the water pressures that are associated with the waves and the wave-induced motions of the ship. Viscous forces, which are found to be relatively unimportant (except for roll damping), are ignored in computing the vertical loads. Within this framework, the vertical fluid forces on the various elements may be subdivided into five categories, as follows, all expressed in force units,

(a) A wave pressure force component computed as though the presence of the ship does not disturb either the incident waves or the dynamic pressure distribution in those waves. This is called the *Froude-Krylov* force,

(b) A wave pressure force component computed from the properties of the diffracted wave system. These waves result from the the reflection and distortion of the incident waves when they impinge upon the ship. This force represents a correction to the Froude-Krylov force for the disturbance introduced into the wave system by the presence of the ship.

(c) A term proportional to the instantaneous vertical displacement of the element of the ship from its mean position, as if in calm water. This is called the hydrostatic restoring force and is equal to the change in the mean static buoyancy of the element.

(d) A term proportional to the instantaneous vertical velocity of the element called a damping force.

(e) A term porportional to the instantaneous vertical acceleration of the element. This is called an added mass force.

The first two of these forces, when added together, comprise the total wave-induced exciting force, computed as though the ship moves steadily forward through the waves but experiences no oscillatory motion response to the wave forces. The last three forces are computed as though the ship is undergoing its oscillatory wave-induced motion while moving at steady forward speed through calm water.

In addition to forces (a) through (e) above, there must be added the inertial reaction force of that portion of the mass (weight/g) of the ship that is contained in the strip, dx. If the ship's mass per unit length is denoted by $m(x)$, this reaction force is given, according to D'Alembert's principle, by $-m(x)a_y dx$, where a_y is the component of the absolute acceleration of the section at x in the direction parallel to the ship y-axis. If we now denote the sum of the five component fluid forces acting on the strip, dx, by $f(x)dx$, then the total force at any instant is the sum of the fluid forces and the inertial reaction, denoted $q(x)dx$, given by

$$q(x)dx = [f(x) - m(x)a_y]dx \qquad (4)$$

At any instant of time, the shear force, $V(x_1)$, at a section whose x-coordinate is x_1 is obtained by integrating $q(x)$ from the after end of the ship, $x = 0$, up to the station at $x = x_1$. The bending moment at x_1 is obtained, in turn, by integrating the shear force, $V(x)$, from $x = 0$ to $x = x_1$,

$$V(x_1) = \int_0^{x_1} q(x)dx \qquad (5)$$

$$M(x_1) = \int_0^{x_1} V(x)dx \qquad (6)$$

Fig. 8 illustrates the different components of the load distribution at fixed time for an example ship moving in a simple sinusoidal wave of unit amplitude. In this figure we see that the total loading consists of a number of terms of somewhat similar magnitude which may differ in sign and phase. There may be cancellation or reinforcement among the different components, with the result that the total loading may be larger or smaller than any individual component. This cancellation or reinforcement varies along the ship length and also varies with the frequency of wave encounter.

In the foregoing discussion of the force on a ship section, we have described a procedure in which the total force is subdivided into several components, each of which may be computed independently of the others. In consequence of the assumed linearity underlying strip theory, it is possible to calculate shear and bending moment in regular waves of any desired amplitude and frequency. Most ship motion programs contain a module to perform this computation at oblique headings to the waves as well as the head sea case discussed here.

It is shown in Chapter VII that the component regular wave forces depend upon the wave frequency, hull shape, ship speed and heading. The hydrodynamic coefficients of damping and added mass depend upon the hull shape, the ship speed and wave encounter frequencies. The wave forces act upon the ship at a frequency equal to the encounter frequency and, as a consequence of the linear representation of the ship motion response, the motions and motion-related loads will occur at this same frequency. In general, each motion or load response variable can be divided into a component that is in phase with the encountered waves and a component out of phase with the waves. We see, therefore, that the structural load components $q(x)$, $V(x)$ and $M(x)$ at a specific location, x, along the ship length, are sinusoidally varying quantities whose frequency equals the frequency of wave encounter and whose amplitude and phase vary with frequency.

When we consider a ship sailing through a realistic, irregular seaway it is fortunate that linearity also applies approximately in the description of the seaway. As shown in Chapter VII, the seaway can be broken down into a theoretically infinite number of wave components of various amplitudes, frequencies (lengths) and directions. Since by the linearity assumption the load responses of the ship to any regular wave component can be assumed to be directly proportional to the amplitude of that wave, the response of the ship to the random sea can be computed by linear superposition of the responses to the various seaway components present. As a result, the computations of the force, motion and loading components are initially performed for a series of elementary regular wave components each of unit wave amplitude and having a frequency equal to one of the components of the random seaway. The resulting unit responses are then weighted by the actual component wave amplitudes and added together in order to obtain the corresponding response spectrum in the real random seaway. This process of linear superposition forms the basis for virtually all computations of ship responses to realistic random seaways, and the details may be found in Section 4 of Chapter VII, including an example of wave bending moment. Results are in the form of rms values of shear and bending moment, from which various short-term statistical properties of the response can be derived, as discussed in Sections 2.7 and 2.8. Extension to the computation of the extreme loadings to be expected during the ship's lifetime is discussed in Sections 2.9 and 2.10 of the present chapter.

If, during its lifetime, the ship operates at several different conditions of load distribution and draft, then there will be a different set of the functions $q(x)$, $V(x)$ and $M(x)$ associated with each loading. In order for the naval architect to design for the most severe structural loading that the ship will experience during its operating lifetime, the strip theory computations must be carried out for the full range of frequencies, wave heights and wave headings expected to be encountered. These computations must be repeated for all of the combinations of speed and loading conditions at which the ship will operate. Many of the available computer programs contain the means of performing these multiple computations easily and efficiently. The more sophisticated of them also contain provisions for performing the superposition that yields the response to random seas, together with the probabilistic analysis leading to estimates of extreme motions and loads.

2.5 Transverse Distribution of Wave Loads. In order to compute the secondary or tertiary response of structural components such as panels of stiffened or

STRENGTH OF SHIPS

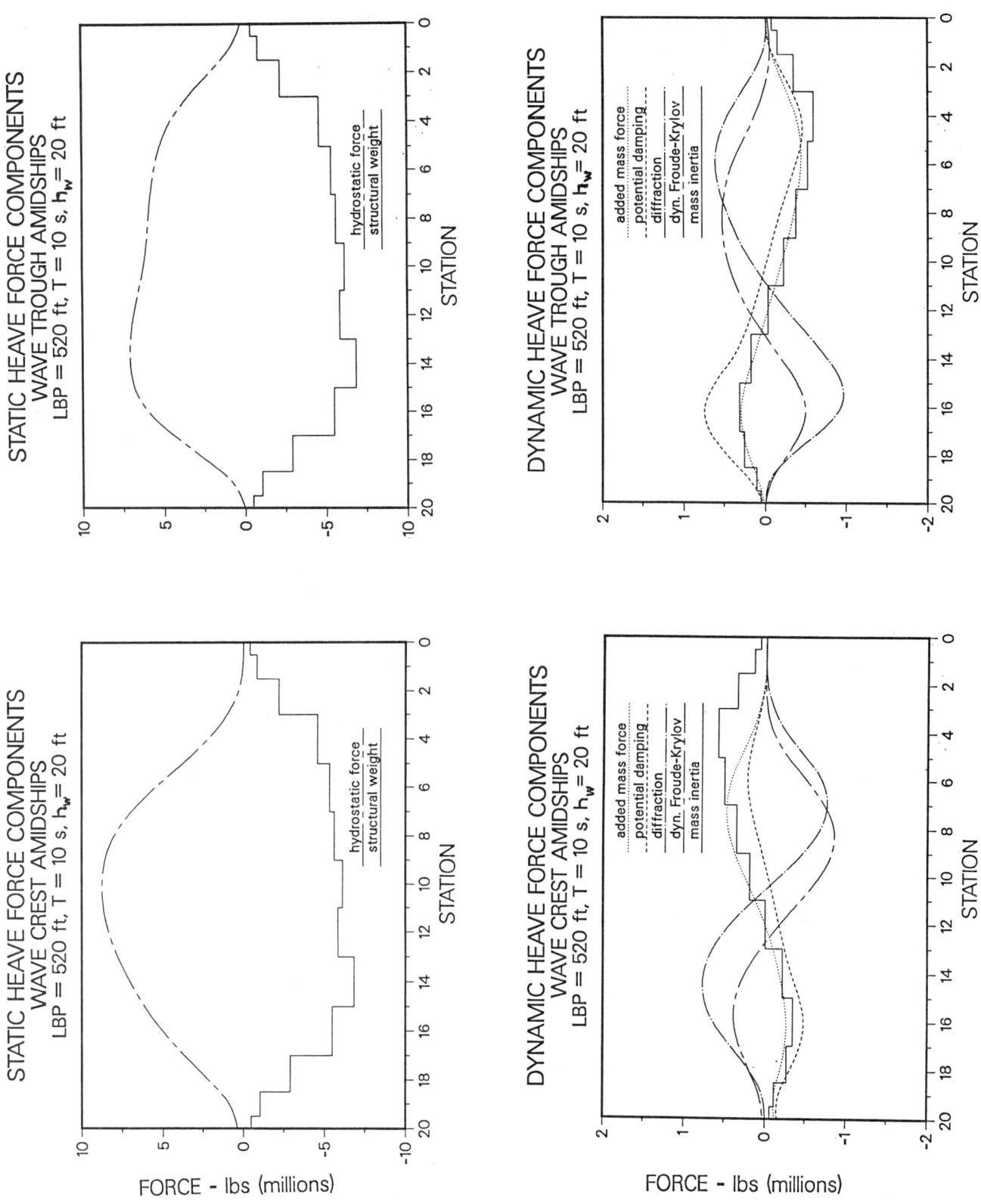

Fig. 8 Calculated components of typical wave-induced vertical loads for *Mariner* class cargo ship (English units)

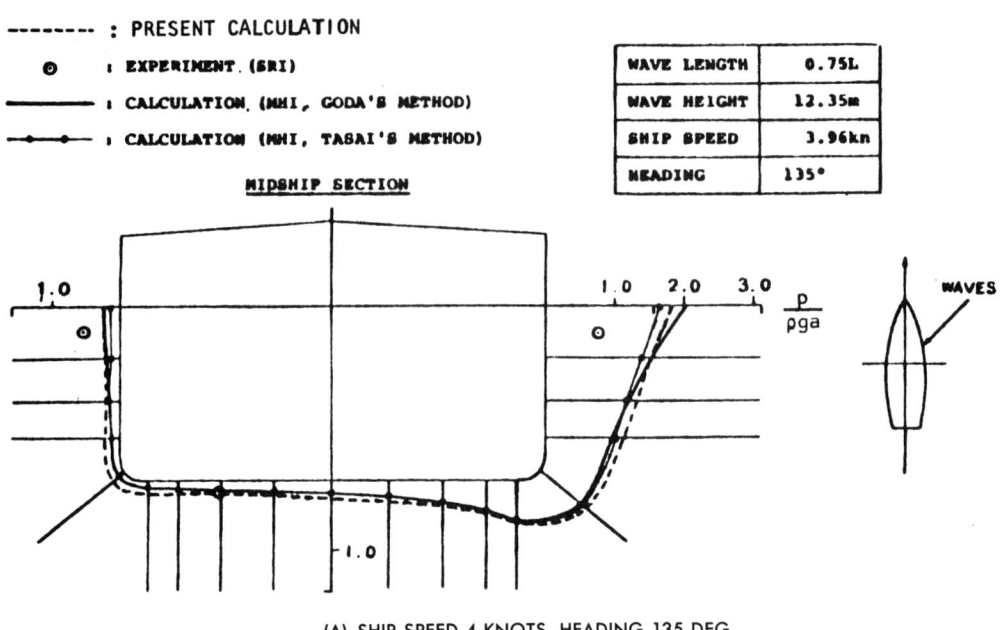

(A) SHIP SPEED 4 KNOTS, HEADING 135 DEG

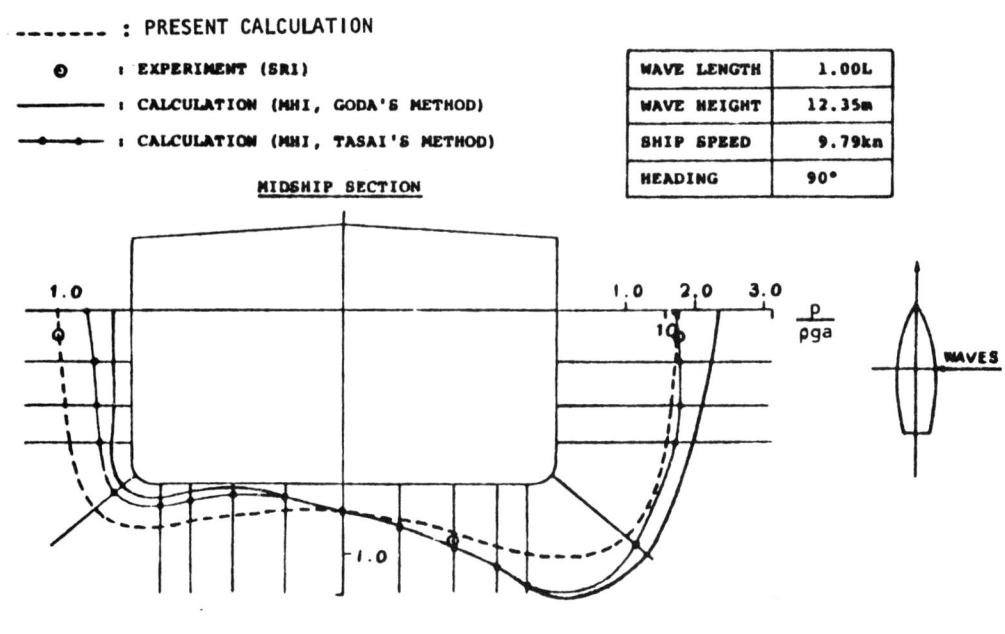

(B) SHIP SPEED 9.8 KNOTS, HEADING 90 DEG

Fig. 9 Typical transverse wave load distributions: resultant hydrodynamic pressure amplitudes (Kim, 1982)

unstiffened plating, it is necessary to know the distribution of fluid pressures and acceleration loads over the surface of the panel. For the purpose of analyzing transverse strength, therefore, the distribution of loads transversely around the ship section is required. Note that the sectional force per unit length used in computing the longitudinal shear and bending moment in the preceding section is the resultant or integral of this distributed load around the section. A computation of the wave and motion-induced distribution of pressure is given by Kim (1982) and some of his results are shown in Fig. 9. The two plots shown in this figure

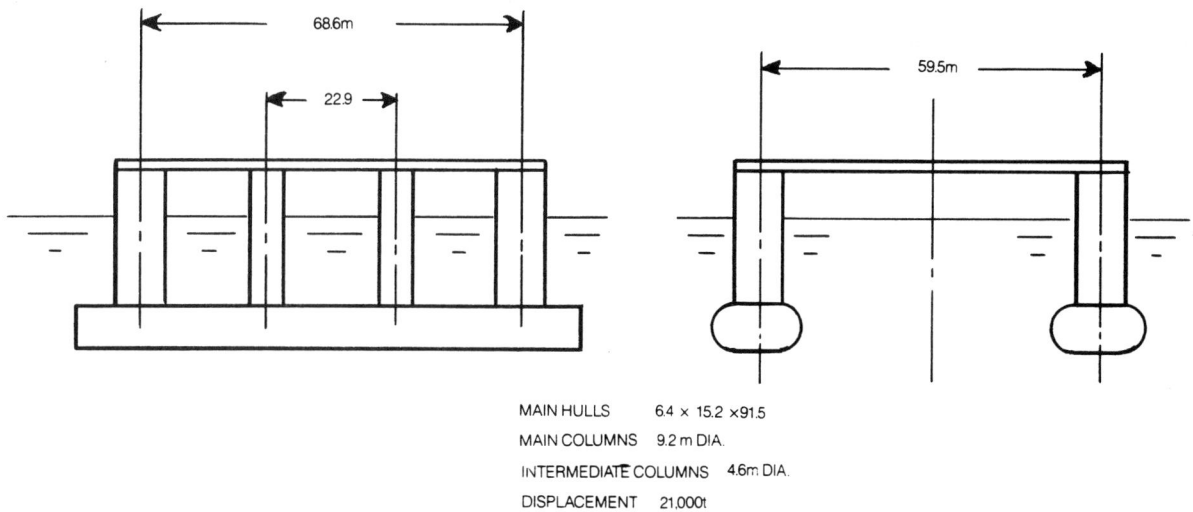

Fig. 10 Example of twin-hull semi-submersible platform

give the amplitude (but not the phase) of the dynamic pressure variation around the midship section, and this includes the effects of both ship and wave motions. It will be noted that the pressures in the wave *above* the still waterline are not obtained, since the linear hydrodynamic theory computes the fluid force on only the mean immersed portion of the ship. The pressure on this area must therefore be estimated separately.

The greatest dynamic pressure variation is found near the waterline in all cases, and in beam or bow seas the amplitude is greatest on the side facing the incoming waves. The lowest amplitude of pressure variation is found in the vicinity of the keel, which will, however, experience the highest static pressure.

The more complete ship motion programs used for the computation of the wave-induced shear and bending moment also provide the pressure distribution over the hull surface.

2.6 Wave Loads on Offshore Platforms. Many offshore oil drilling and production platforms are decidedly nonship-like in geometric form, but are closely approximated as a space-frame assembly of slender tubular members. The structural analysis of such a platform is carried out approximately by three-dimensional frame analysis methods, and the applicable loads require computational methods appropriate to the geometry of the structure. Such a platform is illustrated in Fig. 10. A reasonably good estimate of the wave forces may be obtained by computing the force on each member as though there were no hydrodynamic interference between members and the member in question were exposed to the flow field due to the combined wave and platform motion.

Using a formula first proposed by Morison, et al (1950), the force on one member, is illustrated in Fig. 11, and the force per unit length, l, may be approximated by,

$$\frac{dF}{dl} = -\int p\,\vec{n}\,ds + \frac{\rho}{2} D C_D \vec{u}_n |u_n| + \rho \pi \frac{D^2}{4} C_M \vec{a}_n \quad (7)$$

where

p = fluid pressure on the surface of the member computed as though the member does not disturb the flow,
$\vec{n}$ = surface outward normal vector on the member,
D = diameter of member,
C_D = drag coefficient,
C_M = added mass coefficient,
$\vec{u}_n$ = resultant velocity normal to centerline,
$\vec{a}_n$ = resultant acceleration normal to centerline,
ds = element of circumferential arc length on member.

Through the application of the method of equivalent linearization, the quadratic drag force term may be replaced by a linear drag force. The resulting analysis for the platform motions and structural loading now becomes linear and the principles of load decomposition and superposition that were previously applied in the ship case may be applied here also. The equivalent linear drag coefficient is found to be dependent upon the amplitude of the relative motion between member and fluid and, since the member motion depends upon the resultant platform motion, an iterative solution procedure must be employed. An application of the method of equivalent linearization to problems of fluid forces on cylindrical members is given in Paulling (1982).

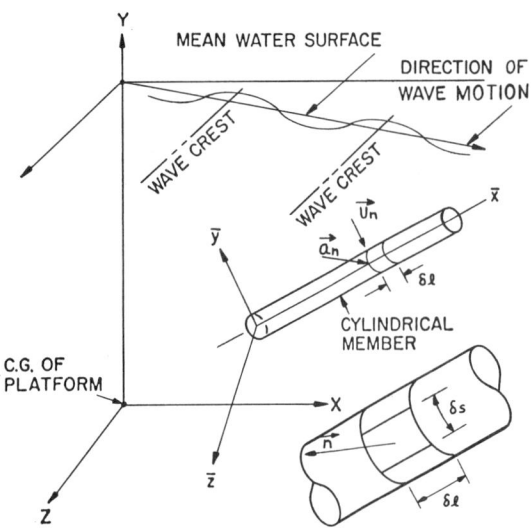

Fig. 11 Nomenclature for fluid forces on a cylinder

The fluid force per unit length given by Equation (7) may be integrated over the length of each member and replaced by statically equivalent forces and moments at the member ends in a manner consistent with the space frame method of structural analysis. The structural response is then obtained through the use of a standard finite-element or space frame computer program. Further discussion may be found in Paulling (1974).

Standards for the structural design of offshore platforms are given in ABS (1987b), *Rules for Building and Classing Mobile Offshore Drilling Units* (MODUs), as well as in the International Maritime Organization (IMO) *Code for the Construction and Equipment of Mobile Offshore Drilling Units*.

2.7 Probabilistic Estimate of Wave-induced Loads in Random Seas. The Preface to the 1967 edition of this book, *Principles of Naval Architecture*, contains the following statement: "Chapter IV has been rewritten to reflect the many developments since 1939, and to recognize the trend toward separate consideration of the static still-water bending moment and the dynamic wave bending moment, which is essential to the eventual possibility of evaluating bending moments in a real sea by statistical analysis." This eventual possibility appears now to have been realized. Statistical analysis, augmented by model experiments, full-scale measurements and computational means, is now acknowledged to form an accepted and important part of the process of evaluating the bending moments and other loads acting on a ship in a seaway.

Many years ago Wendell P. Roop (1932) attempted a limited statistical approach to the problem of longitudinal bending loads and stresses on a naval tanker. Since then statistical analyses have been applied to full-scale naval ship stress data, as in Birmingham (1971). The latter was a probabilistic study in which the author derived bending moment amplitudes in different regular waves for seven ships from trials in which wave records were made. He calculated short-term bending moments for wave spectra representing different sea states, as discussed in Chapter VII, and then predicted lifetime loads for 14 years' operation in the North Atlantic on the basis of a number of assumptions regarding effects of ship heading, speed, wave spectra, etc. More recently probability methods have been applied in the design of unusual naval craft, as in Sikara, Dinsenbacher and Beach (1983).

Meanwhile, a continuing effort in many countries has been directed toward probabilistic methods of estimating wave loads on merchant ships, sponsored and/or carried out mainly by the classification societies, and by the Ship Structure Committee in the U.S. This work has been regularly summarized in the *Proceedings* of the International Ship Structures Congress (ISSC). As explained in Section 2.4, the structural responses to random waves—as well as the motions—are usually predicted by the linear superposition of the responses to a series of elementary regular waves. In applying superposition, it is necessary to assume that the loads are linearly related to the wave amplitude even when extrapolated to the higher sea states. There have been several experimental investigations, for example, Dalzell (1963), intended to test this assumption by examining experimentally determined midship bending moments in waves of progressively increasing wave height. As shown in the Dalzell work, the assumption of linearity is usually found to be conservative, i.e., the measured midship bending moments increase at a rate less than the wave amplitude in the range of the steeper waves. Nevertheless, it is important to bear in mind the assumptions and limitations of linear analysis when using loads that are predicted by means of standard linear ship motions programs. See Section 4 of Chapter VII, Validity of linear theory.

Within the framework of linear ship motion theory, all of the wave-induced components of the hull loading, including the pressure at a point on the hull surface, the hydrodynamic load per unit length, the inertial load per unit length, the shear force and the bending moment, are linearly related to the wave amplitude. When the principle of superposition is used in predicting the response to a random seaway, the methods developed in Chapter VII may be applied in estimating the short-term probabilities of response. If this estimate is of a load component—for example, the midship wave bending moment—there is usually a nonzero mean still-water moment upon which the wave-induced moment is to be superimposed. The load probabilities can then be computed by taking into consideration the combination of the wave-induced and the still water moment.

The statistical quantities that are usually of concern in ship or platform strength investigations are divided into three categories:
- Short-term mean and extreme values.
- Long-term mean and extreme values.

- Cumulative cyclic values.

The short-term refers to a period of time of approximately one to four hours during which it has been found that the sea state remains nearly uniform (i.e., statistically stationary) under normal climatic conditions. A discussion of the general principles of short-term probabilities and their application to the problem of hull loads is covered in the following Section 2.8.

The second category refers to a longer time period, which may be measured in days or years, during which the sea state may vary widely from a calm to severe storm conditions, but the rate at which the conditions change is sufficiently low that the assumption of uniformity is a good approximation over the period of a few hours. Thus, the long-term response may be thought of as an accumulation of short-term responses to different sea states, each having uniform or statistically stationary characteristics. This collection of short-term experiences would extend, in the case of ships and ocean structures, over a period of time equal to the useful operating life of twenty to thirty years. This approach is discussed in Section 2.9.

The third item in the above list refers to the phenomenon of long-term cyclic loading that may cause cumulative fatigue damage to the structure. Here the large numbers of cycles of low to moderate levels of bending moment and stress may contribute as much damage as the fewer cycles of extreme stress. Thus the full range of wave-induced loading, including the more prevalent moderate conditions as well as the less frequent severe conditions, must all be considered. It will be shown that valuable information on cyclic loading can be derived from the second category of long-term data, when expressed in the probability of exceedance format. Further discussion of fatigue is contained in Section 4.11 of the present chapter.

2.8 Short-term Estimates of Loads in Random Seas. For structural design purposes we need to have information about the extra large or extreme values of load rather than some sort of average values. In general there are two approaches to the problem:

(a) Exceedance probability, and
(b) Extreme value theory.

The exceedance probability estimate is simpler, for if we know the density function of the loads, this curve can be integrated to give the probability of exceeding different levels of load, hence the value we would expect to be exceeded once in any number of cycles. But if we know the density function, we can also apply the principles of formal extreme value theory to obtain the expected highest value in any number of cycles. Actually the value to be exceeded once and the highest value are almost the same, and either is suitable for design purposes.

It is shown in Section 4 of Chapter VII, Vol III, that the short-term statistics of maxima or peaks of any response, including bending moment or stress, can be determined from the moments of the response spectrum. For practical purposes it is usually possible to assume a narrow-band process, in which case the statistics are defined by a Rayleigh density function (i.e., $\epsilon = 0$). The corresponding Rayleigh (cumulative) distribution function defines the probability of not exceeding different levels of load or stress. But here Q is taken to be the probability of exceeding the various levels. Hence, the distribution function is $(1 - Q)$. Since the Rayleigh density function is:

$$q(x) = [x/m_o] \exp[-x^2/2m_o] \qquad (8)$$

the exceedance probability is,

$$Q(x > x_1) = \int_{x_1}^{\infty} q(x)\,dx = \exp[-x_1^2/2m_o] \qquad (9)$$

Hence, the exceedance probability can be easily calculated as a function of the parameter, m_o. The number of cycles, n, at which one value is expected to equal or exceed x_1 (the *return period*) is,

$$n = 1/Q = \exp[x_1^2/2m_o]$$

The quantity m_o is the statistical variance of the random process, equal to the zero-th moment or area under the spectrum of the process, and $\sqrt{m_o}$ is the rms value.

Sometimes designers prefer to work with the expected highest value of load, which is derived in Chapter VII on the basis of the theory of extremes. In Section 4.5 it is shown that over the short-term the extreme values for the *maxima* of a narrow-band random process are given by an expression of the form,

$$\overline{Y}_n = C_n \sqrt{m_o} \qquad (10)$$

The extreme value in question, $\overline{Y}_n$, is defined as the expected value of the highest single excursion or maximum in a sample of n excursions. C_n is a constant depending upon n, approximately $2\sqrt{ln(n)}$, for large values of n.

As a result of the randomness of the sea and the ship responses, both the expected value of exceedance and the expected maximum value are subject to variation. Thus if many records were taken on many ships in the same seaway, the highest peak of some of the records will be greater than the expected value computed by Equation (10), and some of the values will be less. The one-in-n peak values will, therefore, be characterized by their own probability distribution having a dispersion about a mean value. In terms of the experience of a single ship undergoing a total of n bending moment oscillations, the value given by (10) must be interpreted as an estimate of the mean or *expected value* of the highest peak. In some analyses, the modal or *most probable* value is used instead of the mean and this is given by an expression of the same form as (10) but with modified values of the coefficient, C_n. For n larger than 1000, which is typical of three to four hours of wave experience, the mean and modal values will be nearly identical.

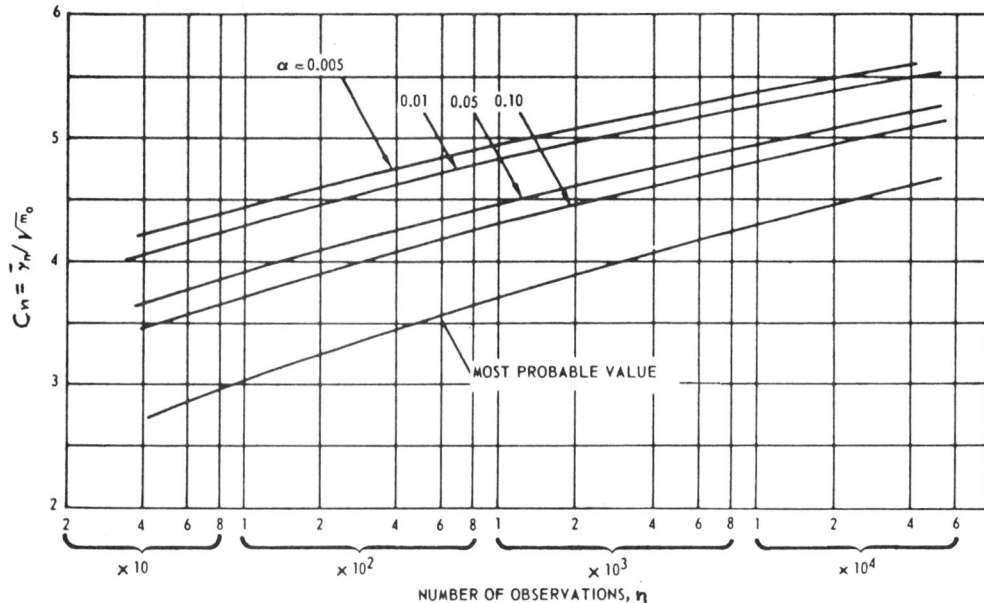

Fig. 12 Extreme values (amplitudes) for the Rayleigh distribution ($\epsilon = 0$) as a function of probability of being exceeded α (Ochi, 1973)

In order to obtain a more useful estimate of the extreme value that reflects this randomness of the peaks, Ochi (1973) applied the probability of exceedance or *risk parameter*, α, described in Chapter VII, Section 4, Vol III. The estimate of the extreme peak value in a sequence of n excursions is still given by an expression of the form of (10), but the constant C_n is now a function of α as well as n. Fig. 12 presents values of this modified C_n for the narrow-band random process as a function of the risk parameter, α. Thus the estimated value of the highst bending moment for a risk parameter of 0.01, is given by Equation (10) with C_n taken from Fig. 12, and this value will apply to all of the ships in a hypothetical fleet of many similar ships. If now we were to examine bending moment recordings for all of the ships in the same seaway and extract from each record the single highest peak, we would find that with $\alpha = 0.01$ the value estimated by the present procedure is exceeded by the actual measured value in approximately one percent of the ships. Note that if we had estimated the *most probable* value for the bending moment peak, we would find the estimated value to be exceeded by the actual value in approximately 63 percent of the ships.

We have seen in Fig. 12 that C_n increases very little as α takes on values less than 0.01. For this reason, a value of $\alpha = 0.01$ is sometimes considered to be suitable for design purposes, so long as we are concerned with a single short-term situation in which the sea condition (spectrum) and its duration are known.

The above theory, applicable to short-term situations, is of limited direct usefulness in ship design, since for most design purposes we must take account of the many different sea conditions to be encountered by a ship during its lifetime. However, if we were able to specify the spectrum of the most severe sea that a structure such as a fixed platform should be able to survive, and its duration (persistence), then we might predict the highest expected load by the short-term techniques.

The idea of designing to a load that corresponds to an extrapolation of extreme waves to the highest expected in a period of 50 or 100 years has been extensively used in civil engineering and recently in some branches of ocean engineering. Current practice in the design of offshore platforms for the North Sea is to design for the "100-year storm." The wave heights and periods corresponding were derived by analyses of wave data using extreme value theory. It should be noted in this context that the value of the extreme wave is a "most probable extreme" and that again there is a significant probability of exceeding this level in the first, or subsequent, samples of length equal to the return period. Accordingly, some design margin must be built in, either in the safety factors subsequently used, or in the definition of an adequate design return period. For example, if it is desired to fix an extreme for design so that there is an 80 percent probability that the value will not be exceeded in a 20-year exposure, the design return period may be chosen to be $[20/(1-0.8)] = 100$ years.

Note also that the above approach assumes that the highest seaway will produce the highest load. Although this may be true for a fixed platform, for ships and other floating structures it may not necessarily be true. The extreme load depends upon the response characteristics of the ship or other structure, as well as on the expected exposure time (probability and duration of the severe sea.)

2.9 Long-term Extreme Values. When we are concerned with the prediction of the most severe loading to be experienced by a ship or structure during the course of its entire useful lifetime, we can no longer assume the sea conditions or the conditions of operation of the ship to remain constant. Sea states of varying severity from flat calm to the most severe storm may be encountered and the occurrence and duration of sea conditions of various degrees of severity will depend upon the geographical and seasonal operational profile of the ship or platform. The conditions of loading, speed and heading will vary from one period of time to another, and the influence of all of these variables must be included in the computation of the long-term extreme loading. As in the short-term case, it is generally not possible to obtain a single precise value for the highest load. Instead, the answer must be expressed in the form of long-term probabilities.

There are two basic approaches—just as in the short-term case—exceedance probabilities and extreme value theory. Although the latter approach is not suitable for providing cyclic loading data for fatigue problems, it has provided a useful method of extrapolating observed extreme wave or load data to the long-term case. See the following Section 2.10. The most common approach involves the estimation of probabilities of exceedance, or what have been called *long-term distributions* (Lewis, 1967), which are synthesized initial distributions of response maxima, or probabilities of exceedance per cycle of load.

There are currently a half dozen methods in use world-wide to obtain such long-term distributions, as summarized in Lewis and Zubaly (1981) and Stiansen and Chen (1982). The basic assumptions are the same in all these approaches, which are broadly similar to the "collection analyses" mentioned in Section 4.10, Chapter VII. In addition to the assumed linearity of response, a fundamental assumption made in all methods is that in the short term the maxima of the response have a probability structure defined by the Rayleigh distribution ($\epsilon = 0$); that is, that in the short term the response is a zero-mean stationary Gaussian narrow-band process with maxima defined by the single parameter, $m_o = \sigma^2$. The short-term probabilities are conditional, i.e., they assume different values for each value of m_o. To be specific, the density may be written,

$$f(x|m_o) = [x/m_o] \exp[-x^2/2m_o] \quad (11)$$

where m_o is the zeroth spectral moment (as treated in Section 2.8) representing the mean-square response of each short-term exposure. This single parameter, m_o, defining a Rayleigh distribution is considered to be a random variable depending upon the sea condition and the speed, loading and heading of the ship during each short-term interval. Because of the assumptions stated above the following development is believed to be conservative.

To obtain a long-term distribution it is necessary to consider many different short-term intervals, in each of which the response is defined by m_o and can be calculated by the methods of Chapter VII. This is the way by which the geometry and characteristics of the ship are injected into the analysis. The factors affecting the value of m_o include speed, V, heading, μ, condition of loading, a measure of wave height, H, some measure of wave period, such as T_m, the modal period, and sometimes another measure of wave spectral form or shape.

In order to develop a long-term distribution in practice, some simplifications are needed. The speed, V, can be eliminated by recognizing that it is not an independent variable and specifying that for each time interval the speed is that which is appropriate for the prevailing sea condition, loading and ship-to-wave heading on which it depends. (Note that bending moment is not greatly affected by ship speed, in any case.) The heading variable, μ, cannot be eliminated, but it is customary to assume that there is an equal probability of all headings, statistically independent of the other variables. This may not be true on any one voyage, but it tends to average out over many round trips. Finally, loading can be handled by assuming that the number of conditions is limited (as outbound loaded, ballasted return), each with a different mean (still water) bending moment, and that completely independent calculations can then be carried out for each. Assuming that two parameters, H and T, suffice to describe the sea state, we are left with the parameters, μ, H, and T.

If the above factors are assumed to be random variables, then the spectral moment, m_o, is also, and its probability density is conditional. Recognizing the conditional relationships between the response, x, and m_o, and the conditional relationship between m_o and all the factors enumerated above, a joint long-term probability density of all the variables may be assumed in the following form:

$$q(x, m_o, \mu, H, T) = f(x|m_o)\, q(m_o|\mu, H, T)\, q(\mu, H, T) \quad (12)$$

where the first factor is the short-term conditional density of x, Equation (11), and the second is the conditional density of m_o, given the operational and environmental factors. Finally, the last factor is the joint probability density of the parameters influencing m_o.

In order to obtain the long-term probability density of the response, x, Equation (12) must be integrated with respect to m_o, μ, H and T. However, the long-term probability of interest is that of the response peaks or maxima of x exceeding some level, say x_1. This is obtained by integrating the long-term probability density of x with respect to x over the interval

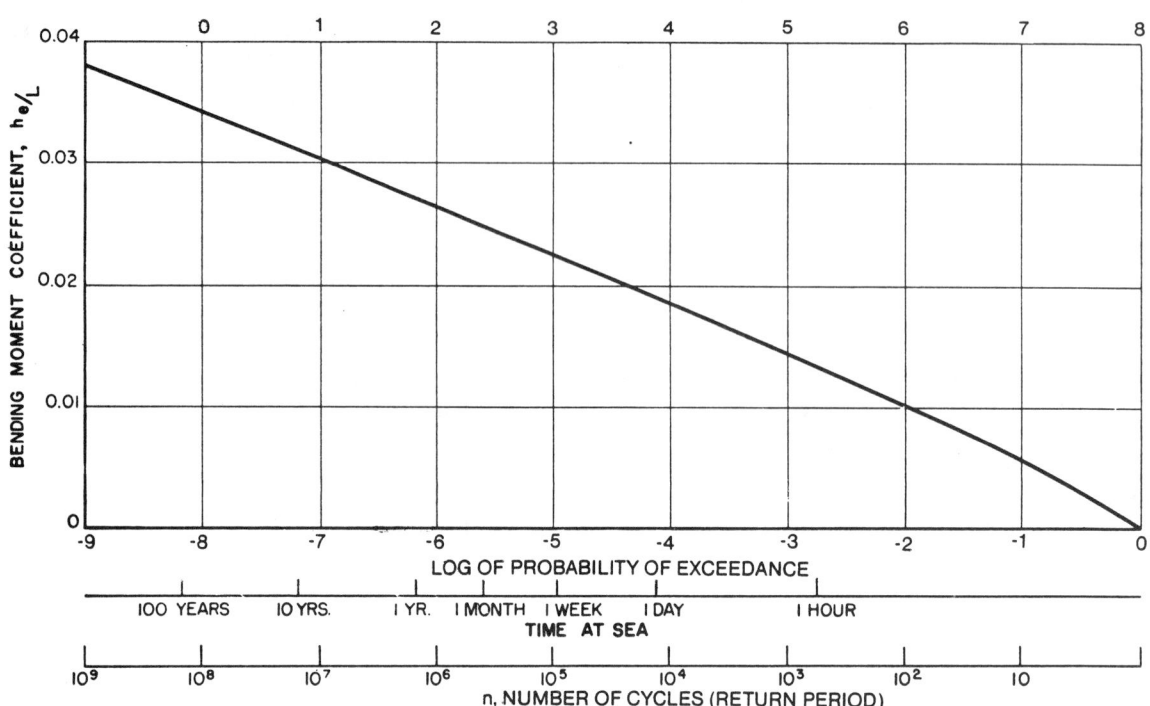

Fig. 13 Long-term probability distribution for wave-induced bending moment in S.S. *Wolverine State* corrected to average North Atlantic weather (based on Band, 1966)

x_1 to infinity. Under the Rayleigh assumption, Equation (11), this last integration may be done formally, so that finally the long-term probability that a response exceeds x_1 may be written,

$$Q(x > x_1) = \int_{m_o} \int_\mu \int_H \int_T \exp[-x_1^2/2m_o] \cdot$$

$$q(m_o \mid \mu, H, T) \cdot q(\mu, H, T) \, dm_o \, d\mu \, dH \, dT \quad (13)$$

After this point the development of a usable engineering procedure requires the synthesis of the conditional and joint probabilities and their integration, and it is here that the divergence in the several detailed approaches begins. We do not know the exact probability densities of any of the factors involved, much less the form of their joint density. The developers of the various methods have had to interpret historical wave data, devise means of utilizing it in the required integrations, devise means of handling the speed, heading and loading factors, and finally have had a choice of the order of integration. The result is that the various methods appear to differ more than they fundamentally do. Details of a number of these methods are given in Bennet, et al (1962), Band (1966), Nordenstrom (1973), Fukuda (1970), Soeding (1974), and Mansour (1972). Ochi (1978) adopts a somewhat different mathematical concept, which leads, however, to similar numerical results.

Whatever the method, the final result is a numerically defined initial distribution of the maxima in the great many short-term Rayleigh distributions, which are in effect superimposed in the synthesis. Some choose to plot the results on some form of probability paper. Fig. 13 from Band (1966), is a somewhat more common form of presentation. The ordinate is the level x_1, in the equation above. The abscissa is a logarithmic scale of $Q(x > x_1)$, or its reciprocal, n, known as the return period (both are shown in the figure). It is important to note that the formulation results in the probability that the peak of a response excursion, chosen at random, will be greater than some level (in the example the probability is about 10^{-8} that the bending moment coefficient will exceed about 0.035). Roughly, the formulation involves probability per cycle of response. This is the reason that the computations are carried out to such low probabilities. The auxiliary scale in the figure indicates, on the basis of average encounter periods for the ship, approximately the corresponding exposure time, and as can be noted it is necessary to consider probabilities less than 10^{-7} in order to represent ship lives and multiples thereof.

In order to illustrate the considerations involved in the long-term prediction methods we will make a particular simplification of Equation (13). Suppose the ranges of all the variables μ, H and T are systematically divided into discrete intervals of width such that the value of the variable at the center of the interval is representative of the variable anywhere in the interval. With this assumption Equation (13) may be approximated as a summation,

$$Q(x > x_1) \approx \sum_\mu \sum_H \sum_T \int_{m_o} \exp[-x_1^2/2m_o] \cdot$$
$$q(m_o|\mu, H, T) \cdot p(\mu, H, T) \, dm_o \quad (14)$$

where it is understood that the discrete central values of the variables are intended. The function, $p(\mu, H, T)$ denotes the probability that the variables μ, H and T lie simultaneously in their respective intervals. In this form the conditional probability of the spectral moment, m_o, accounts for statistical variation of this parameter about the value that would be estimated by considering each of the central values fixed and applying the methods of Chapter VII, Vol III.

Now for present purposes we may make a further simplifying assumption. Suppose that the intervals into which the ranges of the variables, μ, H and T, are divided are sufficiently small that the spectral moment which can be estimated by considering the central values of the variables fixed is also representative of the moment which would be obtained so long as the variables μ, H and T were anywhere within the bounds of their respective intervals. This amounts to saying that the spectral moment, m_o, is a deterministic function of ship heading and sea state. With this assumption the conditional probability in Equation (14) tends toward a delta function, and when the integration with respect to m_o is performed the expression becomes

$$Q(x > x_1) \approx \sum_\mu \sum_H \sum_T \exp[-x_1^2/2m_o] \, p(\mu, H, T) \quad (15)$$

where the notation has been streamlined to facilitate further development. In particular, the summations are over all the discrete intervals previously defined, and the arguments of the joint probability symbolize the central values defined in conjunction with Equation (14). Most importantly, the value of m_o in the exponential is taken as a deterministic function of the central values of the operational and environmental variables. In this form the synthesis problem comes down to constructing a suitable discrete representation of the joint probability.

In order to illustrate how this joint probability might be constructed we can argue as follows. We imagine the ship lifetime to be subdivided into a large number of short-term intervals, for example, 4 hours each, during which the conditions of loading, ship speed, heading, and sea state remain constant. Now, assume that the values of all of the parameters μ, H and T are known for each of these time intervals. This, in fact requires that we be able to predict the operational profile of the ship in terms of its loadings, speeds and headings, and the weather conditions that it will encounter. The forecasting of the ship's cargo loading, speed and routing is normally performed by the ship-owner or designer as a part of the preliminary design process. The forecasting of sea and weather conditions that the ship will encounter depends upon the availability of a sea state data base of suitable form and extent.

In Equation (15) the sea state is represented by two parameters, measures of height, H and of period T. A discussion of the various means, including standardized formulas, currently in use for representing spectral areas and shapes is given in Chapter VII, Vol III. In some spectral formulas, only one parameter, for instance significant height, $H_{1/3}$ is used to characterize the sea state. In others, the additional parameter, T_z, the zero-crossing period, or T_m, the modal period, is employed, giving a somewhat more flexible means of representing a wide range of spectra of similar shape or form (i.e., "families.") Ideally, variation in the shape of the spectra should also be taken into account, since actually measured ocean wave spectra show considerable variety—including double peaks resulting from superposition of waves from two or more storms. But for simplicity it is often assumed that two parameters will suffice.

The long-term frequency of occurrence of sea states of different severity but of similar spectral form can then be expressed in terms of a joint probability density function for $H_{1/3}$ and T_m. This joint probability density of $H_{1/3}$ and T_m may be presented, for a given ocean area, either in the form of a table, as in Hogben and Lumb (1967), given in Chapter VII, or a contour plot or *scatter diagram*, an example of which is shown in Fig. 14. The values in the tables, or contours of the plot, are equal to the probability (fraction of time) that the sea state is characterized by the simultaneous occurrence of values of H falling within the interval H_1 to H_2 and T falling within the interval T_1 to T_2. This diagram would assume a single form for the spectrum (e.g. Bretschneider or JONSWAP). If the ship operates in different geographical areas having differing sea state characteristics, such diagrams would be required for each area of operation.

A composite sea state distribution may be constructed if we have such tables or diagrams for all parts of the ship's route of operation, together with information stating the fraction of time that the ship spends in each area. To illustrate this synthesis, consider a tanker trading between Europe and the Persian Gulf via the Cape of Good Hope. The tropical portions of the route would be characterized by a preponderance of low sea states while the Cape region would have a greater portion of high sea states. A series of diagrams similar to Fig. 14, or a tabulation of the frequency of occurrence of pairs of values of $H_{1/3}$ and T_m, would be necessary to represent these extremes as well as the gradations of sea climates typical of other portions of the route. Now, since each scatter diagram is, in fact, a bivariate probability density function representing the relative frequency of occurrence

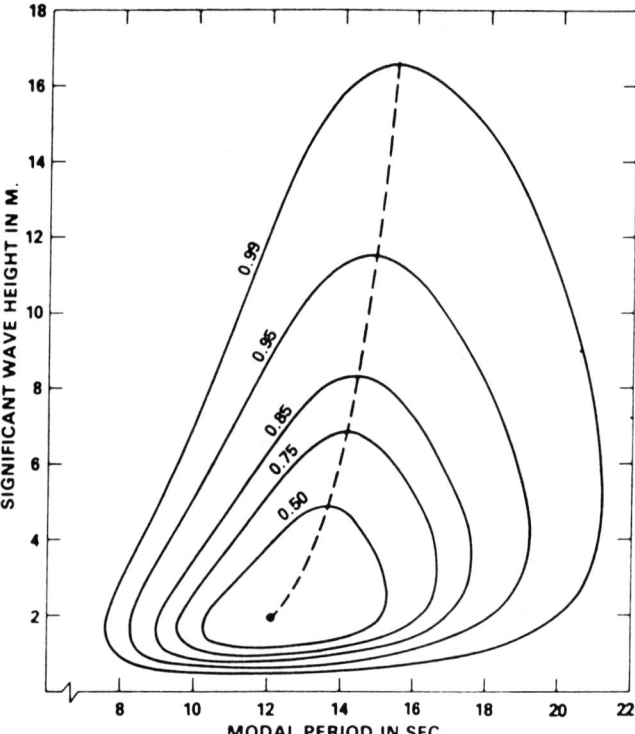

Fig. 14 Confidence domains of significant wave height and modal period for the (mean) North Atlantic (Ochi, 1978)

of sea states of all degrees of severity in the applicable ocean area, the integral of the diagram over all values of $H_{1/3}$ and T_m must equal unity. The ship will, however, be exposed to different geographical areas for different periods of time during each fraction of its voyage. The ordinates of the scatter diagram for each area along the route must, therefore, be multiplied by the fraction expressing the proportion of time that the ship spends in that area. A composite diagram or tabulation for the entire voyage is then constructed by adding together such weighted ordinates of all the diagrams representing voyage segments. Available data are discussed in Section 2 of Chapter VII.

For each set of values of the variables, μ, $H_{1/3}$, T_m, we now may determine the ship load response and compute a value of the spectral moment, m_o, of the loading. This computation would normally be performed using one of the standard ship motions and loads programs referred to earlier. The summation in Equation (15) is then approximated by

$$Q(x > x_1) \approx \sum_\mu \sum_{H_{1/3}} \sum_{T_m} \exp\left(-\frac{x_1^2}{2m_o}\right) p(\mu)\ p(H_{1/3}) p(T_m) \quad (16)$$

The simple product form of Equation (16) is based upon the assumption that all of the remaining random variables are statistically independent. We are in effect making a superposition of many Rayleigh distributions, weighted by the expected frequency of occurrence of all combinations of the three variables. In order to use the results provided by Equation (16) in selecting a design load value, we must choose a load having an acceptably low probability of exceedance and, in concept, we may proceed somewhat as follows. The average period of wave encounter at sea is about ten seconds. In a twenty to twenty-five year lifetime, a ship will encounter approximately 10^8 waves, the exact value depending upon the operating profile and the portion of the time spent at sea. It is reasonable to design the structure so that the ship will be able to survive the highest single peak excursion of the bending moment or other loading to be expected in that lifetime, thus the loading having a probability of exceedance of once in 10^8 cycles would appear to be a reasonable target value. Equation (16) expresses the probability that any one oscillatory peak will exceed the value x_1.

However, it should be noted that the load corresponding to this probability is subject to variation, just as in the short-term case. Thus, during one ship lifetime, there may be one peak value that exceeds the value corresponding to the 10^{-8} probability, there may be none or there may be several. If the ship sails for another twenty years in the same service, or if we consider a second identical ship, this 10^{-8} value may be exceeded one or more times again. If it is exceeded again, the second exceedance may be by a margin much greater than that experienced the first time or it may be less. The problem is seen to be similar to that of the short-term experience in which the risk parameter was introduced to quantify the probabilistic estimates of the extreme events. This random behavior of the expected highest peak load may be taken into consideration by using a procedure suggested by Karst in an Appendix to Hoffman and Lewis (1969). He formulates a problem that can be restated thus: to determine a bending moment (or stress) x_L such that the probability that a ship will exceed it in its lifetime has a specified value $P(x_1 > x_L)$, where x_1 is the expected bending moment corresponding to a lifetime of n_L maxima (or cycles) at $Q(x > x_1)_L = 1/n_L$. The specified probability $P(x_1 > x_L)$ is a risk factor or confidence level analogous to Ochi's α, previously discussed. On the basis of a Poisson model, the approximate result is that the design bending moment x_L can be read at

$$Q(x > x_1) = P(x_1 > x_L) \cdot Q(x > x_1)_L$$
$$= P(x_1 > x_L) / n_L \quad (17)$$

The procedure is illustrated in Fig. 15, which shows a graph of the function $Q(x > x_1) = 1/n$ plotted in the usual manner. At a probability level of 10^{-8}, the ordinate of the curve is the expected value of load (midship bending moment) having this probability of exceedance in any one cycle. The small graph plotted along the vertical axis at this probability level is, sche-

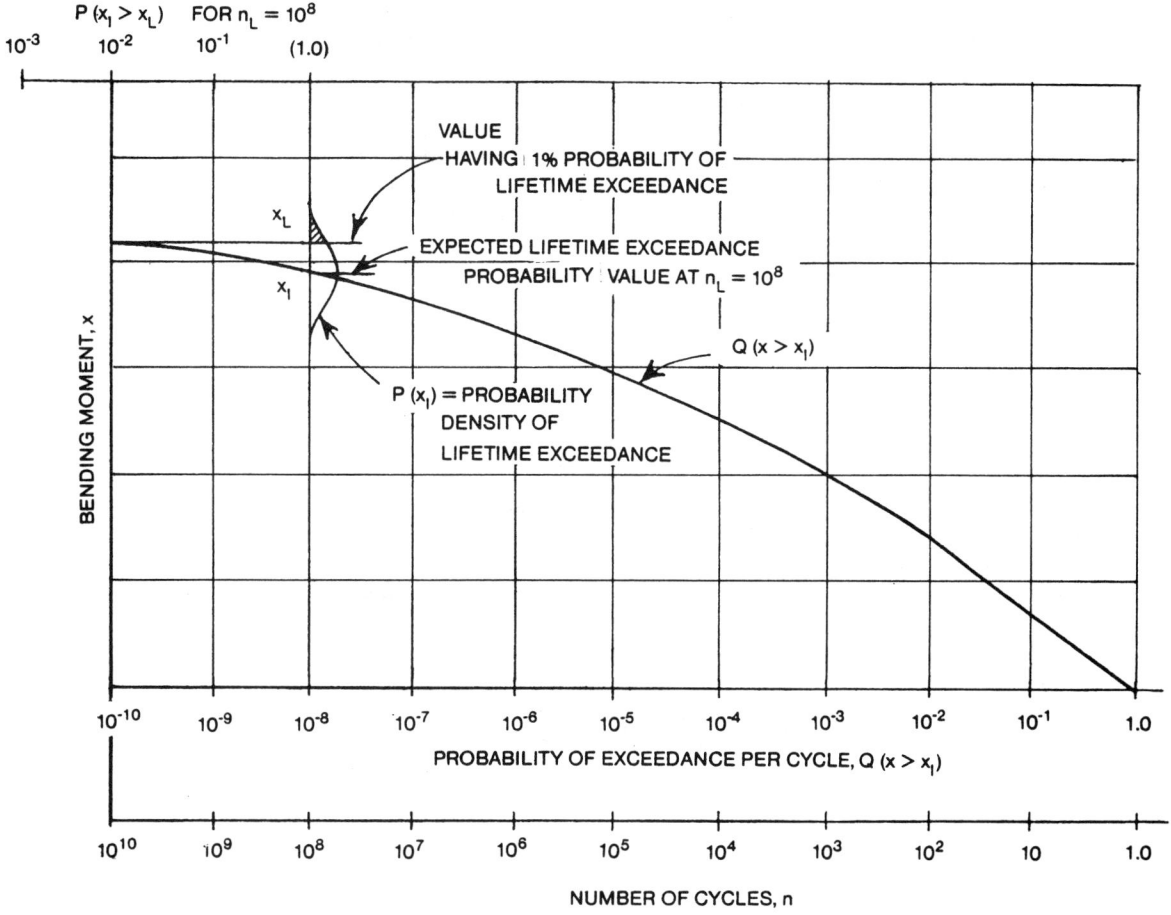

Fig. 15 Typical long-term exceedance probability of wave-induced bending moment

matically, the probability density function for the lifetime $1/10^8$ peak values. This density function expresses the random behavior of the highest peaks occuring in many records containing n_L peaks each. The boundary of the shaded portion of this probability density function corresponds to the one percent value, i.e., the shaded area is equal to one percent of the total area under the curve. Hence, $P(x_1 > x_L) = 0.01 = 10^{-2}$. If we extrapolate this one percent level horizontally to the left, it is found to intersect the $Q(x > x_1)$ curve very nearly at the probability of 10^{-10}, which is the value given by Equation (17). This provides a convenient way of obtaining $Q(x > x_L)$ for a $P(x_1 > x_L) = 0.01$ by extrapolating the long-term curve to lower values of probability. The significance of the result is that it defines a design load that would be expected to be exceeded once in any one ship in a fleet of 100 similar vessels in the same service.

For example, if we set $P(x_1 > x_L)$ equal to 0.01 and n_L equal to 10^8, we find that, to a close approximation, the design load x_L can be read at $Q(x > x_L) = 10^{-2}/10^8 = 10^{-10}$. Thus it is seen that the one percent level of confidence is obtained by using a design load x_L corresponding to the probability of exceedance $Q(x > x_1)$ having a value equal to the product of the risk factor (or confidence level) and the expected probability of exceedance in a ship's lifetime of n_L cycles. Alternatively, the lifetime probability (risk factor) is

$$P(x_1 > x_L) = n_L \cdot Q(x > x_L) \qquad (18)$$

Note that Equations (17) and (18) break down, however, when $x_L = x_1$, or $Q(x > x_L) = 1/n_L$. In this limiting case $P(x_1 > x_L)$ is 0.667 instead of $P(x_1 > x_L) = n_L \cdot 1/n_L = 1.0$, as given by Equation (18). See Fig. 15. Equation (18) is useful in calculating probability of failure (Section 5.3), for the risk factor $P(x_1 > x_L)$ can also be considered the probability that any one ship in a fleet of many similar ships will exceed the design load x_L in its lifetime. Hence, it is sometimes referred to as the *lifetime probability* to distinguish it from the probability per cycle, $Q(x > x_1)$, or $1/n$. See further discussion in Faulkner and Sadden (1978).

Note also that a curve of $P(x_1 > x_L)$ can be obtained for any value of n_L by adding a new scale to the plot of $Q(x > x_1)$. For example, a scale of $P(x_1 > x_L)$ has been added at the top of Fig. 14 for $n_L = 10^8$.

The foregoing analysis yields an estimate of the probable exceedance value to be expected during the long period of exposure in which the full range of variation of speeds, headings and other variables is experienced. Equation (16) is based upon the Rayleigh distribution for the peak values of the random process, and, as pointed out by Ochi (1978), it does not explicitly contain the time of exposure. Instead, it should be thought of as applying to the large number of peak excursions roughly estimated for the ship's lifetime. In order to calculate accurately the number of oscillatory peaks in the time period, we proceed as follows. During one of the short-term intervals in which the variables μ, $H_{1/3}$, T_m remain constant, the mean number of zero-crossings in unit time is given by

$$n_z = \frac{1}{2\pi} \sqrt{\frac{m_2}{m_0}} \qquad (19)$$

Here, m_2 is the second moment of the spectrum of the response, as defined in Section 4 of Chapter VII. It is consistent with the Rayleigh assumption that each zero crossing corresponds to one peak of the random process, so that Equation (19) will also give the number of peaks per unit time, n_z. Now, if T equals the long-term period in hours, the total time during which a specific set of the variables, μ_i, H_j, T_k, will prevail is given by,

$$T_i = p(\mu)\, p(H) p(T) T \qquad (20)$$

and the total number of oscillations during the time T by,

$$N = T \,\Sigma\, n_z p(\mu) p(H) p(T) \qquad (21)$$

To incorporate this number into the calculation of the probability, Q, each term in Equation (16) must now be multiplied by a weighting function which is the ratio of the number of oscillatory peaks corresponding to each of the intervals of summation, to the total number of peaks given by Equation (21),

$$n_z^* = \frac{n_z(\mu, H, T)}{\Sigma n_z p(\mu) p(H) p(T)}$$

instead of 1.0, as assumed in Equation (16). Hence, the final result, modified to express the probability in terms of time, is

$$Q(x > x_1) \qquad (22)$$
$$= \sum_\mu \sum_H \sum_T n_z^* \, \mathrm{Exp}\left(\frac{x_1^2}{2m_0}\right) p(\mu) p(H) p(T)$$

Lewis and Zubaly (1981) evaluated an example which indicates that the result obtained by (16) and (22) are nearly the same for the example containership. This is explained by noting that the higher bending moment values generally occur in head seas corresponding to the greater number of peaks, or higher mean frequencies of oscillation. The loadings in following seas are generally much lower and make very little contribution to the overall probabilities.

The foregoing procedure leads to an estimate of the long-term exceedance values that takes into consideration all of the different sea conditions the ship may encounter during its lifetime. The weighted summation includes contributions from low sea states that occur frequently but individually have low probabilities of causing extreme events and from high sea states, each of which has a high probability of causing an extreme event but which occur relatively infrequently. As a result of the wide range of conditions to be included in a computation of this nature, it is seen that considerable computational resources will be required.

For the problem of determining cyclic loading for fatigue design, it is important to note that the long-term distribution discussed here can provide basic information. Fig. 13 shows such a distribution, with a scale of probability and number of cycles at the bottom. If the life of the ship for fatigue studies is assumed to correspond to 10^8 cycles, a new reversed scale is constructed, as shown at the top, giving the number of cycles expected to reach any specified level of bending moment.

2.10 Extrapolation of Observed Extremes. An alternate approach to the problem of design loads considers only the largest maxima (extremes) in short-term samples—instead of all maxima. Applications of the *theory of extreme values* are presented in Sections 2 and 4 of Chapter VII, to situations in which the initial distribution is known in closed form, as in the case of a stationary zero-mean Gaussian process such as a wave record or a record of short-term ship response to a seaway. The theory can also be applied and extended to the case of extrapolation of long-term historical data in the form of records of either waves or ship responses.

If historical data on the extremes (i.e., highest values) in many samples from a population are available, the classical approach of Gumbel (1958) may be utilized to forecast extremes likely to occur in the future. Because the density function of extremes tends to be more concentrated than the initial density, asymptotic forms of the density of extremes may be derived that depend more upon the general properties of the initial density than the detail. Jasper (1956) and Yuille (1963) are among the early marine examples of application of this approach, where historical data are fitted to an assumed distribution of extremes. Fig. 16 from Yuille (1963) is included to illustrate the method. In this particular case the data were the daily readings of a mechanical maximum reading strain gauge installed in a warship. Each reading is the maximum stress range experienced in a day at sea, and thus may be interpreted as the extreme in a one day sample. To analyze a set of N such observations they are arranged in order of increasing magnitude and numbered $m = 1, 2, \ldots N$, thus converting the original sequential observations to *Order Statistics*. The fractions: $m/(N+1)$ for $m = 1, N$ are computed and these represent an estimate of the probability distribution corresponding to each level of observation. The next step is to

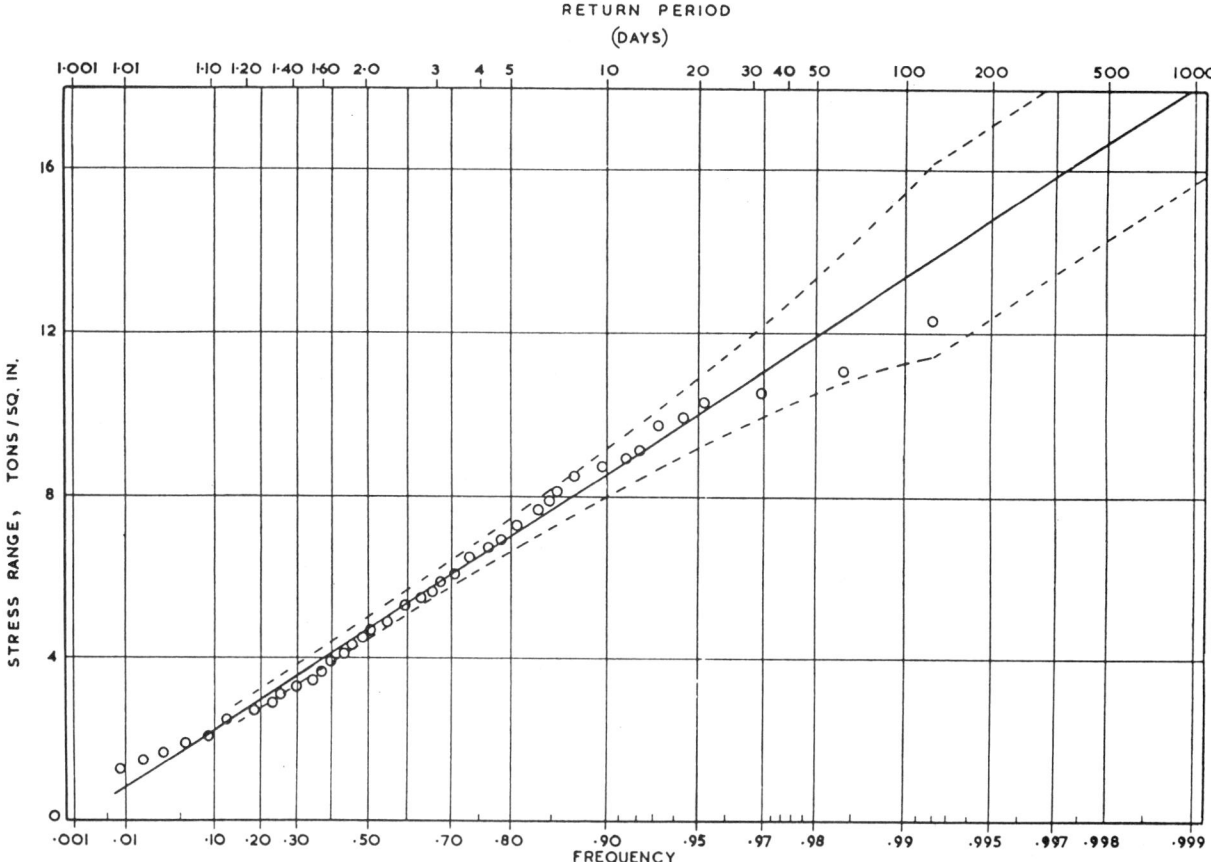

Fig. 16 Results from typical maximum reading strain gage (Yuille, 1963)

assume an analytical distribution and fit the data. The distribution assumed in the example, Fig. 16, was the first asymptotic distribution of extremes (Gumbel, 1958). The usual way of fitting the data is to use a probability paper appropriate to the assumed distribution.

The fractions are plotted upon such a paper in Fig. 16, where the stress range is the ordinate and the distorted frequency or probability scale is the abcissa. The probability scale in any probability paper is distorted in such a way that the theoretical distribution plots as a straight line. Thus fitting the data to the assumed distribution amounts to fitting the best straight line shown in the example. The dashed lines in the figure are *confidence limits*. In the example they were derived so that if the scatter of the data points about the fitted straight line be due to random errors, 68 percent of the data points may be expected to fall within the dashed lines, and these lines provide a statistical criterion for the adequacy of the fit of observation to the assumed theoretical distribution. Extrapolation of the data so as to forecast extremes that are likely to occur in future samples is done simply by extending the fitted straight line to higher probability levels. It should be noted with respect to the detail of the fit in Fig. 16 that the data appear to deviate from the straight line systematically. This raises the point that no one of the several analytical distributions that may be assumed is guaranteed to be best relative to a given set of data. Indeed, better fits to this particular set of data have been made (Ochi and Bolton, 1973). The art in the approach is to find the best analytical approximation. The Weibull distribution has been found to be useful for this purpose (Nordenström, 1973).

An interesting hybrid approach to the design load problem is discussed in detail in a paper by Ochi (1978). Instead of integrating over all conceivable sea states, the most severe conditions likely to occur and their persistence (duration) are derived from historical data and the short-term extreme theory of Section 2.8, with appropriate risk factor, is applied only to the responses to these severe seas. In Ochi (1978) some comparisons made with results obtained by the long-term weighted summation method show good agreement. The method depends for its success on the user's covering a sufficient number of severe seas, considering their probabilities and persistences, to clearly establish the "most extreme" response. For most ship design problems the long-term approach described previously is

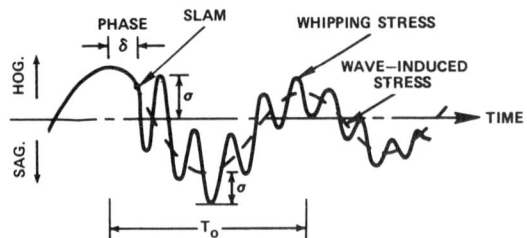

Fig. 17 Explanatory sketch of whipping stresses

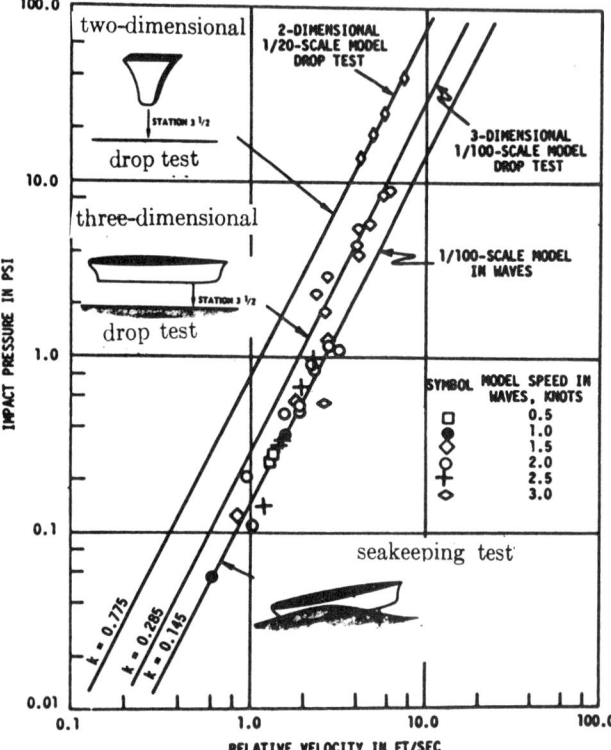

Fig. 18 Comparison of pressure-velocity relationships obtained in three different types of experiment

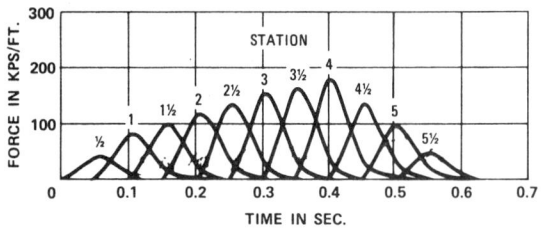

Fig. 19 Calculated force applied at various stations as a function of time; Mariner, Sea State 7, significant wave height 25 ft, ship speed 7.4 knots, light draft (all above Ochi and Motter 1973)

recommended, since it provides more complete information, such as cyclic loading data for fatigue. The further use of probability-based wave loads in design is discussed in Section 5 on Reliability of Structures.

2.11 Dynamic Loads. (a) *Springing.* An important effect of sea waves on some ships is the excitation of random hull vibration that may continue for extended periods of time. This phenomenon, known as *springing*, has been noticed particularly in Great Lakes bulk carriers (Matthews, 1967), but it has also been reported on large ocean-going ships of full form (Goodman, 1971). The explanation is that long ships of shallow draft and depth are comparatively flexible in longitudinal bending and consequently have unusually long natural periods of vertical hull vibration (two-noded periods of 2 *sec* or longer). Experimental and theoretical studies (Hoffman and van Hooff, 1973, 1976) confirmed that when such a ship is running into comparatively short waves that give resonance with the natural period of vibration, significant vibration is produced. A corresponding fluctuation in stress amidships is therefore superimposed on the quasi-static wave bending stress.

The well-developed strip theory of ship motions has been applied to springing in short waves (Goodman, 1971). Although motions of a springing ship may then be very small, the theory provides information on the exciting forces acting on the ship in the short waves that produce springing. Hence, when these forces are applied to the ship as a simple beam the vibratory response can be predicted. Despite the fact that strip theory is not rigorously applicable to such short waves, results for one ocean-going ship were found to agree quite well with full-scale records. Further coordination between theory and experiment has been attempted for Great Lakes ships, including model tests where idealized wave conditions can be provided (Hoffman and van Hooff, 1973, 1976; Stiansen, et al, 1977; Troesch, 1984a).

If the waves that excite springing were regular in character, the springing could be avoided by a small change in speed. But in a real seaway containing a wide range of frequencies the springing varies in a random fashion and a speed change may have little effect. The springing excitation and response can then be treated as stochastic processes that can be handled by the techniques discussed in Section 4 of Chapter VII. However, it has been shown by Kumai (1974) and Troesch (1984b) that longer waves in the spectrum can also excite the hull vibration. This introduces non-linear aspects that are important to consider in relation to structural responses. (See Section 4.10).

(b) *Slamming Loads.* When a ship operates at high speeds, especially in head seas, the bow may occasionally emerge from one wave and re-enter the next wave with a heavy impact or *slam* as the bottom forward comes in contact with the water. Related phenomena are associated with the impact of large waves on the bow topsides having pronounced flare and with green water on deck coming in contact with the front of a deckhouse or superstructure. In each case the phenomenon involves the impact at high relative velocity between the free surface of the nearly incompressible water and a portion of the ship's structure.

Two noticeable effects may be caused by bottom slamming. There may be localized structural damage in the area of the bottom that experiences the highest impact pressure. This may include set up plating and buckled internal frames, floors and bulkheads. A second effect of slamming is a transient vibration of the entire hull in which the principal contribution comes from the fundamental two-noded vertical mode. This slam-induced vibration is termed *whipping*, and it may result in vibratory stress intensities that are equal in magnitude to the wave-induced low-frequency bending stresses. See Fig. 17.

Ochi (1964) has concluded that slamming is possible if two conditions are fulfilled simultaneously:

1. The forefoot must emerge above the surface of the waves.
2. At the time of re-entry, the relative velocity between the ship bottom and the water must exceed some threshold value. From model experiments, the threshold velocity was found to be 3.6m per sec (12 ft per sec) for the *Mariner* class cargo ship.

The intensity of whipping depends upon the magnitude of the force resulting from the impulsive pressure of the slam, on its longitudinal location and on the duration of the force pulse. Much of the available information on the intensity of pressure, p, resulting from slamming has been obtained from model tests. These have been performed with two-dimensional models in calm water, with full three-dimensional models in calm water, and with full models moving through waves. Fig. 18, from Ochi and Motter (1973), contains a compilation of such model test data. These results are usually fitted to a curve of the form given by,

$$p = kv^2 \qquad (23)$$

where v is the relative vertical velocity. The coefficient k is shown for the three types of tests in the figure.

The impact pressure is distributed over an area of the ship bottom in the immediate vicinity of the point of re-entry, and is typically a maximum on the centerline at any instant of time. Higher pressures are found to occur where the bottom is nearly flat. The total force is then given by the integral of this pressure over the area of bottom on which it acts. As the ship forefoot re-enters the water, the point of maximum pressure tends to move toward the bow. At a given station along the length, the duration of the pressure pulse is typically a few tens of milliseconds, but as a result of the movement of the re-entry location, the pressure pulse moves also, meanwhile maintaining its peak intensity. The total duration of the force pulse that the ship experiences will, therefore, be several times as great as the pulse duration at a single station. This space-time behavior of the force is illustrated in Fig. 19 from Ochi and Motter (1973).

The prediction of the occurrence of slamming, and of the resulting pressures on the hull, are considered in greater detail in Section 5 of Chapter VII, Vol III. The determination of local structural response and whipping stress is discussed in Section 4.10 of this chapter.

(*c*) *Inertial loads on components.* The accelerations resulting from the motions of a ship in a seaway produce forces (or inertial reactions) on local components of the ship, as well as on personnel, cargo and liquids in tanks. Consequently the magnitudes of these forces are often needed for the design of local structure, foundations, lashings, securing devices, etc. For such purposes the estimated maximum values of the forces may usually be considered as static design loads, because of the relatively long periods of ship motion amplitudes. See Section 4 of Chapter VII (Forces due to ship motions) in Vol III.

Hull vibrations caused by machinery or propeller action may in some cases cause resonant response of structural components, Chapter VI, Vol. II.

Section 3
Analysis of Hull Girder Stress and Deflection

3.1 Nature of Ship Structural Reactions. The reactions of structural components of the ship hull to external loads are usually measured by either stresses or deflections. Structural performance criteria and the associated analyses involving stresses are referred to under the general term of *strength* while deflection-based considerations are referred to under the term *stiffness*. The ability of a structure to fulfill its purpose may be measured by either or both strength and stiffness considerations. The strength of a structural component would be termed inadequate and structural failure would be deemed to have occurred if the material of which the component is constructed experiences a loss of load-carrying ability through fracture, yield, buckling, or some other failure mechanism in response to the applied loading. Excessive deflection, on the other hand, may also limit the structural effectiveness of a member, even though material failure does not occur, if that deflection results in a misalignment or other geometric displacement of vital components of the ship's machinery, navigational equipment, or weapons systems, thus rendering the system ineffective.

The present section will be concerned with the determination of the response, in the form of stress or deflection, of structural members to the applied loads. Once these responses are known it is necessary to determine whether the structure is adequate to withstand the demands placed upon it, and this requires consideration of the several possible *failure modes*, as discussed in detail in Section 4.

In analyzing the response of the ship structure it is convenient to subdivide the structural response into categories logically related to the geometry of the structure, the nature of the loading and the expected response. Appropriate methods are then chosen to analyze each category of structural component or response, and the results are then combined in an appropriate manner to obtain the total response of the structure.

As noted previously, one of the most important characteristics of the ship structure is its composition of an assemblage of plate-stiffener panels. The loading applied to any such panel may contain components in the plane of the plating and components normal to the plane of the plating. The normal components of load originate in the secondary loading resulting from fluid pressures of the water surrounding the ship or from internal liquids, and in the weights of supported material such as a distributed bulk cargo and the structural members themselves. The in-plane loading of the longitudinal members originates mainly in the primary external bending and twisting of the hull. The most obvious example of an in-plane load is the tensile or compressive stress induced in the deck or bottom by the bending of the hull girder in response to the distribution of weight and water pressure over the ship length.

The in-plane loads on transverse members such as bulkheads result from the edge loads transmitted to these members by the shell plate-stiffener panels and the weights transmitted to them by deck panels. In-plane loads also result from the local bending of stiffened panel components of structure. For example, a panel of stiffened bottom plating contained between two transverse bulkheads experiences a combined transverse and longitudinal bending in response to the fluid pressure acting upon the panel. This panel bending, in turn, causes stresses in the plane of the plating and in the flanges of the stiffening members. Finally, the individual panels of plating contained between pairs of stiffeners undergo bending out of their initial undeformed plane in response to the normal fluid pressure loading. This results in bending stresses parallel to the plane of the plate, and the magnitudes of these stresses vary through the plate thickness.

In order to perform an analysis of the behavior of a part of the ship structure, it is necessary to have available three kinds of information concerning the structural component:

- The dimensions, arrangement and material properties of the members making up the component.
- The boundary conditions on the component, i.e., the degree of fixity of the connections of the component to adjacent parts of the structure.
- The applied loads.

It is possible, in principle, using a computer-based method of analysis known as the finite-element method, to analyze the entire hull at one time without the necessity of such subdivision into simpler components. There are at least two reasons, however, for retaining the subdivision into simpler components:

- By considering the structural behavior of individual components of structure and their interactions with each other, a greater understanding is developed on the part of the naval architect of their functions, and this leads to improved design.
- Many of the problems facing the practicing naval architect involve the design or modification of only a limited part of a ship, and a full-scale analysis would be neither necessary nor justified.

A brief introduction to the finite element procedure is given in Taggart (1980), Chapter VI. It is a powerful tool that is widely and routinely used in most aspects of modern structural analysis, and standard computer programs are available from computer service bureaus and a number of other sources.

As noted in the previous section, it is convenient to subdivide the structural response into primary, secondary and tertiary components, and we shall here examine these components in detail. Note that the primary and secondary stresses in plate members are membrane stresses, uniform (or nearly uniform) through the plate thickness. The tertiary stresses, which result from the bending of the plate member itself, vary through the thickness, but may contain a membrane component if the out-of-plane deflections are large compared to the plate thickness.

From this we see that the resultant stress at a given point in the ship structure is composed of several parts, each of which may arise from a different cause. In many instances, there is little or no interaction among the three (primary, secondary, tertiary) component stresses or deflections, and each component may then be computed by methods and considerations entirely independent of the other two. The resultant stress, in such cases, is obtained by a simple superposition of the three component stresses. An exception occurs if the plate (tertiary) deflections are large compared to the thickness of the plate. In this case the primary and secondary stresses will interact with this tertiary deflection and its corresponding stress, so that simple superposition may no longer be employed to obtain the resultant stress.

Fortunately for the ship structural analyst, such cases rarely occur with the load magnitudes and member scantlings used in ships, and simple superposition of the three components may usually be performed to

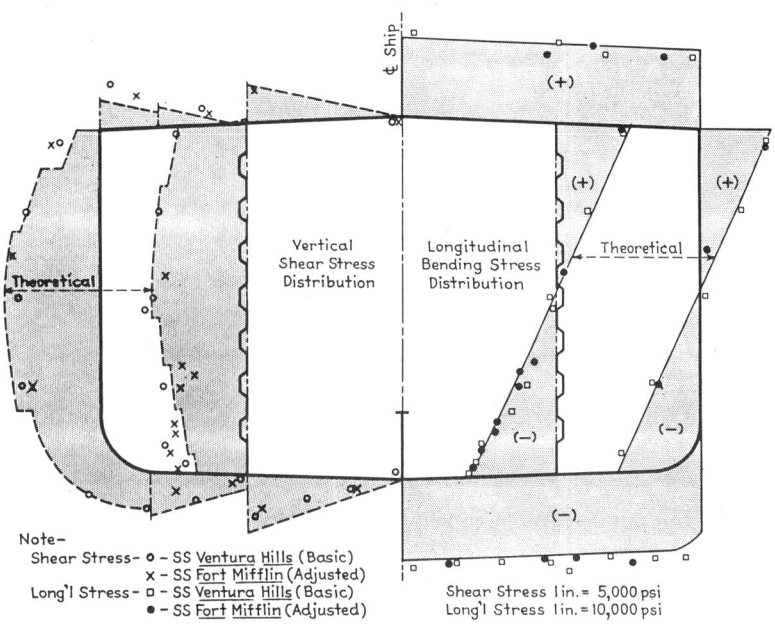

Fig. 20 Shear and bending moment study on tanker Ventura Hills (Vasta, 1958)

obtain the total stress. In performing this superposition, the relative phasing in time of the components must be kept in mind if the components represent responses to time-varying loads such as those caused by waves. Under such circumstances, for a particular location in the structure such as a point in the bottom plating, the maximum value of the primary stress may not necessarily occur at the same instant of time as the maximum of the secondary or tertiary stress at that same location.

3.2 Primary Direct Stress. The structural members involved in the computation of primary stress are, for the most part, the longitudinally continuous members such as deck, side, bottom shell, longitudinal bulkheads, and continuous or fully effective longitudinal primary or secondary stiffening members. By our original definition, however, primary stress also refers to the in-plane stress in transverse bulkheads due to the weights and shear loads transmitted into the bulkhead by the adjacent decks, bottom and side shell.

Elementary Bernoulli-Euler beam theory is usually utilized in computing the component of primary stress or deflection due to vertical or lateral hull bending loads. In assessing the applicability of this beam theory to ship structures, it is useful to restate the underlying assumptions:

- The beam is prismatic, i.e., all cross sections are the same.
- Plane cross sections remain plane, and merely rotate as the beam deflects.
- Transverse (Poisson) effects on strain are neglected,
- The material behaves elastically, the moduli of elasticity in tension and compression being equal.
- Shear effects (stresses, strains) can be separated from and do not influence bending stresses or strains.

Many experiments have been conducted to investigate the bending behavior of ships or ship-like structures as noted, for example in Vasta (1958). The results, in many cases, agree quite well with the predictions of simple beam theory, as shown in Fig. 20 taken from the Vasta paper, except in the vicinity of abrupt changes in cross section.

In way of deck openings, side ports, or other changes in the hull cross sectional structural arrangements, stress concentrations may occur that can be of determining importance in design of the structural members. Proper design calls for, first, the avoidance where possible of abrupt change in geometry, and, second, the introduction of compensating structural reinforcements such as doubler plates where stress concentrations cannot be avoided. In most cases, serious stress raisers are associated with local features of the structure, and the design of such features is treated in greater detail in Chapter VII of Taggart (1980).

Since stress concentrations cannot be avoided entirely in a highly complex structure such as a ship, their effects must be included in any comprehensive stress analysis. Methods of dealing with stress concentrations are presented in Section 3.14 of this chapter.

The derivation of the equations for stress and deflection under the assumptions of elementary beam theory may be found in any textbook on strength of materials, for instance, Timoshenko (1955). The elastic curve equation for a beam is obtained by equating the resisting moment to the bending moment, M, at section x, all in consistent units,

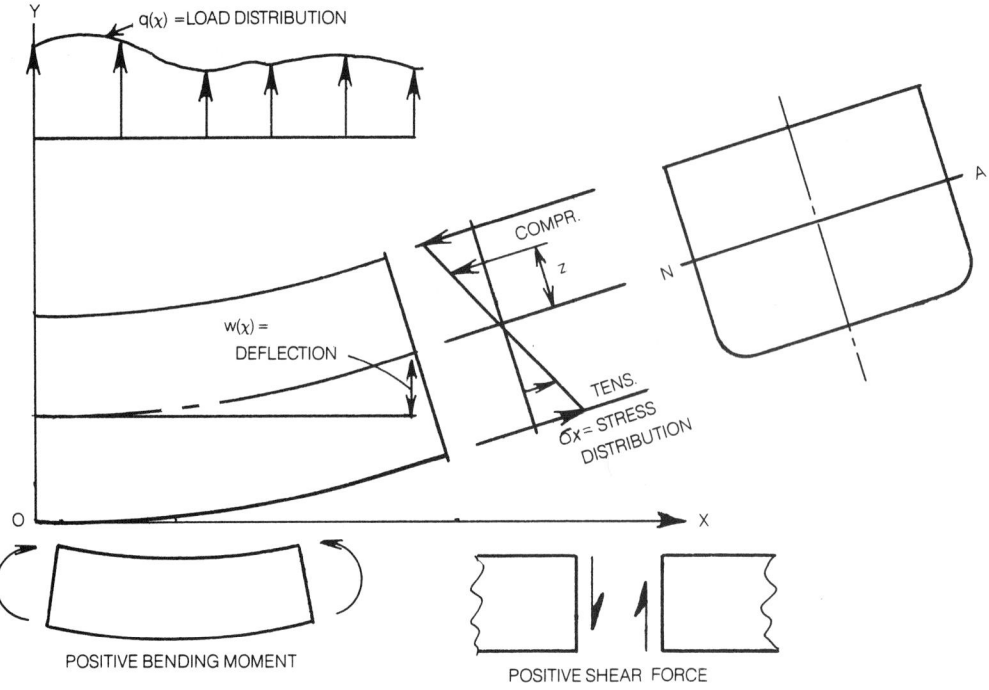

Fig. 21 Nomenclature for shear, deflection and loading of elementary beam

$$EI \frac{d^2w}{dx^2} = M(x) \quad (24)$$

Where:

w is deflection (Fig. 21),
E is modulus of elasticity of the material,
I is moment of inertia of beam cross section about a horizontal axis through its centroid.

This may be written in terms of the load per unit length, $q(x)$, as

$$EI \frac{d^4w}{dx^4} = q(x) \quad (25)$$

The deflection of the ship's hull as a beam is, obtained by the multiple integration of either of Equations (24) or (25). It may be seen that the deflection—hence stiffness against bending—depends upon both geometry (moment of inertia, I) and elasticity (E). Hence, a reduction in hull depth or a change to a material such as aluminum (E approximately 1/3 that of steel) will reduce hull stiffness.

Since flexibility is seldom a problem for hulls of normal proportions constructed of mild steel, primary structure is usually designed on the basis of strength considerations rather than deflection. However, classification society rules deal indirectly with the problem by specifying a limit on L/D ratio of 15 for oceangoing vessels and 21 for Great Lakes bulk carriers (which experience less severe wave bending moments.) Designs in which L/D exceeds these values must be "specially considered." There is also a lower limit on hull girder moment of inertia, which likewise has the effect of limiting deflection—especially if high-strength steels are used. An all-aluminum alloy hull would show considerably less stiffness than a steel hull having the same strength. Therefore, classification societies agree on the need for some limitation on deflection, although opinions differ as to how much.

Coming to strength considerations now, we note that the plane section assumption together with elastic material behavior results in a longitudinal stress, σ_x, in the beam that varies linearly over the depth of the cross section. The condition of static equilibrium of longitudinal forces on the beam cross section is satisfied if σ_x is zero at the height of the centroid of the area of the cross section. A transverse axis through the centroid is termed the neutral axis of the beam and is a location of zero stress and strain. Accordingly, the moment of inertia, I, in Equations (24) and (25) is taken about the neutral axis.

The longitudinal stress in Section x is related to the bending moment by the following relationship, as illustrated in Fig. 21.

$$\sigma_x = -\frac{M(x)}{I} z \quad (26)$$

It is clear that the extreme stresses are found at the top or bottom of the beam where z takes on its numerically largest values. The quantity SM $= I/z_o$, where z_o is either of these extreme values, is termed the section modulus of the beam. The extreme stress, deck or bottom, is given by[2]

$$\sigma_{x_0} = -\frac{M(x)}{\text{SM}} \qquad (27)$$

The sign of the stress, either tension or compression, is determined by the sign of z_0. For a positive bending moment the top of the beam is in compression and the bottom is in tension.

The computation of the section modulus for a ship hull cross section, taking into consideration all of the longitudinally continuous, load carrying members, is described in Section 3.3.

Two variations on the above beam equations may be of importance in ship structures. The first concerns beams composed of two or more materials of different moduli of elasticity, for example, steel and aluminum. In this case, the flexural rigidity, EI, is replaced by $\int_A E(z) z^2 dA$,

where:

A is cross sectional area,
$E(z)$ is modulus of elasticity of an element of area dA located at distance z from the neutral axis.

The neutral axis is located at such height that

$$\int_A E(z) z \, dA = 0 \qquad (28)$$

A second related modification may be described by considering a longitudinal strength member composed of thin plate with transverse framing. This might, for example, represent a portion of the deck structure of a transversely framed ship. Let us consider one module of a repeated system of deck plate plus transverse frame, as shown in Fig. 22, that is subject to a longitudinal stress, σ_x, from the primary bending of the hull girder. As a result of the longitudinal strain, ϵ_x, which is associated with σ_x, there will exist a transverse strain, ϵ_s. For the case of a plate that is free of constraint in the transverse direction, the two strains will be of opposite sign and the ratio of their absolute values, given by $|\epsilon_s/\epsilon_x| = \nu$, is a constant property of the material. The quantity ν is called *Poisson's Ratio* and, for steel, has a value of approximately 0.3.

In the module of deck plating shown in Fig. 22, the transverse beams exert some restraint against this transverse strain, with the result that stresses of opposite sign are set up in both the beam and plate. Equilibrium of the transverse force resultants of these stresses for the module is expressed as

$$\bar{\sigma}\bar{A} + \sigma_s A_p = 0 \qquad (29)$$

where:
$\bar{A}$ = cross sectional area of one stiffener

A_p = cross sectional area of one module of plating.
$\bar{\sigma}$ and σ_s are defined in Fig. 22.

Hooke's Law, which expresses the relation between stress and strain in two dimensions, may be stated in terms of the plate strains:

$$\begin{aligned}\epsilon_x &= \frac{1}{E}(\sigma_x - \nu \sigma_s) \\ \epsilon_s &= \frac{1}{E}(\sigma_s - \nu \sigma_x)\end{aligned} \qquad (30)$$

In the plate-stiffener field, there will be equality of strain at the joint of plate to stiffener. If the stiffeners are closely spaced, this is assumed to be applied, on the average, to the entire plate field and the procedure is sometimes referred to as a *smearing* of the effect of the stiffeners. The stiffeners themselves are assumed to behave as one-dimensional elastic members. The above equality of strain, therefore, requires

$$\bar{\sigma} = \sigma_s - \nu \sigma_x \qquad (31)$$

if the plate and bar are of the same material.

If $\bar{\sigma}$ and σ_s are eliminated from Equations (29), (31) and the first of (30), the relation between longitudinal stress and strain in the plate may be written

$$\frac{\sigma_x}{\epsilon_x} = E \frac{1 + r}{1 + r(1 - \nu^2)} \qquad (32)$$

Here $r = \bar{A}/A_p$ is the ratio of stiffener to plate area per module.

Thus, even though the entire cross section may be constructed of material having the same E, the section behaves as though it were a composite section if the stiffener size and/or spacing varies around the section. Note that the stress in the cross section is affected by the above phenomenon through a change in the position of the neutral axis compared to the unstiffened plate. The stiffness against bending is seen to be affected by both the geometry, through the moment of inertia, and the apparent modulus of elasticity. While these effects are usually insignificant in the static analysis of ship structures, they may be of importance in vibration analysis.

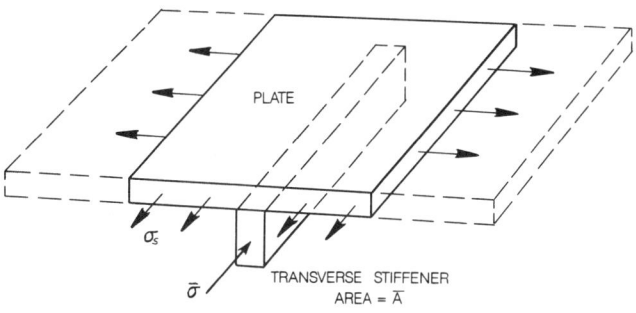

Fig. 22 Module of stiffened deck plate

---

[2] First stated by C. A. de Coulomb, French physicist, based on beam theory of Jakob Bernoulli and L. Euler.

Table 2—Section Modulus Calculation (English Units)

Moment of Inertia of Midship Section Cargo Vessel 563 ft 7¾ in. × 528 ft 6 in. × 76 ft 0 in. × 44 ft 6 in.
Assumed Neutral Axis = 20 ft 0 in. WL   Design Draft = 31 ft 6 in.

Member	Scantling	a	d_n	ad_n	ad_n^2	h	i_0
Granwale angle	8 × 8 × 1	15.00	24.79	371.85	9,218.08		
Main deck plating	276 × 1.125	310.50	24.92	7,737.66	192,820.49		
Main deck strap	13.5 × 1.125	15.19	25.25	383.55	9,684.64		
Second deck plating	276 × 0.5625	155.25	15.54	2,412.59	37,491.65		
Sheer strake	60 × 0.8125	48.75	23.00	1,121.25	25,788.75	5	101
Side shell	246.5 × 0.7187	175.36	10.28	1,802.27	18,527.34	20.54	6,169
					293,530.95		6,270.00
				Σi_0	6,270.00		
Total above 20 ft 0 in. WL		720.05		13,829.17	299,800.95		
Side shell	123 × 0.7187	88.40	5.24	463.22	2,427.27	10.25	775
Bilge strake	195 × 0.8125	158.44	17.25	2,733.09	47,145.80	9.65	1228
Bottom shell	315.5 × 0.8125	256.34	20.04	5,137.05	102,946.48		
Flat keel	26.5 × 1.00	26.50	20.04	531.06	10,642.44		
I.B. margin	53 × 0.5937	31.43	15.03	472.34	7,100.02		
I.B. center strake	26.5 × 0.5937	15.71	15.03	236.12	3,548.88		
I.B. plating	365.5 × 0.50	182.75	15.06	2,752.22	41,448.43		
C.V. keel	59 × 0.5937/2	17.49	17.50	306.08	5,356.40	4.91	33
Inboard longitudinal girder	59 × 0.5312	31.34	17.50	548.45	9,597.88	4.91	67
Outboard longitudinal girder	59 × 0.4062	23.97	17.50	419.48	7,340.90	4.91	48
					237,554.55		2,051
				Σi_0	2,051.00		
Total below 20 ft 0 in. WL		830.37		13,598.12	239,605.55		

$A = \Sigma a = 1{,}550.42$ $\Sigma ad_n = 231.05$ $I_n = 539{,}406.50$

$d_g = \dfrac{231.05}{1550.42} = 0.149$ $A \times d_g^2 = 1550.42 \times 0.149^2 = 34.43$

$I/2 = 539{,}406 - 34.43 = 539{,}372$ $I = 539{,}372 \times 2 = 1{,}078{,}744$

Top $C = 24.59 - 0.149 = 24.44$ Bottom $C = 20.08 + 0.149 = 20.23$

Top $\dfrac{I}{C} = \dfrac{1{,}078{,}744}{24.59} = 43{,}869$ Bottom $\dfrac{I}{C} = \dfrac{1{,}078{,}744}{20.23} = 53{,}323$

3.3 Calculation of Section Modulus. An important step in routine ship design is the calculation of midship section modulus. As defined in connection with Equation (27), it indicates the bending strength properties of the primary hull structure. The standard calculation is described in ABS (1987a), Section 6: "The section modulus to the deck or bottom is obtained by dividing the moment of inertia by the distance from the neutral axis to the molded deck line at side or to the base line, respectively." See Fig. 23.

"In general, the following items may be included in the calculation of the section modulus, provided they are continuous or effectively developed.

- Deck plating (strength deck and other effective decks).
- Shell and inner-bottom plating.
- Deck and bottom girders.
- Plating and longitudinal stiffeners of longitudinal bulkheads.
- All longitudinals of deck, sides, bottom and inner bottom.
- Continuous longitudinal hatch coamings."

The designation of which members should be considered as effective is subject to differences of opinion. The members of the hull girder of a ship in a seaway are stressed alternately in tension and compression, and certain of them will take compression even though deficiency in end connection makes them unable to take full tension, while other members, perhaps of light plating ineffectively stiffened, may be able to withstand tension stresses to the elastic limit, but may buckle under a moderate compressive stress. In general, however, only members which are effective in both tension and compression are assumed to act as part of the hull girder.

The section-modulus calculation for the cargo ship shown in Fig. 23 is carried out in Table 2. This calculation is based on the following formula for the moment of inertia of any composite girder section:

$$I = 2[I_n - A\,d_g^2] = 2[\Sigma(i_0 + ad_n^2) - A\,d_g^2] \quad (33)$$

where

I is moment of inertia of the section about a line par-

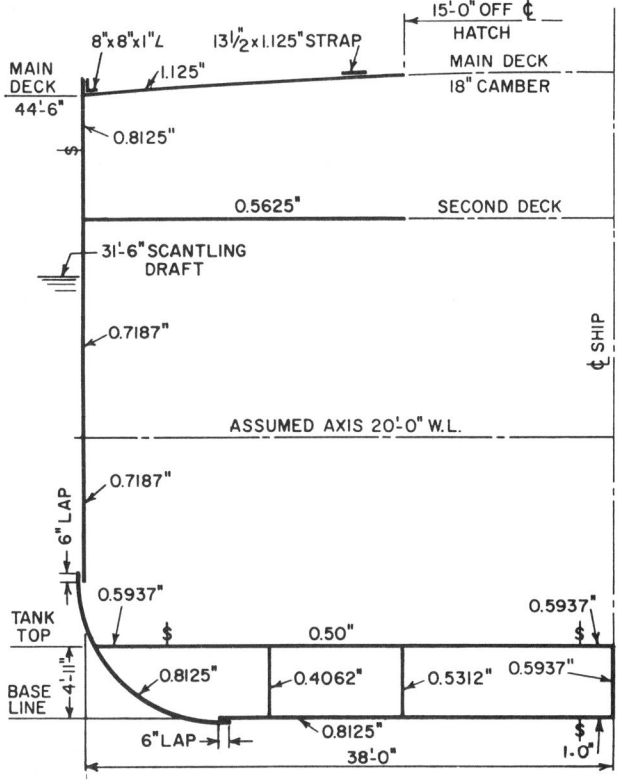

Fig. 23 Cargo ship midship section

allel to the base through the true neutral axis (center of gravity), expressed in cm²-m² (in²-ft²).

I_n is moment of inertia of the half-section about an assumed axis parallel to the true neutral axis, $= \Sigma(i_0 + ad_n^2)$.

A is total half-section area of effective longitudinal strength members, $= \Sigma a$, in cm² (in²). Generally no deduction is made for rivet holes.

d_g is distance from the assumed axis to the true axis, m (ft).

i_0 is vertical moment of inertia (about its own center of gravity) of each individual plate or shape effective for longitudinal strength.

a is area of each such plate or shape, in cm³ (in)².

d_n is distance of the center of gravity of each such plate or shape from the assumed axis, in m (ft).

Owing to symmetry it is necessary to include in the calculation only the structural parts on one side of the centerline, the result of the calculation being multiplied by 2 as indicated in Equation (33). Accordingly, quantities listed in Table 2 are for one side of the ship shown in Fig. 23.

If the assumed axis be assigned an arbitrary location, the known or directly determinable values are i_0, a, and d_n; hence I_n may be obtained. The value of A is also known and $d_g = \Sigma a d_n / \Sigma a = \Sigma a d_n / A$; therefore $A d_g^2$ is determinable. The baseline may be used for the assumed axis. There is, however, some advantage in using an assumed axis at about mid-depth in that lever arms are decreased. In that case the assumed axis should be located at about 45 percent of the depth of the section above the baseline, the actual position of the neutral axis being normally at less than half-depth because the bottom shell plating has greater sectional area than has deck plating (except in such cases as tankers). This condition is accentuated when an inner bottom is fitted.

After I had been calculated as outlined and as indicated in Table 2, the section modulus I/c may be obtained to both top and bottom extreme fibers.

For the sake of convenience and uniformity, the following conventions are usually observed:

• Since the moments of inertia i_0 of individual horizontal members are negligible, they are omitted from the calculations.

• The top c is taken from the neutral axis to the deck at side, the bottom c to the baseline.

3.4 Distribution of Shear and Transverse Stress Components. The simple beam theory expressions given in the preceding section permit us to evaluate the longitudinal component of the primary stress, σ_x. In Fig. 24 we see that an element of shell or deck plating may, in general, be subject to two other components of stress, a direct stress in the transverse direction and a shearing stress. Fig. 24 illustrates these as the *stress resultants*, defined as the stress multiplied by plate thickness. The stress resultants have dimensions of force per unit length and are given by the following expressions:

$N_x = t\sigma_x, N_s = t\sigma_s$ stress resultants
$N = t\tau$ shear stress resultant or *shear flow*
σ_x, σ_s stresses in the longitudinal and girth-wise directions
τ shear stress
t plate thickness

Here σ_s designates the transverse direct stress parallel to the vertical axis in the ship's side and parallel to the transverse axis in the deck and the bottom.

Through considerations of static equilibrium of a triangular element of plating, it may be shown that the plane stress pattern described by the three component stresses σ_x, σ_s, τ may be reduced to a pair of alternative direct stresses, σ_1, σ_2. The stresses σ_1, σ_2 are called *principal stresses* and the directions of σ_1 and σ_2 are principal stress directions. The principal stresses are related to σ_x, σ_y and τ by

$$\sigma_1, \sigma_2 = \frac{\sigma_x + \sigma_s}{2} \pm \sqrt{\left(\frac{\sigma_x - \sigma_s}{2}\right)^2 + \tau^2} \quad (34)$$

The two angles, θ, between the x-axis and the directions of σ_1, and σ_2 are

$$\tan 2\theta = -\frac{2\tau}{\sigma_x - \sigma_s} \quad (35)$$

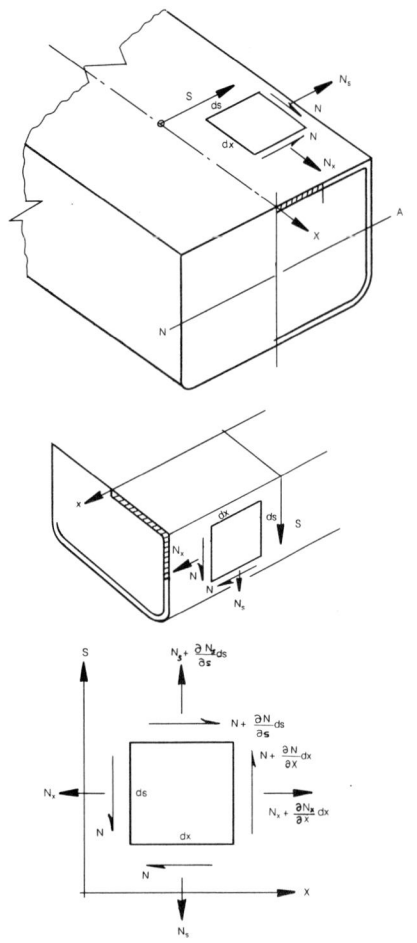

Fig. 24 Element of plate structure in deck or side shell, illustrating components of bending stress resultants

Detailed derivation of these expressions may be found, for example, in Timoshenko (1955).

In many parts of the ship, the longitudinal stress, σ_x, is the dominant component. There are, however, locations in which the shear component becomes important and under unusual circumstances the transverse component may, likewise, become important. A suitable procedure for estimating these other component stresses may be derived by considering the equations of static equilibrium of the element of plating illustrated in Fig. 24. In case the stiffeners associated with the plating support a part of the loading, this effect may also be included.

The static equilibrium conditions for the element of plate subject only to in-plane stress (i.e., no bending of the plate) are

$$\frac{\partial N_x}{\partial x} + \frac{\partial N}{\partial s} = 0$$
$$\frac{\partial N_s}{\partial s} + \frac{\partial N}{\partial x} = 0 \qquad (36)$$

In these expressions, s, is the girthwise coordinate measured on the surface of the section from the x-axis as shown in Fig. 24.

The first of Equations (36) may be integrated in the s-direction around the ship section in order to obtain the shear stress distribution. For this purpose, we assume, as a first approximation, that the longitudinal stress, σ_x, is given by the beam theory expression (27). Then, if we assume that the hull girder is prismatic (or that I changes slowly in the x-direction) so that only $M(x)$ varies with x, the derivative of N_x is given by differentiating Equation (26),

$$\frac{\partial N_x}{\partial x} = -\frac{tz}{I}\frac{dM(x)}{dx} \qquad (37)$$
$$= -\frac{tz}{I} V(x)$$

where $V(x)$ = shear force in the hull at x.

The shear flow distribution around a section is given by integrating the first of Equations (36) in the s-direction,

$$N(s) - N_0 = \int_0^s \frac{\partial N}{\partial s} ds$$
$$= -\int_0^s \frac{\partial N_x}{\partial x} ds \qquad (38)$$
$$= \frac{V(x)}{I} \int_0^s tz\,ds$$

Here, N_0, the constant of integration, is equal to the value of the shear flow at the origin of integration, $s = 0$. By proper choice of the origin, N_0 can often be set equal to zero.

For example, in a section having transverse symmetry and subject to a bending moment in the vertical plane, the shear stress must be zero on the centerline, which therefore, is a suitable choice for the origin of the girthwise integration. The shear flow distribution around a single-walled symmetrical section is then given by

$$N(s) = \frac{V(x)}{I} m(s) \qquad (39)$$

with $N_0 = 0$ in the case of such symmetry.

The quantity $m(s) = \int_0^s tz\,ds$ is the first moment about the neutral axis of the cross sectional area of the plating between the origin at the centerline and the variable location designated by s. This is the shaded area of the section shown in Fig. 24.

If a longitudinal frame or girder that carries longitudinal stress is attached to the plate, as shown in Fig. 25, there will be a discontinuity in the shear flow, $N(s)$, at the frame corresponding to a jump in $m(s)$. This may be seen by considering the equilibrium of forces in the x-direction of the system of plate shears

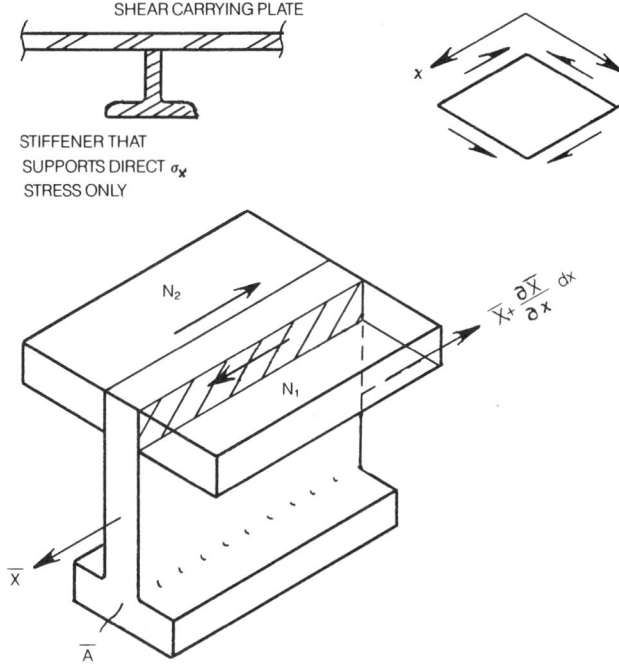

Fig. 25 Free body diagram of plate-frame joint

on each side of the frame plus the x-stress in the frame. These are shown in the lower part of Fig. 25. Immediately adjacent to the frame on either side, the plate shear flows are N_1 and N_2. The stress in the frame, $\overline{\sigma}_x$, is given by Equation (26), and the total force in the frame, $\overline{X}$, is obtained by integrating this expression over the cross-section area, $\overline{A}$, of the frame.

$$\overline{X} = \int_{\overline{A}} \overline{\sigma}_x \, dA \qquad (40)$$
$$= -\frac{M(x)}{I} \overline{m}_A$$

where $\overline{m}_A = \int_{\overline{A}} z \, dA$ is the moment of the section area of the frame about the ship neutral axis.

Equilibrium of forces in the x-direction requires that

$$\overline{X} + \frac{\partial \overline{X}}{\partial \overline{X}} dX - \overline{X} + (N_2 - N_1) \, dx = 0 \qquad (41)$$

or, the stepwise change in shear flows at the frame will be given by

$$N_2 - N_1 = -\frac{\partial \overline{X}}{\partial x} \qquad (42)$$
$$= \frac{V(x)}{I} \overline{m}_A$$

Thus, in evaluating the moment, $m(s)$, in Equation (39) a finite increment equal to $\overline{m}_A$ is added, and there is a corresponding jump in $N(s)$, as the path of integration encounters a frame.

We note that the moment of the stiffener cross section may be written as,

$$\overline{m}_{A_i} = \overline{A}_i \overline{z}_i \qquad (43)$$

where:

$\overline{A}_i$ is sectional area of frame i
$\overline{z}_i$ is distance from neutral axis to centroid of $\overline{A}_i$.

Equation (39) then becomes,

$$N(s) = \frac{V(x)}{I} \left[m(s) + \sum_i \overline{A}_i \, \overline{z}_i \right] + N_0 \qquad (44)$$

For a rectangular cross section having no longitudinal frames, the shear flow distribution corresponding to Equation (39) will have a linear variation in the deck and parabolic variation in the topsides as shown in Fig. 26.

The transverse stress, σ_s, or equivalently, the stress resultant, N_s, may be found by integration of Equation (45),

$$\frac{\partial N_s}{\partial s} = -\frac{\partial N}{\partial x} \qquad (45)$$

Substituting Equation (39) for N, we obtain

$$N_s(s) - N_{so} = -\int_0^s \frac{\partial}{\partial x} \left[\frac{V(x)}{I} m(s) \right] ds \qquad (46)$$

Assuming a prismatic section so that only $V(x)$ depends upon x, and noting that $\frac{\partial V}{\partial x} = q(x)$, the vertical load per unit length, we obtain

$$N_s(s) - N_{so} = -\frac{q(x)}{I} \int_0^s m(s) \, ds \qquad (47)$$

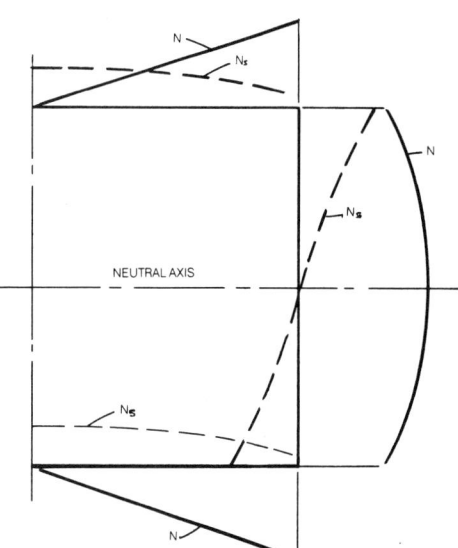

Fig. 26 Shear flow and girthwise stress around a rectangular ship cross section

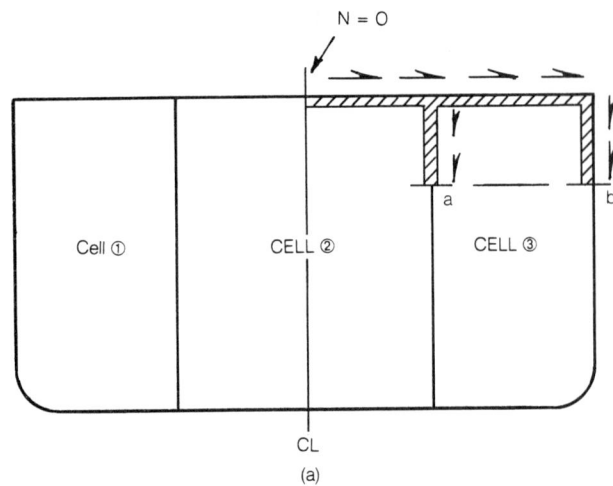

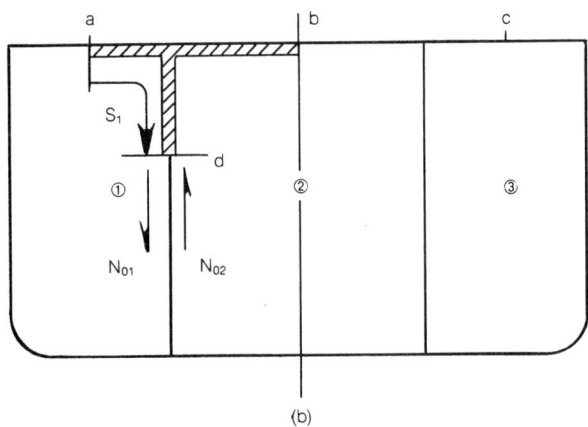

Fig. 27 Shear flow in multiple-bulkhead tanker section

For a rectangular section, this expression gives a parabolic variation of N_s in the horizontal members and a cubic variation in vertical members. The constant of integration, N_{so}, is adjusted so that N_s is equal to the girthwise stress resultant at some location where this stress can be determined. Examples of such locations are the edge of a deck, side or bottom panel.

In the side plating at the deck (upper) edge, the girthwise resultant, N_s, must be equal to the weight per unit length of deck cargo plus deck structure for one side of the ship. At the bottom (bilge) it would be one half of the total buoyancy per unit length less the downward force due to structural weight and internal cargo, liquids, etc. that are supported directly by the bottom plating. In the bottom plating the girthwise resultant is in the transverse direction and will equal the resultant transverse hydrostatic force per unit length. Fig. 26 illustrates the distribution of N_s for a rectangular barge section carrying deck cargo and subject to a hydrostatic load on sides and bottom. It is assumed that the transverse framing transmits all of the water pressure load from the bottom plate panel to the edge where it is transmitted as in-plane stress into the side plating and similarly for the hydrostatic pressure on the sides and weight load on deck. With few exceptions, for example a ship carrying very large localized deck load, such vertical and horizontal in-plane stresses are usually negligibly small in comparison to the longitudinal and shear stresses.

3.5 Shear Flow in Multicell Sections. If the cross section of the ship shown in Fig. 24 is subdivided into two or more closed cells by longitudinal bulkheads, tank tops, or decks, the problem of finding the shear flow in the boundaries of these closed cells is statically indeterminate. In order to visualize this, refer to Fig. 27 showing a typical tanker midship section which is subdivided into three cells by the two longitudinal bulkheads. Equation (39) may be evaluated for the deck and bottom of the center tank space since the plane of symmetry at which the shear flow vanishes, lies within this space and forms a convenient origin for the integration. At the deck-bulkhead intersection, the shear flow in the deck divides, but the relative proportions of the part in the bulkhead and the part in the deck are indeterminate. The sum of the shear flows at two locations lying on a plane cutting the cell walls, e.g., at points a and b in Fig. 27 (a), will still be given by (39), with $m(s)$ equal to the moment of the shaded area. However, the distribution of this sum between the two components in bulkhead and side shell requires additional information for its determination.

This additional information may be obtained by considering the torsional equilibrium and deflection of the cellular section. In order to develop the necessary equations, we first consider a closed, single cell, thin-walled prismatic section subject only to a twisting moment, M_T, which is constant along the length as shown in Fig. 28. The resulting shear stress may be assumed uniform through the plate thickness and is tangent to the mid-thickness of the material. Under these circumstances, the deflection of the tube will consist of a twisting of the section without distortion of its shape, and the rate of twist, $d\theta/dx$, will be constant along the length.

Now consider equilibrium of forces in the x-direction for the element $dx\,ds$ of the tube wall as shown in Fig. 28(b). Since there is no longitudinal load, there will be no longitudinal stress, and only the shear stresses at the top and bottom edges need be considered in the expression for static equilibrium, giving,

$$-t_1\tau_1 dx + t_2\tau_2 dx = 0 \qquad (48)$$

The shear flow, $N = t\tau$, is therefore seen to be constant around the section.

The magnitude of the moment, M_T, may be computed by integrating the moment of the elementary force arising from this shear flow about any convenient axis. If r is the distance from the axis, 0, perpendicular to the resultant shear flow at location s, as shown in Fig. 28(c) we find,

$$M_T = \oint rN\,ds$$
$$= N\oint r\,ds \qquad (49)$$
$$= 2N\Omega$$

Here the symbol $\oint$ indicates that the integral is taken entirely around the section and, therefore, Ω is the area enclosed by the mid-thickness line of the tubular cross section. The constant shear flow is then related to the applied twisting moment by,

$$N = M_T/2\Omega \qquad (50)$$

We now consider the deformation of the element $ds\,dx$ which results from this shear. Let u, v be the displacements in the axial and tangential directions respectively of a point on the surface of the tube. This is shown in Fig. 28(d). As a result of the constant twisting moment and prismatic geometry, u is seen to be a function of s only and v will be given by a rigid body rotation of the cross section,

$$\frac{\partial v}{\partial x} = r\frac{d\theta}{dx} \qquad (51)$$

where r is defined as before.

From elementary elasticity, the shear strain, γ, is related to the displacements by,

$$\gamma = \frac{\partial u}{\partial s} + \frac{\partial v}{\partial x} \qquad (52)$$

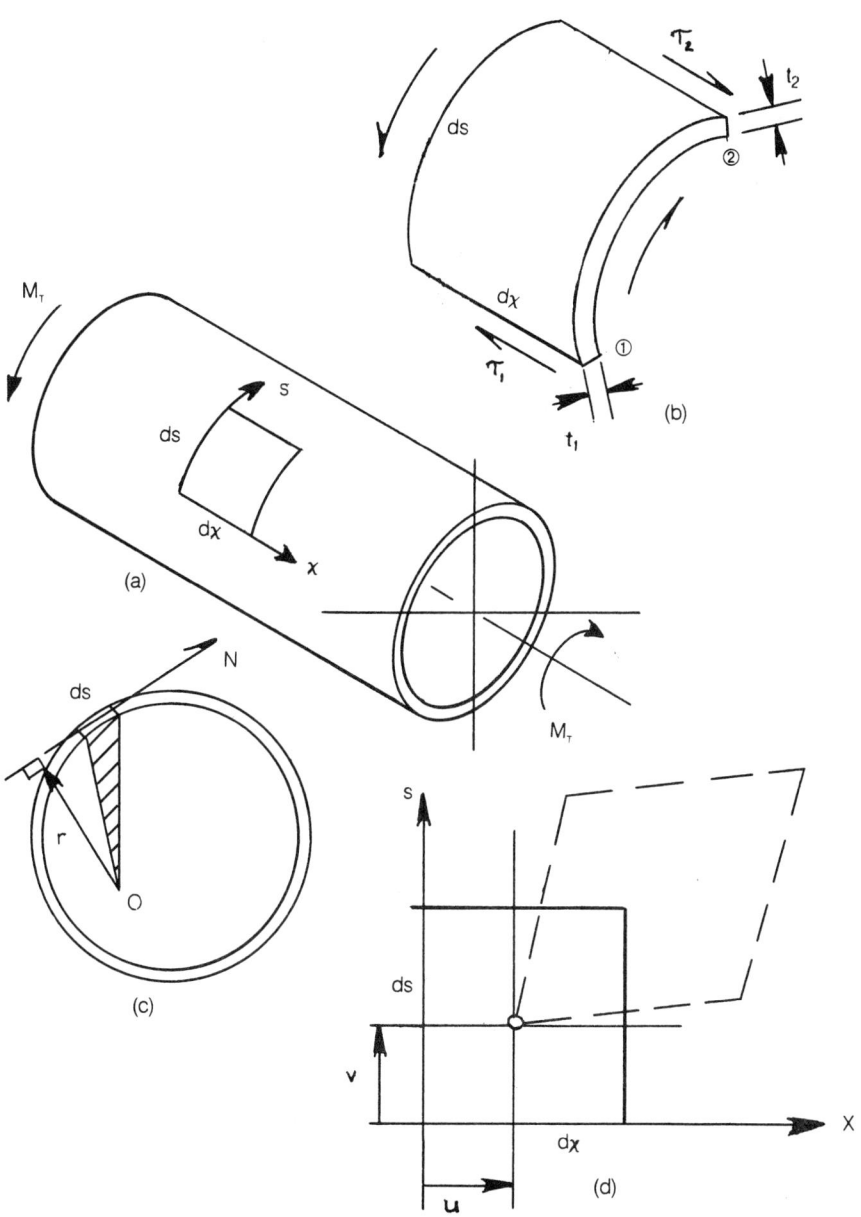

Fig. 28 Twist of closed prismatic tube

also, $\gamma = \dfrac{\tau}{G}$, where $G = \dfrac{E}{2(1+\nu)}$ is the shear modulus of elasticity. Substituting Equation (51) into (52) and rearranging, we obtain,

$$\frac{\partial u}{\partial s} = \frac{\tau}{G} - r\frac{d\theta}{dx} \qquad (53)$$

Since u depends only on s, this may be integrated to give

$$\begin{aligned} u(s) &= \int \frac{\tau}{G}\,ds - \int r\frac{d\theta}{dx}\,ds + u_0 \\ &= \frac{1}{G}\int \frac{N}{t}\,ds - \frac{d\theta}{dx}\int r\,ds + u_0 \end{aligned} \qquad (54)$$

where u_0 is a constant of integration.

The quantity, $u(s)$, termed the *warp*, is seen to be the longitudinal displacement of a point on the cell wall, which results from the shear distortion of the material due to twist. If the section is circular the rotation will take place without warping, but for other shapes the warping will be nonzero and will vary around the perimeter of the section.

For a closed section, the differential warp must be zero if the integral in Equation (54) is evaluated around the entire section, i.e., two points on either side of a longitudinal line passing through the origin of the s-integration cannot be displaced longitudinally with respect to each other. This is expressed by

$$\frac{1}{G}\oint \frac{N}{t}\,ds - \frac{d\theta}{dx}\oint r\,ds = 0 \qquad (55)$$

Noting that N is constant around the section and recalling that the second integral was previously represented by 2Ω, the relationship between shear flow and rate of twist is given by the Bredt formula,

$$\frac{d\theta}{dx} = \frac{N}{2\Omega G}\oint \frac{ds}{t} \qquad (56)$$

Substituting (50) for the twisting moment, this gives

$$\frac{d\theta}{dx} = \frac{M_T}{4\Omega^2 G}\oint \frac{ds}{t} \qquad (57)$$

Equation (54) for the warp, $u(s)$, applies to any thin-walled prismatic tube if it can be assumed that the tube twists in such a way that cross sections rotate without distortion of their shape.

Let us now write the shear flow in the tanker section, Equation (39), as the sum of two parts, $N(s) = N_1(s) + N_0$, where N_0 is an unknown constant shear flow. Under a pure vertical loading, the twist of the section, $d\theta/dx$ must be zero, and for this case, Equation (55) becomes

$$\begin{aligned} 0 &= \frac{1}{G}\oint (N_1 + N_0)\frac{ds}{t} \\ &= \frac{1}{G}\oint N_1 \frac{ds}{t} + \frac{N_0}{G}\oint \frac{ds}{t} \end{aligned} \qquad (58)$$

This may be solved for the unknown constant,

$$N_0 = -\frac{\oint N_1 \dfrac{ds}{t}}{\oint \dfrac{ds}{t}} \qquad (59)$$

Equation (59) provides the means of evaluating the constant of integration in Equation (39) or (44) in the case of sections of general shape for which a location of zero shear flow cannot be determined by inspection.

A physical significance may be attached to the quantity N_0 by the following reasoning. Assume that the

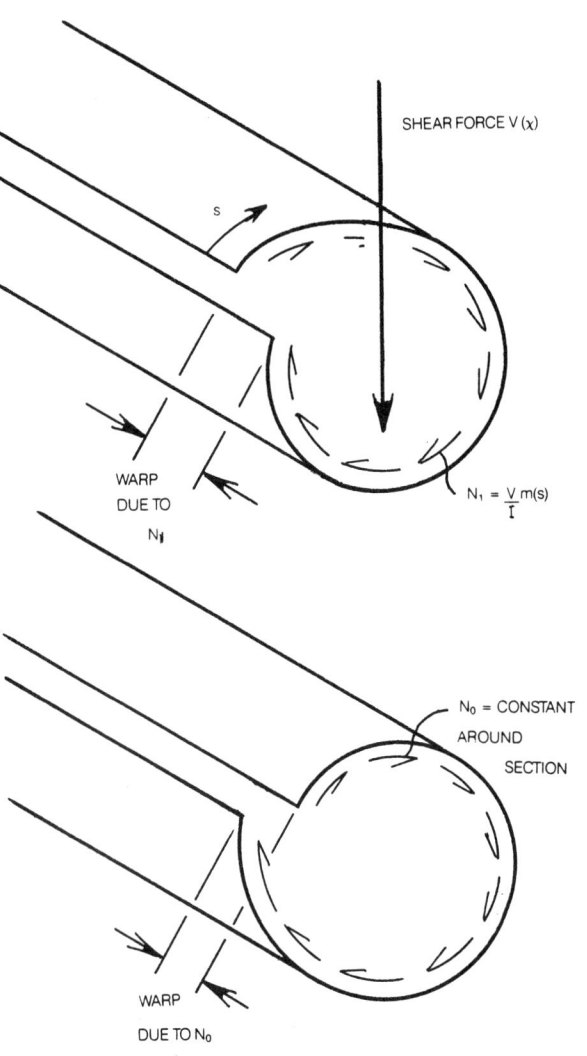

Fig. 29 Warp associated with N_1 and N_0

closed tubular section subject to a vertical loading without twist is transformed into an *open* section by an imaginary longitudinal slit, and the edge of this slit is taken as the origin of the *s*-coordinate. This is shown in Fig. 29. Corresponding to the shearing strains, the two edges of the slit will displace longitudinally relative to each other. If we compare Equations (54) and (58) we see that, since $\frac{d\theta}{dx} = 0$, the two terms in (58) represent *warping* displacements of one edge of the slit relative to the other corresponding to N_1 and N_0, respectively.

The first term is the warping displacement caused by the statically determinate shear flow N, which is given by Equation (44). The second term is the warp due to the constant shear flow N_0. Equation (58) is the statement that N_0 is of such magnitude that the net warp must be zero, i.e., it is of such magnitude that the slit is closed up.

We will now apply a similar procedure to the multicell section shown in Fig. 27. Results will be given for the general case of several closed cells and later specialized for a case of symmetrical section as illustrated.

We first imagine each cell to be cut with longitudinal slits at points *a*, *b*, *c* in Fig. 27(b). Let $N_i(s)$ be the shear flow in cell *i* obtained by evaluating Equation (44) with the origin of *s* located at the slit in that cell. Let N_{i0} be the constant of integration for cell *i*. Note that, when we compute $m(s)$ in Equation (44) for a location in the bulkhead such as point *d*, we must include the area of deck plating and frames up to the imaginary slit in the adjacent cell 2.

The relative warp at slit *a* due to $N_1(s)$ is given by

$$u_{1a} = \frac{1}{G} \oint_1 N_1(s) \frac{ds}{t} \quad (60)$$

where $N_{i(s)}$ is given by Equation (44).

Equations similar to these may be written for each of the remaining cells. For each cell there will be an additional constant shear flow corresponding to the constant of integration. The differential warp in the adjacent edges of one of the slits will include the effects of these constant shear flows for the present cell as well as the constant shear flows for adjacent cells acting along any interior walls that are common to the present and the adjacent cell. For cell 1, this additional warp at slit *a*, resulting from the constant shear flows acting on the boundaries of cell 1, is

$$u_{0a} = \frac{1}{G} N_{01} \oint_1 \frac{ds}{t} - \frac{1}{G} N_{02} \int_{1-2} \frac{ds}{t} \quad (61)$$

The second integral in this expression is evaluated only over the bulkhead dividing cells 1 and 2 and is negative since the constant shear flows of the two cells oppose each other in the bulkhead.

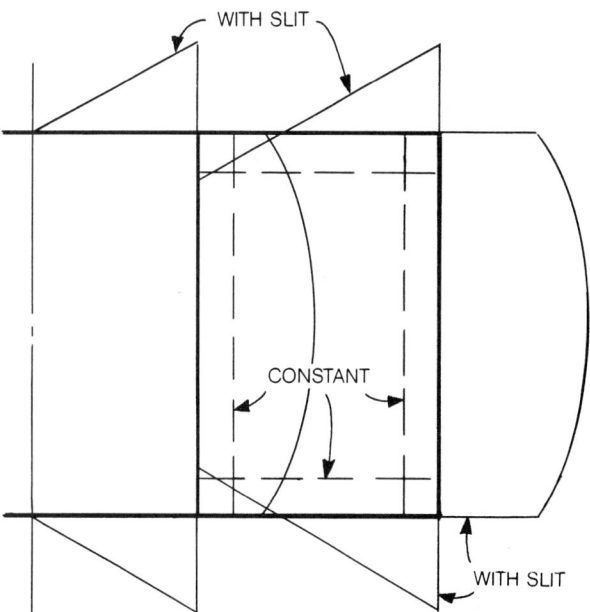

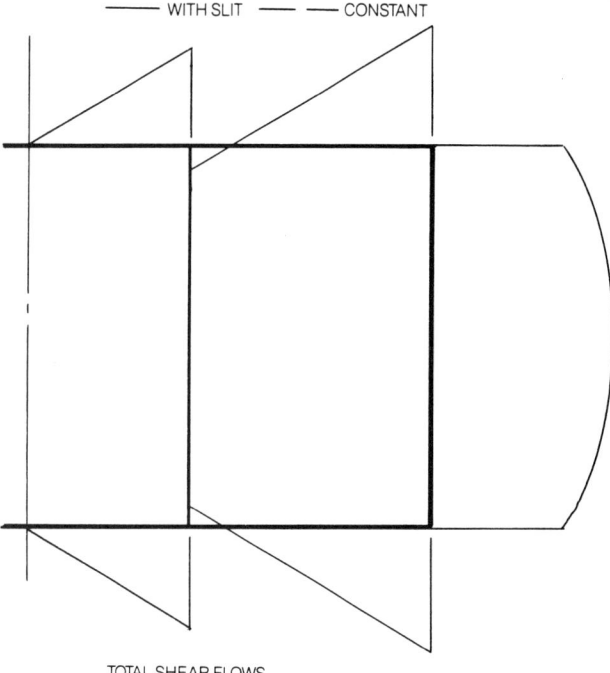

Fig. 30 Component and total shear flows in tanker midship section

We now require that the total warp at slit *a* must vanish and this is given by the condition

$$\oint_1 N_1(s) \frac{ds}{t} + N_{01} \oint_1 \frac{ds}{t} - N_{02} \int_{1-2} \frac{ds}{t} = 0 \quad (62)$$

Similar equations may be written for the remaining cells. For the middle cell, 2:

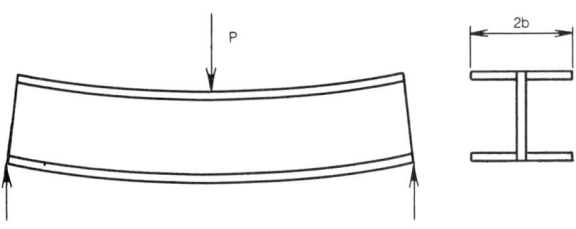

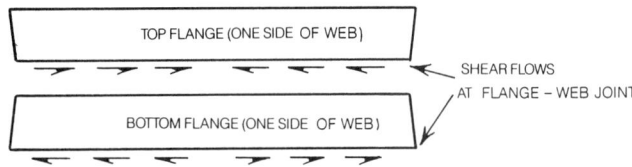

Fig. 31 Loading and resulting strain in flanges of simple beam

$$\oint_2 N_2(s)\frac{ds}{t} + N_{02}\oint_2 \frac{ds}{t} - N_{01}\int_{2\text{-}1}\frac{ds}{t}$$
$$- N_{03}\int_{2\text{-}3}\frac{ds}{t} = 0 \quad (63)$$

For cell 3:

$$\oint_3 N_3(s)\frac{ds}{t} + N_{03}\oint_3 \frac{ds}{t} - N_{02}\int_{3\text{-}2}\frac{ds}{t} = 0 \quad (64)$$

Now observe that the first term in each of these equations may be evaluated and is, therefore, a known quantity. The three equations may, therefore, be solved simultaneously for the three constants of integration N_{01}, N_{02}, N_{03}. Note also that the moment term $m(s)$ in the expressions for N_1, N_2 and N_3 must include the moment of the area of longitudinal stiffeners as well as plating, as shown in Equation (44), while the integrals in Equations (62), (63) and (64) include only the plating. The explanation for this is that the latter integrals evaluate the warping displacements and these result from the shear deformation of the plating only.

In the case of a three-cell, twin-bulkhead tanker that is symmetrical about the centerline plane, we may place slit b on the centerline, in which case N_{02} is zero. Equations (62) and (64) now contain only one unknown constant of integration each, and each may be solved explicitly. Furthermore, by symmetry, it is necessary to solve for only one of the two constants of integration and this is given by

$$N_{01} = -\frac{\oint_1 N_1(s)\frac{ds}{t}}{\oint_1 \frac{ds}{t}} \quad (65)$$

The resulting total shear flow in cell 1 (and cell 3 by symmetry considerations) will be given by

$$N(s) = \frac{V(x)}{I}\left[m(s) - \frac{\oint_1 m(s)\frac{ds}{t}}{\oint_1 \frac{ds}{t}}\right] \quad (66)$$

The shear flows in the deck and bottom of cell 2 are, of course, statically determinate and may be computed directly from Equation (44) with the constant of integration set equal to zero. The resulting component and total shear flows in the typical tanker section are shown in Fig. 30.

3.6 Shear Lag and Effective Breadth. In many ship structural components, the loading is resisted, on a local level, by web-type members before being transformed into a beam bending load. A simple illustration of this, shown in Fig. 31, is a wide-flange girder supporting a concentrated weight. At the local level, the weight is first transmitted into the web of the girder where it is resisted by vertical shears in the web. As a result of the concentrated load, the web tends to bend as shown, resulting in a compressive strain on the upper edge and tensile strain along the lower. The flanges are, of course, required to have a strain equal to the strain in the web at the joint between web and flange, and this results in a shear loading being applied to this edge of each flange member. Each of the four flange members may therefore be viewed as a rectangular strip of plating having a shearing load applied along one of the long edges. The resultant deformation is illustrated in Fig. 31.

The loading situation described above is not confined to isolated beams that support concentrated loads, but appears in many parts of the ship structure in association with distributed as well as concentrated loads. Consider for example the hull girder as a whole. Water pressure on the bottom is resisted locally by the plating that transmits the pressure load to the surrounding frames and floors. Longitudinal framing transmits the force system into the transverse bulkheads or web frames, which in turn transmit their loads into the longitudinal bulkheads or side shell. In the case of transverse frames, these resultant loads are transmitted directly into the side shell by each frame individually. Thus, the water pressure forces and, similarly, the weight forces are ultimately transmitted into the side shell or vertical web of the hull girder as concentrated shearing loads at each bulkhead or transverse frame. The bending tendency of the web (side shell) and the shear loading of the edges of the flange (deck plating) are directly analogous to the behavior of the simple girder illustrated in Fig. 31.

An important effect of this edge shear loading of a plate member is a resulting nonlinear variation of the longitudinal stress distribution. This is in contrast to the uniform stress distribution predicted in the beam flanges by the elementary beam Equation (27). In many practical cases, the departure from the value predicted by Equation (27) will be small, as shown in Fig. 20. But in certain combinations of loading and structural geometry, the effect referred to by the term *shear lag* must be taken into consideration if an accurate estimate of the maximum stress in the member is to be made. This may be conveniently done by defining an *effective breadth* of the flange member. The nomenclature used in the definition of this quantity is illustrated in Fig. 32, where the effective breadth, ρb, is defined as the breadth of plate that, if stressed uniformly at the level σ_B across its width, would sustain the same total load in the x-direction as the nonuniformly stressed plate. Hence,

$$\rho b = (1/\sigma_B) \int_0^b \sigma_x(y) \, dy \qquad (67)$$

The quantity ρ is called the plate effectiveness.

The solution for ρ is seen to require the determination of the plane stress distribution in a plate field under the described edge shear loading. In addition to this edge shear loading, there are kinematic boundary conditions that must be satisfied, appropriate to the physical conditions that prevail on the other edges of the plate. Such conditions would include zero direct and shear stress on a free edge, and zero displacement in the direction normal to the edge if the plate-bar combination is part of a repeating arrangement.

A solution procedure for the stress distribution and the plate effectiveness may be developed making use of the *Airy Stress Function*, for which the fundamental considerations are given in Timoshenko and Goodier (1970). Let us define a function $\phi(x, y)$ such that the direct stresses, σ_x, and σ_y, are given by the following expressions,

$$\sigma_x = \frac{\partial^2 \phi}{\partial y^2}$$
$$\sigma_y = \frac{\partial^2 \phi}{\partial x^2} \qquad (68)$$

From the conditions of static equilibrium of a plane stress element, Equations (36), the shear stress will be given by,

$$\tau = -\frac{\partial^2 \phi}{\partial x \, \partial y} \qquad (69)$$

If Equations (68) and (69) are substituted into the stress-strain Equations (32), the following relations are found:

$$\epsilon_x = \frac{1}{E}\left(\frac{\partial^2 \phi}{\partial y^2} - \nu \frac{\partial^2 \phi}{\partial x^2}\right)$$
$$\epsilon_y = \frac{1}{E}\left(\frac{\partial^2 \phi}{\partial x^2} - \nu \frac{\partial^2 \phi}{\partial y^2}\right) \qquad (70)$$
$$\gamma = -\frac{1}{G}\frac{\partial^2 \phi}{\partial x \, \partial y}$$

A relationship between the direct strains and the shear strain is given by the condition of compatibility

$$\frac{\partial^2 \gamma}{\partial x \, \partial y} = \frac{\partial^2 \epsilon_x}{\partial y^2} + \frac{\partial^2 \epsilon_y}{\partial x^2} \qquad (71)$$

If Equations (70) are now substituted into (71) the result is

$$\frac{\partial^4 \phi}{\partial x^4} + 2\frac{\partial^4 \phi}{\partial x^2 \partial y^2} + \frac{\partial^4 \phi}{\partial y^4} = 0 \qquad (72)$$

Equation (72) is recognized as the *biharmonic equation*, and is the field equation to be satisfied by the stress function, $\phi(x, y)$, at all points in the interior of the plate field. The complete solution of the stress distribution problem requires that we first obtain an expression for $\phi(x, y)$ by solving Equation (72), subject to a set of appropriate boundary conditions on the edges of the plate field. Having this solution for $\phi(x, y)$, we may then obtain the stresses by substitution in

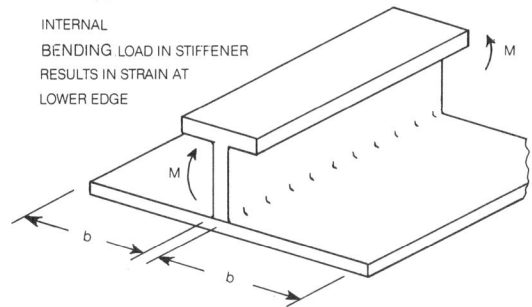

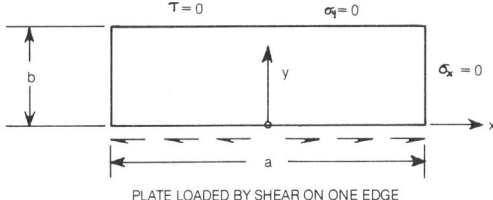

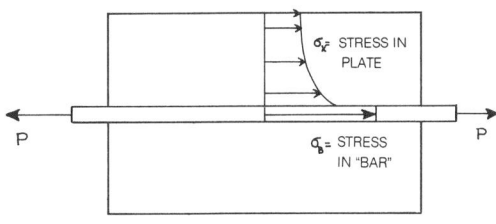

Fig. 32 Nomenclature for shear lag analysis

Equations (68). The plate effective breadth is then found by evaluating the integral in Equation (67).

Examples of typical boundary conditions that may be encountered in ship structural configurations are shown in Fig. 33.

A solution to Equation (71) was obtained by Schade (1951, 1953) using the method of separation of variables. In the Schade solution, the stress function is expanded in a Fourier Series in the longitudinal coordinate,

$$\phi(x,y) = \sum_{n=1}^{\infty} f_n(y) \sin \frac{n\pi x}{L} \qquad (73)$$

Here L is some characteristic length in the x-direction, and $f_n(y)$ is an unknown function of the transverse variable, y.

Upon substituting this expression into the biharmonic equation (72) an ordinary differential equation is obtained for the unknown function $f_n(y)$. Details of the solution need not be repeated here but may be obtained by referring to the Schade papers.

The form of the solution is such that there is an insufficient number of constants of integration of (72) to allow the satisfaction of all of the necessary boundary conditions, and the solution must, therefore, be considered as incomplete. By an appropriate choice of the partial boundary conditions, however, in combination with the application of *St. Venant's Principle*,[3] the solution may be considered accurate for most practical cases.

In order to illustrate this principle, consider the pair of plate panels loaded by the welded bar at the middle, which is illustrated in Fig. 32. The boundary conditions for one of the plate panels are shown in Fig. 33. Integration of the fourth-order differential Equation (42) yields four constants of integration that are found by introducing boundary conditions on all of the plate edges, plus the condition of static equilibrium under the applied load system. With the solution form that was assumed, three of these conditions plus the load equilibrium condition may be satisfied. In many practical cases it is sufficient to satisfy the boundary conditions on the edges parallel to the x-axis, and to neglect the conditions at the ends of the plate panel. According to St. Venant's Principle, the results will then be in error near the plate ends, since the neglect of the boundary conditions is equivalent to the substitution of an equivalent but unspecified end loading. The error will, however, be acceptably small in the interior of the panel if the length is large compared

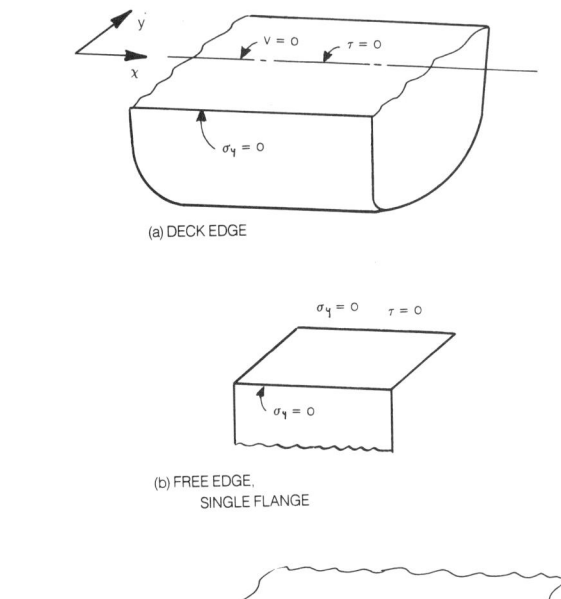

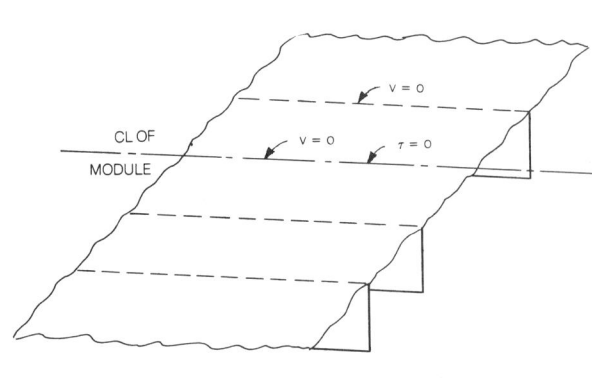

Fig. 33 Example of boundary conditions for shear lag problems in ship structures

to the width. Many of the plate panels in ship structures are relatively slender, as for example, the flanges of beams and girders, and the location of maximum stress is often at mid-length; thus this partial solution still provides results that are valid and useful in these cases. For built-in beams having the maximum stress at the end, the principle of reflection about the built-in end allows this location to be treated as though it were at the mid-length, thus allowing the present interior solution to apply to this important situation.

The Schade solutions have been expressed in the form of the effective breadth ratio, ρ, and this is, in general, a function of the geometry of the plate panel and the form of the applied load. The results are presented in a series of design charts which are especially simple to use. An example of one of the charts is given in Fig. 34, and the remainder may be found in the Schade papers.

Examination of the chart reveals the following properties of the effective breadth:

(*a*) For long slender plate panels, the effective breadth is nearly 100 percent.

(*b*) For a given panel aspect ratio, the effective

---

[3] St. Venant's Principle may be stated as follows; "Two different but statically equivalent force systems that act on a small region of an elastic body will produce the same stress distribution in the portion of the body that is at a large distance from the loading in comparison with the dimensions of the region on which the load acts."

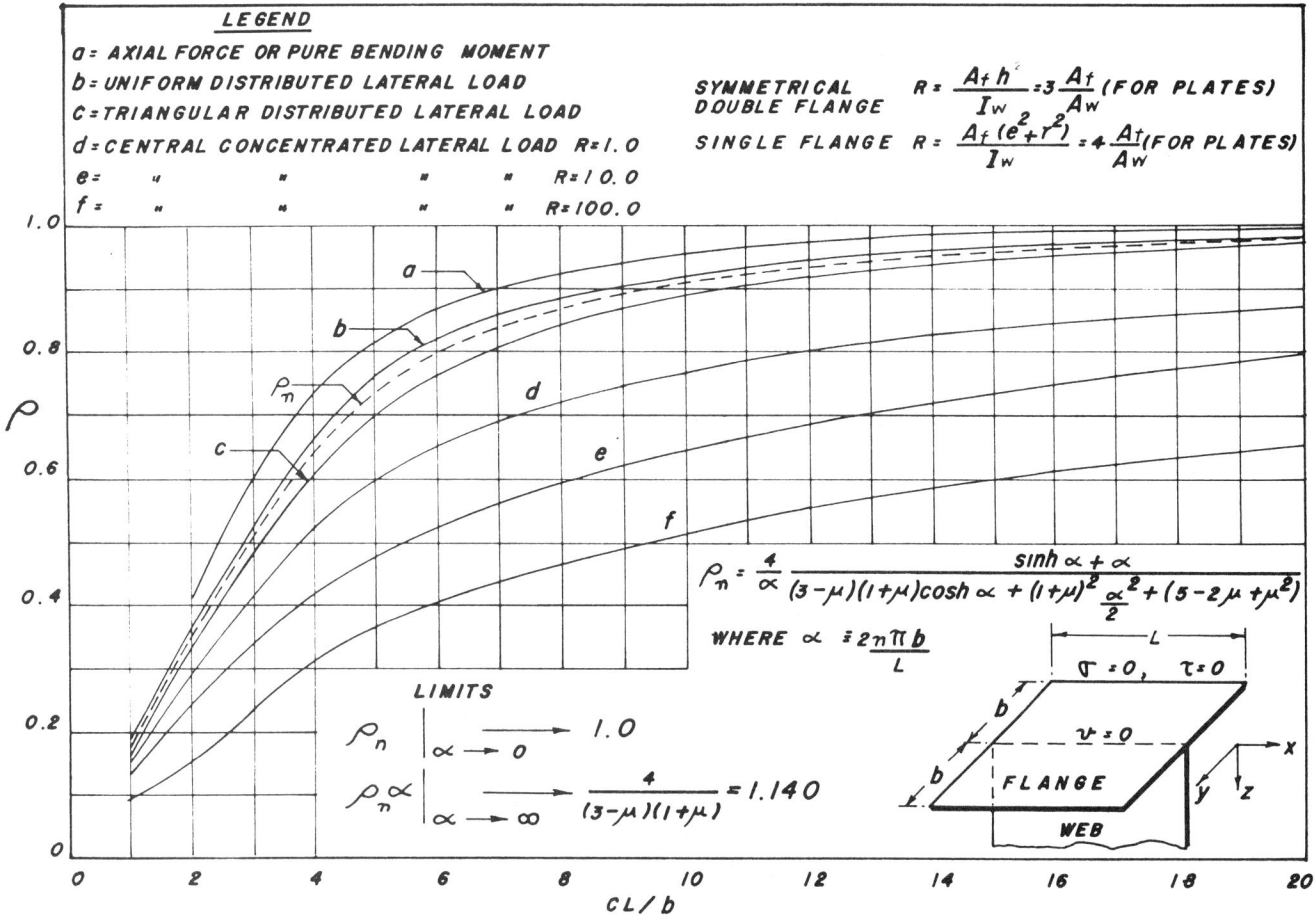

Fig. 34 Effectiveness ratio, ρ, at $x = L/2$

breadth is less for loads that are concentrated or vary abruptly than it is for loads that are evenly distributed over the length.

In Fig. 34 the factor R expresses the relative importance of the plate member, in which shear lag occurs, in comparison to the web or stiffener (assumed not to experience shear lag) in determining the section modulus of the composite. In Fig. 35 are illustrated three typical combinations involving this symmetrical flange member:

(a) H-section having two identical flanges.
(b) T-section or single flange.
(c) Plate flange with stiffener consisting of standard structural shape.

The corresponding expressions for R are as follows:

$$(a) \quad R = 3 A_f/A_w$$
$$(b) \quad R = 4 A_f/A_w \qquad (74)$$
$$(c) \quad R = A_f(r_s^2 + e_s^2)/I_s$$

In all cases: A_f is total flange cross-sectional area
$\qquad = 4bt_f$ in (a)
$\qquad = 2bt_f$ in (b) and (c)
A is web cross sectional area
$\qquad = 2htw$ in (a), (b).

In case (c): I_s is stiffener moment of inertia about its own NA.
r_s is radius of gyration of stiffener = $\sqrt{I_s/A_s}$ where A_s = stiffener cross sectional area.
e_s is distance from stiffener NA to midthickness of plate flange.

The factor CL in Fig. 34 is the total span of the beam between locations of zero bending moment, and CL/b is the aspect ratio, AR. In the case of the constant bending moment (pure axial load) it is the physical length of the beam.

Consider, for example, a symmetrical section, case (a) having a flange area that is 3.33 times the web area; thus $R = 10$. Now, assume that this is the cross section of a simply supported beam having a central, concentrated load and assume the span of the beam to be 10 times the width, b, of one-half the flange. From Fig. 34 curve e ($R = 10$) at an aspect ratio $CL/b = 10$, we see that the effective breadth of the flange is only 65 percent of the physical breadth.

For cases other than the concentrated load, the effective breadth is found to be relatively insensitive to R and, therefore, only a single curve corresponding to a composite value of R is shown in these load cases.

The effect of shear lag in a ship is to cause the stress distribution in the deck, for example, to depart from the constant value predicted by the elementary beam Equation (27). A typical distribution of the longitudinal deck stress in a ship subject to a vertical sagging load is sketched in Fig. 36.

An extreme example of shear lag was observed in an experiment conducted by Glasfeld (1962). The experiment was conducted using a rectangular, thin-walled steel box-girder model to represent the midship portion of a longitudinally framed ship. The vertical loading was applied to the model by means a series of individual pneumatic pressure cells between the bottom of the model and the bed plate of the supporting testing frame. Each cell applied a uniform pressure over a short portion of the length of the model, and the pressure in each cell could be adjusted individually. In the experiment in question, the pressure was adjusted in such a way that adjacent cells applied alternately positive and negative loads. The resulting bending moment was also found to show an alternating form, with relatively short spacing between zero points.

The longitudinal stress distribution measured at amidships of the model is shown in Fig. 37, together with the corresponding stress computed by (a) the finite element method, and (b) a stress function technique similar to that employed in deriving the effective breadth charts. It is remarkable to observe that both the observed and computed stress distributions display such a pronounced shear lag effect that there is a complete reversal in the sign of the longitudinal stress between the centerline of the model and the edge of both the deck and bottom plating. This is obviously an extreme departure from simple beam theory, which predicts a constant longitudinal stress in these members. The longitudinal stress in the sides also shows a departure from beam theory in the S-shaped variation of this quantity from deck edge to bilge.

A real situation in which such an alternating load distribution may be encountered is a bulk carrier loaded with a dense ore cargo in alternate holds, the remainder being empty. An example computation of the effective breadth of bottom and deck plating for such a vessel is given in Chapter VI of Taggart (1980).

3.7 Lateral Bending and Torsional Effects. Up to this point, our attention has been focused principally upon the vertical longitudinal bending response of the hull. As the ship moves through a seaway encountering waves from directions other than directly ahead or astern, it will experience lateral bending loads and twisting moments in addition to the vertical loads. (See Section 2.3). The former may be dealt with by methods that are similar to those used for treating the vertical bending loads, noting that there will be no component of still water bending moment or shear in the lateral direction. The twisting or torsional loads will require some special consideration. Note, however, that under

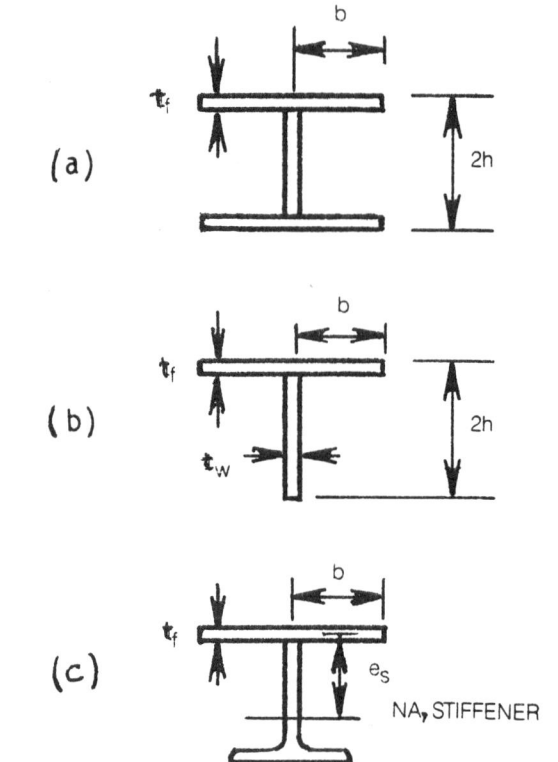

Fig. 35 Composite plate-stiffener beam sections, to accompany Fig. 34

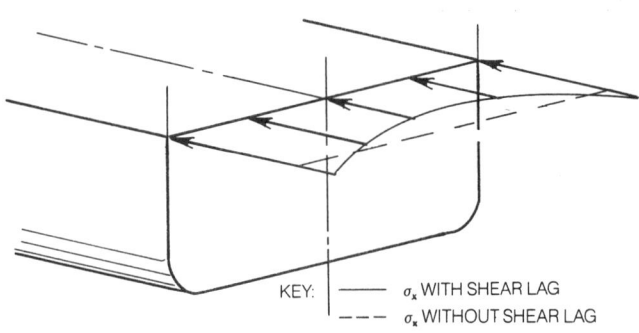

Fig. 36 Deck longitudinal stress, illustrating the effect of shear lag

our subdivision of the loads and response into primary, secondary and tertiary components, the response of the ship to the overall hull twisting loading should be considered a primary response.

The equations for the twist of a closed tube (50) and (56), presented in Section 3.5, are applicable only to the computation of the torsional response of closed thin-walled sections. In ship structures, however, it is found that torsional effects (stresses, deflections) are most often found to be of importance in ships that have large deck openings separated, perhaps, by narrow-transverse strips of structure and closed ends.

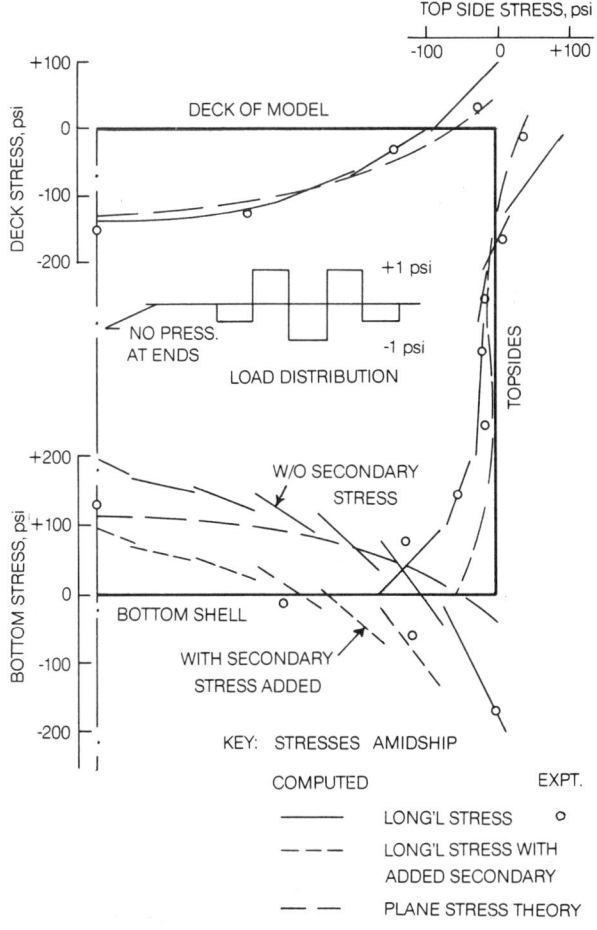

Fig. 37 Longitudinal stresses at midship section of box girder, by experiment, theory and finite element analysis (saddle bending moment)

has a longitudinal slit over its full length as in Fig. 38. The closed tube will be able to resist a much greater torque per unit angular deflection than the open tube because of the inability of the latter to sustain a shear stress across the slot. The only resistance to torsion in the case of the open tube without longitudinal restraint is provided by the twisting resistance of the thin material of which the tube is composed. This is illustrated in the lower part of the figure. The resistance to twist of the entirely open section is given by the St. Venant torsion equation,

$$M_{T1} = GJ \frac{d\theta}{dx} \qquad (75)$$

where $\frac{d\theta}{dx}$ is twist angle per unit length,

G is shear modulus of the material,

J is torsional constant of the section,

For a thinwalled open section, $J = \frac{1}{3} \int_0^S t^3 ds$

If warping resistance is present, i.e., if the longitudinal moment of the elemental strips shown in Fig. 38 is resisted, another component of torsional resistance is developed through the shear stresses that result from this warping restraint. This is added to the torque given by Equation (75). In ship structures, warping resistance comes from four sources:

- The closed sections of the structure between hatch openings.
- The closed ends of the ship.
- Double wall transverse bulkheads.

Such construction is typical of a modern containership, as shown in Fig. 40 of Chapter I.

Experience with the design of such partially *open deck* ships has indicated that the torsional stresses, alone, have seldom been of a serious magnitude. When considered in conjunction with the primary bending stresses, however, they can result in significant localized increases in the combined primary stresses. A more serious structural problem, requiring special attention in the design of such ships, is found at the transition from the torsionally weak open sections to the relatively stiff closed sections that are required to provide torsional rigidity to the hull. The abrupt change in structural properties may result in high stress concentrations in such areas, requiring special attention to the design of details. The principal design objective here is to select material and structural details that are appropriate for regions subject to stress concentrations.

The relative torsional stiffness of closed and open sections may be visualized by means of a very simple example. Consider two circular tubes, one of which

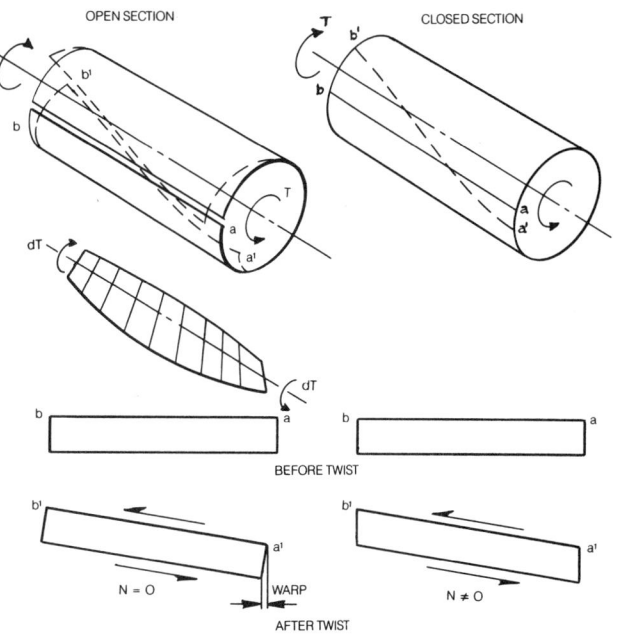

Fig. 38 Twist of open and closed tubes

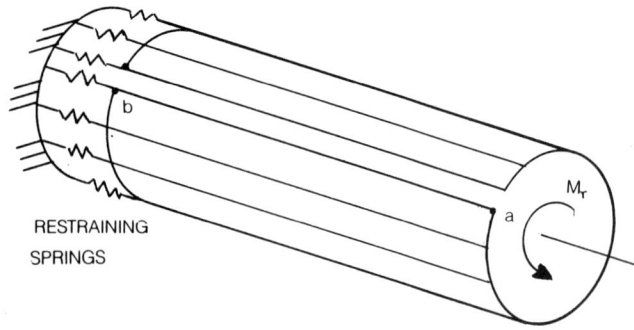

RESTRAINING SPRINGS

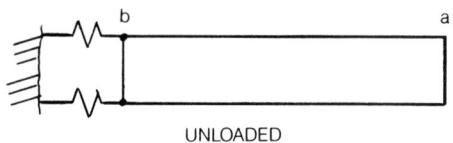

UNLOADED

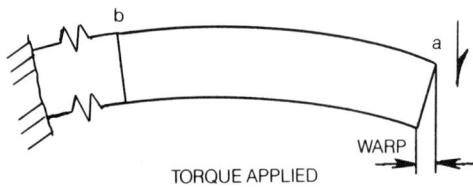

TORQUE APPLIED

Fig. 39 Open tube with warping restraint

- Closed, torsionally stiff parts of the cross section (longitudinal torsion tubes or boxes, including double bottom).

In order to understand the mechanism by which warping resistance leads to a component of torsional stiffness, we refer to Fig. 39 which shows the open tube having a partially built-in condition at end b. The end condition may be visualized by imagining each of the elementary strips of which the tube is composed to be attached to a rigid wall through a set of springs. In this case the tangential deflection of a point, a, on the right-hand end of the tube is due primarily to the in-plane bending of the elementary strip and this is shown in the lower part of the figure. Shear deformation of the strip is relatively unimportant in determining the resulting moment except in the case of a very short tube. The longitudinal displacement of a point on the surface of the tube, or warp is then given by Equation (54) with the shear deformation set equal to zero,

$$u(s) = -\frac{d\theta}{dx} \int_0^s r(s)\, ds \tag{76}$$

$$= -2\omega(s)\frac{d\theta}{dx} + u_0$$

The quantity $\omega(s) = \frac{1}{2}\int_0^s r(s)\, ds$ is called the *sectorial area* and u_0 is the warping displacement at the origin of the coordinate s. If the origin of the s-integration is placed at the open edge of the section, point a in Fig. 39, then $u(s) - u_0$ is the warp of a point on the section measured with respect to the warp at this origin of s.

Now, let us compute the average warp, which is, given by the integral of $u(s)$ around the entire section periphery, S, divided by S,

$$\begin{aligned}\bar{u} &= \frac{1}{S}\int_0^S u(s)\, ds \\ &= \frac{1}{S}\left[\int_0^S u_0\, ds \right.\\ &\quad \left. -\frac{d\theta}{dx}\int_0^S 2\omega(s)\, ds\right] \\ &= u_0 - \frac{S_\omega}{S}\frac{d\theta}{dx}\end{aligned} \tag{77}$$

The quantity $S_\omega = 2S\omega_0$ is called the first sectorial moment with respect to the origin of s. There will in general, be one or more points on the contour which, if used as origins for the S-integration, will result in a zero value for S_ω or ω_0. These points are referred to as sectorial centroids, and for a symmetrical section, the intersection of the plane of symmetry and the contour is a sectorial centroid.

Now, let us measure the warp, $u(s)$, from the plane of the mean warp,

$$\begin{aligned}u(s) &= u - \bar{u} \\ &= \frac{d\theta}{dx}(2\omega_0 - 2\omega(s))\end{aligned} \tag{78}$$

If the origin of s were chosen as a sectorial centroid, the term containing ω_0 would vanish.

The x-strain is,

$$\epsilon_x = \frac{\partial u}{\partial x} = 2\frac{d^2\theta}{dx^2}(\omega_0 - \omega(s)) \tag{79}$$

Neglecting the transverse (Poisson) effect, the x-stress is

$$\begin{aligned}\sigma_x &= E\epsilon_x \\ &= 2E\frac{d^2\theta}{dx^2}(\omega_0 - \omega(s))\end{aligned} \tag{80}$$

From the condition of equilibrium of an element, the shear flow, N is related to the x-stress resultant, N_x, by Equation (36)

After substituting Equation (80) into the first of (36)

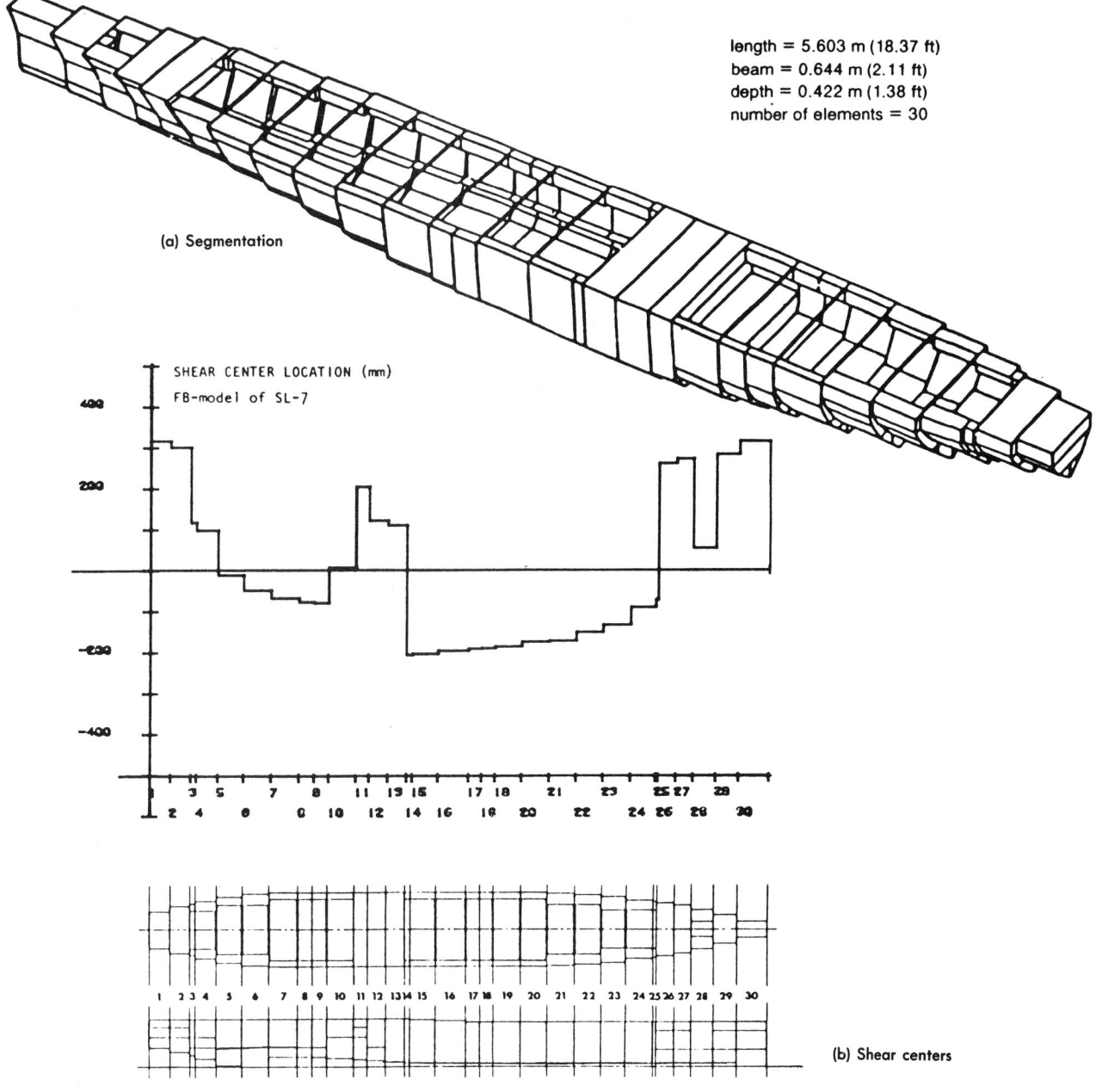

Fig. 40 SL-7 hull structure idealized with finite element method (Westin, 1981)

and integrating, we obtain for the shear flow,

$$N(s) = -2E \frac{d^3\theta}{dx^3} \int_0^s t(s)[(\omega_0 - \omega(s)]\,ds \quad (81)$$

where the origin of the s-integration is now taken on a free edge of the contour for which the shear stress is zero. Note that by defining separate s-origins for the present and earlier s-integrations, the ω_0 term may be made to vanish from this expression. The twisting moment on the end of the section is obtained by integration of the moment of $N(s)$ around the entire contour,

$$T_2(x) = \int_0^S N(s)\,r(s)\,ds \quad (82)$$

where r is defined as in Equation (76).

After substituting (81) into (82) and integrating by parts, we obtain for the twisting moment due to restrained warping

$$T_2(x) = -E\frac{d^3\theta}{dx^3}\int_0^S 4(\omega_0 - \omega(s))^2 t(s)\,ds \quad (83)$$
$$= -E\Gamma\frac{d^3\theta}{dx^3}$$

$\Gamma = 4\int_0^S (\omega_0 - \omega(s))^2 t(s)\,ds$ is called the warping constant of the section.

If $T(x) = T_1 + T_2$ is the total twisting moment at station x the differential equation of twist, taking into consideration unrestrained warping, Equation (75), and restrained warping effects (83) is

$$E\Gamma\frac{d^3\theta}{dx^3} - GJ\frac{d\theta}{dx} = -T(x) \quad (84)$$

In a ship made up of some closed and some open sections, the analysis leading to this equation is assumed to apply only to the open sections.

In order to solve Equation (84) we shall assume that the twisting moment at longitudinal position x may be expressed in the form of a Fourier Series over the length of an open prismatic section of length L.

$$T(x) = \sum_{n=1}^{\infty}(T_{nc}\cos p_n x + T_{ns}\sin p_n x) \quad (85)$$

The solution of the differential equation of the deflection is,

$$\theta(x) = A_0 + A_1 \sinh kx + A_2 \cosh kx \quad (86)$$
$$+ \Sigma(\alpha_n \cos p_n x + \beta_n \sin p_n x)$$

Here,
$$p_n = \frac{\pi n}{L}$$
$$k^2 = \frac{GJ}{E\Gamma}$$

A_0, A_1, A_2 are integration constants of the homogeneous solution and are to be determined by boundary conditions at the ends of the segment of length L.

$$\alpha_n = \frac{T_{cn}}{GJ(p_n^3 + k^2 p_n)} \quad (87)$$
$$\beta_n = -\frac{T_{sn}}{GJ(p_n^3 + k^2 p_n)}$$

The *Bredt Formula*, Equation (56) of Section 3.5, is applicable to the torsional deflection of a closed prismatic tube and is therefore applied to the decked-over sections of the ship between hatch openings. Since this is a first-order equation, there will be one constant of integration in its solution.

By subdividing the ship into a series of open or closed sections and applying the appropriate torsional deflection equations, the result is a system of algebraic equations, containing three unknown constants of integration for each open section and one unknown constant for each closed section. These constants are solved for by imposing requirements of continuity of internal reactions and deflections across the junctions of the closed and open sections.

A model of a large containership subdivided into a series of such prismatic segments is shown in Fig. 40a, taken from Westin (1981). The matching conditions at the junctions state that there is compatibility of twist, compatibility of warp and continuity of the internal loads across the junction between the two types of sections. Example analyses of this type which treat in detail the problem of matching closed and open sections may be found in Haslum and Tonnesen (1972), De Wilde (1967) and Westin (1981).

An inherent difficulty in establishing suitable matching conditions lies in determining the axis of rotation or center of twist of the two types of section. For a beam of uniform section, the center of twist coincides with the shear center of the cross section, which is also the center about which the moment of external loads is computed. For the nonuniform beam, the center of rotation is no longer at the shear center, which itself is at a different vertical location for the closed and open sections.

For a closed ship section, the shear center will be in the vicinity of mid-depth, but for an open section it may be below the keel. Fig. 40b shows the height of the shear center for the structural idealization of the containership of Fig. 40a.

In general, the torsion analysis described above, when applied to the computation of the actual stress distribution in a real ship under torsional loading, can be expected to give results somewhat less exact than were obtained when applying simple beam theory to the vertical bending of the ship structure. This is primarily a result of attempting to apply two separate theoretical procedures, each one of which is based upon an assumption of uniformity of cross section along the ship length, to a problem in which the variation in cross sectional shape is itself, of fundamental importance. The effect of the concentration of stiff and soft sections is to result in a distortion pattern in the ship deck that is somewhat as shown, to an exaggerated scale, in Fig. 41. The term *snaking* is sometimes used in referring to this behavior.

Fortunately, ship structures designed to withstand normal bending loads do not appear to experience large primary hull girder stresses as a result of the normal torsional loads experienced in service. As previously noted, significant stresses may, however, be induced at specific locations such as hatch corners as a result of stress concentrations due to discontinuities in structure. The analysis of structures in which discontinuity plays an essential role is best handled by the finite-element technique, which is described in Taggart (1980) and many text books. Such an analysis of a large

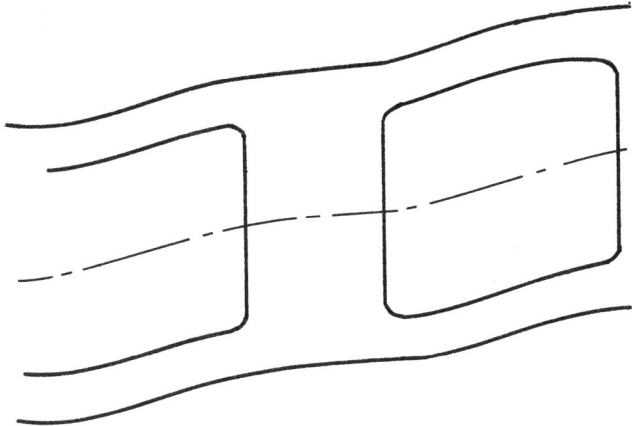

Fig. 41 Torsional distortion of containership deck

containership and comparisons with model experiments are described by El Batouti, Jan and Stiansen (1976).

3.8 Secondary Structural Response. In the case of secondary structural response, the principal objective is to determine the distribution of both in-plane and normal loading, deflection and stress over the length and width dimensions of a panel of stiffened plating. We recall that the primary response involves the determination of only the in-plane load, deflection, and stress as they vary over the length of the ship. The secondary response, therefore, is seen to be a two-dimensional problem while the primary response is essentially one-dimensional in character.

A panel of structure, as used in the present context, usually consists of a flat or slightly curved section of plating with its attached stiffeners. There may be two sets of stiffeners arranged perpendicular to each other. Usually the stiffeners comprising the set of parallel members in one direction will be of equal size and spacing, and the stiffeners in the other direction will also be of equal size and spacing but different from the first set. In some cases, the central stiffener in a rectangular stiffened panel is made somewhat larger than the remaining parallel stiffeners, as in the case of the center keel girder. In some cases, there may be stiffeners in one direction only. There may be a single panel of plating with stiffeners attached to one side, as in decks, side shell and bulkheads; or there may be two parallel panels with the stiffeners between them, as in double bottom construction. The plating may be absent, in which case the module is a grid or grillage of beam members only, rather than a stiffened plate panel.

In most cases the boundaries of a panel are attached to other panels, either in the same plane or perpendicular to the original panel. As an example, we may consider a section of the double bottom structure of a typical dry cargo ship. The forward and after boundaries of this double bottom panel are formed by transverse bulkheads, which are perpendicular to the bottom panel, and by the continuing bottom structure beyond the bulkheads, which is in the same plane as the present panel and of similar construction. The outboard edges of the double bottom panel are bounded by the plate and frame panels comprising the side structure of the ship. The bottom panel consists of two plate members, bottom shell and inner bottom plating, with an orthogonal system of transverse floors and longitudinal girders between them. The center keel girder is typically somewhat heavier, thus stiffer than the other longitudinal girders. The transverse bulkheads and side shell are usually single plate panels with the stiffeners welded to one side. The loading on the bottom panel consists of the external fluid pressures, the distributed weights or pressures of liquids in the inner bottom space and combinations of distributed and concentrated weights of cargo, machinery, and the structural material itself.

In principle, the solution for the deflection and stress in the secondary panel of structure may be thought of as a solution for the response of a system of orthogonal intersecting beams. Interactions between the two beam systems arise from the physical connections between the stiffeners, requiring equality of normal deflection of the two beam systems at the points of intersection. A second type of interaction arises from the two-dimensional stress pattern in the plate, which may be thought of as forming a part of the flanges of the stiffeners. The plate contribution to the beam bending stiffness arises from the direct longitudinal stress in the plate adjacent to the stiffener, modified by the transverse stress effects, and also from the shear stress in the plane of the plate. The maximum secondary stress may be found in the plate itself, but more frequently it is found in the free flanges of the stiffeners, since these flanges are at a greater distance than the plate member from the neutral axis of the combined plate-stiffener.

At least four different procedures have been employed for obtaining the structural behavior of stiffened plate panels under normal loading, each embodying certain simplifying assumptions:

- Orthotropic plate theory.
- Beam-on-elastic-foundation theory.
- Grillage theory.
- The finite element method.

Orthotropic plate theory refers to the theory of bending of plates having different flexural rigidities in the two orthogonal directions. In applying this theory to panels having discrete stiffeners we idealize the structure by assuming that the structural properties of the stiffeners may be approximated by their average values, which are assumed to be distributed uniformly over the width or length of the plate. The deflections and stresses in the resulting continuum are then obtained from a solution of the orthotropic plate deflection equation,

$$a_1 \frac{\partial^4 w}{\partial x^4} + a_2 \frac{\partial^4 w}{\partial x^2 \partial y^2} + a_3 \frac{\partial^4 w}{\partial y^4} = p(x,y) \quad (88)$$

where

a_1, a_2, a_3 express the average flexural rigidity of the orthotropic plate in the two directions.

$w(x,y)$ is the deflection of the plate in the normal direction.

$p(x,y)$ is the distributed normal pressure load per unit area.

Note that the behavior of the isotropic plate, i.e., one having uniform flexural properties in all directions, is a special case of the orthotropic plate problem.

It is not appropriate to go into the detailed derivation of this equation nor its solution, both of which have been presented in detail by Schade (1938, 1940, 1941). The results of Schade's solution have been presented in a series of easily used charts, and their use will be discussed later. The orthotropic plate method is best suited to a panel in which the stiffeners are uniform in size and spacing and closely spaced. The Schade design charts have been developed in such a way that a centerline stiffener that is heavier than the other stiffeners may be included.

The beam on elastic foundation solution is suitable for a panel in which the stiffeners are uniform and closely spaced in one direction and more sparse in the other. One of the latter members may be thought of as an individual beam having an elastic support at its point of intersection with each of the closely-spaced orthogonal beams. An average elastic modulus or spring constant per unit length may be determined by dividing the force per unit deflection of one of these closely spaced members by the spacing. Using this average spring constant per unit length, the effect of the closely spaced members is then represented as an elastic support that is distributed evenly along the length of the widely spaced members. Each of these members is then treated individually as a beam on an elastic foundation, for which the differential equation of deflection is,

$$EI \frac{d^4 w}{dx^4} + kw = q(x) \quad (89)$$

where
w is deflection,
I is sectional moment of inertia of the longitudinal stiffener, including adjacent plating.
k is average spring constant per unit length of the transverse stiffeners,
$q(x)$ is load per unit length on the longitudinal member.

Michelsen and Nielsen (1965) have developed a solution method for this equation, based upon use of the *Laplace Transform*, which is particularly well adapted to machine computation. Various realistic

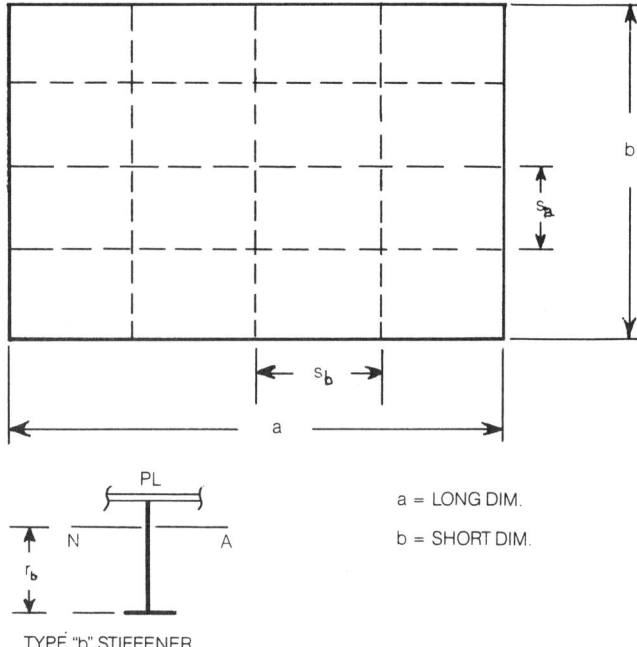

Fig. 42 Stiffened plate nomenclature

boundary conditions may be taken into account, and the solution can also consider several intersecting panels of structure. This procedure has been incorporated into a computer-based scheme for the optimum structural design of the midship section as described by St. Denis (1970).

In the grillage method of Clarkson et al (1959), each stiffener in the two orthogonal sets of members is represented as a simple beam. The external loading may be applied as a set of equivalent point forces at the intersections of the two beam systems. At these points of intersection conditions of equilibrium of the unknown reaction forces between the two beams, together with conditions of equal deflection, are required to be satisfied. The result is a system of algebraic equations to be solved for the deflections. From the solutions the forces in each set of beams and the resulting stresses may be obtained.

The versatile finite element technique may model the structure in a number of different ways. For example, each segment of stiffener between intersection points may be represented by a short beam, and the plating may be represented as a membrane capable of supporting in-plane stress as in the grillage technique. Conditions of equality of deflections and equilibrium of internal and external forces are then required to be satisfied at the points of intersection leading to the formulation of a system of simultaneous algebraic equations relating external loads to deflections. Machine computation is necessary in order to formulate and solve the large number of equations that are necessary in a practical situation. This procedure is the most general of the four, being virtually unrestricted in the degree to which complex structural geometry, variable member sizes, boundary conditions and load distributions can be represented.

In the first three of the methods described above, the shear lag behavior of the plating in a plated grillage is not automatically included, but must be considered by the user when computing the bending stiffness of the orthotropic plate or grillage model. The finite-element technique is inherently capable of including this effect, provided the proper choice is made for the plate element type and the mesh size in the representation.

For hand computations of secondary stress, the Schade design charts based upon the orthotropic plate solution provide the most practical method of those described above. However, Clarkson (1959) has presented a limited number of design charts based upon the discrete grillage solution which are useful in many cases.

Two of the charts from Schade (1941) are reproduced here as Figs. 44 and 45, after the following explanation of terminology and preliminary discussion. Referring to Fig. 42,

p = uniform unit pressure loading.
$a(b)$ = length (width) of rectangular panel,
$s_a(s_b)$ = spacing of long (short) stiffeners,
$I_{na}(I_{nb})$ = moment of inertia, including effective breadth of plating, of long (short) repeating stiffener (as distinguished from central stiffener, which may be different),
$I_{pa}(I_{pb})$ = moment of inertia of effective breadth of plating working with long (short) repeating stiffeners,
$I_a(I_b)$ = moment of inertia of central long (short) stiffener, including effective breadth of plating,
$A_a(A_b)$ = web area of central long (short) stiffener,
$r_a(r_b)$ = distance from its neutral axis to extreme fiber of central long (short) stiffener.

The effective breadth of plating to be used in computing the I's can be estimated by use of the effective breadth charts given in Fig. 33. In most cases, the effective breadth is 100 percent of the stiffener spacing, in which case the moment of inertia should be computed by using a modified thickness obtained by multiplying the actual plate thickness by the factor $1/(1 - \nu^2)$.

Four types of stiffening are shown, together with the definitions of certain additional parameters, in Fig. 43. The four types are as follows:

Type A Cross-stiffening. Two sets of intersecting stiffeners; the middle stiffener of either or both sets may be stiffer than the other stiffeners of the set.
Type B One set of repeating stiffeners and a single central stiffener in the other direction. The

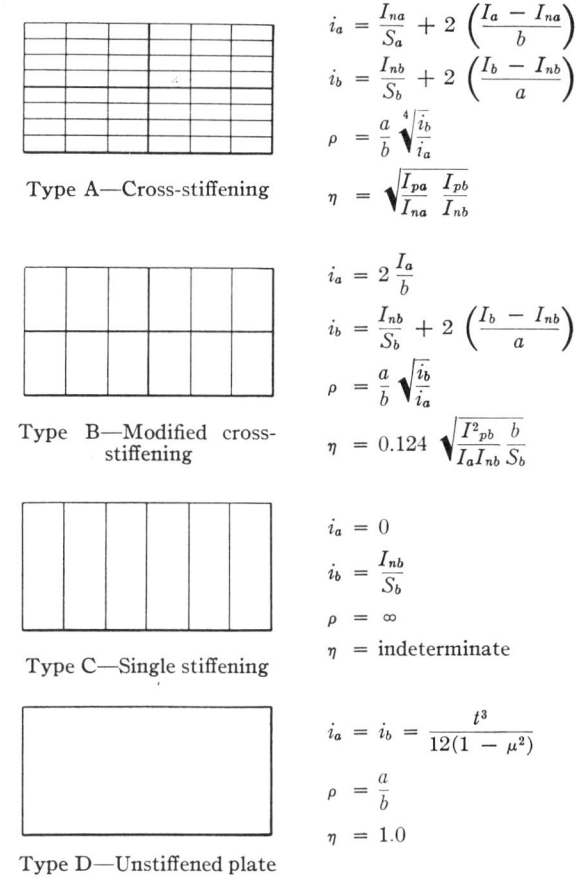

Type A—Cross-stiffening

$$i_a = \frac{I_{na}}{S_a} + 2\left(\frac{I_a - I_{na}}{b}\right)$$
$$i_b = \frac{I_{nb}}{S_b} + 2\left(\frac{I_b - I_{nb}}{a}\right)$$
$$\rho = \frac{a}{b}\sqrt[4]{\frac{i_b}{i_a}}$$
$$\eta = \sqrt{\frac{I_{pa}}{I_{na}}\frac{I_{pb}}{I_{nb}}}$$

Type B—Modified cross-stiffening

$$i_a = 2\frac{I_a}{b}$$
$$i_b = \frac{I_{nb}}{S_b} + 2\left(\frac{I_b - I_{nb}}{a}\right)$$
$$\rho = \frac{a}{b}\sqrt{\frac{i_b}{i_a}}$$
$$\eta = 0.124\sqrt{\frac{I_{pb}^2}{I_a I_{nb}}\frac{b}{S_b}}$$

Type C—Single stiffening

$$i_a = 0$$
$$i_b = \frac{I_{nb}}{S_b}$$
$$\rho = \infty$$
$$\eta = \text{indeterminate}$$

Type D—Unstiffened plate

$$i_a = i_b = \frac{t^3}{12(1-\mu^2)}$$
$$\rho = \frac{a}{b}$$
$$\eta = 1.0$$

Fig. 43 Types of stiffening, with applicable formulas for parameters (Schade, 1941)

middle stiffener of the repeating set may be stiffer than the others as in Type A.
Type C One set of repeating stiffeners only.
Type D Plating without stiffeners (isotropic plate).

For the first three types, there may be stiffeners without plating, there may be one panel of plating with stiffeners on one side, or there may be two courses of plating with stiffeners in between. The full range of possibilities for a rectangular panel stiffened in two directions is therefore covered. Type D, the case of plate alone without stiffeners may be used in computing the tertiary stress.

There are many possible combinations of edge fixity and boundary support for the panels used in ship structures. The solution has been found, and results are given for the following four combinations of built-in and simple support. These may usually be used as limiting cases of the actual, but usually indeterminate, boundary conditions that are found in actual ship structures.

Case 1 All four edges simply supported, i.e., rigidly supported against normal deflection but

(*Continued on page 260*)

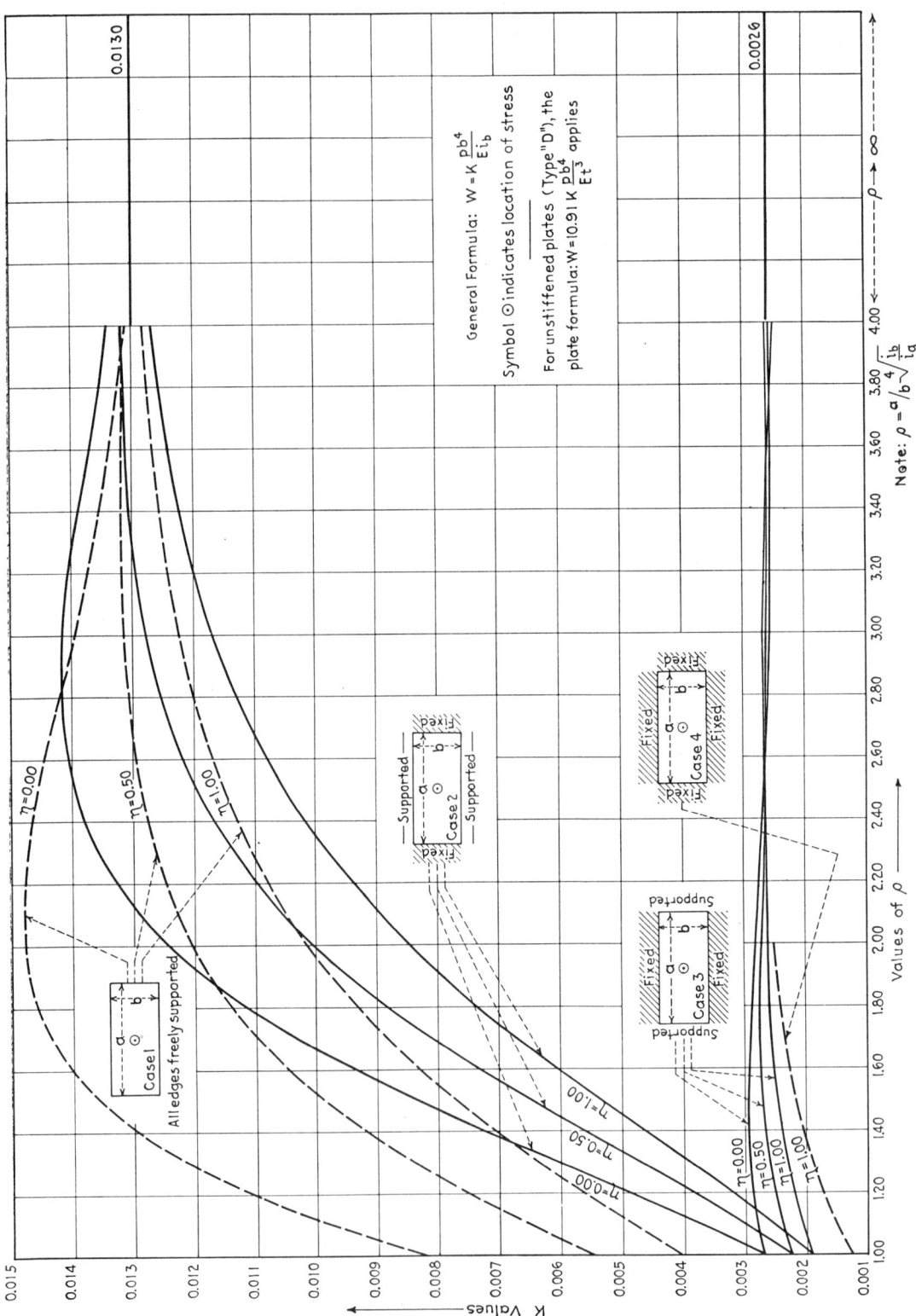

Fig. 44 Plate deflection at center of panel (Schade, 1941)

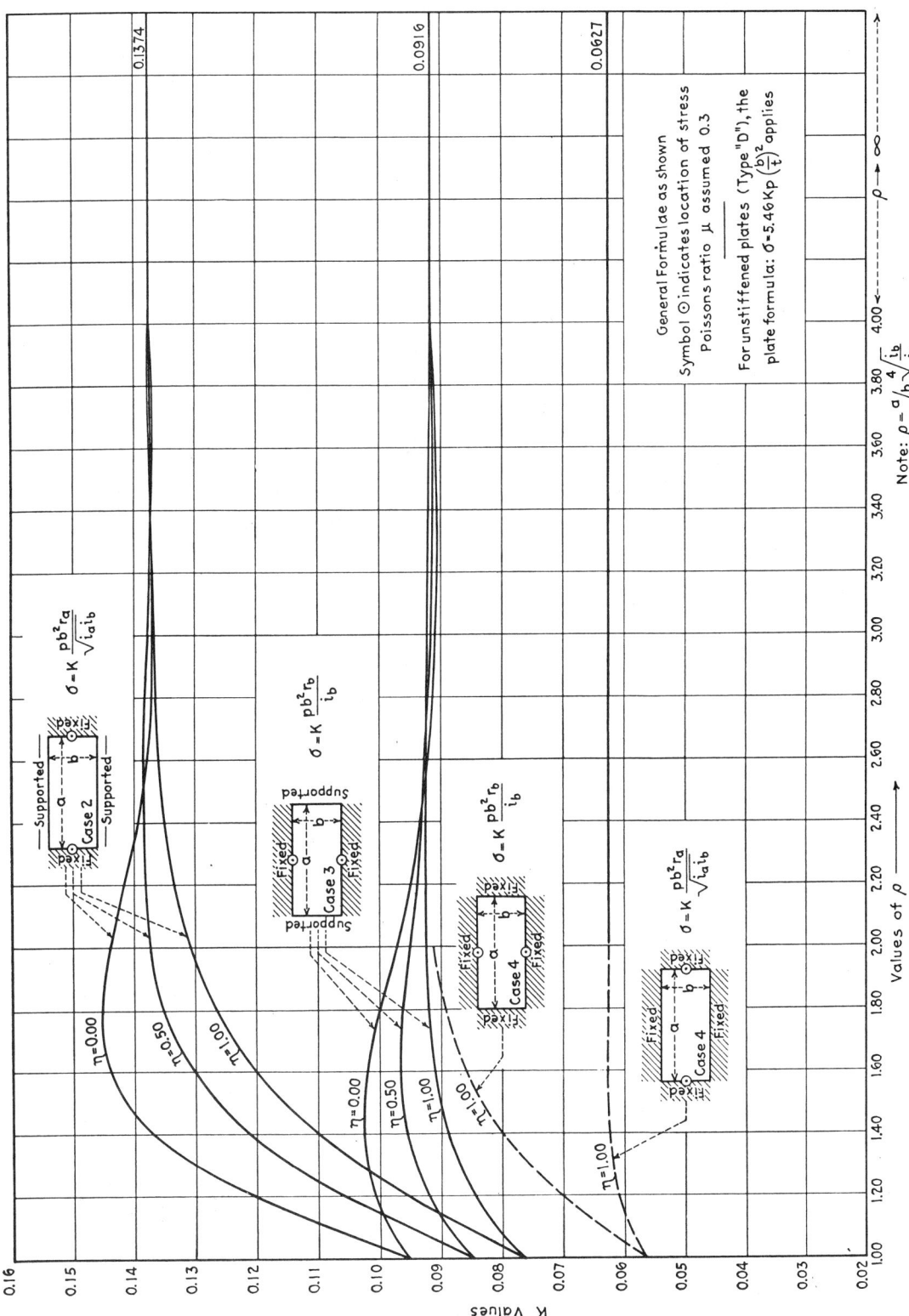

Fig. 45 Support bending stress in plating panel (Schade, 1941)

(*Continued from page 257*)
without edge moment restraint.

Case 2 Both short edges fixed, i.e., with both normal and rotational restraint, both long edges simply supported.

Case 3 Both long edges fixed, both short edges simply supported.

Case 4 All four edges fixed. (Only partial results are given for this case.)

In using the charts, several special parameters are required which are defined as follows:

Unit stiffnesses,

$$i_a = \frac{I_{na}}{S_a} + 2\left(\frac{I_a - I_{na}}{b}\right) \qquad (90)$$

$$i_b = \frac{I_{nb}}{S_b} + 2\left(\frac{I_b - I_{nb}}{a}\right) \qquad (91)$$

Torsion coefficient,

$$\eta = \sqrt{\frac{I_{pa}I_{pb}}{I_{na}I_{nb}}} \qquad (92)$$

Virtual aspect ratio,

$$\rho = \frac{a}{b}\sqrt[4]{\frac{i_b}{i_a}} \qquad (93)$$

Expressions for these parameters are given for each stiffener configuration in Fig. 43.

The charts from Schade (1941) contain the deflection at the center of the panel and the stress in the plating at the panel boundary. Charts containing other results, for example the stress at the panel mid-point, may be found in the original reference. In general, the charts give a nondimensional parameter, k, which may be substituted into a formula, given on the chart, for the corresponding stress or deflection.

3.9 Diffusion of Vertical Loads into Structure. The description of the computation of vertical shear and bending moment (Sections 2.2 and 2.4) by integration of the longitudinal load distribution implies that the external vertical load is resisted directly by the vertical shear carrying members of the hull girder such as the side shell or longitudinal bulkheads. In the case of a ship framed by a closely spaced transverse framing system without the support of longitudinal girders this condition is approached. In such case, the fluid pressure loads on the bottom as well as the weight of cargo and other deck loads are largely transmitted to the side shell by the transverse frames. The double bottom structure, however, has longitudinal as well as transverse supporting members and the 'tween deck beams are usually supported at their inboard ends by longitudinal girders. Such longitudinal structural members, therefore, transmit part of the weight or pressure load to the transverse bulkheads which, in turn, transfer the resultant loads into the side shell or longitudinal bulkheads in the form of localized shear forces. In a longitudinally framed ship, such as a tanker, the bottom pressures are transferred principally to the widely spaced transverse web frames or the transverse bulkheads where they are transferred to the longitudinal bulkheads or side shell, again as localized shear forces. Thus, in reality, the loading, $q(x)$, applied to the side shell or the longitudinal bulkhead will consist of a distributed part due to the direct transfer of load into the member from the bottom or deck structure, plus a concentrated part at each bulkhead or web frame. This leads to a discontinuity in the shear curve at the bulkheads and webs. Fig. 46 provides a simple means, based upon orthotropic plate theory, for estimating the proportion of total load transmitted to each edge of a panel of such structure. In this figure, the symbols are defined as in the preceeding Section 3.8.

In the derivation of the figure it is assumed that the stiffening members are numerous and closely spaced in each direction so that the bending stiffness of the total panel is well represented by the average values of combined plate and stiffener. The curves may, therefore, be reasonably applied, for example, to a double bottom structure in which there are closely spaced transverse floors and several longitudinal girders.

Fig. 47 illustrates the components of the load that are transferred to the side structure and to the transverse bulkheads, which form the boundaries of the panel of bottom structure. It is observed that the load transferred into the transverse bulkhead at its lower edge is resisted by vertical shearing forces in the bulkhead-side shell joint. These latter shearing forces have the effect of concentrated loads insofar as the primary hull girder is concerned. This effect of the concentrated bulkhead edge loads on the primary hull girder shear force is illustrated in the lower part of Fig. 47.

3.10 Tertiary Structural Response. Tertiary response refers to the bending stresses and deflections in the individual panels of plating that are bounded by the stiffeners of a secondary panel. In most cases the load that induces this response is a fluid pressure from either the water outside the ship or liquid or dry bulk cargo within. Such a loading is normal to and distributed over the surface of the panel. In many cases, the proportions, orientation, and location of the panel are such that the pressure may be assumed constant over its area.

As previously noted, the deflection response of an isotropic plate panel is obtained as the solution of a special case of the earlier orthotropic plate equation, and is given by,

$$\frac{\partial^4 w}{\partial x^4} + 2\frac{\partial^4 w}{\partial x^2 \partial y^2} + \frac{\partial^4 w}{\partial y^4} = \frac{p(x,y)}{D} \qquad (94)$$

Here

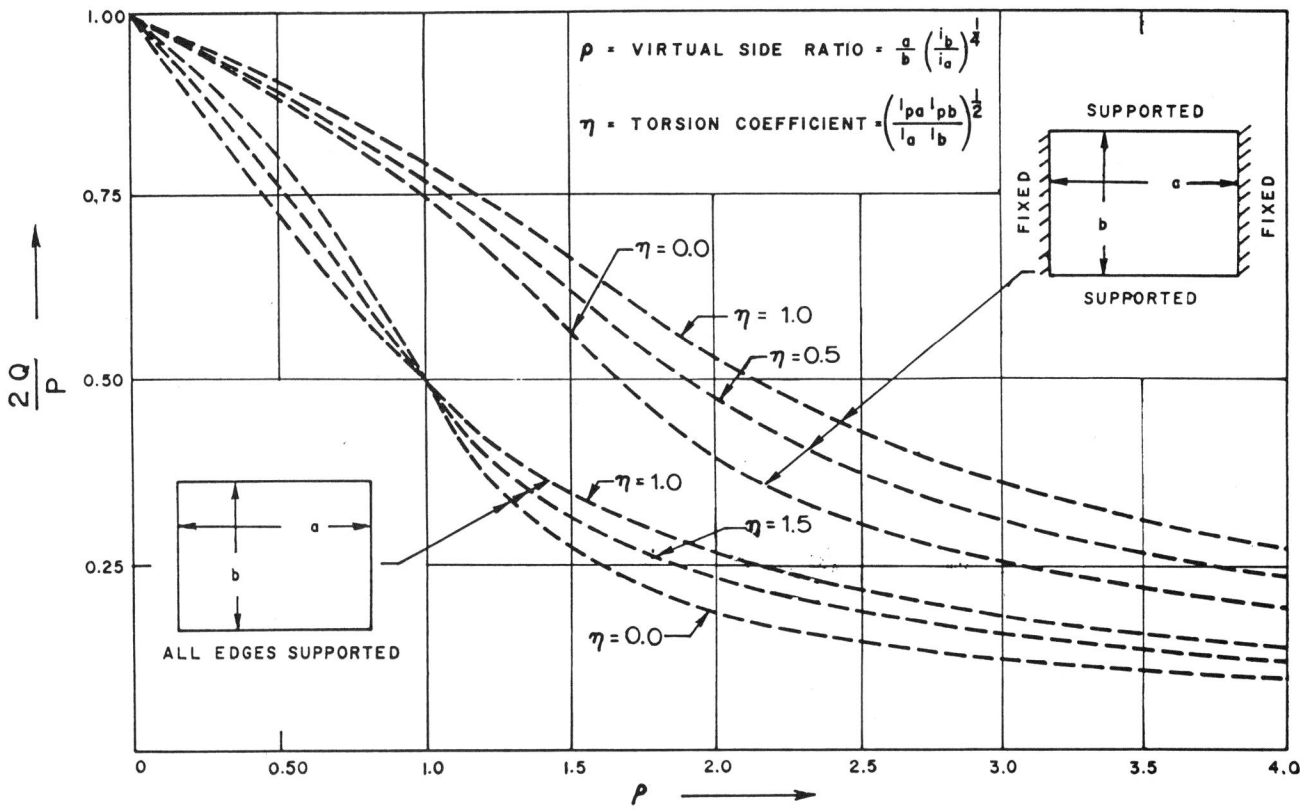

Fig. 46 Proportion of total load, $2Q/P$, carried by transverse boundaries (uniform load, $P = pab$) (D'Arcangelo, 1969)

D is plate flexural rigidity $= \dfrac{Et^3}{12(1-\nu^2)}$,

t is plate thickness (uniform),

$p(x,y)$ is distributed unit pressure load.

Appropriate boundary conditions are to be selected to represent the degree of fixity of the edges of the panel. The stresses and deflections obtained by solving this equation for rectangular plates under a uniform pressure distribution are contained in Figs. 44 and 45 and are labelled Type D in those figures.

A special case of some importance which is not covered in these charts is that of a plate subject to a concentrated point load. Such loads occur when wheeled vehicles such as fork lift trucks are used for cargo handling. Information on plating subject to such loads may be found in Hughes (1983) and in classification society rules.

3.11 Superposition of Stresses. Since the methods of calculation of primary, secondary, and tertiary stress all presuppose linear elastic behavior of the structural material, the stress intensities computed for the same member may be superimposed in order to obtain a maximum value for the combined stress. In performing and interpreting such a linear superposition, several considerations affecting the accuracy and significance of the resulting stress values must be borne in mind. First, the loads and theoretical procedures used in computing the stress components may not be of the same accuracy or reliability. The primary loading, for example, may be obtained using a theory that involves certain simplifications in the hydrodynamics of ship and wave motion, and the primary bending stress may be computed by simple beam theory, which gives a reasonably good estimate of the mean stress in deck or bottom but neglects certain localized effects such as shear lag or stress concentrations.

Second, the three stress components may not necessarily occur at the same instant in time as the ship moves through waves. The maximum bending moment amidships, which results in the maximum primary stress, does not necessarily occur in phase with the maximum local pressure on a midship panel of bottom structure (secondary stress) or panel of plating (tertiary stress).

Third, the maximum values of primary, secondary,

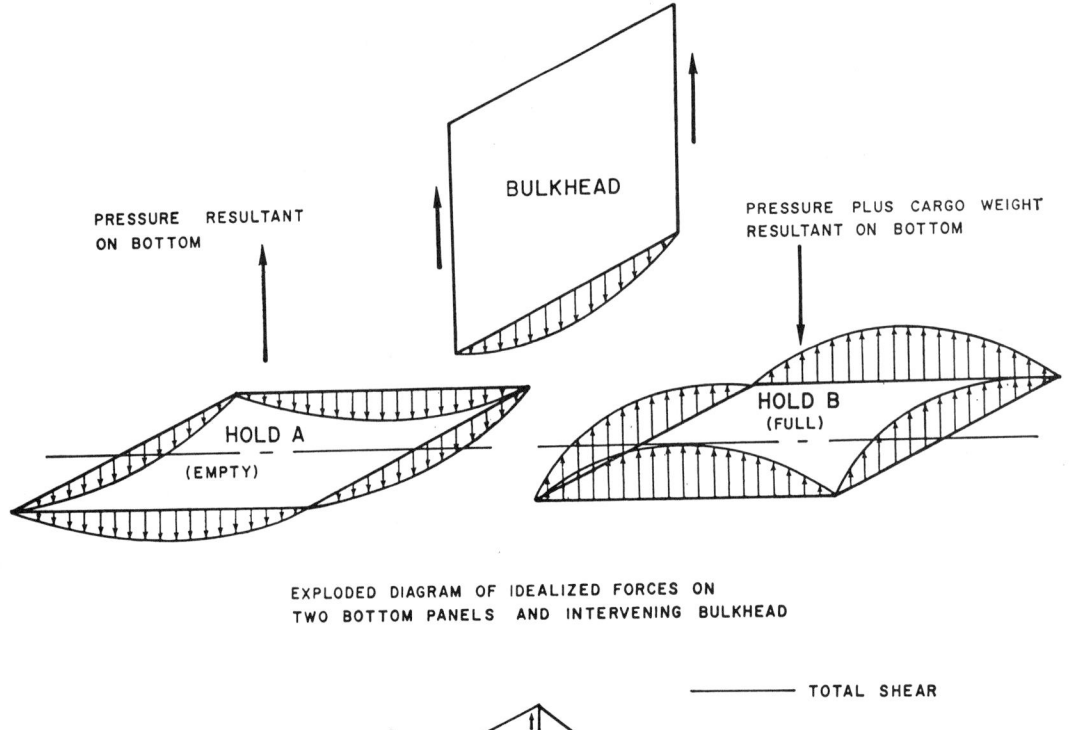

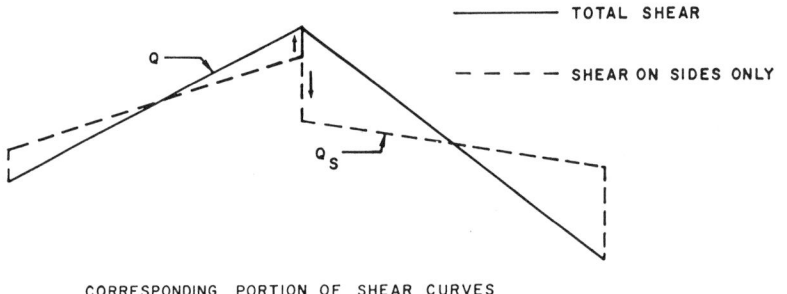

Fig. 47 Force systems on bottoms of two adjacent holds and on intervening bulkhead, with corresponding shear curves

and tertiary stress are not necessarily in the same direction nor even in the same part of the structure. In order to visualize this, consider a panel of bottom structure with longitudinal framing. The forward and after boundaries of the panel will be at transverse bulkheads. The primary stress will act in the longitudinal direction, as given by Equation (27). It will be nearly equal in the plating and the stiffeners, and will be approximately constant over the length of a midship panel. There will be a small transverse component in the plating, given by Equation (47), and a shear stress given by Equation (44). The secondary stress will probably be greater in the free flanges of the stiffeners than in the plating, since the combined neutral axis of the stiffener-plate combination is usually near the plate-stiffener joint. Secondary stresses, which vary over the length of the panel, are usually subdivided into two parts in the case of normal tanker bottom structure. The first part, σ_2, is associated with bending of a panel of structure bounded by transverse bulkheads and either the side shell or the longitudinal bulkheads. The principal stiffeners, in this case, are the center and any side longitudinal girders, and the transverse web frames. The second part, σ_2^*, is the stress resulting from the bending of the smaller panel of plating plus longitudinal stiffeners that is bounded by the deep web frames. The first of these components, as a result of the proportions of the panels of structure, is usually larger in the transverse than in the longitudinal direction. The second is predominantly longitudinal. The maximum tertiary stress is, of course, in the plate, but in the case of longitudinal stiffeners the long dimension of the panel is fore and aft and, consequently, the maximum panel tertiary stress will act in the transverse direction (normal to the framing system) at the mid-length of a long side.

In certain cases, there will be an appreciable shear stress component present in the plate, and the proper interpretation and assessment of the stress level will require the resolution of the stress pattern into principal stress components.

From all these considerations, it is evident that, in

many cases, the point in the structure having the highest stress level will not always be immediately obvious, but must be found by considering the combined stress effects at a number of different locations and times.

3.12 Transverse Strength. Transverse strength refers to the ability of the ship structure to resist those loads that tend to cause distortion of the cross section. When it is distorted into a parallelogram shape the effect is called *racking*. We recall that both the primary bending and torsional strength analyses are based upon the assumption of no distortion of the cross section. Thus, we see that there is an inherent relationship between transverse strength and both longitudinal and torsional strength. Certain structural members, including transverse bulkheads and deep web frames, must be incorporated into the ship in order to insure adequate transverse strength. These members provide support to and interact with longitudinal members by transferring loads from one part of a structure to another. For example, a portion of the bottom pressure loading on the hull is transferred via the center girder and the longitudinal frames to the transverse bulkheads at the ends of the frames. The bulkheads, in turn, transfer these loads as vertical shears into the side shell. Thus some of the loads acting on the transverse strength members are also the loads of concern in longitudinal strength considerations.

The general subject of transverse strength includes elements taken from both the primary and secondary strength categories. The loads that cause effects requiring transverse strength analysis may be of several different types, depending upon the type of ship, its structural arrangement, mode of operation, and upon environmental effects.

Typical situations requiring attention to the transverse strength are:

• Ships out of water—on building ways or on construction or repair dry dock.
• Tankers having empty wing tanks and full centerline tanks or vice versa.
• Ore carriers having loaded centerline holds and large empty wing tanks.
• All types of ships: torsional and racking effects caused by asymmetric motions of roll, sway and yaw.
• Ships with structural features having particular sensitivity to transverse effects, as for instance, ships having largely open interior structure such as auto carriers and RO-RO ships.

As previously noted, the transverse structural response involves pronounced interaction between transverse and longitudinal structural members. As an example, consider a transverse frame at the mid-length of a cargo hold of a transversely framed ship as shown in Fig. 48. The principal loading consists of the water pressure distribution around the ship, and the weights and inertias of the structure and hold contents. As a first approximation, the transverse response of such a frame may be analyzed by a two-dimensional frame

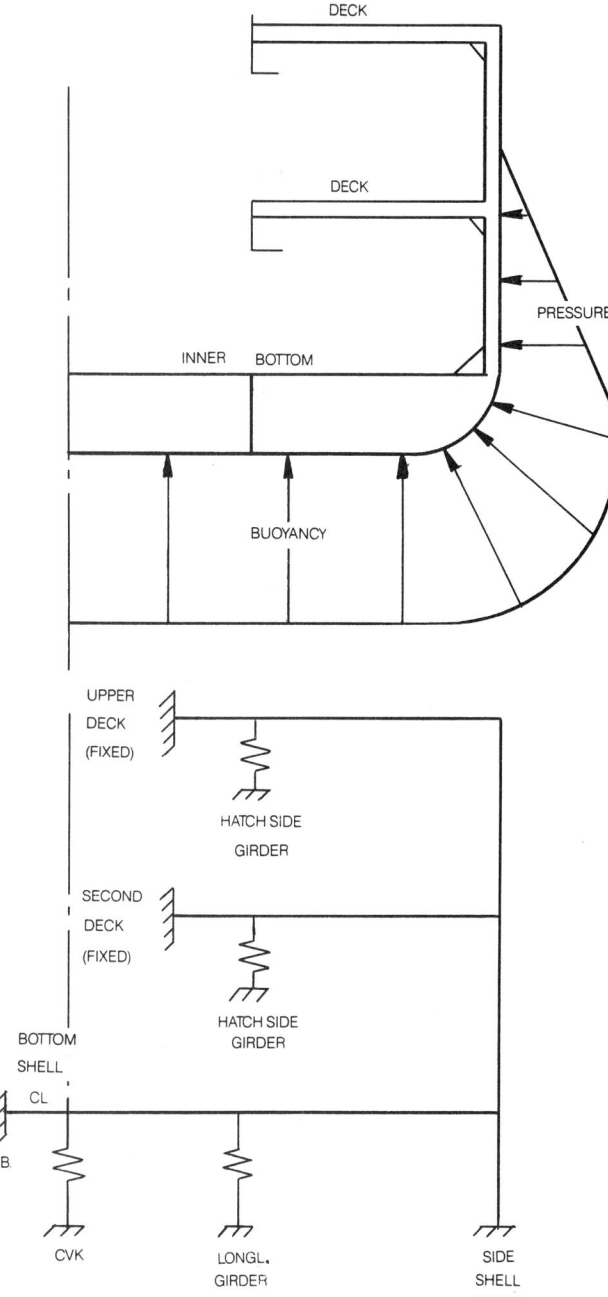

Fig. 48 Ring-frame analysis of transverse frame in a cargo ship

response procedure that may or may not allow for support by longitudinal structure. In the past, the *Hardy-Cross* moment-distribution or a related method might have been used for the manual implementation of this analysis. Today, a computer-based plane frame or plane finite-element analysis would invariably be used.

In a conventional cargo ship, having transverse bulkheads at a spacing approximately equal to the beam, the transverse response of the frame plus plating is

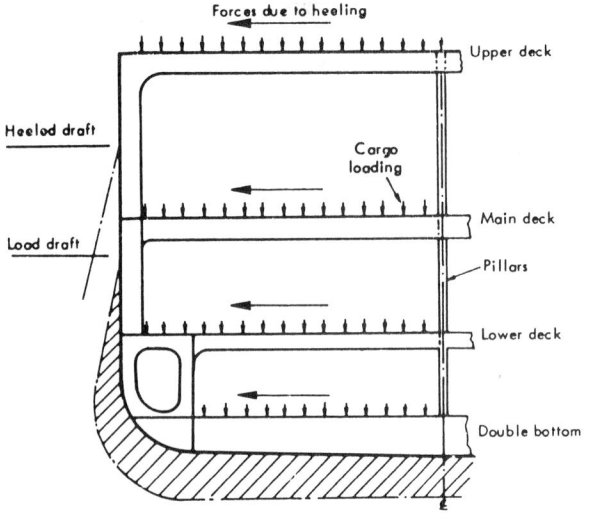

TYPICAL TRANSVERSE FRAME

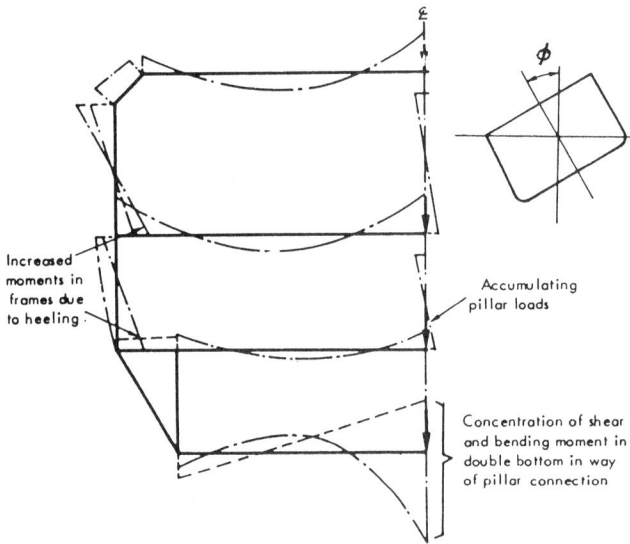

Fig. 49 Racking of a transverse frame

strongly influenced by any significant longitudinal members of the structure to which it may be connected. In the two-dimensional frame computation, these effects can be and are normally taken into consideration by applying appropriate boundary restraint conditions at the intersection of the frame with such members. In the example shown here, the most important of these members are the side and bottom shell, decks, tank top and longitudinally continuous girders. These plate members are assumed to be completely rigid with respect to deflections in their own planes; thus the appropriate boundary conditions in the direction of the plane of the deck or shell at the intersection with the frame, would be a condition of complete fixity. However, rotational restraint of the joint is provided only by the flexural stiffness of the deck beams and side frames. Longitudinal girders, as a result of their smaller cross-sectional dimensions, are less stiff in resisting deflection, and their influence on the frame would be represented by elastic attachments having finite spring constants. The correct value of the spring constant would be determined by evaluating the load vs. deflection characteristics of the girder, including the effective breadth of the deck or other plating to which it is attached, and assuming appropriate support conditions at the transverse bulkheads at its two ends.

Modern RO-RO ships present a particularly severe transverse strength problem since the demands of their mode of operation requires a minimum of obstruction to longitudinal access within the ship. In some cases, transverse bulkheads are absent over the major portion of the middle length of the ship, and vertical support of the decks is accomplished by deep transverse web frames and vertical pillars. As a result of lateral shear deflection, the decks may no longer provide complete transverse fixity at their intersection with the frame ring and large racking stresses and deflections in the frame must be accounted for. These result in large bending moments at the frame-deck beam intersections. An illustration of the midship section of such a vessel and the resulting moments, shears and deflections are shown in Fig. 49 from ISSC (1979). It should be noted also that the principal load component in this case is associated with the rolling (heel) of the ship and includes important contributions from the transverse component of the gravity force as well as the inertia of the structure and contents. Both of these forces experience their maximum value at the same instant during the roll cycle when the angle of roll is at its maximum value, but they are in phase above the roll center and 180 deg out of phase below. See Volume III, Chapter VII, Section 4.2.

Oil tankers and bulk carriers present entirely different problems of transverse strength because of their structural arrangements and distribution of loading. In both types of ship, normal conditions of loading are characterized by pronounced discontinuities of the loads in the transverse plane. In tankers, this comes about from the arrangement of cargo and ballast spaces in which it is customary to utilize a pair of wing tanks as clean ballast tanks with the corresponding centerline tank used for cargo. In the loaded condition the ballast tanks are empty; thus there is an excess of buoyancy over weight in the wings and excess weight on centerline. Forward and aft of such a ballast space, there will usually be fully loaded tanks across the entire width of the ship. In the bulk carrier, particularly one used for heavy ore cargo, the centerline cargo hold will be relatively narrow and will have an excess of weight in the loaded condition when the wing tanks are empty.

Structurally, such ships are invariably longitudinally framed with wide-spaced webs and may be thought of as consisting of several modules abutting each other. A module consists of several panels of stiffened plating

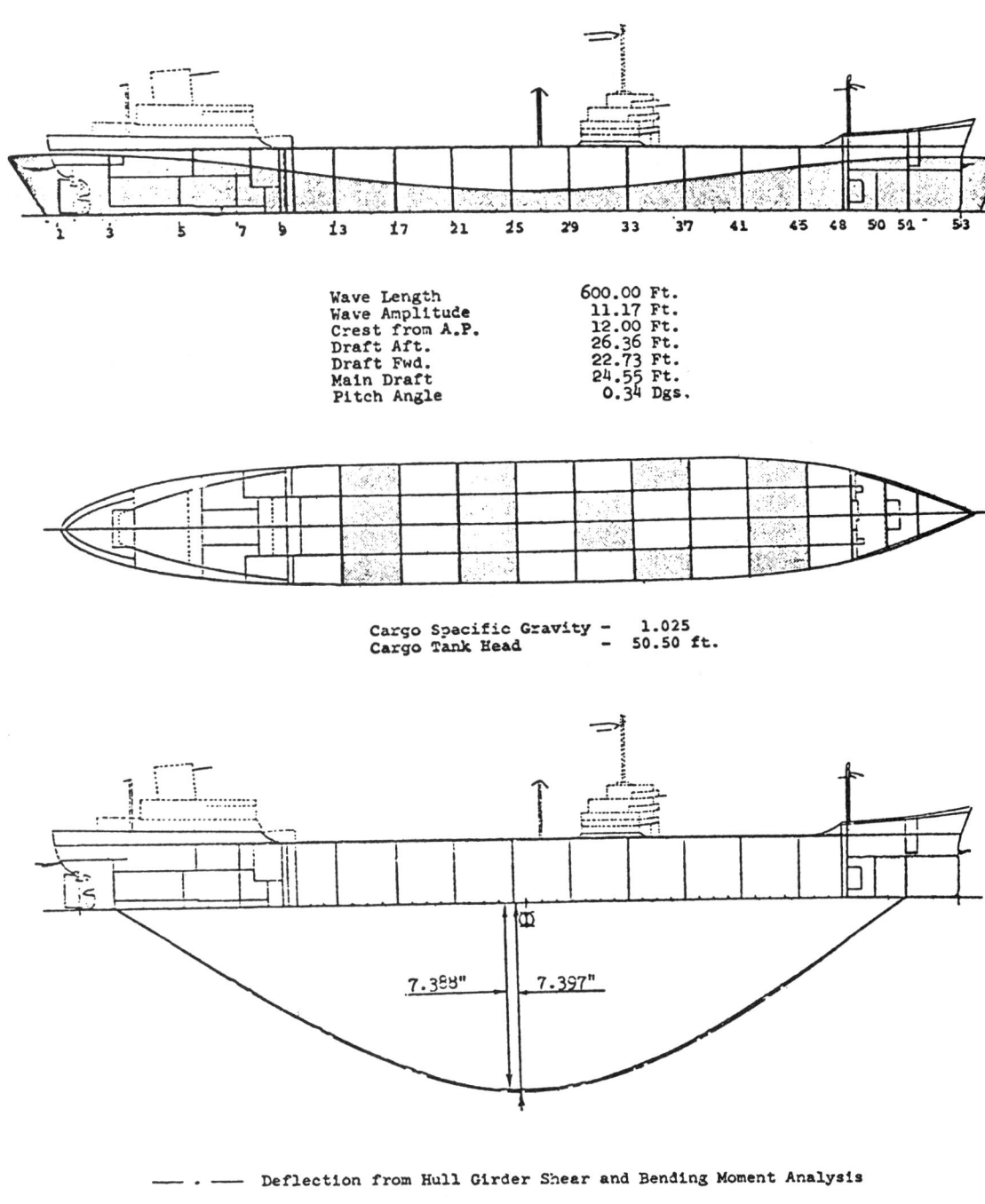

Fig. 50 Typical tanker structural arrangement, with module for three-dimensional finite element analysis

including decks, bottom structure, longitudinal and transverse bulkheads and side shell. Since the spacing of transverse bulkheads is usually of the same order of magnitude as the beam of the ship, a major module, consisting of the section of the ship length contained between two transverse bulkheads, will be approximately square when viewed from above. Within each module, there will be a secondary system of stiffening girders and webs oriented longitudinally and transversely, and these will be of about equal strength to each other. The principal transverse strength members are the bulkheads and transverse webs, but, because of the proportions of the hull module and the stiffness characteristics of the longitudinal and transverse members, there will be strong interactions between the transverse and longitudinal structure and a two-dimensional transverse strength analysis will seldom yield reliable results. For this reason, a three-dimen-

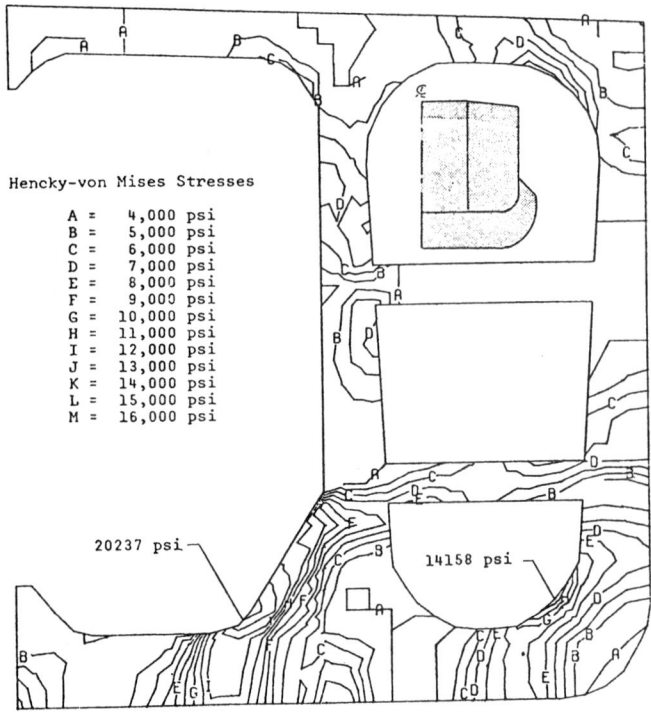

(a) STRESS PLOT, FRAME 23

(b) COMPARISON OF SHEAR STRESS DISTRIBUTION, FRAME 31

Fig. 51 Results of finite element analysis of tanker in Fig. 50 (Liu and Bakker, 1981)

sional analysis is usually performed in order to obtain results that are useful for more than comparative purposes. Since many of the important strength members in such an analysis consist of deep thin plate structure with bracketed intersections, it is not possible to establish an accurate space-frame model. Consequently, finite-element computations utilizing membrane and bar elements must be used in order to achieve an acceptable degree of accuracy in the modelling of the structural behavior.

Ideally, the entire ship hull should be included in such a finite-element (FE) model. The preparation and checking of input data for such an elaborate model would, however, require excessive time and effort, and results of acceptable accuracy can usually be obtained by modelling a portion of the ship length equal to about three tank lengths. If longitudinal symmetry of loading and structural arrangement prevails fore and aft of the space in question, the three-tank space can be reduced, through symmetry considerations, to one and one-half tank spaces. Transverse symmetry can, of course, also be assumed in most cases; so the total FE model would represent one and one-half tank spaces on one side of the centerline. A typical arrangement of the cargo and ballast spaces of a tanker, together with a suitable module for a three-dimensional FE analysis is illustrated in Fig. 50, from Liu and Bakker (1981).

The results of such a three-dimensional FE analysis for the tanker are illustrated in Fig. 51 (Liu and Bakker, 1981), which illustrates the degree of detail to be expected from such an analysis. Skaar (1974), has shown a comparison of a FE calculation similar to that of Figs. 50 and 51, with the results of a three-dimensional space frame analysis. The members used in the latter were specially developed beam members having the capability of representing the neutral axis eccentricity and shear carrying capacity of the deep webs. The stresses in the middle portion of the web members was represented reasonably well, but the space-frame model did not contain the detailed description of stresses in the brackets and other members of varying geometry.

The preparation and checking of input data for a finite-element analysis of the extent described above represents a major expenditure of time and effort. There are available, however, computer-based data generation procedures and graphics packages for the presentation of results, which can be utilized to keep the overall cost and labor of such an analysis at an acceptable level. Such an analysis is now often routinely required by classification societies for the purpose of confirming the transverse structural design of large tankers and bulk carriers.

3.13 Deckhouses and Superstructures. The terms *deckhouse* and *superstructure* refer to a structure usually of shorter length than the entire ship and erected above the strength deck of the ship. If its sides are coplanar with the ship's sides it is referred to as a superstructure. If its width is less than that of the ship, it is called a deckhouse. We shall use the latter as an inclusive term to mean both types of structures since the superstructure may be considered as a special case of a deckhouse.

As the ship hull bends in response to the applied seaway and other external loads, the deckhouse will bend also in response to the loads transmitted to it through its connection to the main hull. These loads will consist of distributed longitudinal shears and vertical loads acting at the lower edges of the sides of the deckhouse. Since there will be equal and opposite reactions applied to the hull, the presence or absence of the deckhouse is seen to affect the structural behavior of the hull. The combined stiffness may be appreciably greater than that of the hull alone if the deckhouse is of substantial length and the two are of the same material, effectively connected together.

In addition to these effects felt in the overall bending stiffness and the corresponding stress patterns, local stress concentrations may be expected at the ends of the house, since here the structure is transformed abruptly from that of a beam consisting of the main hull alone to that of hull plus deckhouse. Particular care is needed in designing the structural details and reinforcement in this region of both the main hull and the deckhouse in order to avoid localized structural problems.

The horizontal shears and vertical loads between hull and deckhouse will tend to produce opposing structural effects as may be seen by considering two extreme cases. Let us consider the ship to be in a hogging bending condition, corresponding to a wave crest amidships. The deck of the ship will be in a state of positive or extensional strain. Let us first assume that the condition of longitudinal strain compatibility is satisfied between the deck and the lower side of the deckhouse, and that there is no interference or vertical resistance force between hull and deckhouse. In this case, the deckhouse will tend to bend in a concave upward mode as a result of the extensional strain of its lower edge; thus the deflection of the deckhouse will be opposite the deflection of the hull.

Now, as a second case, assume that the condition of strain compatibility is not satisfied, i.e., that the deckhouse longitudinal strain is independent of the strain in the hull at the connection between the two. Let the deck of the hull in this case be very stiff, however, so that the bending deflection of the deckhouse is forced to follow that of the hull. Since the bending deflection of the deckhouse is now concave downwards, the lower edge of the deckhouse will be in compression, which is opposite to the tensile strain in the deck of the hull.

In the actual case, except for a small effect of shear lag, the longitudinal shear connection between the ship hull and deckhouse will be nearly completely effective and the condition of longitudinal strain compatibility will be satisfied. The vertical loads between hull and deckhouse will be associated with the relative vertical

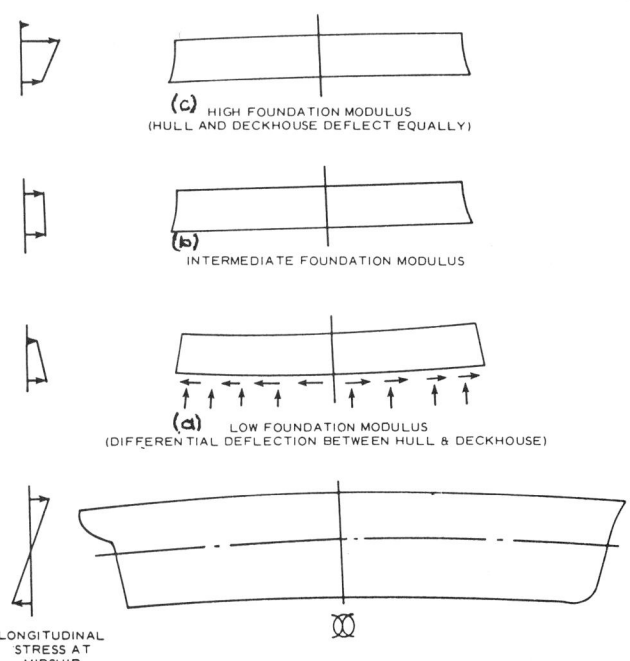

Fig. 52 Bending behavior of deckhouse and hull (Taggart, 1980)

deflection between the two members, and this will, in turn, depend upon the rigidity or *foundation modulus* of the deck structure upon which the deckhouse rests. Fig. 52 illustrates three possible modes of hull-deckhouse interaction, depending on the relative stiffness of the deck. In Fig. 52(a), the deck structure is very flexible (low foundation modulus) allowing nearly unrestricted vertical deflection between hull and deckhouse. For the hogging condition illustrated, the lower edge of the deckhouse is in extensional strain as a result of strain equality at the hull joint. As a result of the flexible deck, this results in differential deflection of the hull and deckhouse.

Fig. 52(c) illustrates a case of very high foundation modulus in which the deckhouse is constrained to deflect with nearly the same curvature as the hull. This is approximately the case of the superstructure, where—if sufficiently long, so that end effects are confined to a short portion of the length—the middle part of the house acts merely as an extension of the structure of the main hull. The longitudinal stress distribution shown in the left-hand part of the figure is colinear with that in the hull, assuming that the two are constructed of the same material. The intermediate case of finite foundation modulus, in which there is some differential deflection between hull and deckhouse is illustrated in Fig. 52(b).

In the analysis of the hull-deckhouse interaction, we assume that the hull and deckhouse each behave as a simple beam undergoing bending deflection. Under such simplifying assumptions, a complete analysis of the stress and deflection, particularly of the stress

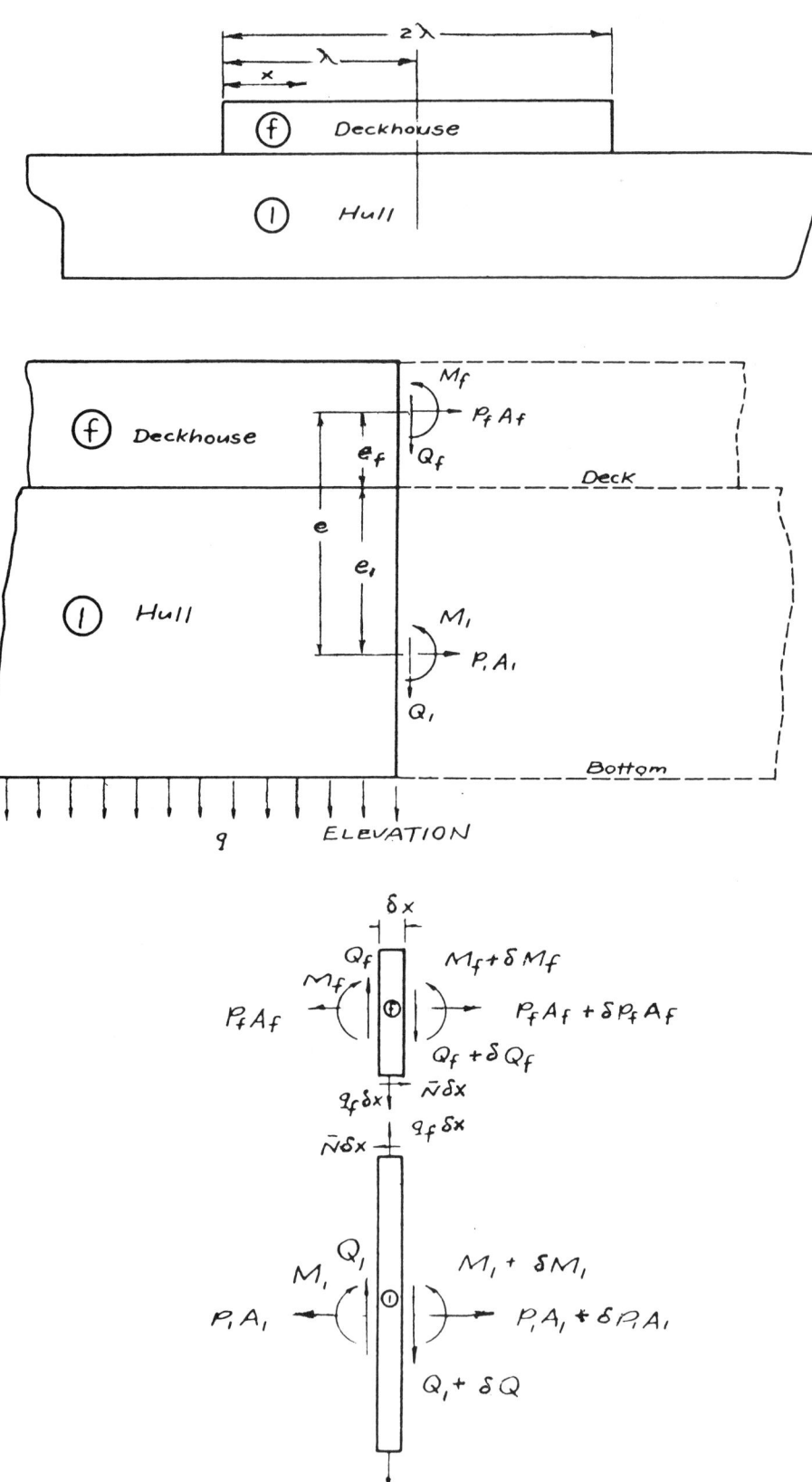

Fig. 53 Identification sketches for hull and deckhouse

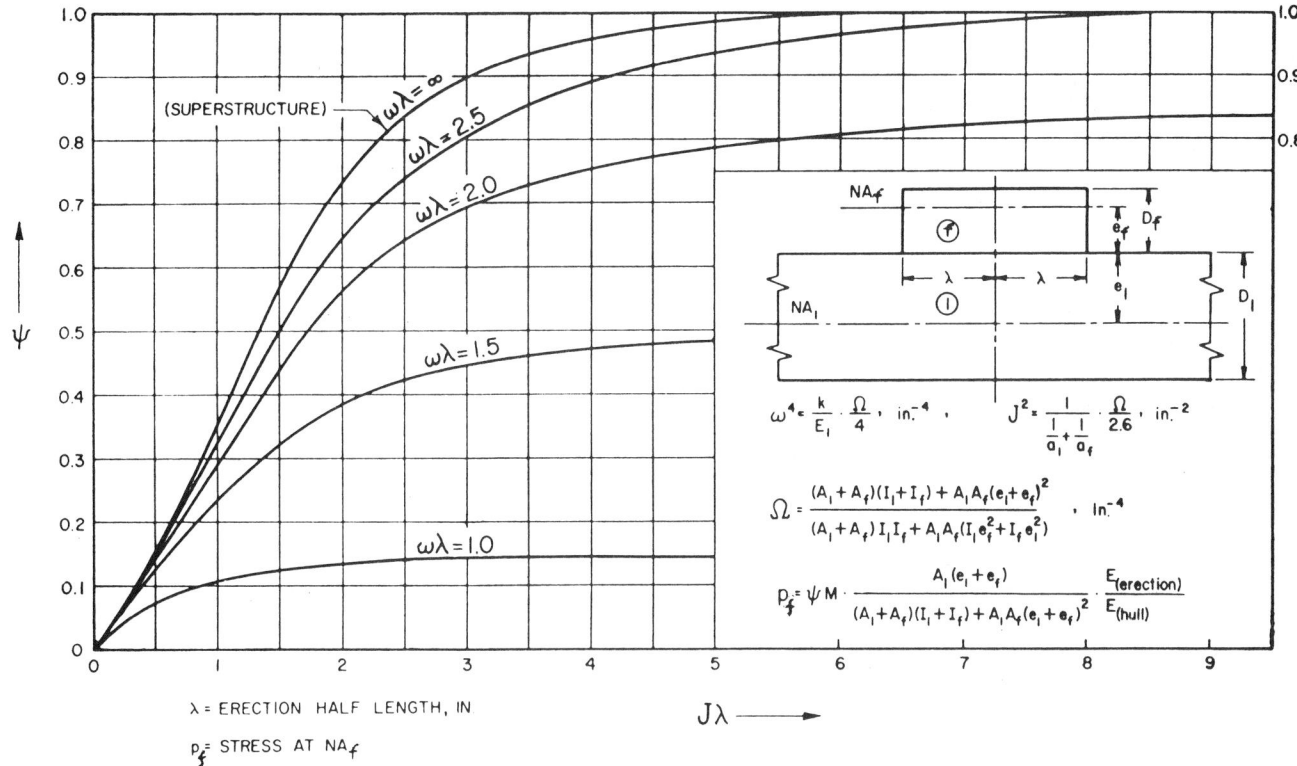

Fig. 54 Trends of erection efficiency Ψ (Schade, 1965)

concentrations at the ends of the house, is not possible. However, it is possible to obtain meaningful results that describe the combined behavior of hull and deckhouse in the middle portion of the length of the house.

In Fig. 53 we assume a vertical cut through the hull and deckhouse at location x, measured from the left hand end of the house. The deckhouse and hull, when each bend as separate beams, will have individual neutral axes NA_1 and NA_f, separated by the vertical distance e. The forces of interaction consist of a horizontal shear flow, N, and a vertical distributed loading, q_n, each having the dimensions of force per unit length. The vertical load is assumed proportional to the relative vertical deflection of the hull and deckhouse, and the proportionality constant, k, or *foundation modulus* has the dimensions of a spring constant per unit length (force per unit length/unit deflection). The separate bending moments in deckhouse and hull are M_1 and M_f, and the net longitudinal stresses in each are p_1 and p_f, respectively.

We now write equations of equilibrium of longitudinal forces and equilibrium of moments about the respective neutral axes for hull and deckhouse separately,

$$A_1 p_1 + A_f p_f = 0$$
$$M_1 + M_f - A_f e p_f = M \qquad (95)$$

The longitudinal stresses at the joint of deckhouse and hull are given by the sum of the average stress due to p and the bending stress due to the moment,

$$p_1 - \frac{M_1}{Z_1} = \frac{1}{r}\left(p_f + \frac{M_f}{Z_f}\right) \qquad (96)$$

where r is a factor included to allow for the effect of shear lag in the deck.

Here A_1 and A_f are the cross-sectional areas of hull and deckhouse, respectively. Z_1 and Z_f are the sectional moduli, and the other parameters are shown in Fig. 53.

As noted earlier, equality of longitudinal strain is required at the joint since the deckhouse is assumed attached to the hull continuously along its length. Neglecting transverse strain effects as is the case in simple beam theory, this reduces to stress equality, or $\sigma_f = r\sigma_1$. Combining Equations (95) and (96), and then solving for M_f we obtain,

$$M_f = \frac{rZ_f}{rZ_f - Z_1}\left\{p_f\left[A_f e + Z_1\left(\frac{rA_f + A_1}{rA_1}\right) + M\right]\right\} \qquad (97)$$

The above considerations have given three equations in the four unknowns, p_1, p_f, M_1, and M_f. A fourth consideration involves the vertical interaction between hull and deckhouse, which is assumed proportional to the differential vertical deflection, $w_2 - w_1$. Equilibrium of the slice of hull and deckhouse shown in Fig. 53 leads to the condition,

$$k(w_1 - w_f) = \frac{rZ_f}{Z_f - rZ_f}\left(\frac{d^2p_f}{dx^2} + \frac{d^2M}{dx^2}\right) \quad (98)$$

Noting that the deflection of each member must be related to the bending moment on that member by the equation of simple beam theory, we obtain Equation (99). Note that in this equation the total deflection, w_1 or w_2, has been corrected for the shear deflection in hull or deckhouse by deducting the shear deflection given by the second term in the parentheses on the right-hand side.

$$M_1 = -E_1 I_1\left(\frac{d^2w_1}{dx^2} + \frac{q_1}{a_1 G_1}\right)$$

$$M_f = -E_f I_f\left(\frac{d^2w_2}{dx^2} + \frac{q_2}{a_f G_f}\right) \quad (99)$$

Here a_1 and a_2 are the vertical shear-carrying areas of the hull and deckhouse, respectively, made up principally of the side plating and longitudinal bulkhead members. G_1 and G_f are the shear moduli of elasticity of hull and deckhouse, respectively.

By combining Equations (95) through (99) we may obtain a fourth-order differential equation for the mean stress in the deckhouse, p_f. The solution has been condensed (Schade, 1965), into a single design chart suitable for most practical ship structural applications and given in Fig. 54. In using this chart, it is necessary to compute the following three parameters:

- Section geometry parameter,

$$\Omega = \frac{(A_1 + A_f)(I_1 + I_f) + A_1 A_f (e_1 + e_f)^2}{(A_1 + A_f) I_1 I_f + A_1 A_f (I_1 e_f^2 + I_f e_1^2)} \quad (100)$$

- Foundation modulus parameter,

$$\omega^4 = \frac{k}{E_1}\frac{\Omega}{4} \quad (101)$$

- Shear stiffness parameter,

$$J^2 = \frac{1}{\frac{1}{a_1} + \frac{1}{a_f}}\frac{\Omega}{2(1+\nu)} \quad (102)$$

It is assumed that Poisson's ratio, ν, is the same for the material in the hull and deckhouse. The shear-carrying areas a_1 and a_f may, however, be modified for any difference in the modulus of elasticity E as, for example, in the case of an aluminum deckhouse on a steel hull.

If the deckhouse were fully effective, corresponding to Fig. 52(c), the mean stress would be given by,

$$\tilde{p}_f = \frac{MA_1(e_1 + e_f)}{(I_1 + I_f)(A_1 + A_f) + A_1 A_f (e_1 + e_f)^2} \quad (103)$$

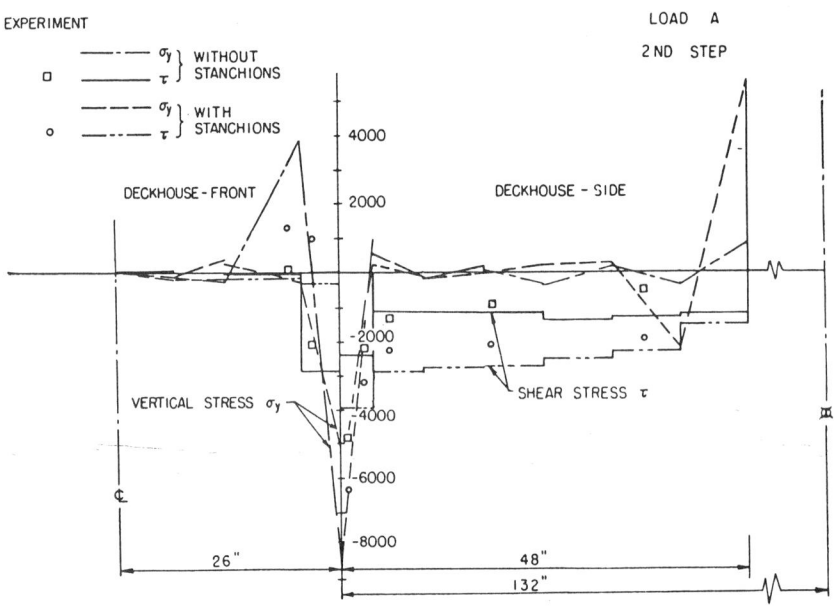

Fig. 55 Stresses in lowest members of deckhouse—second step, showing influence of foundation modulus (Paulling and Payer, 1968)

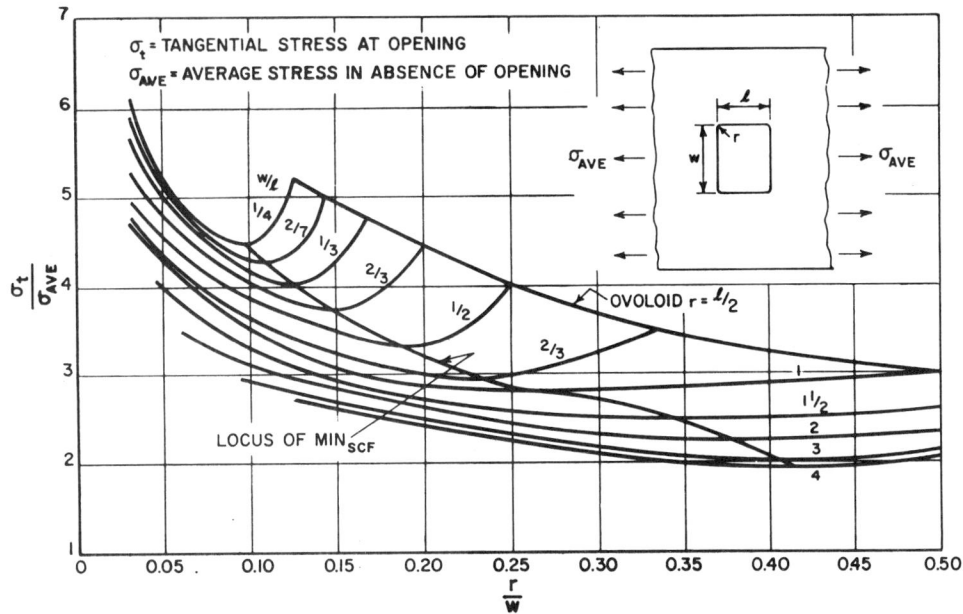

Fig. 56 Stress concentration factor for square hole, tension parallel to side (Heller, et al, 1959)

Fig. 54 expresses the efficiency of the deckhouse in terms of the ratio, ψ, of the actual mean stress to this ideal value. The actual stress is, therefore,

$$p_f = \psi \tilde{p}_f \qquad (104)$$

Having p_f, the mean stress in the hull, p_1, and the bending moments in hull and deckhouse may be obtained from Equations (95), (96) and (97). The stresses at top and bottom of hull and deckhouse may then be computed by Equations (27) of elementary beam theory, using the respective bending moments, and added to the mean stresses, p_1 and p_f.

A somewhat similar analytic solution to that described above was developed by Bleich (1953), which however does not include the effect of shear deflection and shear lag. Both the Schade solution and the Bleich solution may be used to obtain the loads in the middle portions of the deckhouse, but such solutions do not apply near the ends where, as noted, large localized loads may occur. Such solutions are useful for assessing the extent to which the deckhouse contributes to the overall bending strength of the hull, or in deciding whether to design the deckhouse to participate in the hull bending strength or to design it simply as an appendage having no contribution to the strength of the hull.

In order to design the structural details needed to withstand the concentrated loads near the ends of the deckhouse, an analysis method suitable for revealing the high stress concentrations in an area of abrupt changes in geometry is required. The most suitable means presently available for this purpose is the finite element method. Fig. 55, taken from Paulling and Payer (1968), illustrates the computation of the high vertical and shear stresses in the vicinity of the corner of the deckhouse. Also shown on this graph are experimental values of the respective stresses, illustrating the extremely high stress gradients in the vicinity of the house corner. These experiments were conducted using the same experimental apparatus, Glasfeld (1962), on which the shear lag phenomenon illustrated in Fig. 36 was measured.

3.14 Stress Concentrations. Stresses such as those discussed previously are *average* stresses. In general, any discontinuity in a stressed structure results in a local increase in stress at the discontinuity. The ratio of the maximum stress at the discontinuity to the average stress that would prevail in the absence of the discontinuity is called a stress-magnification factor.

Discontinuities in ship structures range from the gross discontinuities formed by the ends of superstructures and by large hatches to the minute corners and notches that may occur in attachments to stressed structure. The most numerous class of discontinuities is the many openings required for cargo handling access, and engineering services. Every such opening causes stress concentration. The importance of these stress concentrations is shown by the fact that the great majority of fractures in ship structures originate at the corners of openings.

At the ends of superstructures the smoothest possible transfer of the stress in the superstructure into the structure below is required. Part (3) of Vasta (1949), reporting a full-scale investigation on the *President Wilson* shows stress-concentration factors at various points near the ends of the superstructure varying from 2.4 to 4.6.

Similar care is needed at the corners of large hatches. The great amount of study that has been

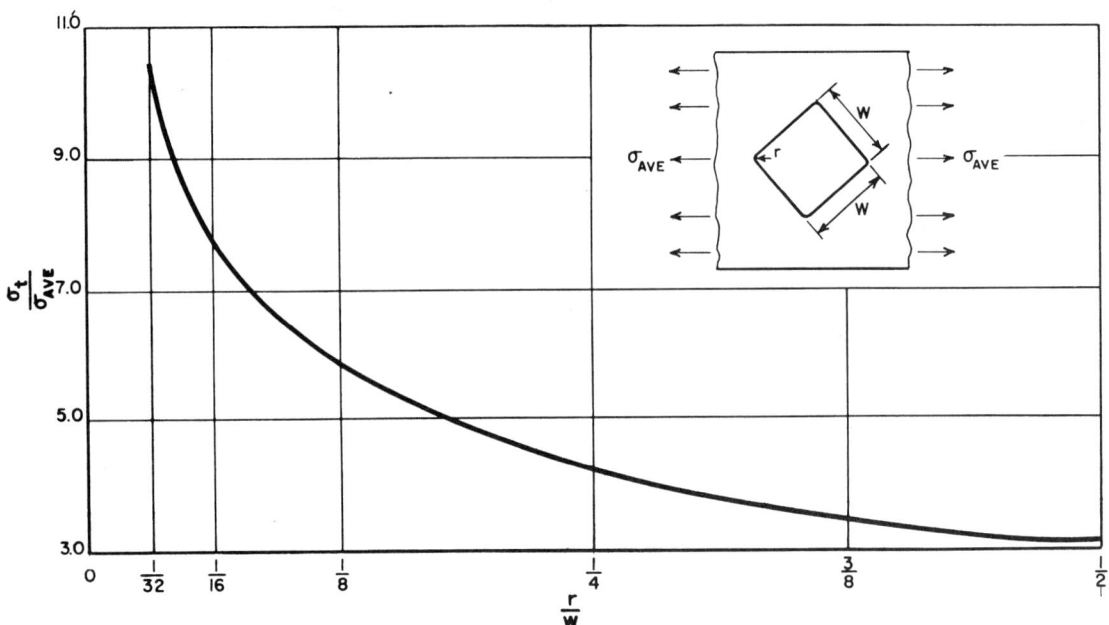

Fig. 57 Stress concentration factor for square hole, tension parallel to diagonal (Brock, 1957)

devoted to the design of hatch corners is summarized in SSC (1952). Square corners in hatches in strength decks are an invitation to fractures.

Openings or cuts in structure present two problems; the reduction of the strength of the member in which the hole is cut, and the stress concentrations adjacent to the opening. The stress-concentration factor increases with the width of the opening relative to the width of the member, and with the acuity of the corners of the opening.

If the size or location of an opening is such as to impair unduly the strength of a member, measures must be taken either to reduce the stress in way of the opening or to compensate for the loss of material, or both. This may be done by changing the location of the opening, e.g., in the case of a beam or girder under flexural loading, locating the opening at or near the neutral axis of the member, changing the shape of the opening, providing an insert or doubling plate at the area of the member circumscribing the opening, or by providing a reinforcing ring around the opening.

The purpose of a reinforcing ring is to stabilize the edge of the cut member around the opening and to concentrate added material around the opening close enough to the plane of the loaded plate so that the ring will deform with the loaded plate, absorb energy and reduce strains at the periphery of the opening.

Large openings in strength members may require the use of an insert plate abreast the opening, thicker than the material in which the hole is cut, or a doubling plate to compensate for the material lost at the opening and possibly a reinforcing ring at the periphery of the opening.

When insert or doubling plates are used as reinforcements, they must be tapered off in the direction of loading and well beyond the opening to minimize stress concentrations due to abrupt changes in thickness and to insure that the cross-sectional area of the insert or doubler will become fully effective in way of the opening. Insert plates are preferable to doublers, since they eliminate lamination and insure effectiveness of the reinforcing material at the approaches to and in way of the opening.

The design of compensation and reinforcement at openings is discussed in various sections of Chapter VII of Taggart (1980). The reader's attention is also invited to the latest NAVSEA Design Data Sheet for methods for determination of reinforcements based on the energy-absorption principle, which are applicable to openings in strength structures of U.S. naval surface ships.

Rectangular Openings. The stress-magnification factors at the corners of a rectangular opening in an axially loaded plate depend on the size of the opening and the radius of the corners. Fig. 56 from Heller, et al (1959) is a family of curves which shows the maximum values of the boundary stress as a function of corner radius and aspect ratio of the openings in an infinite plate subjected to uniaxial stress. Theory shows that when the plate is at least approximately four times the width of the opening, the concentration factors for an infinite width of plate are sufficiently accurate.

For a sharp corner, the maximum stress occurs at the corner, and for finite radii the maximum stress is located between the midpoint of the fillet and the point of tangency.

Fig. 56 also gives the locus of minimum stress-con-

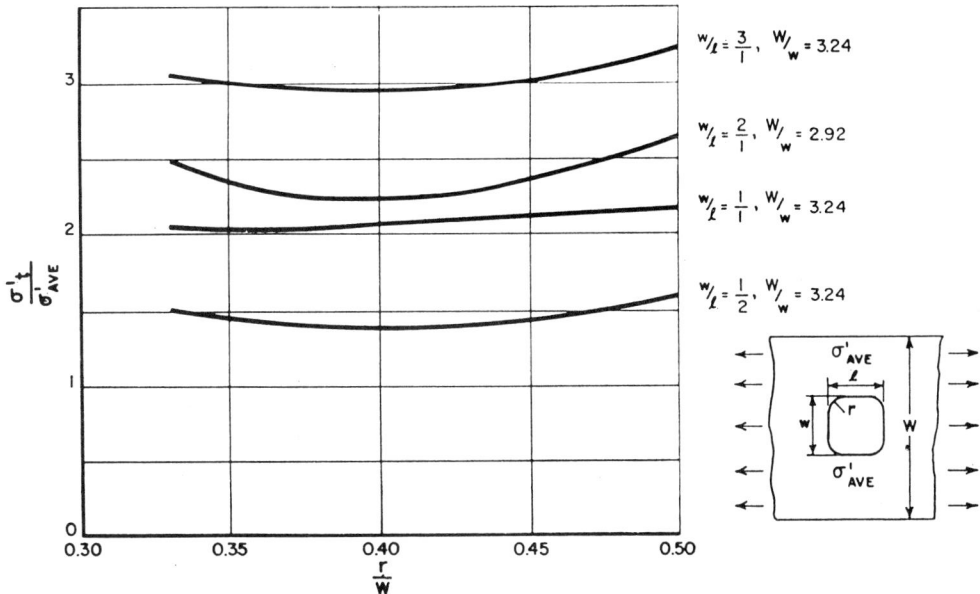

Fig. 58 Stress concentration factor in a finite plate with a square hole (D'Arcangelo, 1969)

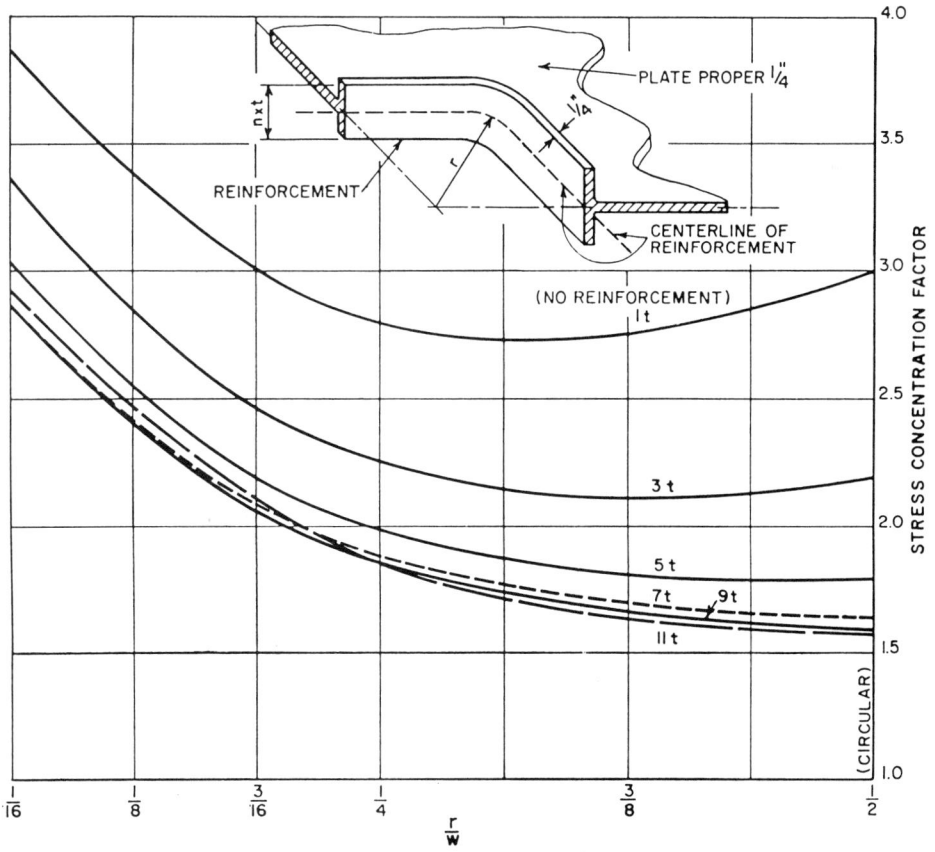

Fig. 59 Stress concentration factor for face-bar reinforced square openings with rounded corners (Kuntsmann and Umberger, 1959)

centration factors and the factors for an ovoloid when $r = l/2$.

Fig. 57 (from Brock, 1957) gives stress-concentration factors for a square hole with the applied load parallel to the diagonal of the hole. The maximum stresses are considerably higher under this condition of loading than when the applied load is parallel to the side of the opening as in the case of Fig. 56.

Fig. 58 (from D'Arcangelo, 1969) gives the stress-concentration factor for rectangular openings in a finite plate. The nomenclature is the same as for Figs. 56 and 57, except the additions of W = width of plate, and σ'_{AVE} = average stress at minimum section.

Increasing the length of the opening, for a constant width, reduces the stress-concentration factor. This is shown in Fig. 58 where the stress-concentration factor is halved as the w/l ratio is reduced from 3/1 to 1/2. Generally the stress concentration is smaller when the long dimension of the opening is parallel to the load.

Fig. 59 gives the maximum stress-concentration factors versus r/w ratio for a face-bar-reinforced square opening with rounded corners. The minimum possible factor is 1.56. For a reinforcement thickness equal to the plate thickness, the maximum effective reinforcement height, above which additional height has little effect, is seven times the plate thickness.

The specimens from which the data were obtained were of sufficient size so that the loading arrangements and stress patterns were independent of the specimen boundaries.

Circular Openings. Stress-concentration factors for non-reinforced and reinforced circular openings in an infinite plate are obtained from Fig. 59 when r/w is equal to 1/2.

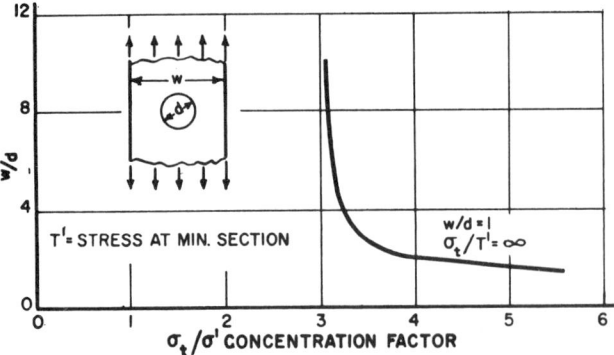

Fig. 60 Stress concentration factor in a finite plate with a circular hole

Fig. 60 (from Coker, et al 1919-20) is a plot of photoelastic studies of tangential stress-concentration factors for openings, centrally located in the width of the plate, of various ratios of plate to opening width.

3.15 Unconventional Craft. There are many types of unconventional craft that require special consideration in their structural design, either because of the nature of the loads encountered, unusual structural configurations, or both. Since most of them are intended for high-speed operation in rough seas, dynamic loads may be of much greater importance than for conventional ships. (Some of these dynamic loads are discussed in connection with other problems of motion in waves in Chapter VII, Vol III.) Unusual structural arrangements are found in such craft as hydrofoil boats, where the hull is supported in air on struts by the foils immersed in water, and in catamarans, where the structure connecting the twin hulls is not found in other types. A general treatment of the problem of impact loads on high-speed V-shaped hulls is given by Stavovy and Chuang (1976).

(a) *Planing Craft.* For low, non-planing speeds the longitudinal hull structure of planing craft can be designed in the same manner as for conventional displacement hulls. But at planing speeds high slamming pressures are generated on the forward bottom as a combined result of relative vertical motions between boat and water surface and forward speed with an angle of attack to the wave surface. These impact loads are of very short duration and at any instant cover a relatively small area. For design purposes an estimate is usually made of the steady load that would produce the same effect as the transient dynamic load. A summary of empirical methods available, based on theory, experiments, and full-scale trials is given by Silvia (1978), covering work of Heller and Jasper (1960), Allen and James (1977), Spencer (1975), and others. Silvia states that after extreme local pressure values have been determined, empirical reduction factors should be applied to allow for averaging or spreading of load over the area of interest and for the longitudinal location of the component being designed. He states that, "on the assumption that the design pressure selected reflects a worst possible case and that it is very unlikely to be exceeded, very low safety factors (1.1 against yield or 1.4 against ultimate strength) can be used." But fatigue and weld-zone effects must also be considered, and he recommends a reduction of allowable stress by 50 percent for aluminum and 65 percent for fiberglass construction.

Useful guidelines are provided by ABS *Rules for Building and Classing Steel Vessels Under 61 Meters (200 ft.) in Length* (1983) and *Provisional Rules for Building and Classing or Certifying Reinforced Plastic Vessels* (1978).

(b) *Catamarans.* Twin-hulled ships represent an unusual type that involve distinctive structural problems, specifically the design of the structure connecting the two hulls. Hadler, et al (1974) found from their studies of the oceanographic research ship, USNS *Hayes*, that the most critical loadings on this structure were the transverse bending moment and vertical shear. Calculation procedures based on the basic theory of Chapter VII, Vol. III, developed for determining motions in waves, were applied to the calculation of these loads in beam seas (Lee, et al, 1973). Model tests (Wahab, et al, 1971) indicated good agreement for design purposes.

Another structural problem, however, arose during the early operation of the *Hayes* in the North Atlantic. It was found that relatively large bow motions resulted in severe slamming of the cross-structure, requiring significant reductions in speed. This was solved by the addition of a hydrofoil between the two hulls forward, which provided significant reduction in relative bow motion (as discussed in Section 6 of Chapter VII). Hadler, et al (1974) present a method for determining the loads on the hydrofoil as a basis for design.

The small waterplane twin-hull (SWATH) is a special type of catamaran in which the two buoyant hulls are completely immersed. It has received considerable attention in regard to structural loads and design. Methods of calculating motions and loads on cross-structure are presented by Lee and Curphey (1977). Active or fixed control surfaces are usually provided both forward and aft to reduce motions and hence to avoid excessive slam impact loads.

(c) SES and ACV. Of the vehicles supported mainly by an air cushion, those with fixed sidewalls are usually called *Surface Effect Ships* (SES) and those with flexible skirts only are designated *Air Cushion Vehicles* (ACV). Because of their relatively simple structural configuration, the only serious structural design problem with these craft is to provide adequate strength against bow impacts—as for other high-performance craft. See Kaplan (1981) and Mantle (1975).

(d) Hydrofoil Craft. Since at high speeds the hulls of these craft are supported in air by foils immersed in water, the distinctive structural design problem is that of the hydrofoils and supporting struts. Such a wide variety of foil arrangements have been developed that they cannot be discussed in detail here. But in all cases the foil and strut design problem can be minimized by making sure that the hull itself is as light as possible. These problems lead to the adoption of principles that are more closely allied to aircraft than to ship design. Multiple production is usually an important consideration also in hydrofoil craft design, as discussed by Bullock and Oldfield (1976). Hulls must, of course, be designed to withstand bottom slamming damage, as in the case of other vehicles discussed here.

Section 4
Load Carrying Capability and Structural Performance Criteria

4.1 The Nature of Structural Failure. As noted in the introduction, ship structural failure may occur as a result of a variety of causes, and the degree or severity of the failure may vary from a minor esthetic degradation to catastrophic failure resulting in loss of the ship. In the report of the Committee on Design Procedures of the International Ship Structures Congress (ISSC, 1973), four contributing failure mechanisms or modes were defined:
- Tensile or compressive yield of the material.
- Compressive instability (buckling).
- Low-cycle fatigue.
- Brittle fracture.

The first mode of failure occurs when the stress in a structural member exceeds a level that results in a permanent plastic deformation of the material of which the member is constructed. This stress level is termed the material *yield stress*. At a somewhat higher stress, termed the *ultimate stress*, fracture of the material occurs. While many structural design criteria are based upon the prevention of any yield whatsoever, it should be observed that localized yield in some portions of a structure is not necessarily serious and may, in case of non-reversing loads, result in a more favorable redistribution and equalization of stress throughout the structure.

Instability failure of a structural member loaded in compression may occur at a stress level that is substantially lower than the material yield stress. The load at which instability or buckling occurs is a function of member geometry and material modulus of elasticity rather than material strength. The most common example of an instability failure is the buckling of a simple column under a compressive load that equals or exceeds the *Euler Critical Load*. A plate in compression will also have a critical buckling load whose value depends on the plate thickness, lateral dimensions, edge support conditions and material modulus of elasticity. In contrast to the column, however, exceeding this load by a small margin will not necessarily result in complete collapse of the plate but only in an elastic deflection of the central portion of the plate away from its initial plane. After removal of the load, the plate will return to its original undeformed configuration. The ultimate load that may be carried by a buckled plate is determined by the onset of yielding at some point in the plate material or in the stiffeners, in the case of a stiffened panel. Once begun, yield may propagate rapidly throughout the entire plate or stiffened panel with further increase in load.

Fatigue failure occurs as a result of a cumulative effect in a structural member that is exposed to a stress pattern alternating from tension to compression through many cycles. Conceptually, each cycle of stress causes some small but irreversible damage within the material and, after the accumulation of enough such damage, the ability of the member to withstand loading is reduced below the level of the applied load. Two categories of fatigue damage are generally recognized and they are termed *high-cycle* and *low-cycle* fatigue. In high-cycle fatigue, failure is initiated in the form of small cracks, which grow slowly and which may often be detected and repaired before the structure is endangered. High-cycle fatigue involves several millions of cycles of relatively low stress (less than yield) and is typically encountered in machine parts rotating at high speed or in structural components exposed to severe and prolonged vibration. Low-cycle fatigue involves higher stress levels, up to and beyond yield, which may result in cracks being initiated after several thousand cycles.

The loading environment that is typical of ships and ocean structures is of such a nature that the cyclical stresses may be of a relatively low level during the greater part of the time, with occasional periods of very high stress levels caused by storms. Exposure to such load conditions may result in the occurrence of low-cycle fatigue cracks after an interval of a few years. These cracks may grow to serious size if they are not detected and repaired.

In the fourth mode of failure, brittle fracture, a small crack suddenly begins to grow and travels almost explosively through a major portion of the structure. The originating crack is usually found to have started as a result of poor design or manufacturing practice, as in the case of a square hatch corner or an undetected weld flaw. Fatigue is often found to play an important role in the initiation and early growth of such originating cracks. The control of brittle fracture involves a combination of design and inspection standards aimed toward the prevention of stress concentrations, and the selection of steels having a high degree of notch toughness or resistance to the growth of cracks, especially at low temperatures. *Crack arrestors* are often incorporated into the structure to limit the travel of a crack if it should occur (fail-safe design).

In designing the ship structure, the analysis phase is concerned with the prediction of the magnitude of the stresses and deflections that are developed in the structural members as a result of the action of the sea and other external and internal causes. Many of the failure mechanisms, particularly those that determine the ultimate strength and total collapse of the structure, involve nonlinear material and structural behavior that are beyond the range of applicability of the linear structural analysis procedures described elsewhere in this chapter, which are commonly used in design practice. Most of the available methods of nonlinear structural analysis are beyond the scope of the present work and are often limited in their applicability to a narrow class of problems. It is one of the difficulties facing the structural designer that he or she must often use linear analysis tools in predicting the behavior of a structure in which the ultimate capability is governed by nonlinear phenomena. This is one of the important sources of uncertainty referred to in the Introduction.

After performing an analysis, the adequacy or inadequacy of the member and/or the entire ship structure must then be judged through comparison with some kind of criterion of performance. The conventional criteria that today are commonly used in ship structural design are usually stated in terms of acceptable levels of stress in comparison to the yield or ultimate strength of the material, or as acceptable stress levels compared to the critical buckling strength of the structural member. Such criteria are, therefore, intended specifically for the prevention of the first two of the four types of failure—tensile yield or compressive buckling.

Design criteria stated expressly in terms of fatigue damage resistance are seldom employed in ship structural design although cumulative fatigue criteria are used in offshore structure design. Fatigue considerations are especially important in the design of details such as hatch corners and reinforcements for openings in structural members. Since the ship loading environment, consisting in large part of alternating loads, is highly conducive to fatigue-type failures, it may be assumed that fatigue resistance is implicitly included in the conventional safety factors or acceptable stress margins based on past experience.

The prevention of brittle fracture, as noted previously, is largely a matter of material selection and proper attention to the design of structural details in order to avoid stress concentrations. Quality control during construction and in-service inspection form key elements in a program of fracture control.

4.2 Material Physical Properties and Yield Criteria. The physical strength properties of shipbuilding materials are normally obtained from standardized testing procedures conducted under closely controlled conditions, as described in Chapter VIII of Taggart (1980). The yield strength of the material is defined as the measured stress at which appreciable nonlinear behavior accompanied by permanent plastic deformation of the material occurs. The ultimate strength is the highest level of stress achieved before the test specimen fractures. For most shipbuilding steels, the yield and tensile strengths in tension and compression are assumed to be equal.

The material properties described above are expressed in the form of simple uniaxial stresses, i.e., the test from which they are obtained is conducted in such a way that the test specimen is subject to stress in the longitudinal direction only, and the transverse stresses are zero. Few ship structural members, chiefly the flanges of slender stiffeners or slender columns, experience pure uniaxial stress. In these cases, however, the computed member stresses may be compared directly with the uniaxial test data in order to ascertain the adequacy of the member.

In the plate members of the hull structure, the stresses do not, in general, form a simple unidirectional pattern. The maximum values of the primary, secondary, and tertiary stresses may not coincide in direction or time, and therefore combined stresses should be calculated—as discussed in Section 3.11—at various locations and times. Furthermore, the stress at each point will not be a simple unidirectional tension or compression but, in most cases, will be found to form a biaxial stress pattern. See Section 3.4.

The stress criterion that must be used in this situation is one in which it is possible to compare the actual multiaxial stress with the material strength expressed in terms of a single value for the yield or ultimate stress. For this purpose, there are several theories of material failure in use, of which the one usually considered the most suitable for ductile materials such as ship steel is referred to as the *Distortion Energy Theory*. (This is also called the *Octahedral Shear Stress Theory* or *Hueber-von Mises-Hencky Theory*).[4]

To illustrate the application of this theory, consider a plane stress field in which the component stresses are σ_x, σ_y, τ and the corresponding principal stresses are σ_1, σ_2. The distortion energy theory states that failure through yielding will occur if the equivalent stress, σ_e, given by

$$\sigma_e = (\sigma_x^2 + \sigma_y^2 - \sigma_x\sigma_y + 3\tau^2)^{1/2}$$
$$= (\sigma_1^2 + \sigma_2^2 - \sigma_1\sigma_2)^{1/2} \qquad (105)$$

exceeds the equivalent stress corresponding to yielding of the material test specimen. The standard material test data, however, are obtained from a uniaxial stress pattern for which $\sigma_x = \sigma_{yield}$, $\sigma_y = 0$, and $\tau = 0$. The material yield strength may, therefore, be expressed through an equivalent stress at failure, σ_0, obtained by substituting the above values into Equation (105) or

$$\sigma_0 = \sigma_{yield} \qquad (106)$$

The margin against yield failure of the structure is, therefore, obtained by a comparison of the structure's σ_e against σ_0 giving the result,

$$\sigma_0^2 = \sigma_1^2 + \sigma_2^2 - \sigma_1\sigma_2 \qquad (107)$$

Equation (107) is the equation for an ellipse in the σ_1 σ_2-plane and is illustrated in Fig. 61. Pairs of values (σ_1, σ_2) lying outside the ellipse correspond to failure by yielding according to this theory.

4.3 Ultimate Strength of Box Girder in Yield. As we have noted earlier, the initial occurrence of local yielding does not necessarily signal the total collapse of the structure. Depending upon the importance and func-

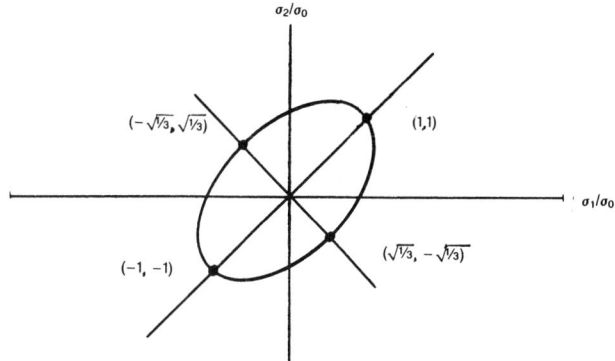

Fig. 61 Failure contour according to the distortion energy theory

tion of the structural member in question, the appropriate design criterion may be severe, requiring the avoidance of any occurrence of yield, or it may be relaxed to require only the prevention of total collapse of the member. In the latter case, yield would be allowed to a limited extent under the maximum design load. Strength deck plating for example, might be designed so as never to experience yield under any circumstances. Subdivision bulkhead stiffeners in a dry cargo hold could be designed to yield under conditions of flooding due to damage, so long as the bulkhead retains its watertight integrity, enabling the ship to remain afloat.

As an illustration of the differences in design that may be introduced by the use of an ultimate strength versus a yield criterion, consider the maximum bending moment sustainable by a simple box girder with the thin-walled rectangular cross section shown in Fig. 62a. Let us assume that the material stress-strain curve may be approximated, as shown in Fig. 62b, by two straight line segments. The initial, elastic part is given by a straight line having a slope equal to the modulus of elasticity. At a stress equal to yield, it is assumed that the strain increases indefinitely without further increase in stress. This is sometimes referred to as elastic-perfect-plastic behavior of the material. The behavior in compression is assumed to be similar to that in tension but with reversal of sign.

The next Fig. 62c shows the stress distribution in the side of the box girder for the cases:

- Stress is everywhere below yield stress, σ_0.
- Stress has just reached yield at deck and bottom.
- Stress equals yield everywhere across the section.

In the first case, assuming the deck stress is just equal to yield, the moment supported by the section (resultant of the stress distribution) is:

$$M_1 = \sigma_0\left(4bdt + \frac{4d^2t}{3}\right) \qquad (108)$$

In the third case, with the stress everywhere equal to yield,

---
[4] This theory, as noted in Timoshenko (1956), was first propounded by J. C. Maxwell in 1856.

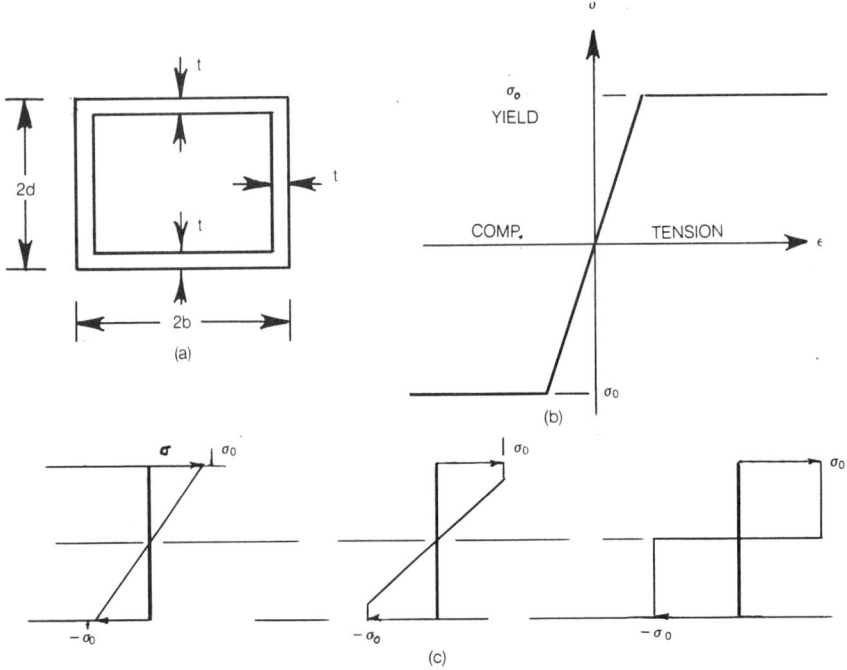

Fig. 62 Ultimate strength of simple box girder

$$M_3 = \sigma_0 \left(4bdt + \frac{4}{2} d^2 t\right) \quad (109)$$

If the cross section is square, $b = d$, then the ratio of the ultimate moment to the moment at yield is

$$\frac{M_3}{M_1} = \frac{9}{8}$$

This shows that the beam is capable of supporting a bending moment at total collapse that is 12.5 percent higher than the moment causing the initial occurrence of yield.

In a complex structure such as a ship the occurrence of progressive failure is seldom as simple as in this example since some parts may experience failure by buckling, and others by yield. The failure of one member in turn results in a redistribution of the load over the cross section and may drastically modify the load on other members. The analysis of the ultimate bending moment, including both yield and buckling, has been carried out by Caldwell (1965). Results are presented in the form of charts that clearly illustrate the effect of sectional geometry and buckling on the ultimate moment.

The simple example given here illustrates an important characteristic of real structures in which there often exists the capability of sustaining a total collapse load considerably in excess of that which initiates yield. The computation of the ultimate load capability of real ship structures by incremental finite element procedures is discussed in Section 4.9.

4.4 Instability Failure and Its Prediction. The fundamental characteristics of plate buckling were discussed in Section 4.1. The loads that may cause buckling are the primary hull bending loads in the secondary panels of stiffened plating in deck and bottom, and the combined primary and secondary loads in the individual tertiary panels of plating between stiffeners. These in-plane stresses alternate between tension and compression as the ship moves through waves. Tensile or compressive yield failure, which was discussed in the previous section, occurs if the combined stresses in a member exceed the load carrying capability of the material of which the structure is fabricated. Buckling constitutes a different mode of failure which is possible as a result of instability of the members, and in some cases it may take place at a compressive load substantially less in magnitude than that necessary to cause material yielding. Instability failure depends upon the material modulus of elasticity and member geometry. Its initial occurrence does not, however, depend upon either the material yield or ultimate strength. Such an instability or buckling failure may occur in a single panel of plating between stiffeners in a transversely framed ship. The buckling may also be more extensive, involving the stiffeners as well as the plating to which they are attached, and this is the more probable mode in the case of longitudinal framing.

Two important modifications to the originally uniform, unidirectional stress pattern in a plate are found as a result of buckling. First, tensile membrane stresses in the transverse direction are set up which tend to resist the out-of-plane deformation. Second, the initially uniform longitudinal compressive stress distribution in the plate changes. The compressive stress in the central, deflected portion of the plate is reduced below the average stress, and the compressive stresses near the restrained edges becomes greater than the average compressive stress in the panel. This is illustrated in Fig. 63. If the total compressive load on the panel is increased sufficiently, the stress near the panel edges may reach the yield point of the material, resulting in permanent localized deformation. If the load is increased further, this will ultimately lead to total collapse of the panel as the yield zone continues to grow. If, on the other hand, the load is relaxed before yield stress is attained at any location, the panel will return to its original undeformed shape. Stiffeners, which are nearly always present in ship structures, will modify the behavior of the plate as they participate in carrying a part of the load, as discussed subsequently.

By the above reasoning, it is clear that the maximum load carried by the panel at the time of collapse may be substantially greater than the critical load at the onset of buckling. The ratio of ultimate failure load to critical load is higher for thin plates, i.e., those whose thickness is small compared to their lateral dimensions, than it is for thick plates. In panels of plating having the proportions found in ship structures, this load ratio seldom exceeds two. In the following sections, we shall consider first the elastic buckling of simple plate panels, and then proceed to a discussion of the post-buckling behavior and ultimate strength of stiffened panels.

4.5 Elastic Buckling of Rectangular Plates. The *critical stress* is defined as the highest value of compressive stress in the plane of the initially flat plate for which a nonzero out-of-plane deflection of the middle portion of the plate can exist. For values of stress lower than the critical, the plate may be compressed in length but no deflection out of the initial plane occurs. The theoretical solution for the critical buckling stress in the elastic range has been found for a number of cases of interest, and is given by the *Bryan Formula*, Equation (110). For a rectangular plate subject to a compressive in-plane stress in one direction,

$$\sigma_c = k_c \frac{\pi^2 E}{12(1-\nu^2)} \left(\frac{t}{b}\right)^2 \qquad (110)$$

where the plate nomenclature is shown in Fig. 63.

Here k_c is a function of the plate aspect ratio, $\alpha = a/b$, the boundary conditions on the plate edges and the type of loading. If the load is applied uniformly to a pair of opposite edges only, and if all four edges are simply supported, then k_c is given by

$$k_c = \left(\frac{n}{\alpha} + \frac{\alpha}{n}\right)^2 \qquad (111)$$

where

n is the number of half-waves of the deflected plate in the longitudinal direction.

Fig. 64 presents, k_c vs. a/b for rectangular plates with uniform compressive stress in one direction. Wah (1960), Chapter 5, presents graphs for a number of other cases including shear loads, linearly varying edge stress and elastic restraint on the plate edges.

If $\alpha < 1$ (wide plates), the critical stress will correspond to $n = 1$ and a more convenient expression for σ_c is given by

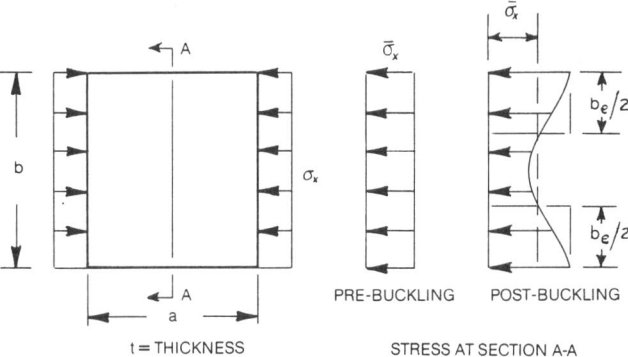

Fig. 63 Plate buckling nomenclature

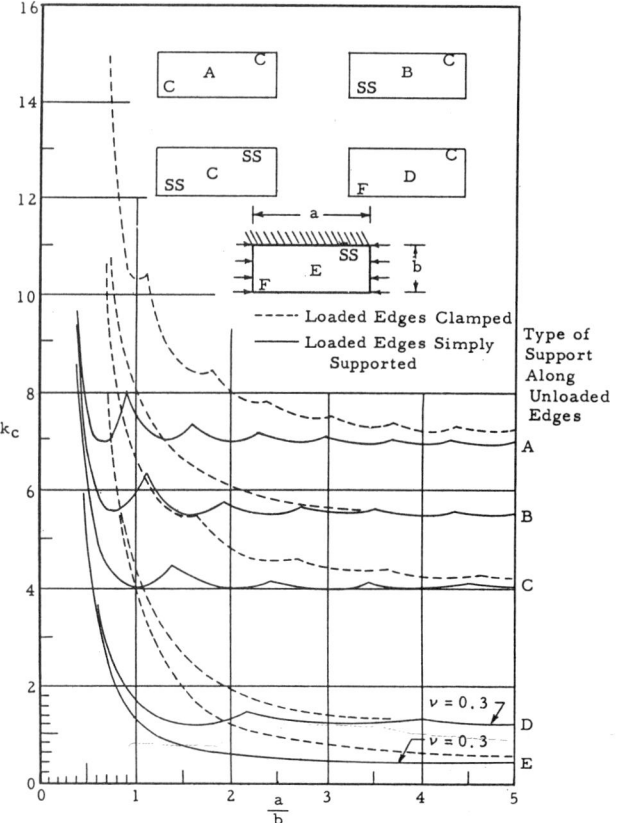

Fig. 64 Compressive buckling coefficients for plates

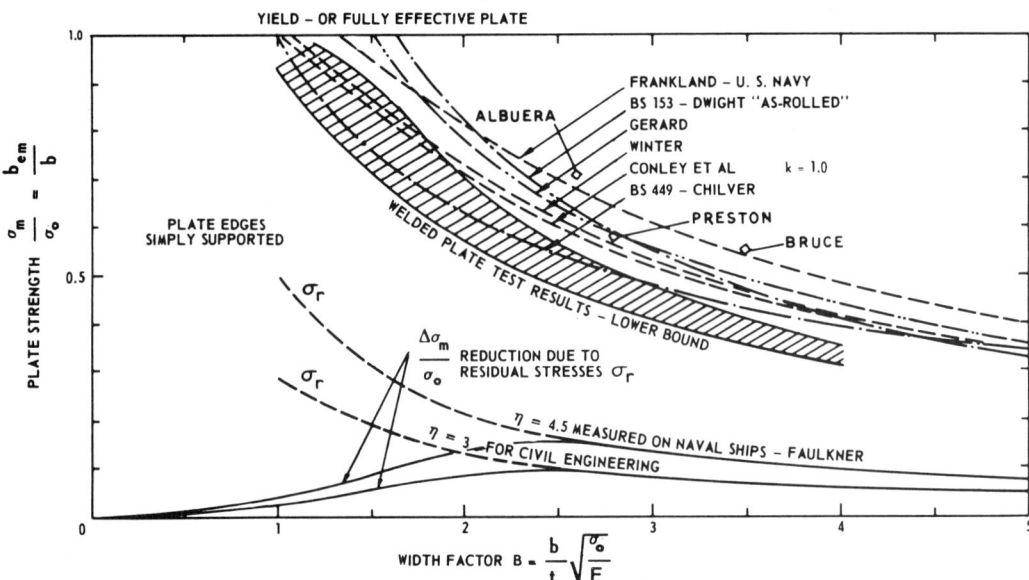

Fig. 65 Plate strength curves for simply supported edges (Faulkner, 1975)

$$\sigma_c = \frac{\pi^2 E}{12(1-\nu^2)} \left(\frac{t}{a}\right)^2 (1+\alpha^2)^2 \quad (112)$$

where t is the plate thickness.

For very wide plates, $\alpha \ll 1$, this may be replaced by the limiting value,

$$\sigma_c = \frac{\pi^2 E}{12(1-\nu^2)} \left(\frac{t}{a}\right)^2 \quad (113)$$

For narrow, simply-supported plates, $\alpha > 1.0$, the coefficient, k_c, of Equation (110) is approximately 4 and the elastic critical stress is given by,

$$\sigma_c = \frac{\pi^2 E}{3(1-\nu^2)} \left(\frac{t}{b}\right)^2 \quad (114)$$

It should be emphasized that these formulations for critical stress do not describe the strength of the structural member in question, nor do they even give a stress at which the load-carrying ability of the structure can be expected to be impaired in any substantial way. Elastic material behavior is assumed even after buckling, and therefore, upon removal of the load, the buckled member will return to its original undeflected shape. So long as the support of the unloaded edges of the plate remains intact, the only effect of buckling will be a visible deformation of the surface of the plate and an increase in the apparent rate of panel overall strain versus stress.

4.6 Postbuckling Behavior and Ultimate Strength of Simple Plates. In predicting the strength of a plate element, the objective is to determine the maximum average stress that the plate can sustain before the stress at some point reaches the yield stress of the material, at which point plastic deformation or panel collapse occurs. A complete theoretical solution for this problem is lacking at the present time, and the designer in most cases must rely heavily upon empirical rules and data. The plate strength data in most common use are based upon the concept of an *effective width*, which follows directly from the typical plate stress distribution in the postbuckling regime, as shown in the right-hand part of Fig. 63. The effective width, b_e, is defined as the reduced width of plate that would support the same total load as the buckled plate, but at a uniform stress equal to the maximum stress, σ_e, at the plate edge

$$b_e = (1/\sigma_e) \int_0^b \sigma_x(y) dy \quad (115)$$

where

$\sigma_x(y)$ is stress in x-direction,
b is plate width.

Note that the definition of effective *width*, Equation (115), is equivalent to the definition of effective *breadth*, Equation (67). Some authorities reserve the term effective *breadth* to refer to the shear lag phenomenon and effective *width* to refer to the postbuckling phenomenon. This distinction in terminology is not universal, however.

A comprehensive review of the numerous formulations proposed and in use for estimating b_e or related quantities for narrow unstiffened plates has been given by Faulkner (1975). Fig. 65 adapted from this reference gives the effective breadth ratio, b_{em}/b, or equivalently, the ratio of maximum mean stress at failure to plate yield stress σ_m/σ_o. The width parameter, β, against which these curves are plotted is

$$\beta = (b/t)\sqrt{\sigma_o/E} \quad (116)$$

It is derived as follows. Consider a simply supported plate under uniaxial compression. If the aspect ratio, a/b is greater than 1.0, the coefficient, k_c, of Equation (110) is approximately 4 and the elastic critical stress is given, as in Equation (114), by

$$\sigma_c = 4\,\frac{\pi^2 E}{12(1-\nu^2)}\left(\frac{t}{b}\right)^2 \qquad (117)$$

For steel, where ν is constant,

$$\sigma_c \propto E\,(t/b)^2.$$

If the plate width, b_0, is chosen in relation to the thickness, t, so that σ_c equals the yield stress, σ_0, we obtain:

$$b_0 = \pi t\,\sqrt{E/3(1-\nu^2)\,\sigma_0} \propto t\sqrt{E/\sigma_0} \qquad (118)$$

The factor β in Fig. 65 is proportional to the ratio of the actual width to this nominal width, given by,

$$\beta = \frac{b}{t}\sqrt{\sigma_0/E} \qquad (119)$$

This width parameter is often referred to as the plate slenderness ratio.

For steel, where the value of Poisson's ratio is $\nu = 0.3$, the actual value of b_0 in Equation (118) becomes,

$$b_0 = 1.9\,t\,\sqrt{E/\sigma_0} \qquad (120)$$

For typical mild steel the yield strength, σ_0, is about 241 MPa (35,000 psi) and b_0, from Equation (119), is approximately 55 t. On the basis of experimental data, a somewhat lower value of 50t is sometimes stated as an appropriate effective width of plating to be used in design.

A suitable design formula for expressing the effective width of simply supported rectangular plates without residual stress is suggested,

$$\frac{b_{em}}{b} = \frac{\sigma_m}{\sigma_0} = \frac{2}{\beta} - \frac{1}{\beta^2} \qquad (121)$$

where

b_{em} = minimum effective width
σ_m = maximum average plate stress

The effectiveness of plating in compression will also be modified by three factors, in addition to plate dimensions and material properties:

- Initial deformation, due principally to welding distortion.
- Residual stresses also resulting from welding.
- Normal pressure.

It is suggested that if the maximum initial panel deflection is less than 0.3t, buckling strength is not affected. On the other hand, Faulkner (1975) reports the results of numerous measurements of ships in drydock which indicate that larger deformations may be expected in lightly plated naval vessels.

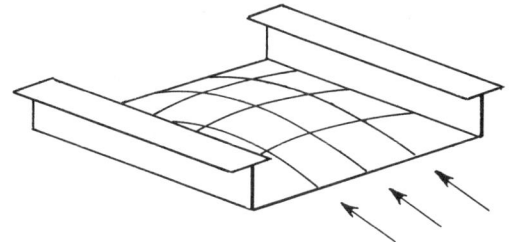

MODE 1. PLATE ALONE BUCKLES BETWEEN STIFFENERS.

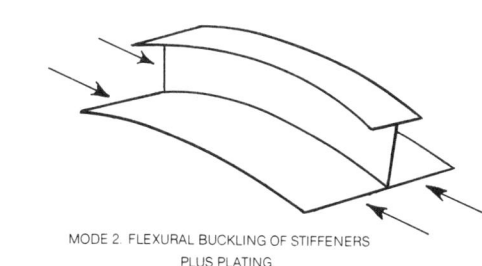

MODE 2. FLEXURAL BUCKLING OF STIFFENERS PLUS PLATING.

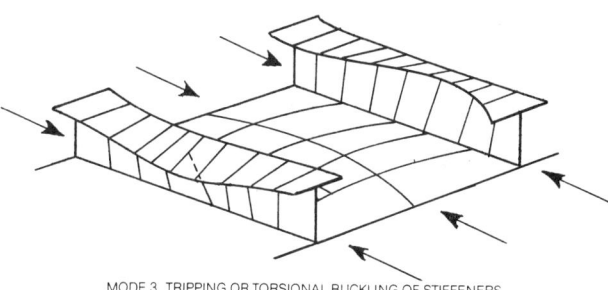

MODE 3. TRIPPING OR TORSIONAL BUCKLING OF STIFFENERS.

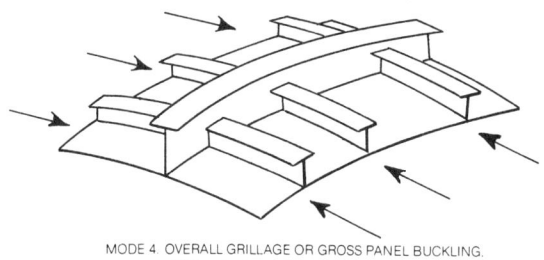

MODE 4. OVERALL GRILLAGE OR GROSS PANEL BUCKLING. BOTH LONGL. AND TRANS. STIFFENERS.

Fig. 66 Four modes of stiffened panel buckling

The effect of residual stress also is to reduce the maximum mean stress that the plate can sustain. For values of $\beta > 2$, it is suggested that σ_m be merely reduced by σ_r, where σ_r is the value of residual stress. For lower values of β, the reduction is dominated by inelastic effects and results may be found in Bleich (1952). These results are plotted, along with experimental data, in Fig. 65.

Bleich (1952) and others have indicated that the effect of normal pressure is to increase the critical buckling stress of flat panels. However, for the plate dimensions and pressures encountered in ship plating, this effect is small and may usully be neglected. Theoretical and experimental studies reported by H. Okada et. al. (1980) have shown that the effect of normal pressure on the ultimate strength of ship plating is usually negligible, as well.

4.7 Elastic Buckling of Stiffened Panels. In the previous section, we have considered the stability and strength of individual simple plate panels with various types of edge restraint. Stiffened plate panels are of greater practical interest, in particular when subject to in-plane loading resulting from the primary bending of the hull girder. This type of buckling behavior, as was the case with the bending behavior treated in Section 3.8, will involve the interaction of plates and stiffeners in response to the applied loading. In the case of longitudinally-framed ships, the stiffened panel behavior plays a more important role than the individual plate panel in determining the ultimate strength of the ship's structure since, in this case, catastrophic buckling of the simple panel is nearly impossible without the involvement of its associated stiffeners. On the other hand, for transversely-framed ships, individual panel buckling with the frames forming nodal lines is the most probable buckling mode. For a comprehensive review of the buckling behavior of stiffened plate panels, the reader is referred to Mansour (1977).

Four different modes of buckling are usually recognized in describing the behavior of a stiffened plate panel and these are illustrated in Fig. 66.

• Mode 1 is the simple buckling of the plate panel between stiffeners and was discussed in the previous section.

• Mode 2 consists of flexural buckling of the individual stiffener together with its effective breadth of plating in a manner analogous to a simple column. For a panel with the edges simply supported, Mansour gives the following expression for the critical stress:

$$\sigma_{cr} = \frac{\pi^2 EI}{Al^2} \left[\frac{1}{1 + \frac{\pi^2 EI}{l^2 GA_s}} \right] \quad (122)$$

Here, I is the effective moment of inertia of the stiffener plus associated plating, where

A is the total cross section area,
A_s is the shear area,
l is length of the stiffener,
G is shear modulus.

The effect of shear deformation is seen to be included in this expression. A treatment of panels subject to other boundary conditions is given in Bleich (1952).

• Mode 3 is referred to as the lateral-torsional or tripping mode. In this mode, the stiffener is relatively weak in torsion, and failure is initiated by twisting of the stiffener in such a way that the joint between stif-

(a) INTERFRAME FLEXURAL BUCKLING (MODE 2)

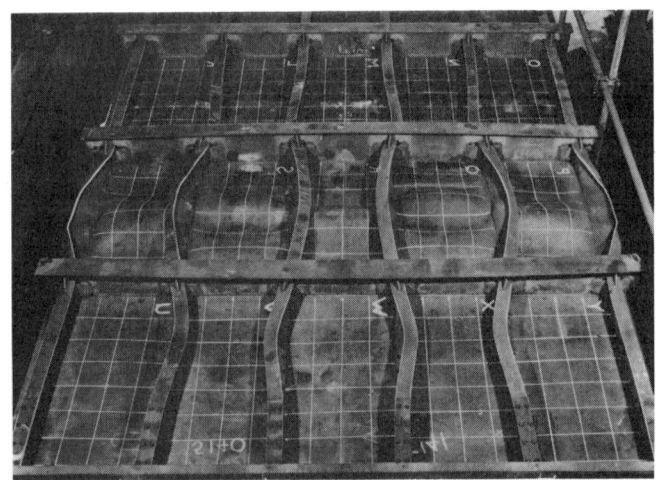

(b) INTERFRAME LATERAL-TORSIONAL BUCKLING (MODE 3)

(c) OVERALL GRILLAGE BUCKLING (MODE 4)

Fig. 67 Examples of test panel failure (Smith, 1975)

fener and plate does not move laterally. A portion of the adjacent plate may participate in the twisting, and the flange of the stiffener may twist together with the web, or the two may twist differentially. The critical stress for the lateral-torsional mode 3 is given by a formula from Bleich (1952), p. 138,

$$\sigma_{cr} = \frac{\pi^2 E}{(l/r_e)} \qquad (123)$$

Here, is l is stiffener length,
  is r_e is effective radius of gyration of the cross section.

Bleich gives expressions and charts from which to estimate the effective radius of gyration for a variety of stiffener cross sections.

In general, the means of predicting the occurrence of this mode of buckling are somewhat less reliable than the other three modes. Fortunately, mode 3 may usually be avoided by fitting tripping brackets to the web of the stiffener, and it does not appear to play an important role in determining ultimate strength of the hull.

• Mode 4, overall grillage buckling, has been treated by Mansour (1976, 1977) using orthotropic plate theory. The following expressions, taken from the latter reference may be used in this case if the number of stiffeners in each direction exceeds 3. For gross panels under uniaxial compression the critical buckling load is given by,

$$\sigma_{xcr} = k \frac{\pi^2 \sqrt{D_x D_y}}{h_x B^2} \qquad (124)$$

where B is gross panel width, h_x is effective thickness, k is given by different expressions, depending on the boundary conditions,

• For simply supported gross panels,

$$k = \frac{m^2}{\rho^2} + 2\eta + \frac{\rho^2}{m^2} \qquad (125)$$

• For gross panels with both loaded edges simply supported and both of the other edges fixed,

$$k = \frac{m^2}{\rho^2} + 2.5\eta + 5\frac{\rho^2}{m^2} \qquad (126)$$

where

  m is number of half-waves of buckled plate, η and ρ are defined in Equations (92) and (93).

The Mansour (1976) reference contains an extensive treatment of the behavior of orthotropic plate panels in the buckling and elastic postbuckling range. Design charts are given which contain the mid-panel deflection, critical buckling stress, and bending moment at the mid-length of the edge. The loading conditions include normal pressure, direct in-plane stress in two directions and edge shear stress.

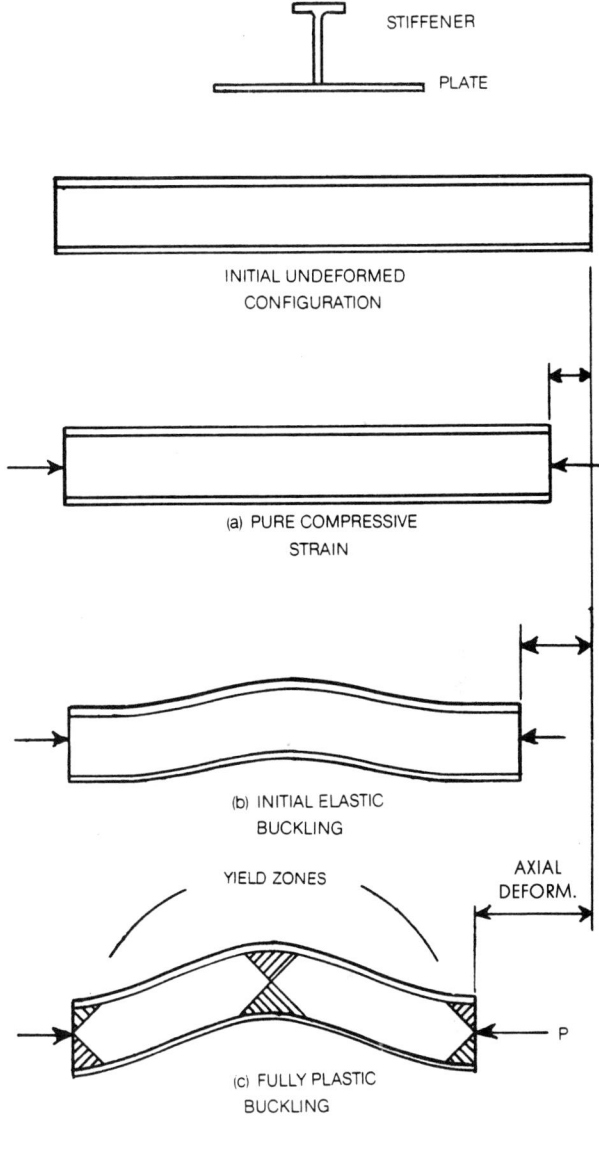

Fig. 68 Failure regimes in compression

4.8 Ultimate Strength of Stiffened Plate Panels. The previous section has dealt with stiffened panels whose behavior remains in the elastic range. The ultimate strength involves deformations in which the material behavior is no longer elastic, and combinations of analytic, numerical and experimental methods have been employed to obtain understanding and design information. Failure is usually observed to occur in one of the four modes defined in the previous section. The aforementioned reference by Mansour (1976) contains charts from which predictions may be made of the large-deflection behavior in Mode 4 up to the initiation of yield.

As noted by Smith (1975), despite a considerable body of research, the understanding of collapse behavior of welded grillages is far from complete, and much reliance is still placed upon experiments, Fig. 67, reproduced from this last reference, shows test panels of stiffened plating in which examples of panel failures in each of the modes 2 to 4 listed above were observed.

If the panel buckling occurs by one or more of the mechanisms shown in Fig. 66, the stress-strain curve for the corresponding part of the deck or bottom structure can no longer be represented by the ideal elastic-perfectly plastic curve of Fig. 4.2.

While the tension side of the curve will level off and remain at the constant stress level corresponding to yield, the compressive behavior is somewhat more complex as may be seen by considering the behavior of a single stiffener and its associated plating under a compressive load. This is illustrated in Fig. 68, where we see three distinct regimes in the behavior of the load versus deformation:

(a) Simple elastic strain or shortening before buckling occurs.
(b) After initial buckling but before the development of extensive yield in the stiffener.
(c) Development of yield over the full depth of the member to form one or more plastic hinges, normally at the ends and mid-span.

From considerations similar to those that underly Equation (109), we see that the fully plastic resisting moment that may be developed in case (c) remains constant if there is further out-of-plane deflection of the stiffener. If the external compressive load, P, were to remain constant, the bending moment at the mid-span point would increase with increase of out-of-plane deflection since the moment is equal to the product of the external load and deflection, δl. Increased out-of-plane deflection is accompanied by a shortening of the distance between the two ends of the stiffener (apparent strain). We see therefore, that in order to maintain moment equilibrium, it is necessary for the external compressive load to decrease with increase in apparent strain of the stiffener beyond the point of formation of the plastic hinges. This effect is termed *unloading* of the member.

An idealized graph of the load versus apparent strain is shown in Figure 69. Here, the tension side is characterized by a nearly linear elastic zone followed by an idealized plastic zone in which the load remains constant with increasing strain. The compression side of the graph exhibits the three zones described above. The first is the elastic compression without buckling. The second corresponds to either compressive plastic yield or elastic buckling, whichever occurs at the lower loading. The third is the unloading regime in which fully plastic hinges have developed in the stiffener.

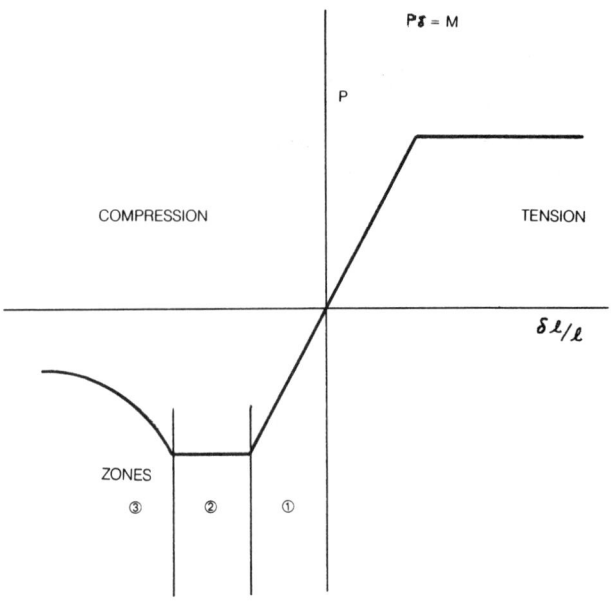

Fig. 69 Idealized load versus strain

4.9 Ultimate Strength of the Hull Girder. In the ship structure, we have numerous plate-stiffener panel members. At any given instant, many of these members are subject to compressive loads and the loads vary in intensity depending upon the location of the member within the ship. Under conditions of extremely severe hull loading, it is apparent that the most highly loaded compression members may experience buckling and plastic yield and go into the zone of unloading as described above. A portion of the load that would otherwise be carried by such members is then shifted to nearby intact members. A further increase in the hull loading beyond this level will result in some of these members becoming so heavily loaded that they now experience yield. Further increase in the hull loading will eventually lead to total collapse of the structure as a complex sequence of interdependent panel collapses.

The ultimate strength of the hull is therefore a composite of the ultimate strength characteristic of all of these panels. The ultimate collapse behavior of the individual panel alone is a complex subject not amenable to a simple analytical description. Since each of the numerous panels making up the ship differs in geometry, loading and boundary conditions, there is obviously considerable difficulty in developing a comprehensive analytic solution for the entire hull. Early investigations of hull ultimate strength have usually been experimental in nature. The most informative of these have been conducted on actual obsolete ships by applying static loads to the ship in drydock through

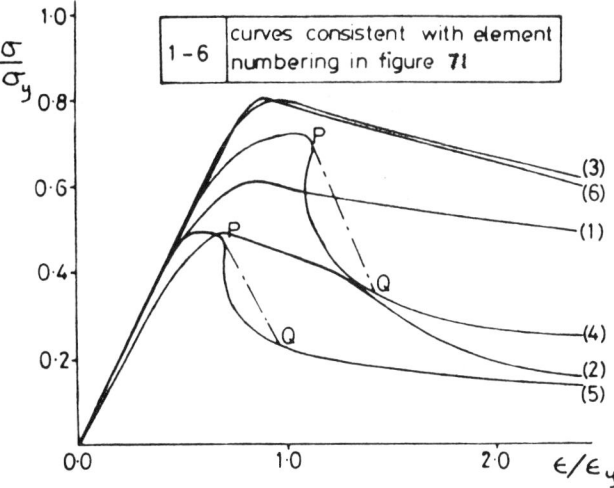

Fig. 70 Compressive stress-strain curves for stiffened panels
(Dow et al, 1981)

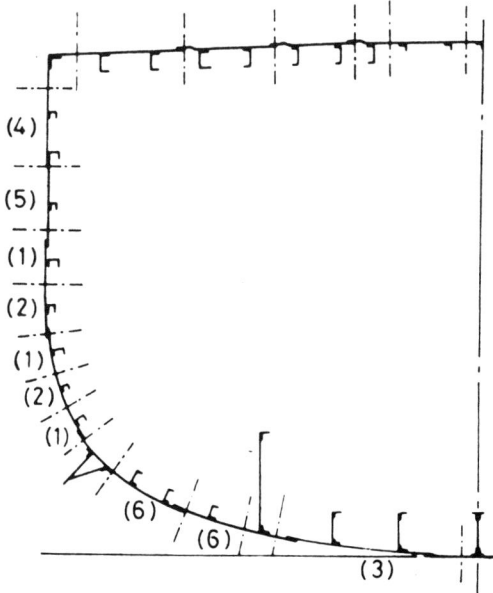

Fig. 71 Midship section of destroyer *Albuera* showing elemental subdivision
(Dow et al, 1981)

ballast shifts and dewatering the dock. The reference by Vasta (1958) contains descriptions of several classic experiments of this type.

At the present time it is possible to perform an approximate numerical analysis of the ultimate strength of the ship hull by using nonlinear finite-element methods. Two examples of such analyses are described by Smith (1977) and Dow, et al (1981). (The latter includes a description of computer program, ULTSTR, intended for design use.) In these procedures, the hull cross-section is first subdivided into a number of panels consisting of plating and the associated stiffeners, and for each such panel the load-shortening curve of approximately the form of Fig. 69 is constructed. This can be accomplished by any of a number of methods. Experimental data could be used or, alternatively, one might perform a nonlinear finite element analysis of the individual panel itself. A third procedure would employ one or more analytic formulations suitable for describing the large amplitude deflection to be experienced by the member in the failure regime. Example stress-strain curves for plate-stiffener panels obtained by Dow, et al (1981) using a nonlinear finite element analysis are shown in Fig. 70. The unloading behavior in compression is clearly observed in this figure.

An appropriate collection of such elementary members is then assembled in order to represent the midship portion of the ship's hull. Some *hard corner* elements that exhibit exceptionally high buckling resistance will be included to represent portions of the structure at shell-deck or bulkhead-shell intersections where buckling is not expected to be the primary failure mode. An example of such a discrete model is shown in Fig. 71.

The hull is then subjected to an incrementally increasing bending deflection pattern in which it is assumed that cross sections that are initially plane remain plane after deflection and experience only rotation about an assumed neutral axis. It is recalled that a similar assumption was made in developing the equations of elementary beam theory. After this rotation, the strain of each longitudinal member is determined to correspond with the assumed position of the neutral axis. By reference to the stress-strain curves of each member the stress, and thus the load on that member, are determined. Next, the loads and their moments are summed over all members making up the cross section. The total load must be zero for longitudinal force equilibrium, and the total moment must be equal to the external hull bending moment at the cross section in question. As a result of the nonlinear stress-strain behavior of the members, the neutral axis will, in general, not be located at the geometric centroid of the cross section. A trial and error procedure must, therefore, be used at each increment of angular deflection in order to find the neutral axis location that results in equilibrium of longitudinal forces. Once this has been obtained, the moment of the longitudinal forces in the member may be computed. See also the papers by Billingsley (1980) and Chen, et al (1983), with discussions.

Typical results for naval ships from Dow, et al (1981) and Smith (1977) are shown in Figs. 72 and 73, respectively. Several interesting characteristics are revealed by such computations. For example, it is quite apparent that, for the lightly-built type of ship investigated here, the ultimate strength of the midship section will be substantially less than the strength corresponding to yield failure alone, as was depicted

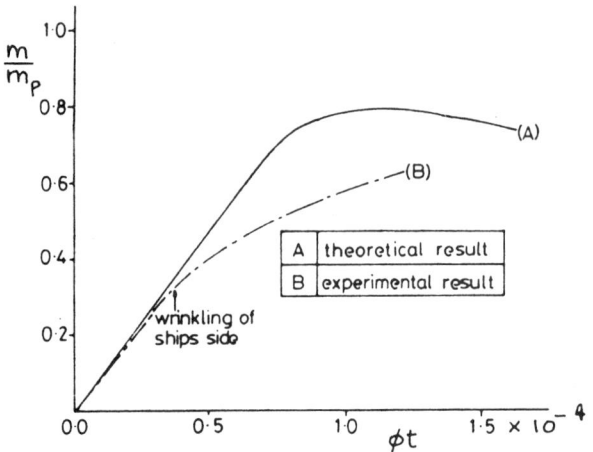

Fig. 72 Midship bending moment-curvature relationship, by theory and experiment (Adamchak, 1982)

in Fig. 62. Furthermore, we observe an unloading effect in the overall hull strength that is similar to the unloading phenomenon in individual plate-stiffener members. If we compare the strength curves corresponding to hogging to those in sagging, an asymmetry is apparent which may be attributed to the difference in the buckling strength of deck and bottom structure.

On the other hand, heavily built ships such as tankers and most longitudinally framed commercial vessels can be designed to avoid buckling up to the load at which full compressive strength is attained. This requires a proper balance among plate thickness, stiffener (frame) scantlings and frame spacing (Faulkner, 1981). When such a balance cannot be obtained, the procedure described previously is available to determine a realistic value of ultimate strength.

4.10 Vibratory Response to Dynamic Loads. Two types of dynamic loading were discussed in Section 2.11, one being a slamming impact on bottom or flare forward, followed by transient vibratory *whipping*, and the other a more-or-less steady-state random vibration or *springing* excited by certain wave frequency components.

Considering the response to slamming first, the determination of the local structural response involving damage requires consideration of the inelastic behavior of the structure. The computation of the overall whipping response may be performed using a linear elastic model. Mansour and D'Oliveria (1975) have developed a procedure for computing the combined rigid-body and elastic dynamic response of a ship to head seas. Their results, as well as full-scale experimental data from Lewis, et al (1973), and others, have shown that the whipping induced stresses amidships may be equal in magnitude to the wave-frequency bending stresses. High-speed craft, which present special problems, are treated in Section 3.15.

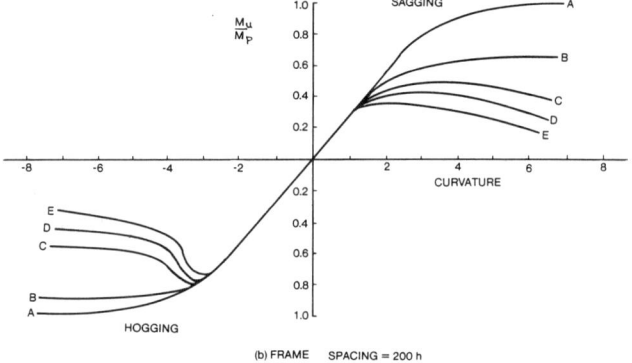

Case A: the entire cross-section is assumed to follow an elastic-perfectly plastic stress-strain curve, buckling effects being ignored.

Case B: element compressive stress-strain curves follow trends predicted by a detailed nonlinear F. E. analysis of stiffened panel behavior up to peak loads and thereafter remain horizontal, ie post-buckling reduction of load is ignored. "hard corners" are assumed to have conservative areas.

Case C: element compressive stress-strain curves are as in Case B except that post-buckling load reductions are included; "hard corners" have double the areas shown in Case B.

Case D: same as Case C except that "hard corners" have the same areas as Case B.

Case E: same as Case C except that "hard corners" are eliminated, the entire cross-section being assumed to follow the computed panel stress-strain curves.

Fig. 73 Midship bending moment-curvature relationships for destroyer hull girder, under various assumptions (Smith, 1977)

The combining of whipping loads or stresses with low-frequency bending loads involves problems of phasing (see Lewis, et al, 1973) and of duration of the high-frequency load peaks. As noted by Dow, Hugill, Clark and Smith (1981), "Whipping of a ship's hull due to slamming or explosive loading can cause large additional hull girder bending moments, but because of their transient nature, these loadings may be less effective in producing plastic collapse than the more slowly varying bending moment due to buoyancy effects." On-going research is needed to determine how significant whipping stresses are in producing hull failures.

Meanwhile, a number of cases of deck buckling have been reported on ships with large flare forward, which seem to have been associated with slamming and whipping (McCallum, 1975). The transient loading associated with flare immersion is characterized by a longer duration of impact than bottom slamming. Elementary beam theory shows that the dynamic load factor will therefore in general be greater. For example, consider the case of a 500-ft cargo ship with natural period of vibration, T, of 0.75 s. Assume that the duration of a bottom slam impact, t_1 is 0.1 sec and of a flare immersion impact is 0.5 sec. Then in the first case $t_1/T = 0.13$ and in the second $t_1/T = 0.67$. Simple theory (Frankland, 1942) assuming triangular or sinusoidal pulses gives a magnification factor of 0.3 in the first case and 1.5 in the second. A theoretical treatment of the important case of flare immersion has been given by Kaplan and Sargent (1972).

One difficulty in estimating slamming response for design purposes is that the relative vertical velocity when a slam occurs depends on the ship's speed and heading, as well as on the severity of the sea. Hence, it is to some extent under the control of the ship master. As shown by Maclean and Lewis (1970), the highest slamming loads actually recorded on a typical cargo ship did not increase with sea severity after a certain level was reached, since speed was gradually reduced voluntarily by the captain. This suggests the need for more data on actual slam loads permitted to be experienced by ships of various types.

For naval ships impact on above-water appendages is a problem. Keane (1978) has pointed out that, "no analytic methods exist for assessing the hydrodynamic loads imposed upon sponsons at various heights above the waterline during the early stages of design." However, the U.S. Navy has developed empirical rules for use in design.

Springing has been found to cause significant increases in wave bending moments and stresses only in the case of long, flexible ships of full form, particularly Great Lakes bulk ore carriers and a few ocean-going bulk vessels. The solution to the problem of vibratory springing involves the assumption of an "Euler" or "Timoshenko" beam on an elastic foundation (the sea). The equations of motion in a vertical plane are set up, balancing the wave excitation against the elastic beam response of the hull, including effects of mass inertia and added mass, and both structural and hydrodynamic damping, as discussed in a general way in Chapter VI, Vol. II. The solution for the simple 2-noded case is of particular interest, and such solutions have been given by Goodman (1971) and Hoffman and van Hooff (1976); Stiansen, et al (1977) describe the ABS computer program SPRINGSEA II for carrying out routine calculations of linear springing response. As shown by model tests (Hoffman and van Hooff, 1976) and confirmed by Troesch [(1984a)], the bending moment response operator (at constant speed in head seas) shows an oscillatory character when plotted against wave length (encounter frequency.) Peaks correspond to wave lengths such that hydrodynamic forces at bow and stern reinforce one another. (Forces along most of the ship length tend to cancel out.)

Bishop, et al (1977) presented a more general theoretical approach, which however does not provide a solution to the important problem of damping. Stiansen (1984) notes that hydrodynamic damping is negligible and "the overall damping primarily consists of the speed correction, proportional to the derivative of added mass, and structural and cargo damping." Analysis of full-scale data on the *Stewart J. Cort* gave results in good agreement with damping coefficients calculated by SPRINGSEA II.

Stiansen (1984) reviews recent research under ABS sponsorship on this and other aspects of dynamic behavior of large Great Lakes bulk carriers. Experiments in waves (Troesch, 1984) on a model jointed amidships measured both wave excitation and springing response. He found that in addition to the response at ω_e, the encounter frequency, there was a measurable springing excitation at $2\omega_e$ and sometimes at $3\omega_e$. Should $2\omega_e$ or $3\omega_e$ equal ω_0, the natural two-noded frequency of the hull, there will be a large increase in the springing response. This non-linear response is quadratic in wave amplitude; if wave amplitude doubles, response increases by a factor of four. The experiments also showed that response in the natural frequency is excited when the sum of the encounter frequencies of two wave components equals the natural frequency. See Chapter VII, Section 5, Vol. III.

Troesch (1984) made use of exciting functions (first and second order inelastic bending moments) determined experimentally on a jointed model to calculate the 2-noded springing response in typical Great Lakes wave spectra. Results showed that at certain speeds the combined first and second order resonant response was significantly greater than the first order alone. Work continues on developing a theoretical basis for calculating the non-linear springing response, following the approach of Jensen and Pedersen (1981).

4.11 Cumulative Fatigue Damage. Fatigue constitutes a major source of local damage in ships and other marine structures, since the most important loading on the structure, the wave-induced loading, consists of large numbers of load cycles of alternating sign.

The effects of fatigue are especially severe in locations of high stress concentration, and fatigue cracks have sometimes proven to be the triggering mechanism for brittle fracture. The prevention of fatigue failure in ship structures is strongly dependent on proper attention to the design and fabrication of structural details in order to reduce stress concentrations. This must be followed by thorough and regular inspection of the structure in service in order to detect and repair any fatigue cracks that do occur before they can grow to such size that the structure is endangered.

Much of the quantitative information on fatigue has been obtained by experiments in which an alternating load is applied to a simple test specimen. In the most usual form of this experiment, the loading varies sinusoidally in time with constant amplitude and frequency. Under such a load, it is found that many engineering materials, including the steels commonly used in shipbuilding, will fracture after a sufficient number of cycles even though the alternating stress amplitude is less than the static yield stress of the material. The number of cycles that results in fracture of the specimen is found to depend on the amplitude of the alternating stress, and this number is less the higher the stress amplitude. For a sufficiently low stress level, some materials are found to be capable of withstanding an (apparently) indefinitely large number of cycles, and this threshold stress level is termed the *endurance limit*. A graph of stress amplitude versus the number of cycles to failure is termed the *S-N curve*, and an example is shown in Fig. 74, where σ is the fracture stress and σ_y is the yield stress. It is clear that the S-N curve is usually well-defined in the high-cycle range (as defined in Section 4.1), but more uncertain in the low-cycle range at stresses near the yield point.

The resistance of a material to fatigue failure depends on a number of factors including the material itself, the surface finish, corrosion extent and the presence of stress concentrations. The simple sinusoidal loading that was described in the previous paragraph is approximated in the load experience of certain machinery parts that are subject to forces caused by rotational unbalance or vibration. Such parts are usually fabricated to close tolerances and operate in a uniform environment, all of which contribute to a relatively predictable fatigue life. The ship hull structure, on the other hand, is exposed to loads that vary randomly in time, the parts are fabricated by welding with much looser tolerances than machine parts, the surface finish is relatively rough and the structure is exposed to a harsh and corrosive environment.

The prediction of the fatigue life of the hull structure is, therefore, much less certain than the corresponding prediction for many other types of members such as machinery components. Consequently, fatigue damage computations have seldom been made for ship structures despite the recognized importance of fatigue as a ship structural failure mechanism. Fortunately, fatigue cracks have been mostly of the nuisance variety occurring in poorly designed brackets and other details, requiring repair at times of overhaul. Cracks in longitudinal strength members can be readily detected and repaired before the safety of the ship is threatened.

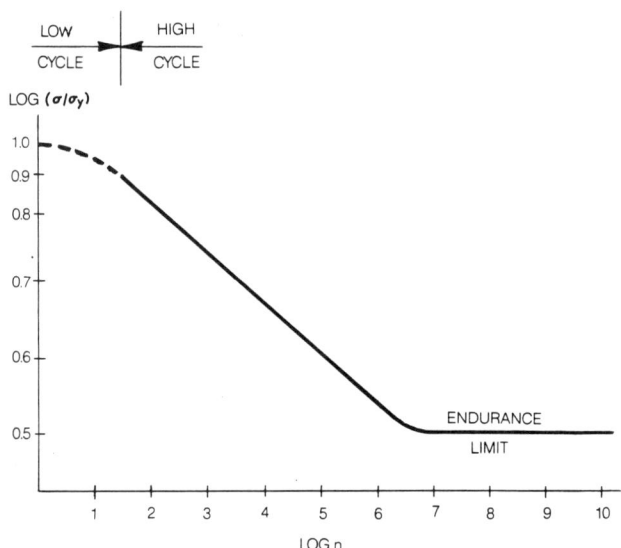

Fig. 74 Typical S-N curve for mild steel

Fatigue damage criteria have, however, been applied in the design of some offshore structures. Here the nature of the loading and the structural response are such that the computations are somewhat more reliable in predicting structural failure than in the case of ships. It is anticipated that the development of such techniques will in time be applied to refining the structural design of ships.

A major source of difficulty in predicting fatigue damage in ships and other marine structures lies in the form of the alternative loading, which is a random rather than simple periodic function. As we have seen in Section 2, the stress at a given location in the ship varies in an irregular fashion, which may have an approximately constant mean value over a short term, but even this mean value changes with changes of sea state and ship loading. The estimation of the fatigue damage or, rather, the probability of fatigue failure in the random loading case is usually performed according to a procedure proposed by Miner (1945). By the Miner hypothesis, it is assumed that one cycle of the randomly varying stress, having an amplitude s_i, causes an amount of fatigue damage in the following proportion:

$$\delta D_i = \frac{1}{N_i} \qquad (127)$$

Here, N_i is the Number of cycles of a sinusoidally varying stress of amplitude s_i required to cause failure.

The cumulative damage due to fatigue during exposure to the random stress environment will then be given by

$$D = \sum_i \frac{n_i}{N_i} \qquad (128)$$

Here n_i = the number of cycles of stress of level s_i during the period of exposure and the summation is taken over all levels of stress experienced during the period of time under consideration. Failure of the structure is then presumed to occur when the length of exposure is sufficient for this sum to equal unity.

For the spectral representation of the seaway and, thus, the stress environment, the fatigue damage summation may be expressed in a more compact form, in the high-cycle case. Let $p(s)$ be the probability density function for the stress. This is defined so that the quantity, $p(s_1) \, ds$ equals the fraction of all of the oscillatory stress peaks whose values lie in the interval ds centered on the mean value s_1. We shall assume that the average frequency of the randomly varying stress is f and that the total time of exposure is T. The incremental damage caused by all of the stress oscillations of amplitude s_1 occurring during the interval T is then given by

$$dD = Tf\, p(s_1) \, ds \Big/ N(s_1) \qquad (129)$$

where $N(s_1)$ is the number of cycles to failure at stress s_1 as obtained from the S−N curve for the material or the structural component.

The expected value of the total damage during the time period T is then given by the integral of Equation (129), or

$$E[D] = Tf \int_0^\infty \frac{p(s)\, ds}{N(s)} \qquad (130)$$

The S−N curve in the high-cycle range is sometimes approximated by the following function, which is piecewise linear in log-log coordinates.

$$NS^b = C \qquad (131)$$

The distribution function, $p(s)$, is often approximated by a Rayleigh distribution

$$p(s) = \frac{s}{m_{os}} \exp\left(-\frac{s^2}{2\, m_{os}}\right) \qquad (132)$$

In this case, the integral in Equation (129) can be evaluated, giving

$$E[D] = \frac{Tf}{c} \sqrt{2} \; m_{os}^2 \; \Gamma(1 + b/2) \qquad (133)$$

Here $\Gamma(x)$ is the Gamma function.

The prediction of fatigue life by the above procedure involves uncertainties that depends upon three general categories or sources of error:

- Uncertainty in the estimated stress levels, including errors of stress analysis and errors in the loading prediction for the ship in the random sea environment.
- Uncertainty in the basic fatigue life data for the structural detail in question, including scatter in the experimental data for the detail and effects of workmanship and fabrication in the real structure.
- Uncertainty in the basic cumulative damage rule (Miner's hypothesis).

Predicting cyclic loading (or stresses) from long-term distributions is discussed in Section 2.9. Of course, this procedure covers only primary bending loads, not local cyclic loads that may also be important for fatigue but are less well understood.

In welded structures such as ships, the fatigue strength is found to be closely related to the geometry and workmanship of the structure. For the steels commonly used in ship construction, fatigue strength is not strongly dependent on the material itself. Fatigue cracks in actual structures are usually found in details, of which examples are hatch corner reinforcements, beam-bracket connections, and stiffener-bulkhead intersections. In addition to attention to the design of such details, a high level of workmanship, including accurate fit-up and alignment of components, and good weld quality are of importance in achieving fatigue strength.

A very comprehensive treatment of the high-cycle fatigue characteristics of ship structural details is presented in a report by Munse (1983). Fatigue design criteria from other fields of engineering are examined with the objective of developing simple criteria for the design of ship structural details. An extensive catalog of structural details has been compiled and fatigue data for these details are presented in tabular and graphical form. Methods of combining these data with information on ship loading are presented and the errors and uncertainties in both the fatigue data and in the computations of fatigue life are examined.

On the basis of an assumed lifetime cyclic loading pattern, it is possible to use these data to determine whether specific structural details can be expected to suffer high-cycle fatigue cracking during the ship's lifetime. If so, that particular detail can be redesigned.

Several difficulties in principle arise in applying the Miner hypothesis to ships and ocean structures. The infinite upper limit of the integral in Equation (130) in conjunction with the use of an analytic expression for the distribution function implies the possibility of occurrence of infinite or very high stress cycles, which would exceed the ultimate strength of the material. Cycles exceeding yield contribute to what was described in Section 4.1 as *low-cycle* fatigue damage. One cycle exceeding the ultimate would, presumably, cause fracture of the part under examination. Because of these and the other sources of uncertainty listed, the estimate of damage by the Miner hypothesis must be considered a probabilistic estimate, to be characterized by its own level of uncertainty.

In the design criteria that are based upon the Miner hypothesis, it is customary to allow for the uncertainties by requiring the structure to have an extended value of the estimated fatigue life or, equivalently, a value less than unity for the damage parameter, $E(D)$ in Equation (133). The American Bureau of Shipping (1983) recommends that steel offshore structures be designed to have an estimated fatigue life of two to three times the design life of the structure. The factor of two is to be used in structures having a sufficient degree of redundancy that failure of the member under consideration will not result in catastrophic failure of the structure. If the degree of redundancy is less or if the redundancy would be significantly reduced by fatigue, the factor of three is to be used. Det norske Veritas (1977) recommends values for $E(D)$ ranging from a low of 0.1 for major structural members that are inaccessible for regular inspection and repair to 1.0 for minor members which are readily accessible.

Low-cycle fatigue failure is less well understood than the high-cycle type discussed above. However, since it is known to occur only when cyclic stresses reach the yield point, conservative design can be based on the avoidance of stresses exceeding that level at points of stress concentration, particularly in longitudinal strength members.

The prevention of fatigue failure is of paramount importance in submarine design, where the designer must deal with the fatigue life of highly stressed details in the 20,000-cycle range. This relatively low fatigue life is in the low-cycle range. Since any fatigue crack will destroy the watertight integrity of the pressure hull or of any other structures, such as tank bulkheads, which experience submergence pressure, and in some cases, may impair the capacity of the structure to resist applied loadings, special attention must be given to the fatigue resistance of the many critically stressed structural members and details of submarines.

Section 5
Reliability of Structures

5.1 General Aspects. As we have seen from the material in the preceding sections, there is a certain degree of randomness or uncertainty in our ability to predict both the loads imposed on the ship's structure (the demand) and of the ability of the structure to withstand those loads (the capability). The sources of these uncertainties include phenomena that can be measured and quantified, but cannot be perfectly controlled or predicted by the designer, and phenomena for which adequate knowledge is lacking. The terms *objective* or *random* uncertainties are sometimes applied in describing the former, and the terms *systematic* or *subjective* are used in describing the latter. In principle, the objective uncertainties can be expressed in statistical terms, using available data and theoretical procedures. The systematic uncertainties, which are known to exist but which cannot be fully quantified as a result of a lack of knowledge, must be dealt with through judgment and the application of factors of safety.

An example of an objective uncertainty is the variability in the strength properties of the steel used in constructing the ship. The magnitude of this variability is controlled to some extent through the practices of specifying minimum properties for the steel, and then testing the material as produced by the steel mill to insure compliance with the specifications. Departures from the specified properties may exist for several reasons. The sampling and testing cannot, for practical reasons, be applied to all of the material going into the ship, but only to a limited sampling of the material. Some of the material, as a result of slight variations in its manufacturing experience, may exhibit different properties from those of material manufactured by supposedly identical procedures. After arrival in the shipyard, the material properties may be altered by the operations such as cutting, forming and welding which are involved in building it into the ship. These variations in properties may be reduced by a more rigorous system of testing and quality control, all of which adds to the final cost of the ship. A compromise must therefore be reached between cost and the level of variation or uncertainty that is considered acceptable and that may be accommodated by the degree of conservatism in the design.

The subjective uncertainty, on the other hand, cannot be quantified on the basis of direct observation or analytical reasoning, but must be deduced by indirect means. The most common source of this uncertainty is a deficiency in the understanding of a fundamental physical phenomenon or incomplete development of the mathematical procedures needed for the purpose of predicting a certain aspect of the structural response. An example of incomplete theoretical knowledge is the small amplitude limitation inherent in the linear theory of wave loads and ship motions as outlined in Section 2.7 of this chapter and in Chapter VII. An analogous limitation exists in the application of linear elastic theory to the prediction of the structural response. Even though there have been important advances in theoretical and computational methods of nonlinear structural analysis, there is still a significant element of uncertainty in predictions of structural behavior in the vicinity of structural collapse. In this region, nonlinear material behavior as well as nonlinear geometric effects are present, and the overall response may involve the sequential interaction of several elementary response phenomena. It is, of course, the goal of ongoing research to change these subjective uncertainties into objective uncertainties.

It is clear, therefore, that the design of ship structures must take into account the uncertainties in the predictions of both demand and capability of strength. In order to arrive at the most efficient structure that will achieve an acceptable degree of reliability, it is necessary to attempt to quantify the uncertainties and to allow for their possible magnitudes and consequences. Three categories or classes of design procedure are now recognized and have been applied as a means of quantifying these uncertainties and incorporating such quantification into the design process. They are termed by Faulkner and Sadden (1979):

- Level 1—*Deterministic* or *safety factor* methods.
- Level 2—*Semi-probabilistic* methods.
- Level 3—*Fully probabilistic* methods.

5.2 Level 1—Safety Factor Methods. Essentially, the traditional safety factor approach assumes that a "worst load" can be evaluated and a load under which the structure will just fail can similarly be determined. The safety factor represents an elementary attempt to quantify the combined effect of the uncertainties in both of these estimates, although in an empirical and indirect manner. The safety factor is just the ratio of the calculated capability, C, (failure load or strength) to the calculated demand, D (worst load), provided that both are expressed in dimensionally similar quantities,

$$F = C/D \qquad (134)$$

The safety factor, F, is normally greater than unity, implying a margin of safety of $1.0 - F$ in the load-carrying capability of the structure to compensate for the designer's inability to predict either demand or capability with absolute certainty. Very often the demand is based upon a highly idealized and presumably conservative computation exemplified by the practice of poising the ship in an attitude of static equilibrium on the crest of a wave of some standard height, then applying elementary beam theory to compute the structural response (Section 2.3). The structural capability is assumed to be expressed by the material yield strength in simple unidirectional, tension or compression. It should be clear, on the basis of the discussion in earlier parts of this chapter, that neither the static wave bending moment nor the simple material yield may resemble very closely the most severe or limiting conditions of the ship's behavior.

A refinement of the above approach, involving partial safety factors, has become an accepted part of a number of civil engineering structural codes. The fundamental difference between this procedure and the elementary safety factor procedure lies in the use of multiple safety factors expressing the variability in loads and demand, and an additional factor expressing the economic, sociological and other consequences of failure of the structure. A typical expression is given in the following equation for a structure in which there may be several sources of loading that combine to form the total demand, D:

$$\gamma_c \sum_i \gamma_{fi} D_i \leq C_j/\gamma_m \qquad (135)$$

For each load component, D_i, (still water load, wave load), there may be a different value of the load factor, γ_{fi} which reflects the variability of that load component and the uncertainty in our ability to predict it. Typical values of the load factors, γ_f, in a formula of the form of (135) will lie in the range of 1.0 to 1.5, while typical values of the material and fabrication factor, γ_m, will be in the range of 1.0 to 1.35.

The consequences of failure factor, γ_c, represents an attempt to quantify the broader effects of potential failure of the structure. This includes certain consequences that may be expressed in very specific quantitative terms, such as the value of the ship and cargo, the loss of earnings if the ship is out of service for repair or replacement, and the cost of restoration of the environment in the event of casualty caused pollution. The potential consequences of loss of life, either of crew or passengers would also be included in the value assigned to this factor. In the case of a passenger ship carrying a large number of persons untrained in seamanship and emergency procedures a higher value would be assumed than in the case of a cargo ship having a small and highly trained crew.

The numerical values of the partial safety factors depend upon the form of the equation in which they are combined and upon the degree of structural reliability that is to be achieved, thus a unique set of values cannot be stated for general use. They have found their widest application in codes for fixed land structures of various types, although they are beginning to be used in the codes for offshore drilling structures. For a specific application, the engineer should consult the relevant code to obtain the formula in which the partial safety factors are applied, the relevant definitions and specific values of partial safety factors to be used. A recent paper discussing the relationship of partial safety factor formulations and the rules and practices for ships and offshore structures may be found in Mansour, et al, (1984).

5.3 Level 3 Methods. In the ideal situation, the designer would have correct and complete means of

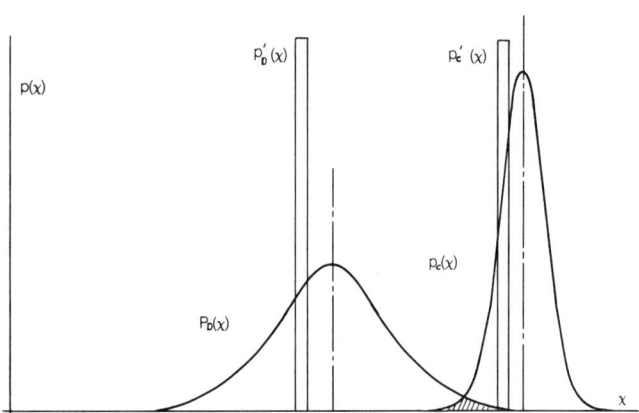

Fig. 75 Probability density functions of demand and capability

predicting both the loads on the structure and the structural response to those loads. Furthermore, the data concerning the objective uncertainties, such as the variation in material properties and the variation in sea conditions that the ship is expected to encounter, would be available in the form of quantitative statistical distributions. Using standard procedures of probability theory, it would then be possible to predict and express in probabilistic terms the reliability of the structure. Such a prediction would take the form of a quantitative estimate of the probability of failure occurring in a specified time interval such as one wave cycle, one year or one ship lifetime of twenty-five years. Alternatively, the prediction could be expressed in reciprocal terms of the probable life expectancy for the ship. In either event it can be incorporated into a structural performance criterion.

The demand and capability can each be expressed in the form of probability density functions as illustrated in Fig. 75. Here, the horizontal scale, x, represents a loading in a simple case, e.g., the midship bending moment. The density function $p_D(x)$ is so defined that $\int_{x_2}^{x_1} p_D(x)\,dx$ equals the probability that the actual maximum or exceedance value of load experienced by the ship during its lifetime will lie between x_1 and x_2. Note that, in general, $\int_{-\infty}^{\infty} p_D(x)\,dx = 1.0$, since there is 100 percent probability that the maximum experienced load will lie in the interval $(-\infty, \infty)$. Contained within the function $p_D(x)$, sometimes referred to as *lifetime probability* (Section 2.9), should be the effects of the random seaway loading, the variable ship and cargo load distribution, other minor loadings and the uncertainties in the prediction methodology.

The effect of the random seaway alone is given by Equation (18) in Section 2.9, where $P(x_1 > x_L)$ is the probability of lifetime exceedance of wave-induced bending moment in one ship among many similar ships. The most important additional loading to include is the still water bending moment. To combine the wave-induced and still water bending moments it is necessary to differentiate $P(x_1 > x_L)$ numerically to obtain the density function, $p(x_w)$, of lifetime probability. The density function of still water bending moment, $p(x_s)$, can be estimated for the loaded or ballasted condition. We now assume that the still water and wave bending moments are statistically independent quantities. We let x_s represent the possible values of the still water load, having a probability density $p_s(x_s)$. Similarly, x_w will represent the possible values of wave loading with a probability density $p_w(x_w)$. If x represents the total loading, $x_s + x_w$, the probability density of total load, or demand, will be given by the product of the densities of still water and wave loading,

$$p(x) = p_s(x_s)\,p_w(x_w)$$
$$= p_s(x_s)\,p_w(x - x_s) \qquad (136)$$

The exceedance probability distribution of the total load may be expressed by integrating Equation 136,

$$P_D(x > x_1) = \int_{x_1}^{\infty} dx \int_{-\infty}^{\infty} p_s(x_s)\,p_w(x - x_s)\,dx_s \qquad (137)$$

An example of this for the cargo ship *Wolverine State* is shown in Fig. 76 from Lewis, et al (1973), in which different normal density functions were assumed for still water bending moment in the loaded and in the ballast conditions. Here the wave bending moment is an effective value that takes account of the effect of lateral as well as vertical longitudinal bending. Further work is needed to permit other minor loads to be combined in a similar manner.

One result of including still water moments is to reveal the significant difference between hogging and sagging loads. The one to be used in any case depends on the mode of failure under consideration. A scale has been added to Fig. 76 as an example of how the exceedance probability, $P_D(x_1 > x_L)$ can be shown directly for the case of a ship lifetime of $n_L = 10^8$ cycles (Section 2.9). This will be used subsequently in developing the probability of failure.

The second function shown in Fig. 75, $p_C(x)$, is the probability density function of the ship structural capability or ultimate strength. In this case, the quantity, $\int_{x_1}^{x_2} p_C(x)\,dx$ is equal to the probability that failure of the structure will occur at any time under a load having a value lying between x_1 and x_2. The function $p_C(x)$ contains information quantifying the extent to which the actual ultimate strength differs from the computed ideal strength. As we have pointed out, this difference is the result of variation in material properties and structural geometry from assumed values, of con-

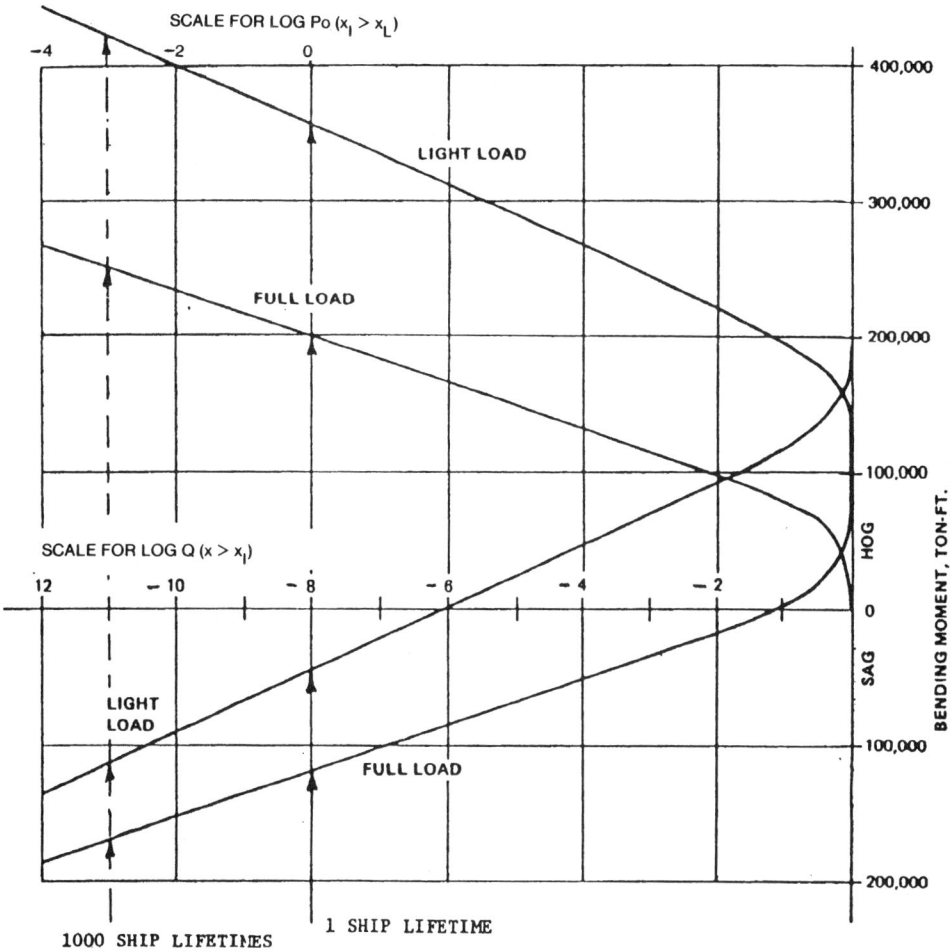

Fig. 76 Long-term distributions of combined bending moments: wave bending (vertical and lateral) and still-water bending (Lewis, et al, 1973)

struction irregularities, of deterioration and of deficiencies in the analytical procedures. Ideal ultimate strengths for different modes of failure are discussed in Section 4.9, but comparatively little information is available on the variability of actual strength. Some data are given by Stiansen, et al (1979) and Goodman and Mowatt (1976), and some assumptions expressed as coefficients of variation (c.o.v.), or ratios of standard deviation to mean value, are given by Mansour and Faulkner (1972) for different failure modes. Thus there would be different curves of $p_c(x)$ for each of these modes, just as there are different curves of $P_D(x_1 > x_L)$ for different conditions of load, and hog or sag. In each case, presumably, the most unfavorable curves would govern.

If the prediction methods for demand and capability were perfect, and if there were no variation in material or construction, and if there were no deterioration in the strength, the designer would then be able to predict with certainty a single value for the maximum load on the structure, and a single value for the capability of the structure. The two probability density functions in such case would reduce to two unit impulse functions and safety against failure would be assured if the demand impulse were infinitesimally displaced to the left of the capability (Fig. 75). Although the Level-1 method appears to be based upon single-value estimates of demand and capability, we see that the separation of the two values as measured by the safety factor acknowledges the uncertainty in the estimates of these quantities.

In any real situation, the two probability density functions have a certain dispersion about a mean value, and the functions are of such a character that there is a small overlap between the tails of the two curves (Fig. 75). This overlap implies that there exists a small probability that the demand equals or exceeds the capability, thus that failure may occur. If the probability densities of demand, p_D, and capability, p_C, are both known in the overlap region, the probability of failure may be computed by straightforward means. It should be noted in advance, however, that the tails of the functions, in general, are much less accurately known than the central part (mean, standard deviation) of the density functions. We assume that the tails are known and that the demand and capability are statistically independent random variables (not an unreasonable assumption). Then the incremental probability of failure under a particular level of loading is given by the product of the probability of capability, $p_C(x_1)\,dx$, and the probability that the lifetime demand is equal to or greater than x_L, $P_D(x_1 > x_L)$.

$$dp_f = P_D(x_1 > x_L)\, p_C(x)\, dx \qquad (138)$$

where the quantity, shown by the upper scale in Fig. 76,

$$P_D(x_1 > x_L) = \int_{x_L}^{\infty} p_D(x_1)\, dx \qquad (139)$$

is known as the exceedance probability of x_1 and its value, $P_D(x_1 > x_L)$, equals the probability that the demand has a value greater than or equal to x_L. The total probability of failure is, therefore, given by the integral of (138) or,

$$p_f = \int_0^{\infty} P_D(x_1 > x_L)\, p_C(x)\, dx \qquad (140)$$

This approach is the only one of the three that is fundamentally sound and able to take into consideration the detailed real variation in the strength and demand on the structure. Its application, however, presents several difficult problems. First, the probability of failure is determined principally by the tails of the two probability density functions, and, as noted earlier, these are the least known parts of the functions. Second, in most cases, neither the demand nor the capability involve simple phenomena that may be characterized by a single function. Instead, there may be several components of loading, each with a different probability density function, and all of these must be taken into consideration simultaneously. Similarly, there may be several different failure mechanisms, some of which may interact with each other, no one of which will clearly dominate the capability characterization over the entire range of demand.

Examples of applications of the procedure to ship structural reliability are given by Mansour (1972),

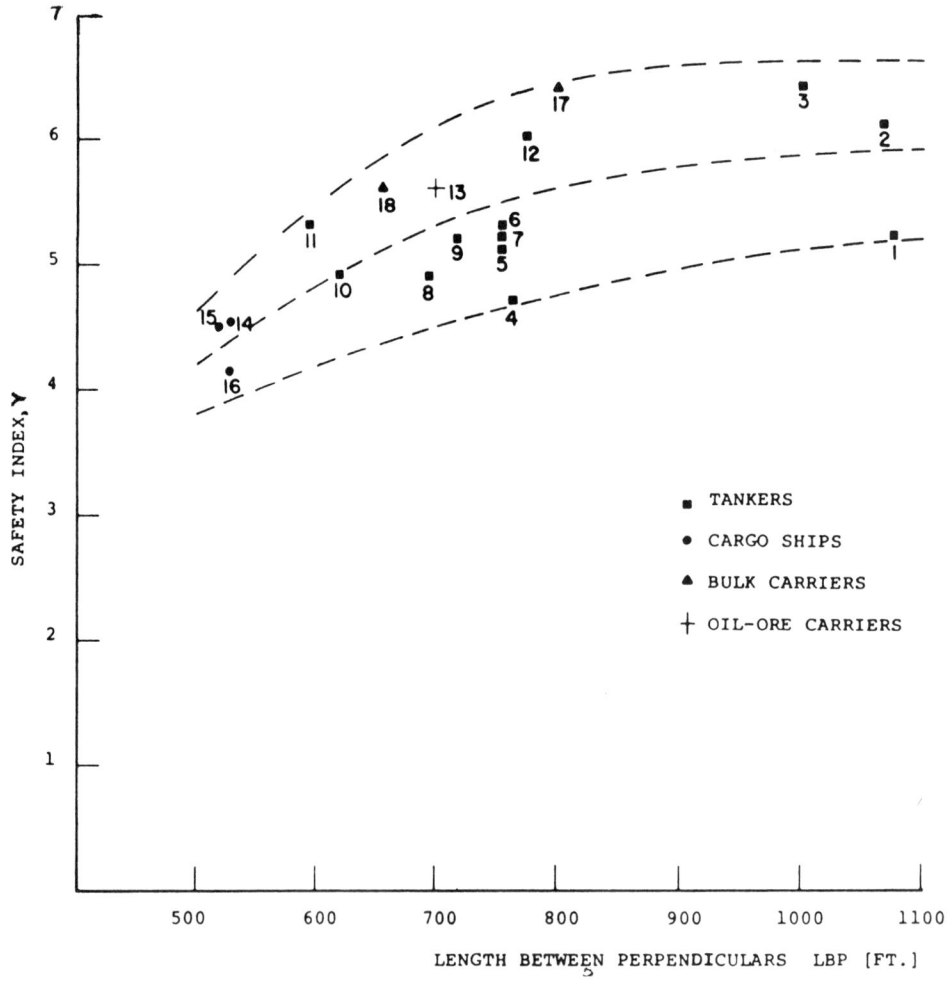

Fig. 77 Safety index for eighteen ships (Mansour, 1974)

Mansour and Faulkner (1973) and Lewis and Zubaly (1981). The latter reference also discusses the problem of establishing acceptable levels of failure probability on the basis of statistical data on actual ship failures, or by minimizing the total *expected cost* as explained by Freudenthal and Gaither (1969). A continuing goal of research is to develop the level-3 method to the point that it is suitable for the design of ships and other complex structures. The current limitations and difficulties of implementation inherent in the procedure have resulted in the development of the semi-probabilistic methods.

5.4 Semi-Probabilistic or Level 2 Methods. These methods of establishing structural performance criteria use similar terminology to that of the level 1 safety factor approach. They attempt to quantify in an approximate way, however, the random nature of the demand and capability. Rather than using the complete probability density functions for these parameters, only the mean and standard deviation of the two functions are employed directly. This leads to the term *second-moment* methods since the mean is equal to the first and the standard deviation is equal to the second moment of the probability density function. An example of a level 2 method as applied to ship structures is the safety index method of Mansour (1974). See also Mansour, et al (1984).

In the safety index method, it is assumed that the capability, C, and demand, D, are uncorrelated random variables having means, m_C and m_D, respectively, and standard deviations, σ_C and σ_D. The margin against failure, M, is given by the difference of C and D.

$$M = C - D \qquad (141)$$

Noting that both C and D are random variables, M will also be random with mean m_M and standard deviation σ_M. The safety index, γ, is defined as follows,

$$\gamma = \frac{m_M}{\sigma_M} \qquad (142)$$

The central safety factor is given by the ratio of the mean values of capability and demand,

$$\theta = \frac{m_C}{m_D} \qquad (143)$$

The safety index may then be expressed in terms of the central safety factor and the coefficients of variation of the capability and demand,

$$\gamma = \frac{\theta - 1}{\sqrt{\theta^2 V_D^2 + V_C^2}} \qquad (144)$$

where

$V_C = \dfrac{\sigma_C}{m_C}$ = *coefficient of variation (cov) of capability,*

$V_D = \dfrac{\sigma_D}{m_D}$ = cov of demand.

The mean of the capability is the design estimate of the strength for the assumed failure mode. The mean of the demand is estimated on the basis of the expected loading of the ship together with the operational profile in terms of sea state, trade route, speeds and headings. The determination of mean and standard deviation of the sea loading is discussed in Section 2 of this chapter. A discussion of the determination of variation of strength may be found in Stiansen, et al (1979).

Fig. 77 from Mansour (1974) illustrates the calculated variation of the safety index with ship size for a representative sample of merchant ships.

REFERENCES

Adamchak, J.C. (1982), "ULTSTR: A program for estimating the collapse moment of a ship's hull under longitudinal bending," David W. Taylor NSRDC Report DTNSRDC-82/076, Oct.

Admiralty Ship Committee (1953), "S.S. *Ocean Vulcan* Sea Trials," Report No. R8.

Allen, R.G., and Jones, R.R. (1977), "Consideration in the Structural Design of High Performance Marine Vehicles," N.Y. Metropolitan Section, SNAME, Jan.

American Bureau of Shipping (1978), *Provisional Rules for Building and Classing or Certifying Reinforced Plastic Vessels.*

American Bureau of Shipping (1983a), *Rules for Building and Classing Steel Vessels under 61 Meters (200 feet) in Length.*

American Bureau of Shipping (1983b), *Rules for Building and Classing Offshore Installations*—Part I, Structures.

American Bureau of Shipping (1987a), *Rules for Building and Classing Steel Vessels.*

American Bureau of Shipping (1987b), *Rules for Building and Classing Mobile Offshore Drilling Units.*

Band, E.G.U. (1966), "Analysis of Ship Data to Predict Long-Term Trends of Hull Bending Moments," ABS Report.

Bennet, R., Ivarson, A., and Nordenström, N. (1962), "Results from Full-Scale Measurements and Predictions of Wave Bending Moments Acting on Ships," Report No. 32, Swedish Ship Research Foundation.

Billingsley, D.W. (1980), "Hull Girder Response to Extreme Bending Moments," Spring Meeting/STAR, SNAME.

Birmingham, J.T. (1971), "Longitudinal Bending Moment Predictions derived from Results of Seven Ship Trials," DTNSRDC Report 3718.

Bishop, R.E.D., et al (1978), "A Unified Dynamic Analysis of Ship Response to Waves," *Trans.* RINA, Vol. 120.

Bleich, F. (1952), *Buckling Strength of Metal*

Structures, McGraw Hill, New York.

Bleich, H.H. (1953), "Nonlinear Distribution of Bending Stresses due to Distortion of the Cross Section," *Journal of Applied Mechanics*, March, pp. 95–104.

Boentgen, R.R. et al (1976), "First Season Results from Ship Response Instrumentation Aboard the SL-7 Class Containership SS *Sealand McLean* in North Atlantic Service," Ship Structures Committee, Report SSC 264.

Brock, J. (1957), DTMB Report 1149, Nov.

Bullock, O.R., and Oldfield, B. (1976), "Production PHM Design-to-Cost Hull Structure," SNAME/AIAA Advanced Marine Vehicles Conference, Sept.

Caldwell, J.B. (1965), "Ultimate Longitudinal Strength," *Trans*. RINA, Vol. 107, pp. 411–430.

Chen, Y.K., Kutt, L.M., Piaszcyk, C.M., and M.P. Bieniek (1983), "Ultimate Strength of Ship Structures," *Trans*. SNAME, Vol. 91.

Clarkson, J., Wilson, L.B., McKeeman, J.L. (1959), "Data Sheets for the Elastic Design of Flat Grillages under Uniform Pressure," *European Shipbuilding*, Vol. 8, pp. 174–198.

Coker, E.G., Chakko, K.C., and Satake, Y. (1919-20), "Photo-Elastic Strain Measurements," *Trans*. IESS.

D'Arcangelo, A.M. (1969), *A Guide to Sound Ship Structures*, Cornell Maritime Press, Inc.

Dalzell, J.F. (1963), "Summary of Investigation of Midship Bending Moments Experienced by Models in Extreme Regular Waves," SSC Report No. SSC-157.

DeWilde, G. (1967), "Structural Problems in Ships with Large Hatch Openings," *International Shipbuilding Progress*, Vol. 14.

DNV (1977), *Rules for the Design, Construction and Inspection of Offshore Structures*, Det Norske Veritas, Hovik, Norway.

Dow, R.S., Hugill, R.C., Clark, J.D., and Smith, C.S. (1981), "Evaluation of Ultimate Ship Hull Strength," Extreme Loads Response Symposium, SNAME/SSC, Arlington, Va.

Elbatouti, A.M.T., Jan, H.Y., and Stianson, S.G. (1976), "Structural Analysis of a Containership Steel Model and Comparison with the Test Results," *Trans*. SNAME, Vol. 84.

Faulkner, D. (1975), "A Review of Effective Plating for Use in the Analysis of Stiffened Plating in Bending and Compression," *Journal of Ship Research*, Vol. 19, No. 1, March, pp. 1–17.

Faulkner, D., Sadden, J.A. (1978), "Toward a Unified Approach to Ship Safety," *Trans*. RINA, Vol. 120.

Faulkner, D. (1981), "Semi-Probabilistic Approach to the Design of Marine Structures," Extreme Loads Response Symposium, SNAME/SSC, Arlington, Va.

Frankland, J.M. (1942), "Effects of Impact on Simple Elastic Structures," D.W. Taylor Model Basin Report, No. 481.

Freudenthal, A.M., and Gaither, W.S. (1969), "Probabilistic Approach to Economic Design of Maritime Structures," XXIInd International Navigation Congress, Paris.

Fukuda, J. (1970), "Long-Term Predictions of Wave Bending Moment," Selected papers from the *Journal of SNA of Japan*.

Gerritsma, J., and Beukelman, W. (1964), "The Distribution of Hydrodynamic Forces on a Heaving and Pitching Ship in Still Water," *International Shipbuilding Progress*, Vol. 11, pp. 506–522.

Glasfeld, R.D. (1962), "Design and Construction of a 42-foot Ship Structural Model Testing Facility," University of California, Institute of Engineering Research, Reports Series 184, Issue 1.

Goodman, R.A. (1971), "Wave Excited Main Hull Vibration in Large Tankers and Bulk Carriers," *Trans*. RINA, Vol. 113.

Goodman, R.A., and Mowatt, G. (1976), "Allowance for Imperfection in Ship Structural Design," Conference on the Influence of Residual Stresses and Distortions on the Performance of Steel Structures, Inst. Mech. Eng. (Great Britain).

Gumbel, E.J. (1958), *Statistics of Extremes*, Columbia University Press, N.Y.

Hadler, J.B., Lee, C.M., Birmingham, J.T. and Jones, H.D. (1974), "Ocean Catamaran Seakeeping Design, based on the Experience of USNS *Hayes*,"*Trans*. SNAME, Vol. 82.

Haslum, K., Tonnessen, A. (1972), "An Analysis of Torsion in Ship Hulls," *European Shipbuilding*, No. 5/6, pp. 67–89.

Heller, S.R., Brock, J.S., and Bart, R. (1959) DTMB Report No. 1290.

Heller, S.R., and Jasper, H.H. (1960), "On the Structural Design of Planing Craft," RINA *Trans.*, Vol. 102.

Hoffman, D., and Lewis, E.V. (1969), "Analysis and Interpretation of Full-Scale Data on Bending Stresses of Dry Cargo Ships," Report SSC-196.

Hoffman, D., and vanHooff, R.W. (1973), "Feasibility Study of Springing Model Tests of a Great Lakes Bulk Carrier," *International Shipbuilding Progress*, March.

Hoffman, D., and vanHooff, R.W. (1976), "Experimental and Theoretical Evaluation of Springing of a Great Lakes Bulk Carrier," *International Shipbuilding Progress*, June.

Hogben, N., and Lumb, F.E. (1967), *Ocean Wave Statistics*, Her Majesty's Stationery Office, London.

Hughes, Owen (1983), "Design of Laterally Loaded Plating-Concentrated Loads," *Journal of Ship Research*, Vol. 27, N. 4, Dec., pp. 252–265.

ISSC (1973), Report of Committee 10, Design Procedure, Fifth International Ship Structures Congress, Hamburg.

ISSC (1979), Report of Com. II.1 Linear Structural Response.

Jasper, N.H. (1956), "Statistical Distribution Patterns of Ocean Waves and of Wave-Induced Ship Stresses and Motions," *Trans.* SNAME, Vol. 64.

Jensen, J., and Pedersen, T. (1981), "Bending Moments and Shear Forces in Ships Sailing in Irregular Waves," *Journal of Ship Research*, Vol. 25, No. 4, Dec.

Kaplan, P., and Sargent, T.P. (1972), "Further Studies of Computer Simulation of Slamming and Other Wave-Induced Vibratory Structural Loadings on Ships in Waves," Ship Structure Committee Report, SSC-231.

Kaplan, P., Sargent, T.P., Cilmi, J. (1974), "Theoretical Estimates of Wave Loads on the SL-7 Containerships in Regular and Irregular Seas," Ship Structures Committee Report SSC-246.

Kaplan, P., Bentson, J., Davis, S. (1981), "Dynamics and Hydrodynamics of Surface-Effect Ships," *Trans.* SNAME, Vol. 89.

Keane, R.G. (1978), "Surface Ships Hydrodynamic Design Problems and R & D Needs," NAVSEC Report No. 6136-78-37.

Kim, C.H. (1975), "Theoretical and Experimental Correlation of Wave Loads for SL-7 Containership," Davidson Laboratory Report SIT-DL-75-1829, June.

Kim, C.H. (1982), "Hydrodynamic Loads on the Hull Surface of a Seagoing Vessel," SNAME Star Symposium, Paper No. 8.

Kumai, T. (1974), "On the Exciting Force and Response of Springing of Ships," International Symposium on the Dynamics of Marine Vehicles and Structures in Waves, Institution of Mechanical Engineers, London.

Kuntsman, C.M., and Umberger, R.C. (1959), "The Stresses Around Reinforced Square Opening with Rounded Corners in an Uniformly Loaded Plate," WINA Thesis, June (Fig 59).

Lee, C.M., and Curphey, R.M. (1977), "Prediction of Motion, Stability, and Wave Loads of Small-Waterplane-Area, Twin-Hull Ships," *Trans.* SNAME, Vol. 85.

Lewis, E.V. (1954), "Ship Model Tests to Determine Bending Moments in Waves," *Trans.* SNAME, Vol. 62, pp. 426–490.

Lewis, E.V. (1967), "Predicting Long-term Distributions of Wave-Induced Bending Moments on Ship Hulls," *Proceedings of Spring Meeting*, SNAME, Montreal.

Lewis, E.V., Hoffman, D., MacLean, W.M., vanHooff, R., and Zubaly, R.B. (1973), "Load Criteria for Ship Structural Design," Ship Structure Committee Report SSC-240.

Lewis, E.V., and Zubaly, R.B. (1981), "Predicting Hull Bending Moments for Design," Extreme Loads Response Symposium, SNAME and SSC, Arlington, Va.

Little, R.S., Lewis, E.V., Bailey, F.C. (1971), "A Statistical Study of Wave-Induced Bending Moments on Large Ocean-Going Tankers and Bulk Carriers," *Trans.* SNAME, Vol. 79, pp. 117–168.

Liu, D., Bakker, A. (1981), "Practical Procedures for Technical and Economic Analysis of Ship Structural Details," *Marine Technology*, Vol. 18, No. 1, Jan.

Liu, D., Chen, H., and Lee, F. (1981), "Application of Loading Predictions to Ship Structure Design: A comparative analysis of methods," Extreme Loads Response Symposium, SNAME/SSC, Arlington, Va.

MacLean, W.M., and Lewis, E.V. (1970), "Analysis of Slamming Stresses on S.S. *Wolverine State*," Webb Institute of Naval Architecture Report No. 10-17 to ABS; also *Marine Technology*, Jan. 1973.

Mantle, P.J. (1975), "A Technical Summary of Air Cushion Craft Development," DTNSRDC Report No. 4727, Oct.

Matthews, S.T. (1967), "Main Hull Girder Loads on a Great Lakes Bulk Carrier," Spring Meeting, SNAME.

Mansour, A.E., and Faulkner, D. (1972), "On Applying the Statistical Approach to Extreme Sea Loads and Ship Hull Strength," *Trans.* RINA., Vol. 114.

Mansour, A.E. (1972), "Probabilistic Design Concepts in Ship Structural Safety and Reliability," *Trans.* SNAME, Vol. 80.

Mansour, A.E. (1974), "Approximate Probabilistic Method of Calculating Ship Longitudinal Strength," *Journal of Ship Research*, Vol. 18, Sept., pp. 203–213.

Mansour, A.E., and D'Oliveira, J.M. (1975), "Hull Bending Moment due to Ship Bottom Slamming in Regular Waves," *Journal of Ship Research*, Vol. 19, No. 2, pp. 80–92.

Mansour, A.E. (1976), "Charts for Buckling and Postbuckling Analysis of Stiffened Plates under Combined Loading," SNAME T & R Bulletin, 2-22, July.

Mansour, A.E. (1977), "Gross Panel Strength under Combined Loading," Ship Structures Committee Report SSC-270.

Mansour, A.E., Jan, H.Y., Zigelman, C.I., Chen, Y.N., and Harding, S.J. (1984), "Implementation of Reliability Methods to Marine Structures," *Trans.* SNAME, Vol. 92.

McCallum, J. (1975). "The Strength of Fast Cargo Ships," *Trans.* RINA, Vol. 117.

Meyers, W.G., Sheridan, D.J., Salvesen, N. (1975), "Manual—NSRDC Ship-Motion and Sea-Load Computer Program," NSRDC Report 3376, Feb.

Michelsen, F.C., Nielsen, R. (1965), "Grillage Structure Analysis through Use of the Laplace Integral Transform," *Trans.* SNAME, Vol. 73, pp. 216–240.

Miner, M.A. (1945), "Cumulative damage in fatigue," *Journal of Applied Mechanics*, Sept., pp. A159–A164.

Morison, J.R., O'Brien, M.P., Johnson, J.W., Schaaf, S.A. (1950), "The Forces Exerted by Surface Waves on Piles," *Trans.* AIME, Vol. 189, pp. 149–157.

Munse, W.H., Wilson, T.H., Tellalian, M.L., Nicoll, K., Wilson, K. (1983), "Fatigue Characterization of Fabricated Ship Details for Design," Ship Structures Committee, Report SSC-318.

Nordenström, N. (1973), "A Method to Predict

Long-Term Distributions of Waves and Wave-Induced Motions and Loads," Det norske Veritas Publication 81, April.

Numata, E. (1960), "Longitudinal Bending and Torsional Moments Acting on a Ship Model at Oblique Headings to Waves," *Journal of Ship Research*, Vol. 4, No. 1, June.

Ochi, M.K. (1964), "Prediction of Occurrence and Severity of Ship Slamming at Sea," Prec. 5th Symposium on Naval Hydrodynamics, ONR ACR 112, pp. 545–596.

Ochi, M.K., and Motter, L.E. (1973), "Prediction of Slamming Characteristics and Hull Response for Ship Design," *Trans.* SNAME, Vol. 81, pp. 144–176.

Ochi, M.K. (1973), "On Prediction of Extreme Values," *Journal of Ship Research*, Vol. 17, March, pp. 29–37.

Ochi, M.K., and Bolton, W.E. (1973), "Statistics for the Prediction of Ship Performance in a Seaway," *International Shipbuilding Progress*, Vol. 22, Nos. 222, 224, 229.

Ochi, M.K. (1978), "Wave Statistics for the Design of Ships and Ocean Structures," *Trans.* SNAME, Vol. 86, pp. 47–76.

Okada, H., Oshima, K., Fukumoto, Y. (1979), "Compressive Strength of Long Rectangular Plates under Hydrostatic Pressure," *Journal of Society of Naval Architects of Japan*, Vol. 146, Dec. (English Translations in *Naval Architecture and Ocean Engineering*, Vol. 18, 1980, SNA Japan, pp. 101–114.)

Paulling, J.R., Payer, H.G. (1968), "Hull-Deckhouse Interaction by Finite-Element Calculations," *Trans.* SNAME, Vol. 76.

Paulling, J.R. (1974), "Elastic Response of Stable Platform Structures to Wave Loading" Int'l. Symp. on the Dynamics of Marine Vehicles and Structures in Waves, IME, London.

Paulling, J.R. (1982), "An Equivalent Linear Representation of the Forces Exerted on the OTEC Cold Water Pipe by Combined Effects of Wave and Current." *Ocean Engineering for OTEC* Vol. 9, ASME.

Peery, D.J. (1950), *Aircraft Structures*, McGraw-Hill, N.Y., p. 493.

Popov, E.P. (1968), *Introduction to Mechanics of Solids* Prentice Hall, Englewood Cliffs, New Jersey.

Raff, A.I. (1970), "Program SCORES—Ship Structural Response in Waves," Ship Structures Committee Report No. SSC-230.

Roop, W.P. (1932), "Elastic Characteristics of a Naval Tank Vessel," *Trans.* SNAME, Vol. 40.

St. Denis, M. (1970), "A guide for the Synthesis of Ship Structures, Part One, the Midship Hold of a Transversely-Framed Dry Cargo Ship," Ship Structures Committee Report SSC-215.

Salvesen, N., Tuck, E.O., Faltinsen, O. (1970), "Ship Motions and Sea Loads," *Trans.* SNAME, Vol. 78, pp. 250–287.

Schade, H.A. (1938), "Bending Theory of Ship Bottom Structure," *Trans.* SNAME, Vol. 46, pp. 176–205.

Schade, H.A. (1940), "The Orthogonally Stiffened Plate under Uniform Lateral Load," *Journal of Applied Mechanics*, Vol. 7, No. 4, Dec.

Schade, H.A. (1941), "Design Curves for Cross-Stiffened Plating," *Trans.* SNAME, Vol. 49, pp. 154–182.

Schade, H.A. (1953), "The Effective Breadth Concept in Ship Structure Design," *Trans.* SNAME, Vol. 61.

Schade, H.A. (1951), "The Effective Breadth of Stiffened Plating under Bending Loads," *Trans.* SNAME, Vol. 59.

Schade, H.A. (1965), "Two Beam Deckhouse Theory with Shear Effects," University of California Institute of Engineering Research Report Na-65-3.

Silvia, P.A. (1978), "Structural Design of Planing Craft," Chesapeake Section, SNAME, March.

Sikara, J.P., Dinsenbacher, A., and Beach, J.E. (1983), "A Method for Estimating Lifetime Loads and Fatigue Lives for SWATH and Conventional Monohull Ships," *Naval Engineers Journal*, May.

Skaar, K.T. (1974), "On the Finite Element Analysis of Oil Tanker Structures," *Maritime Research*, Vol. 2, No. 2, Nov., pp. 2–11.

Smith, C.S. (1975), "Compressive Strength of Welded Ship Grillages," *Trans.* RINA, Vol. 117, pp. 325–359.

Smith, C.S. (1977), "Effect of Local Compressive Failure on Ultimate Longitudinal Strength of a Ship's Hull," PRADS Symp., Tokyo.

Smith, C.S. (1983), "Structural Redundancy and Damage Tolerance in Relation to Ultimate Ship Hull Strength," *Proceedings.* Symp. on Design, Inspection, Redundancy, Williamsburg, Va. National Academy Press, Washington, D.C.

Soeding, H. (1974), "Calculation of Long-Term Extreme Loads and Fatigue Loads of Marine Structures," Symposium on the Dynamics of Marine Vehicles and Structures in Waves, IME, London.

SSC (Ship Structure Committee) (1952), "Considerations of Welded Hatch Corner Design," Report SSC-37, Oct. 1.

Spencer, J.S. (1975), "Structural Design of Aluminum Crew Boats," *Marine Technology*, July.

Stambaugh, K.A., and Wood, W.A. (1981), "SL-7 Research Program Summary, Conclusions and Recommendations," Ship Structure Committee Report, AD-A120599.

Stavovy, A.B., and Chuang, S.L. (1976), "Analytical Determination of Slamming Pressures for High-Speed Vehicles in Waves," *Journal of Ship Research*, Dec. (Errata: *JSR*, Dec. 1977).

Stiansen, S.G. (1984), "Recent Research on Dynamic Behavior of Large Great Lakes Bulk Carriers," *Marine Technology*, Vol. 21, No. 4.

Stiansen, S.G., Mansour, A.E., and Chen, Y.N. (1977), "Dynamic Response of Large Great Lakes Bulk Carriers to Wave-Excited Loads," *Trans.* SNAME, Vol. 85.

Stiansen, S.G., Mansour, A.E., Jan, H.Y., Thayamballi, A. (1980), "Reliability Methods in Ship Structures," *Trans. RINA*, pp. 381–406.

Stiansen, S.G., and Chen, H.H. (1982), "Application of Probabilistic Design Methods to Wave Loads Prediction for Ship Structures Analysis," SNAME T&R Bulletin 2-27.

Taggart, R. (Ed.) (1980), *Ship Design and Construction*, SNAME, New York.

Timoshenko, S. (1956), *Strength of Materials*, Vol. I (3rd Ed.); Vol. II.

Timoshenko, S., Goodier, J.N. (1970), *Theory of Elasticity*, McGraw-Hill, 3rd Ed.

Troesch, A.W. (1984a), "Wave-Induced Hull Vibrations: An Experimental and Theoretical Study," *JSR*, Vol. 28, No. 2, June, pp. 141–150.

Troesch, A.W. (1984b), "Effects of Nonlinearities on Hull Springing," *Marine Technology*, Vol. 21, No. 4, Oct.

Vasta, J. (1949), "Structural Tests on Passenger Ship S.S. *President Wilson*; Interaction between Super-structure and Main Hull Girder, *Trans. SNAME*, Vol. 57.

Vasta, J. (1958), "Lessons learned from full scale structural tests," *Trans. SNAME*, Vol. 66, pp. 165–243.

Wah, T. (1960), *A Guide for the Analysis of Ship Structures*, U.S. Dept. of Commerce, PB181168, Govt. Printing Office.

Wahab, R., Pritchett, C., Ruth, L.C. (1975), "On the Behavior of the ASR Catamaran in Waves," *Marine Technology*, July.

Westin, H. (1981), "Analysis of the Torsional Response of the SL-7 Hull Structure by Use of the Finite-Beam Technique and Comparison with Other Techniques," *Journal of Ship Research*, Vol. 25, No. 1, Mar., pp. 62–75.

Yuille, I.M. (1963), "Longitudinal Strength of Ships," *Trans. RINA*, Vol. 105.

This page intentionally left blank

Volume 1 Nomenclature

The following symbols apply to Volume I only. The phrase "stands for" is understood between the symbol and its definition.

A	stands for area, generally
A_M	area of midship section
A_W	area of waterplane
AP	after perpendicular
A_X	area of maximum section (if not amidships)
B	maximum molded breadth
B	center of buoyancy
B_1	etc., changed positions of center of buoyancy
BL	molded baseline
$\overline{BM}$	transverse metacentric radius, or height of M above B
$\overline{BM}_L$	longitudinal metacentric radius, or height of M_L above B
b	width of a compartment or tank
b	span of a control surface (perpendicular to direction of flow)
C	constant or coefficient
CL	centerline; a vertical plane through centerline
C_B	block coefficient, ∇/LBT
C_D	drag coefficient, $D/\tfrac{1}{2}\rho AV^2$
C_M	midship section coefficient, A/BT; added mass
C_P	prismatic coefficient, $\nabla/A_m L$
C_V	volumetric coefficient
C_{VP}	vertical prismatic coefficient $\nabla/A_w T$
C_{WP}	waterplane area coefficient, A_w/LB
c	chord of a control surface (parallel to direction of flow)
c	distance from neutral axis to extreme fiber
D	molded depth
D	cumulative damage function
D	diameter, generally
D	drag force (resistance)
D	plate flexural rigidity
DWL	designed load waterline
DWT	deadweight
E	Young's modulus of elasticity
E	energy, generally
e	base of Naperian logarithms, 2.7183
F	force, generally
F	factor of safety
F	center of flotation (center of gravity of waterplane)
F_E	effective freeboard, flooded
FP	forward perpendicular
FW	fresh water
f	flow coefficient
f	frequency, hertz
G	center of gravity of ship's mass
G_1	etc., changed positions of the center of gravity
$\underline{G}$	shear modulus, $E/2(1+\mu)$
$\overline{GM}$	transverse metacentric height; height of M above G
$\overline{GM}_E$	effective metacentric height, flooded
$\overline{GM}_L$	longitudinal metacentric height, height of M_L above G
$\overline{GZ}$	righting arm; horizontal distance from G to Z
g	acceleration due to gravity
g	center of gravity of a component
H	head
h	depth of water or submergence
H, h_w	height of a wave, from hollow to crest
I	moment of inertia, generally
I	moment of inertia of hull-girder section about neutral axis
I_L	longitudinal moment of inertia of waterplane
I_T	transverse moment of inertia of waterplane
I_P	polar moment of inertia, generally

i_T	stands for transverse moment of inertia of free surface in a compartment or tank	SM	Simpson's multiplier
		SW	salt water
		s	spacing of ordinates
i_L	longitudinal moment of inertia of free surface in a compartment or tank	T	draft
		T	period, generally
J	torsional constant of a section	T_W	period of a wave
K	any point in a horizontal plane through the baseline	TCG	transverse position of center of gravity
$\overline{KB}$	height of B above the baseline	$TPcm$	tons per cm immersion
$\overline{KG}$	height of G above the baseline	TPI	tons per in. immersion
$\overline{KM}$	height of M above the baseline	t	thickness, generally
$\overline{KM_L}$	height of M_L above the baseline	t	time, generally
k	radius of gyration	V	total vertical shearing force across a section
L	length, generally		
L	length of ship	V	linear velocity in general; speed of ship
L_E	length of entrance		
L_P	length of parallel body	V_k	speed of ship, knots
L_R	length of run	V_c	velocity of a surface wave (celerity)
L_{PP}	length between perpendiculars	V_a	speed of advance (flow through propeller)
L_{OA}	length overall		
L_{WL}	length on designed load waterline	VCB	vertical position of B
L	lift force, perpendicular to direction of flow	VCG	vertical position of G
		vcg	vertical position of g
L_W	length of a wave, from crest to crest	W	displacement/weight of ship, $\rho g \nabla$
LCB	longitudinal position center of buoyancy	WL	any waterline parallel to baseline
		WL_1, etc.	changed positions of WL
LCF	longitudinal position center of flotation	V	volume of an individual item
		V	linear velocity
LCG	longitudinal position center of gravity	w	weight of an individual item
LWL	load, or design, waterline	x	distance from origin along X-axis
l	length of a compartment or tank	y	distance from origin along Y-axis
M	moment, generally	z	distance from original along Z-axis
M_B	bending moment	Z	a point vertically over B, opposite G
M	transverse metacenter		
M_L	longitudinal metacenter	α (alpha)	angle of incidence; angle of attack
M_Q	torsional moment	β (Beta)	width parameter
M_T	trimming moment	Δ (Delta)	displacement mass = $\rho \nabla$
M_W	wave-induced bending moment	δ (delta)	specific volume
$MTcm$	moment to trim 1 cm	γ (gamma)	warping factor
MTI	moment to trim 1 in.	γ (gamma)	safety index
m	mass, generally; (W/g or w/g)	η (eta)	efficiency, generally
m	transverse metacenter of liquid in a tank or compartment	ϑ (theta)	angle of pitch or of trim (about OY-axis)
m_L	longitudinal metacenter of liquid in a tank or compartment	Λ (Lambda)	tuning factor
		λ (lambda)	linear scale ratio
N	shear flow	μ (mu)	permeability
NA	neutral axis	μ (mu)	coefficient of dynamic viscosity
O	origin of coordinates	ν (nu)	Poisson's ratio
OX	longitudinal axis of coordinates	ν (nu)	coefficient of kinematic viscosity; μ/ρ
OY	transverse axis of coordinates		
OZ	vertical axis of coordinates	ρ (rho)	density; mass per unit volume
P	(upward) force of keel blocks	σ (sigma)	direct, or bending, stress
p	pressure per unit area in a fluid	τ (tau)	shear stress
p	probability, in general	ϕ (phi)	angle of heel or roll (about OX-axis)
Q	fore and aft distance on waterplane	ϕ (phi)	velocity potential
R	radius, generally	ψ (psi)	angle of yaw (about OZ-axis)
S	wetted surface of hull	ω (omega)	angular velocity
SM	section modulus, I/c	ω (omega)	circular frequency, $2\pi/T$, radians

Special Naval Architectural Symbols

⌊	denotes baseline
℄	centerline
⊗	midlength, in general
⊗pp	midlength between perpendiculars
⊗wl	midlength on load water line
⊗B	longitudinal distance from amidships to center of buoyancy, B
⊗G	longitudinal distance from amidships to center of gravity, G
⊗F	longitudinal distance from amidships to center of flotation F
∇	vol volume of displacement

Mathematical Symbols

∂	is a partial derivative sign
i	is $\sqrt{-1}$
$\approx$	approximately equal to
$<$	less than
$>$	greater than
$\dot{x}$	(one dot over a variable) is the first derivative of the variable with respect to time
$\ddot{x}$	(two dots over a variable) is the second derivative of the variable with respect to time
$\propto$	means proportional to
∞	infinity
δ (delta)	a finite increment
Σ (Sigma)	summation of
π (pi)	ratio of circumference of circle to diameter
$\int$	integral of
f	some function of
$\rightarrow$	approaches as a limit
$\equiv$	is identical to

Abbreviations for References

ABS	American Bureau of Shipping
ATMA	Association Maritime Technique et Aeronautique
ASNE	American Society of Naval Engineers
ASCE	American Society of Civil Engineers
ATTC	American Towing Tank Conference
BMT	British Maritime Technology (formerly BSRA)
DTNSRDC	David Taylor Naval Ship Research and Development Center (formerly DTMB) (now DTRC, David Taylor Research Center)
IESS	Institute of Engineers and Shipbuilders in Scotland
IMO	International Maritime Organization (formerly IMCO)
ISSC	International Ship Structures Congress
ITTC	International Towing Tank Conference
JSR	SNAME *Journal of Ship Research*
JSTG	*Jahrbuch des Schiffbautechnischen Gesellschaft*
MARIN	Maritime Research Institute Netherlands (formerly NSMB)
NECI	Northeast Coast Institute of Engineers and Shipbuilders
RINA	Royal Institute of Naval Architects
SSC	Ship Structures Committee
SNAME	Society of Naval Architects and Marine Engineers
WEBB	Webb Institute of Naval Architecture

International System of Units (Systeme International d'Unites, or SI) Useful quantities for Naval Architecture

Quantity	SI Unit	Definition	Conversions English to SI	Conversions SI to English
Base Units				
Length	meter, m		1 ft = 0.305 m	1 m = 3.28 ft
Mass	kilogram, kg		1 lb = 0.454 kg	1 kg = 2.20 lb
Time	second, s			
Supplementary Units				
Angle, plane	radian, rad	1 rad = 180°/π		
Density of solids of liquids		kg/cm^3 or t/m^3 kg/L		
Distance	nautical mile, knot	1.852 km		1 knot = 6,080 ft
Force	newton, N kilonewton, kN	1 kg-m/s^2 103 kg-m/s^2	1 lb (force) = 4.45 N	1 N = 0.225 lb
Frequency	hertz, Hz	cycle/sec, cps		
Mass	metric ton, t	10^3 kg	1 long ton (weight) = 1.016 t	1 t = 0.98 long tons
Power	watt, W kilowatt, kW	1 N-m/s 1 kN-m/s	1 hp = 0.746 kW	1 kW = 1.34 hp
Pressure	kilopascal, kPa	10^3 N/m^2	1 lb/in^2 = 6.895 kPa	1 kPa = 0.15 lb/in^2
Specific vol.	1/density	m^3/t		
Stress	megapascal, MPa	MN/m^2 = N/mm^2	1 long ton/in^2 = 15.44 MPa	1 MPa = 0.065 long tons/in^2
Volume of solids of liquids	m^3 liter, L	$10^{-3} m^3 = d^3$		
Velocity	meters/sec knot	m/s 1 nmi/hr = 1.852 km/kr.	1 ft/sec = 0.305 m/s	1 m/s = 3.28 ft/sec

Index

Acceleration
  forces, 209
  inertial reactions, 207
  inertial loads, 209
Added-weight method
  flooding calculation, 150, 159, 166
Afterbody, 2
After perpendicular, 2
Airy stress function, 247
Aluminum
  elasticity of, 236
American Bureau of
  Shipping Rules, 2, 178, 214
Andrea Doria tragedy, 144, 194
Appendages
  displacement, 39
  stability calculation, 83
Archimedes Principle, 16
Areas
  gyradii, 23
  integration of, 22
  moments of, 22
  sectional curve, 6
  water plane, 32
  wetted surface, 47
Arrestors, crack, 207, 276

Ballast
  variable, 104
  liquid, 104
  fuel compensation, 104
  tank capacity, 133
  permanent, 135
Base line, molded, 5
Bending moment
  diffusion of, 260
  hull girder, 211
  lateral, 211, 215
  on a section, 246
  in waves, 215, 216
  response to, 212
  still water, 212
  vertical, 212, 215
Bilge diagonal, 7
Bilge radius, 6
Block coefficient, 18
Body plan stations, 2
Bonjean curves, 36
  construction, 43
  uses of, 44
  for displacement, 115
  for flooding calculations, 152
Bredt formula, 254
Brittle fracture, 207
Bryan formula, 279
Buckling failure
  of plating, 275, 279
  of stiffened panels, 282
  criteria of, 276

Buckling failure (*continued*)
  in compression, 276
  elastic, 282
Bulkhead deck, 179
Bulkheads, subdivision factor
  minimum spacing, 191
  recessed, 191
  stepped, 191
Buoyancy, 147
  intact, 150, 160
  lost buoyancy method, 159
  residual, 177
Buoyancy, center of, 16
  approximation of, 38
  longitudinal (LCB), 35
  vertical (VCB or $\overline{KB}$) 36, 38
Buttock lines, 11

Camber, deck
  curve of 4, 6
Cant frames
  single, 7
  double, 7
  inclined, 7
Cargo
  cubic, 55
  storage factors, 59
  bulk dry, 100
  effect on stability, 103
  liquid, 104
  damping, 287
Capacity
  bale, 54
  consumables, 59
  curve of, 55
  cargo, 54
  fuel and ballast, 60
  heel and trim effects, 58
  soundings, 57
  storage factor, 59
  plan, 51
  ullage of tanks, 59
Capsizing
  longitudinal, 66
  transverse, 67, 143, 147
Center of buoyancy; also see
  Buoyancy, center of, 16, 17
Center of flotation; also see
  Flotation, center of, 33, 74
Center of gravity; also see
  Gravity, center of, 22, 23
Chines, 13
Coefficients, hull form
  block, 18
  midship, 18
  prismatic, 19
  vertical prismatic, 19
  volumetric, 19
  waterplane, 19
  wetted surface, 49

Compensating fuel tanks, 104
Compensation (discontinuity of
  structure), 235
Computers
  applications, 15, 42
  HULDEF, 42, 43
  lines definition, 15
  trim and bending moment, 106
  Navy Ship Hull
    Characteristics Program, 43, 92, 159
Conditions of loading, 76
Conference, MARPOL
  (Marine Pollution), 144
Container systems, 55
Convention SOLAS 1960,
  (Safety of Life at Sea), 144, 146, 181,
  189
Conversion tables, liquids, 57
Criteria
  subdivision, 180
  damage stability, 180, 194
Criteria of service, subdivision
  definition, 144, 180
  formula, 189, 194
Cross-connection (flooding), 175
Cross curves of stability, 78
  appendages, 82
  calculation, 79
  wedges method, 81
  computer methods, 82
Curves of form, 31, 40, 64

Damage
  effect on draft and trim, 143,
  effect on stability, 177,
  extent of, 143, 183
  fundamental effects, 143, 146
  survival, 186
  special requirements 183
  location, 185
  stability criteria, 180
Damage stability
  definition, 149
  regulations, 180
Damage stability calculation
  trim-line-added weight method, 149, 159
  lost-buoyancy method, 159
Deadrise, 6
Deadweight cargo, 51
Deckhouses, superstructure
  forces on, 266
  response of, 267
Definitions for regulations, 178
Deflections
  deckhouse, 266
  structural members, 206, 275
  mid-panel, 282
  hull girder analysis, 233

Density
  weight, 210
  mass, 210
Density of water
  effect of, 18
Deterministic theory
  of reliability, 291
Diagonals
  bilge, 7
Discontinuities (structural)
  compensation for, 235
  stress concentrations, 271
Displacement
  appendages, 39
  and center of buoyancy, 17, 64
  curve, 35
  change per cm of trim
    immersion, 39
  determination of, 51, 71, 116
  effect of weight change on, 102
  molded, 34
Displacement sheet, 32
Displacement vs. weight
  estimate, 18
Diving trim (submarines), 130
Docking, stability during, 136
Draft
  diagram, 121
  keel, 6
  marks, 7
  molded, 6
  navigational, 118
Draft and trim flooded
  by added weight, 166
  by lost buoyancy, 159
Drafts, determination of
  effect of change in loading, 115, 118
  using curves of form, 116
  using Bonjean curves, 115, 116
Drafts
  general, 6
Drag to the keel, 2
Drag, 120
Ducks, 8
Dynamic Stability, 106

Elements (rulings), 13
Entrance, 6
End effect, erections, 270
Equalization (flooding), 175
Equilibrium
  general, 63, 69
  immersed bodies, 16
  static, 16
  submarines, 131
Equilibrium polygon, 132, 133
Erections
  aluminum on steel, 270
Extent of damage, 143, 146

Factor of safety, 291
Factors of subdivision, 180, 189
Failure, modes of, 207
Fairbody line, 13

Fairing, of lines, 7
Fatigue failure, 287
  damage, 288
  endurance limit, 288
  high-cycle, 289
  low-cycle, 290
Finite element analysis
  of a model, 234, 266
Five-eight rule for area, 25
Flare, 6
Floodable length, 149
  approximations, 149
  curve of, 152
  definition, 149, 180
  determination of 150, 152, 176
  direct method, 152
  simplified direct, 152
Flooding
  equalization, 175
  heel after, 175
  partial stages, 175
  residual $\overline{GM}$, 162, 177
  sinkage and trim, 160
  stability, 170
  unsymmetrical, 173
Flooding effect diagram, 106
Flotation, center of, 33
  definition, 115
  uses of, 115
Forces
  acceleration, 209
  damping, 209
  added mass, 214
  wave-induced, 214
  upsetting (heeling) 67, 109
Forebody, 6
Form coefficients
  definition of, 18
Forward perpendicular, 2
Foundations
  design considerations, 209
  forces on 209, 268
  reaction loads, 209, 268
  modulus, 267
Freeboard after damage, 143, 146
Free surface (liquids in tanks), 93
  calculation of, 94
  correction for, 94
  determination of moment
    of inertia, 98
  effect of anti-roll tanks, 100
    on righting arm, 94
    on stability, 94
    of tank size, 99
    on metacentric height 94, 97
    on trim, 99
    of two liquids, 100
Froude-Krylov force, 217
Fuel requirements, 60
Full-scale tests
  bending moment in waves, 222

Gravity, center of, 22
  calculation, 63
  definition, 64

Gravity, center of (*continued*)
  from drafts and trim, 115
  from inclining experiment, 122
  effect of weights, 67
  variation with loading, 71
  of an area, 23
Grounding, stability during, 136

Half-breadth plan, 1
Half siding, 6
Heaving
  forces due to, 209
Heel
  after damage, 146, 147
  angle of steady, 106
  due to flooding, 146
  due to wind, 109
  permissible, 181
Heeling
  acceptable, 146, 173
  arm, 110
  curve of, 106
  forces, 109
  during intermediate flooding, 173
  in a turn, surface ships, 111
Heeling moment
  definition, 65
  representation, 90
Hog
  correction for, 120
  definition, 120
  determination of, 120
Hooke's Law, 237
HULDEF, 42, 43
Hull form
  effect on stability, 79
Hull girder
  beam theory, 236
  bending moment
    longitudinal, 236
    lateral, 250
    torsional, 250
  deflection, 236
  shear flow, 244
  shear lag, 246
  ultimate strength, 284
  vertical, 212
  warp, 244
Hydrostatic curves
  calculations required, 31, 40

Inclining experiment
  conduct of, 122, 124
  in air, 126
  preparation for, 123
  principles of, 122
  report, 126
Induced rolling (sallying), 127
Intact buoyancy
  definition, 178
  effect of, 147
Immersion, 39, 115
Integrating rules
  five-eight rule for area, 25
  for moments, 22

Integrating rules (*continued*)
  polar integration, 28
  Simpson's rules, 24
  Tchebycheff's rules, 26
  trapezoidal rule, 23
Integrator, mechanical, 30, 31, 80
Intergovernmental
  Maritime Consultative
    Organization (IMCO), 107, 144, 147, 181
International Maritime
  Organization (IMO), 107, 144, 147, 181

Keel drafts, 7

Laplace transform, 256
Lateral bending moment, 250
Lead ballast
  submarines, 130
  surface ships, 135
Length
  definition, 180
  floodable, 149, 152
  between perpendiculars, 2
  for displacement, 2
  over-all, 2
  waterline, 2
  permissible, 152, 180
Lines, ships'
  definition, 1
  fairing, 7
  offsets, 11
Liquid loading diagram, 104
Load line (subdivision), 143
Loads, structural
  static, 209
  dynamic, 212
  impact, 233
  internal (sloshing), 233
  inertial, 233
  wave induced, 212, 216, 222
Loading instructions, 136
Longitudinal bending moment
  still water, 208
  in waves, 212
Longitudinal center of buoyancy, 35
Longitudinal metacenter;
  see Metacenter, longitudinal
Lost buoyancy method,
  flooding calculation, 151, 164

Machinery space (flooding)
  definition, 180
Magnification factor (stress), 271
Margin line (flooding) 149, 152
  definition, 179
Margins
  in weight calculations, 70
Midship section coefficient, 18
Midship section lines, 2, 4
Midship section (structural), 238
Metacenter, longitudinal
  effects of trim, 74
  location, 73

Metacenter, longitudinal (*continued*)
  application, 74
  height, 34
Metacenter of submerged bodies, 74
Metacenter, transverse
  approximation for, 71
  definition, 71
  effect of free surface on, 93
  height, 34
Metacentric height
  transverse, 71
  application, 74
  calculation, 72
  definition, 72
  effect on period of roll, 77
  "suitable", 77
  longitudinal, 72
Metacentric height (flooding)
  calculation 150, 168
  change due to, 146, 148, 160, 173
  during intermediate, 173
  negative, 173
  required to limit heel, 150, 160
  required for zero heel, 150
  residual, 177, 180, 188
Metacentric radius, 34
Methodical series of hull forms, 14
Model tests
  bending moment in waves, 214
Modes of structural failure
  brittle fracture, 207, 275
  elastic buckling, 279, 283
  fatigue cracking, 275
  excessive yield, 275
  Euler critical load, 275
Molded surface
  baseline, 1, 5
  dimensions, 1, 5
  form, 1, 5
Moments, righting, heeling, 65
Moments of areas, 23
Moment of inertia
  definition, 22
  of free surface, 96
  of midship section, 239
  of waterplanes, 23
Moment of transference, 94
Moment to change trim calculation, 39, 115
Morrish formula, 38
Morro Castle, Mohawk disasters, 144

Neutral equilibrium, 63

Offsets
  table of, 10
One-minus-prismatic rule, 21

Parallel middle body
  definition, 2, 6
Parallel sinkage, 118
Passenger space (flooding)
  definition, 180
Period of roll, 75, 77, 104

Permeability
  definition, 180
  for stability after damage, 189
  standards for, 190
  values, 190
Permissible heel (flooding), 180
Perpendiculars
  definition, 2
Pitching
  forces due to, 209
Planimeter, 30
Plating, strength of
  elastic buckling, 279
  post buckling behavior, 280
  ultimate strength, 282
Plating stress
  orthotropic theory, 255
  beam on elastic, 255
  grillage theory, 255
  finite element, 255
Poisson's ratio, 270
Polar integration, 28
Polygon, equilibrium (submarines), 132
Posdunine's formula, 39
President Wilson investigation, 271
Primary response
  (structure), 210, 235
Prismatic coefficient, 19
Probabilistic theory
  of reliability, 206, 222, 291

Racking
  transverse strength, 263
Radius of gyration, 75
Range of positive stability
  flooded, 177
  intact, 106
Reaction loads, 209
Recesses (subdivision), 191
Regulations, damage stability, 194
Reliability of structure, 290
  deterministic design, 291
  probabilistic design, 292
Residual metacentric height, 188
Residual righting arm, 188
Response, structural
  primary, 210, 235
  secondary, 210, 255
  tertiary, 210, 260
Righting arm, 78, 79, 89
  effect of free surface, 94
  after damage, 177, 188
  residual, 188
Righting moment
  calculation of, 65
  curve of, 65
  definition, 65
  after damage, 177, 188
Rolling
  induced (sallying), 127
  natural period, 77, 104
  forces due to, 209
Rulings (elements), 13

Safety, factor of, 291

Safety of Life at Sea (SOLAS)
  1960 Convention, 144, 146, 181, 189, 194
Sag
  correction for, 120
  definition, 120
  determination of, 120
Sallying, 127
Scantlings, 207
Secondary response (structure), 255
Section area curve, 6
Section modulus
  calculation of, 238
Shear flow, 242
Shear lag, 246
Shear stress (hull girder), 237
Sheer, deck, 5
Sheer plan
  definition, 1
Shifting boards, 101
Shifting cargo (stability) 93, 103
Shirokauer calculation, 152
Simpson's Rules, 24, 25, 43, 49, 80, 153
Sinkage
  due to added weight, 118
  due to flooding, 146, 150, 164, 180
  salt to fresh water, 18
Slamming, 209
  loads, 232
  stresses, 287
Sloshing, 209, 233
Smearing, 237
Snaking, 254
Sounding tables, 57
Springing, 209, 232, 286
Stability after damage, 146
  calculations, 149, 152
  criteria, 180
  definition, 143
  during intermediate flooding, 170
  limits of heel, 186
  "lost buoyancy" method 150, 164, 160
  requirements, 200, 180, 194
  "trim-line-added weight"
    method, 150, 159, 166
Stability and Trim Booklet, 76
Stability, gravitational, 63
  advanced marine vehicles, 139
  criteria, 106, 107, 108
  cross curves of, 78
  definition, 63
  in depth (submarines), 114, 134
  in drydock, 136
  residual dynamic, 106
  effect of beam on, 85
    cargoes, 100
    changes in weight, 102
    depth, 85
    consumption of
      fuel and stores, 104
    free surface, 93
    icing, 111
    suspended weights, 101
  floating platform, 138
  methods of improving, 135

Stability, gravitational (*continued*)
  range of positive, 106
  regulations, 107
Stable equilibrium, 63
Standards of subdivision
  factorial system, 181
  integer compartmentation, 181
Statical stability, 84
  stranded, 138
  submerged bodies, 113, 134
  unusual ship forms, 138
Statical stability curves
  effect of beam, 84
    depth, 85
    form, 85
  significance of, 87
  work and energy determination, 90
Station ordinates, 4
Stepped bulkheads
  (flooding), 191
Still-water bending moment, 209, 211
Stowage factors (cargo), 59
Strength
  of ships, 205
  of material, 276
  hull girder ultimate 277, 284
  plating, 280, 281
  transverse, 263
  stiffened panels, 282
  classification society rules, 205
Stress components
  resultants, 239
  principal, 239
  dominant, 240
Stress concentration, 271
  around rectangular openings, 272
  around circular openings, 274
Stress distribution, 239, 247
Stress superposition, 261
Stresses, 233
  primary direct, 235
  secondary, 234
  tertiary, 234
Strip theory, 216
Structures
  deterministic design, 291
  reliability of, 290
  probabilistic design, 291
  secondary response, 255
  tertiary response, 260
  unconventional craft, 274
Subdivision
  alternate regulations, 194
  general, 146, 185
  SOLAS requirements, 144, 146, 180
  special requirements, 180, 189
    tankers (Marpol), 187
  U.S. requirements, 183
  standards, 181
Submarines, stability, 128
  diving trim, 69, 130
  metacentric stability, 114
  stability in depth, 85, 134
  weight relationships, 130

Submerged equilibrium, 67
  conditions of, 131
  definition, 128
  the polygon, 132
  the trim dive, 134
Suspended weights
  effect on stability, 101
Superposition of stresses, 261
Superstructure
  aluminum, 236
  stresses in, 267
Swinging stations, 21

Tanks, capacity, 51
Tanks, free surface effect, 93
Tchebycheff's Rules, 26
Tertiary response (structure), 210, 260
Thermal loads, 209
Titanic disaster, 144
Tons per unit immersion, 32
  per inch, 33
  per cm, 33, 115
Torsion, hull girder, 250
Transverse strength, 263
Transverse metacenter, 71
Trapezuidal Rule, 23
Trim
  after damage, 146, 150, 160
  calculations, 116
  effect of free surface, 99
  effect on displacement, 7
  effect on stability, 103
  moment to change, 115
Trim and Stability Booklet, 76
Trim dive (submarines), 134
  report of, 135
Trim-line-added weight
  method (subdivision) 150, 159, 166
Trimming forces, 99, 115
Tumble-home, 6
Turn of bilge, 6

Ullage, 59
Ultimate strength, 277
Unstable equilibrium, 63
Upsetting forces, 67
U.S. Coast Guard Rules
  (subdivision), 178
U.S. Navy
  damage stability standards, 145
  stability criteria, 108
U.S. Coast Guard
  stability criteria, 107
U.S. Maritime Administration
  stability criteria, 106, 107

Variable ballast (submarines), 104
Vertical bending moment, 211, 215
Vertical center of buoyancy
  approximations, 38
  calculations, 36
Vertical prismatic coefficient, 19
Volumetric coefficient, 19

Water, fresh
  service requirements, 59
Waterlines plan, 1
Waterplanes area, 32
Waterplane coefficient, 19
Wave bending moment
  approximations, 212
  in real seas, 212, 225
  transverse, 212, 218
  long-term prediction, 225
Wave induced loads, 212
  short-term, 223
  long-term, 223
  extreme, 225

Wave induced loads (*continued*)
  on offshore platforms, 221
  deterministic evaluation, 216
  probabilistic estimate, 222
Wedges method (cross curves
  of stability), 81
Weight
  effect of changes on the
    center of gravity, 63, 69, 70
    center of buoyancy, 64
    displacement and stability, 102
    trim, 103
  effect of removal, 136
  effect of suspended, 101

Weight (*continued*)
  estimates, 69
  changes (submarines), 103
Wetted surface
  approximations for, 47
  calculation of, 47
  definition of, 47
Whipping, 209, 233
Wind heeling moment, 67
Wolverine State experiments, 292

Yield
  criteria of, 276

This page intentionally left blank